NUCLEAR ACOUSTIC RESONANCE

NUCLEAR ACOUSTIC RESONANCE

Dan I. Bolef
Ronald K. Sundfors

DEPARTMENT OF PHYSICS
WASHINGTON UNIVERSITY
SAINT LOUIS, MISSOURI

ACADEMIC PRESS, INC.
Harcourt Brace & Company, Publishers
Boston San Diego New York
London Sydney Tokyo Toronto

This book is printed on acid-free paper. ∞

Copyright © 1993 by Academic Press, Inc.

All rights reserved.
No part of this publication may be reproduced
or transmitted in any form or by any means,
electronic or mechanical, including photocopy,
recording, or any information storage and
retrieval system, without permission in writing
from the publisher.

ACADEMIC PRESS, INC.
1250 Sixth Avenue, San Diego, CA 92101-4311

United Kingdom Edition published by
ACADEMIC PRESS LIMITED
24-28 Oval Road, London NWI 7DX
'TEX' and '$\mathcal{A}_{\mathcal{M}}$S-TEX' are trademarks of the American Mathematical Society.

Library of Congress Cataloging-in-Publication Data
Nuclear acoustic resonance/Dan Bolef, Ronald Sundfors
       p.   cm.
   Includes bibliographical references and index
   ISBN 0-12-111250-0
   1. Acoustic nuclear magnetic resonance. I. Sundfors, Ronald
II. Title.
QC762.6.A25B65   1993   93-13624
537'.362-dc20   CIP

Printed in the United States of America

93 94 95 96 BB 9 8 7 6 5 4 3 2 1

CONTENTS

	PREFACE	xi
1	**INTRODUCTION**	**1**
	1.1 What is Nuclear Acoustic Resonance?	1
	1.2 Nuclear Magnetic Resonance	3
	1.3 Nuclear Electric Quadrupole Effects	7
	1.4 Line Width Effects	13
	1.5 Nuclear Spin–Lattice Relaxation	14
	1.6 Interaction of Acoustic Waves with Nuclear Spins	18
	1.7 Acoustic Saturation of Nuclear Magnetic Resonance	21
	1.8 Nuclear Acoustic Resonance	24
	1.9 Historical Note	28
	References	29
2	**BASIC THEORY**	**32**
	2.1 Elastic Waves in Solids	33
	2.2 Nuclear Spin Interactions	37
	2.3 Nuclear Spin–Phonon Interactions: Fermi Golden Rule Approach	40
	2.4 NAR Absorption and NAR Dispersion: Müller Approach	43
	References	47
3	**DYNAMIC NUCLEAR ELECTRIC QUADRUPOLE INTERACTIONS**	**48**
	3.1 Multipole Expansions Using Spherical Tensors: The A_{lm} Irreducible Tensor Operators	49
	3.2 Nuclear Electrostatic Multipole Interactions	58
	3.3 Dynamic Electric Quadrupole Interaction	62

3.4	Electric Field Gradient: Symmetry Considerations and the S–Tensor	73
3.5	Fedders' Calculation of Absorption and Dispersion	83
3.6	Application. NAR in Noncubic Metallic Rhenium: Pure Nuclear Electric Quadrupole Resonance	87
3.7	Dynamic Nuclear Electric Hexadecapole Interaction	93
3.8	Is the Hexadecapole Interaction (HDI) Observable by NAR Techniques?	100
3.9	Dynamic Magnetic Dipole–Dipole Interaction	103
3.10	Additional Mechanisms for Multipole Quantum Transitions	107
	References	111

4 DYNAMIC ALPHER–RUBIN (DIPOLAR) INTERACTION — 115

4.1	Nature of the Coupling	115
4.2	Classical Derivation of Dipolar NAR Absorption and Dispersion: I. Method of Quinn–Buttet–Fedders	117
4.3	Classical Derivation of Dipolar NAR Absorption and Dispersion: II. Method of Müller	130
4.4	Experimental Verification in Metals	138
4.5	The 'Low Temperature' Limit	145
4.6	NAR Absorption and Dispersion: Quantum Mechanical Approach	149
4.7	Influence of Electronic Band Structure	158
	References	160

5 LINE BROADENING AND RELAXATION EFFECTS — 163

5.1	Introduction	163
5.2	The Method of Moments Applied to NAR1 and NAR2	165
5.3	Generalized Bloch Equations	172

	5.4	Quadrupole–Split High Field Case: Hamiltonian, Energy Levels, Transitions	173
	5.5	Line Shape of the Quadrupole–Split High Field Line: Intraspin Cross Relaxation	176
	5.6	Spin Relaxation by Dynamic Quadrupole Interaction	183
	5.7	Dipolar and Quadrupolar Exchange Contributions to NAR Line Shapes	192
	5.8	Relaxation by the Intrinsic Direct Process in Van Vleck Paramagnets	197
		References	199
6	**BONDING IN INSULATORS, SEMICONDUCTORS AND METALS: STRAIN–ELECTRIC FIELD GRADIENT TENSORS**		**201**
	6.1	Introduction	201
	6.2	Ionic and Covalent Effects in Insulators: The Sternheimer Antishielding Factor	203
	6.3	Semiconductors	207
	6.4	Antishielding in Metals	215
		References	226
7	**EXPERIMENTAL TECHNIQUES**		**230**
	7.1	Introduction	231
	7.2	Continuous Wave Ultrasonics: Propagating Wave Model	232
	7.3	CW Spectrometers	238
	7.4	Signal Calibration	261
	7.5	Transient NAR Techniques	263
	7.6	The Composite Resonator	272
	7.7	Application: NAR in Rhenium	283
	7.8	Application: Experimental Search for the Dynamic Nuclear Hexadecapole Interaction	287
		References	292

8 ACOUSTIC SATURATION NMR AND DOUBLE RESONANCE — 295

8.1 Introduction — 295
8.2 Acoustic Saturation of NMR — 297
8.3 Continuous Wave ASNMR — 310
8.4 Transient ASNMR — 311
8.5 Double Acoustic–Magnetic Resonance — 319
References — 321

9 MAGNETIC MATERIALS — 323

9.1 NAR in Antiferromagnetic Insulators — 323
9.2 NAR in Ferromagnets — 346
9.3 Enhanced Nuclear Acoustic Resonance — 346
9.4 Effect of Demagnetization on NAR Line Shapes in Bulk Metals — 353
References — 361

10 SQUID DETECTION OF NUCLEAR ACOUSTIC RESONANCE — 363

10.1 Introduction — 363
10.2 The SQUID Acoustomagnetic Effect — 365
10.3 The SQUID Acoustomagnetic Spectrometer — 370
10.4 Detection of Acoustic Composite Resonator Responses — 374
10.5 Nuclear Spin–Phonon Interactions in Tantalum Metal — 378
10.6 Nuclear Spin–Phonon Interactions in Antimony Metal — 385
10.7 SQUID-Detected Acoustic Magnetic Resonance: A Prognosis — 390
References — 393

APPENDICES — 394

A. Selected Physical Constants; Energy Conversion Factors — 395

B. Properties of Selected Stable Nuclei;

	Paired Isotopes Suitable for the Investigation of the Hexadecapole Interaction	397
C.	S-Tensor Components for Various Crystal Structures	402
D.	Dynamic Quadrupole Interaction: Transformation of the S-Tensor for Cubic $\bar{4}3m$ Symmetry	407
E.	T-Tensor Components for Cubic $\bar{4}3m$ Symmetry	416
F.	Dynamic Hexadecapole Interaction: Transformation of the T-Tensor	418

INDEX   **421**

PREFACE

The authors intend this book to serve as an up-to-date introduction to the field of nuclear acoustic resonance. Although nuclear acoustic resonance is closely related to the well-established field of nuclear magnetic resonance, we take pains to distinguish the characteristics and capabilities of the two fields. The distinction is most marked in the nature of the coupling mechanisms, treated in some detail in Chapter 3 (dynamic electric quadrupole coupling) and Chapter 4 (dynamic Alpher–Rubin coupling); in the experimental techniques, treated in Chapter 7; and in the ease of application of nuclear acoustic resonance to the study of conducting media, treated in many of the applications cited throughout the text. An associated technique, acoustic saturation nuclear magnetic resonance, serves as a bridge between nuclear magnetic resonance and nuclear acoustic resonance; it is treated, albeit in summary form, in Chapter 8.

Emphasis in this text is on fundamentals, both in theory and in experimental techniques. A notable feature of the theory presented by us is the detailed treatment of nuclear electrostatic multipole interactions in terms of the A_{lm} irreducible tensor operators and their application, following Fedders, to the calculation of nuclear acoustic resonance absorption and dispersion (Chapter 3), as well as of line width and relaxation effects (Chapter 5). We also treat in detail the alternative approach, building on the concepts of acoustic impedance and susceptibility for calculating absorption and dispersion in nuclear acoustic resonance, developed by Müller and his colleagues. In an extension of the usual treatment of nuclear dipolar and nuclear quadrupolar interactions, we derive the appropriate expressions for nuclear acoustic coupling in solids via the dynamic hexadecapole moment. Our treatment of dynamic Alpher–Rubin coupling emphasizes the classical approach of Quinn–Buttet–Fedders, but also summarizes the alternative classical and quantum mechanical approaches of Müller and his colleagues. The extreme sensitivity of acoustic techniques required to observe most nuclear acoustic resonance effects has been achieved by careful application of standard techniques, described in some detail in Chapter 7, and of the SQUID (Chapter 10).

We have had to be selective in citing results, as well as in mentioning important contributors. Probably the most glaring omission is that of R. L. Melcher, a pioneer in the application of NAR to magnetic materials, both in theoretical and experimental aspects, as well as in the use of reflection CW spectrometers for NAR. We encourage the reader to dig deeper in the literature with the aid of the detailed references at the end of each chapter, as well as by utilizing Physics Abstracts. We also have not explored in detail the applications of nuclear acoustic resonance. The final chapter, on

SQUID nuclear acoustic resonance, we feel, points the way most clearly to future research and to novel applications in this field.

The presentation in this book owes much to the contributions of our students and collaborators at Washington University who have worked in the field of nuclear acoustic resonance and in related areas: Janet Brown, R. G. Leisure, R.L. Melcher, J. B. Merry, J. G. Miller, George Mozurkewich, Charles Myles, K. S. Pickens, H. I. Ringermacher, Rebecca Scholz–Hudson, R. E. Smith, Willis Smith, T. H. Wang and Marjorie P. Yuhas. The criticism and advice of our colleagues, George Mozurkewich, P. A. Fedders, J. G. Miller and Yaotian Fu, have been especially valuable. Detailed and very helpful readings of .portions of the manuscript were made by R. G. Leisure and Volker Müller. George Mozurkewich did heroic duty in reviewing the entire manuscript. We thank them for their time, insight and patient good humor. We thank Pranoat Suntharothok for her fine contributions to the production and editing of the manuscript. We have benefited from the help and advice of Vicki Jennings and her colleagues at Academic Press.

CHAPTER 1
INTRODUCTION

   1.1 What is Nuclear Acoustic Resonance?
   1.2 Nuclear Magnetic Resonance
   1.3 Nuclear Electric Quadrupole Effects
   1.4 Line Width Effects
   1.5 Nuclear Spin–Lattice Relaxation
   1.6 Interaction of Acoustic Waves with Nuclear Spins
   1.7 Acoustic Saturation of Nuclear Magnetic Resonance
   1.8 Nuclear Acoustic Resonance
   1.9 Historical Note
   References

1. INTRODUCTION

1.1 What is Nuclear Acoustic Resonance?

The nuclei of many atoms have magnetic and electrical properties that are associated with the fact that the nucleus as a whole has an intrinsic angular momentum, or *spin*. The nucleus may be characterized, for example, by a magnetic dipole moment and by an electric quadrupole moment. The static interactions between the magnetic dipole moment and external or internal magnetic fields, and between the electric quadrupole moment and electric field gradients, give rise to energy levels whose separations in frequency characteristically fall into the RF or ultra–high frequency ranges. In the well–known technique of nuclear magnetic resonance, an externally generated RF magnetic field is used to induce transitions among the nuclear spin energy levels by coupling to the nuclear magnetic dipole moment. Thus, in nuclear magnetic resonance, the resonance phenomenon characteristically is accompanied by the absorption (or dispersion) of electromagnetic radiation.

In nuclear acoustic resonance, in contrast, *acoustic* radiation is utilized to induce transitions among the energy levels that characterize the orientation of nuclear spins subjected to external or internal magnetic fields and to electric field gradients. Since the frequency of the acoustic wave usually falls in the range of 1 MHz to 100 MHz, the convenient terms *ultrasound* and *ultrasonic* are often used to denote the acoustic radiation. Transitions among the nuclear spin energy levels in nuclear acoustic resonance are induced by the ultrasound through modulation of internal interactions involving the magnetic or electric multipole moments of the nucleus.

At first glance, there appear to be marked similarities between the techniques of nuclear acoustic resonance and nuclear magnetic resonance. The energy levels among which transitions are induced are the same. To display the resonance lines, both techniques utilize the variation of an external variable, most often the DC magnetic field or the frequency. The line spectra and the line widths observed by the two techniques are often closely related. The differences between the techniques, however, are equally notable and allow us to view nuclear acoustic resonance as a valuable adjunct to—and occasional substitute for—nuclear magnetic resonance. Although full discussion of these differences must await the detailed treatment of the theory, experimental techniques and characteristic results to be given in the remaining chapters, we may cite immediately two of the most obvious:

(1) The acoustic wave utilized in nuclear acoustic resonance couples energy to the nuclear spin system by the modulation of an *internal* spin–dependent interaction whereas in nuclear magnetic resonance the coupling is always achieved by means of the interaction of an external RF magnetic field with the magnetic dipole moment. In fact, in nuclear acoustic resonance the internal interactions compete to determine which will best serve to couple energy to the spin system from the exciting acoustic wave. Since the nature of the internal interactions depends upon the type of material (for example, magnetic versus nonmagnetic, conducting versus nonconducting) as well as upon the physical environment (*e.g.*, temperature, pressure), one anticipates that much of the physical understanding to be elucidated derives from detailed studies of the nature and properties of the nuclear spin–acoustic wave interaction itself.

(2) There are several classes of materials—most notably bulk metals and alloys—into which electromagnetic waves cannot easily penetrate, whereas there is no inherent difficulty for the passage of ultrasound.

There exist techniques for observing nuclear spin resonance phenomena that combine nuclear magnetic resonance and the use of ultrasound. Among these are acoustic saturation of nuclear magnetic resonance and, in more sophisticated form, double resonance effects that utilize both electromagnetic and acoustic energy. These *combination* spin resonance techniques sometimes include the use of electron spin resonance as detectors of acoustic absorption, and these techniques always require electromagnetic excitation of one form or another. The latter feature clearly distinguishes them from nuclear acoustic resonance techniques, in which only acoustic energy is utilized. The combination spin resonance techniques are closely related to nuclear acoustic resonance with respect to the physical understanding obtained, however, and will be treated in some detail.

In the remainder of the present chapter, we attempt a more detailed, though still introductory, description of the three techniques: nuclear acoustic resonance, nuclear magnetic resonance and combination acoustic–electromagnetic spin resonance. Since nuclear magnetic resonance (NMR) is by far the best known of these and the subject of a truly extensive literature, we begin with a very brief review of its fundamentals, and we emphasize—because of their importance to nuclear acoustic resonance—an understanding of the nuclear electric quadrupole interaction and of nuclear spin–lattice relaxation. We end the chapter with an overview of the subjects of acoustic saturation of nuclear magnetic resonance (ASNMR) and, finally, of nuclear acoustic resonance (NAR).

1.2 Nuclear Magnetic Resonance

There exists today an extensive literature on the subject of nuclear magnetic resonance. General and comprehensive treatments are given, for example, in books by *Abragam* [1.1] and by *Slichter* [1.2]. Only topics directly relevant to the subjects of nuclear acoustic resonance and of acoustic saturation of nuclear magnetic resonance are treated in the present, brief review. We restrict our discussion, in order to emphasize essentials, to insulating, non-magnetic solids.

The phenomenon of nuclear magnetic resonance follows directly from the magnetic properties of the nucleus. All nuclei with nonzero angular momentum or spin (I) possess a magnetic dipole moment similar to that associated with a classical magnetic dipole. From quantum mechanics we know that the matrix elements of the magnetic moment are proportional to the matrix elements of the total nuclear angular momentum. We can write for the magnetic dipole moment, then,

$$\boldsymbol{\mu} = \gamma \hbar \boldsymbol{I} , \qquad (1.1)$$

where γ is the gyromagnetic ratio, $\hbar$ is Planck's constant and $|\boldsymbol{I}|$ is in units of $\hbar$. In the presence of an applied DC magnetic field $\boldsymbol{H}_0$, the magnetic moment $\boldsymbol{\mu}$ experiences a torque perpendicular to the plane containing $\boldsymbol{\mu}$ and $\boldsymbol{H}_0$. As in the case of the classical gyroscope, the magnetic moment can be thought of as precessing at an angle θ about the about the axis of a magnetic field (Fig. 1.1). The angular frequency of precession (Larmor

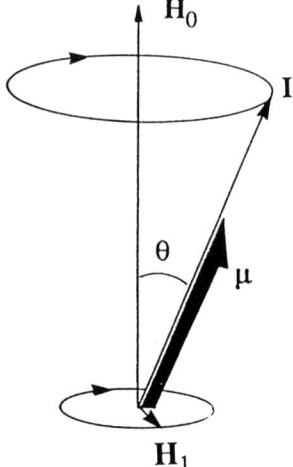

FIGURE 1.1. Gyroscopic precession of a nuclear magnetic moment μ about the axis of an externally applied magnetic field H_0. A rotating magnetic field H_1 tends to increase the angle θ between the magnetic moment vector and the field axis.

frequency) ω_0 is given by

$$\omega_0 = -\gamma H_0 . \tag{1.2}$$

If an RF magnetic field H_1, rotating in the same direction as the nuclear precession, is applied at right angles to H_0 and at right angles to the plane of H_0 and μ, the effect will be to tip μ in a direction antiparallel to H_0. This corresponds to an absorption of energy from the source of the perturbing field H_1. Maximum absorption occurs when the rotational frequency ω of H_1 equals the resonant frequency ω_0. For a typical nucleus, the frequency at which resonance occurs in a field of 1 Tesla (10,000 Oe) is $\nu_0 \equiv \omega_0/2\pi \simeq 10\,\mathrm{MHz}$.

It is more useful and more rigorous to treat the nuclear spins as a quantized system. The magnitude of the spin angular momentum I, a constant characteristic of stable nuclei, is given by $\sqrt{I(I+1)}$, where I is either integer or half–integer.

1.2 NUCLEAR MAGNETIC RESONANCE

The Hamiltonian for a system of isolated spins in a DC magnetic field is given by

$$\mathcal{H}_m = -\boldsymbol{\mu} \cdot \boldsymbol{H}_0 \,. \tag{1.3}$$

The quantities $I(I+1)$ are eigenvalues of the operator $\boldsymbol{I}^2$. The components of $\boldsymbol{I}$ (I_x, I_y, I_z) commute with $\boldsymbol{I}^2$, so that we may specify simultaneously eigenvalues of both $\boldsymbol{I}^2$ and I_z, which we call $I(I+1)$ and m, respectively. A nucleus with spin $\boldsymbol{I}$ may assume only $2I+1$ discrete orientations in the DC magnetic field. These orientations correspond to discrete energy levels characterized by the magnetic quantum number m, corresponding to the projection of $\boldsymbol{I}$ along the direction of the DC magnetic field. The energy levels for isolated nuclear spins in an external DC magnetic field $\boldsymbol{H}_0$ are given by

$$E_m = -mg\mu_n H_0\,, \quad m = I, I-1 \ldots -I\,, \tag{1.4}$$

where g is the nuclear g factor. $\mu_n = e\hbar/2Mc$ is the nuclear magneton, M is the proton mass, e is the electronic charge and c is the velocity of light.

Application of an RF field $\boldsymbol{H}_1$ of frequency $\omega_0/2\pi$ induces transitions among these energy levels. Under the action of a harmonically varying perturbation the probability per unit time that a nucleus in a state initially characterized by the magnetic quantum number m will be found in a state m' is given quite generally by the expression

$$W = \frac{1}{4\hbar^2}|\langle m'|\mathcal{H}|m\rangle|^2 g(\nu)\,, \tag{1.5}$$

where $\langle m'|\mathcal{H}|m\rangle$ is the appropriate matrix element of the perturbing Hamiltonian, and $g(\nu)$ is a shape function such that $\int_0^\infty g(\nu)d\nu = 1$ and $g(\nu)$ is peaked near the resonant frequency ν_0. For the case in which energy is coupled to the nuclear magnetic moments by a time varying RF magnetic field $\boldsymbol{H}_1$, $\mathcal{H} = -\boldsymbol{\mu} \cdot \boldsymbol{H}_1$ and (1.5) becomes

$$W_M = \frac{\gamma^2 H_1^2}{4}|\langle m'|I_\pm|m\rangle|^2 g(\nu)\,, \tag{1.6}$$

where $I_+ = I_x + iI_y$, $I_- = I_x - iI_y$ are the *raising* and *lowering* spin operators, respectively. $|\langle m'|I_\pm|m\rangle|^2$ has a definite value depending upon whether $m' = m+1$ or $m-1$ given by

$$|\langle m+1|I_+|m\rangle|^2 = (I-m)(I+m+1) \tag{1.7a}$$
$$|\langle m-1|I_-|m\rangle|^2 = (I+m)(I-m+1)\,. \tag{1.7b}$$

Substituting (1.7) into (1.6), and allowing for the possibility that H_1 is not at right angles to H_0, we obtain

$$W_{M(\pm 1)} = \frac{1}{4}\gamma^2 H_1^2 g(\nu)(I \mp m)(I \pm m + 1)\sin^2\theta , \qquad (1.8)$$

where $2H_1$ is the peak amplitude of the RF magnetic field, and θ is the angle that H_1 makes with the static magnetic field H_0. The allowed transitions under these circumstances are those for which m changes by ± 1 ($\Delta m = \pm 1$). Applying this selection rule to (1.3), one obtains for the resonance condition

$$\hbar\omega_0 = E_{m+1} - E_m = -g\mu_n H_0 . \qquad (1.9)$$

A pictorial representation of the energy level scheme for a nucleus of spin $I = \frac{1}{2}$ is shown in Fig. 1.2(a), in which N_+ and N_- represent the number of nuclei, respectively, in level $m = +\frac{1}{2}$ and level $m = -\frac{1}{2}$. The absorption of energy from the RF magnetic field as the DC magnetic field (or frequency) is varied is shown schematically in Fig. 1.2(b), in which the center of the absorption line coincides with the resonant field H_0 (or, equivalently, the resonant frequency $\omega_0/2\pi$).

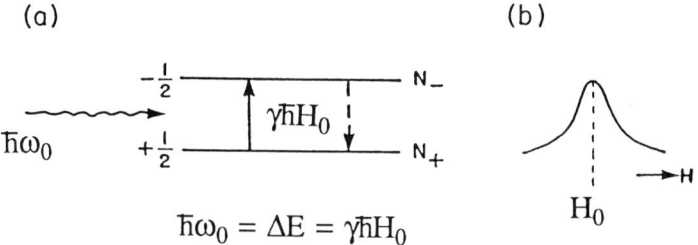

FIGURE 1.2. (a) Energy levels for a nucleus with spin $I = \frac{1}{2}$. (b) A typical nuclear magnetic resonance absorption line.

From quantum mechanics we know that the probabilities for transitions between any two energy levels induced by a time-dependent interaction are equal for transitions up and down. It might appear, then, that for the case illustrated in Fig. 1.2, for example, there would be no net absorption of energy (since a down transition from the higher level, corresponding to a release of energy, might be expected to cancel the absorption of energy in the corresponding up transition). This would indeed be true if the two states were equally populated. If, however, an assembly of nuclei occupying two states ($I = \frac{1}{2}, m = \pm\frac{1}{2}$) is in thermal equilibrium with a reservoir at a temperature T, the upper of the two states is less densely populated according to the Boltzmann distribution factor

$$N_-^o = N_+^o e^{-\Delta E/k_B T} , \qquad (1.10)$$

where k_B is the Boltzmann constant, ΔE is the energy difference between the levels, and N_-^o and N_+^o are the equilibrium populations of the two levels. The population difference $N_+^o - N_-^o$ for DC magnetic fields readily attainable in the laboratory is of the order of parts per million. Because of the greater population of spins in the lower energy state, therefore, there is a net absorption of energy from the source of the RF magnetic field, as shown schematically in Fig. 1.2(b). The mechanism by which the population difference (1.10) is maintained is discussed in Section 1.5. The existence of discrete magnetic energy levels and of a nonuniform distribution of nuclear magnetic moments among these levels leads to a net macroscopic magnetization in the specimen containing these moments. By applying the Boltzmann law, one obtains the expression for M, the net nuclear magnetization per unit volume:

$$M = \frac{N\mu^2}{3k_B T}\left(\frac{I+1}{I}\right) H_0 , \qquad (1.11)$$

where N is the number of spins per unit volume. It is this net magnetization that is often measured in nuclear magnetic resonance experiments.

1.3 Nuclear Electric Quadrupole Effects

Nuclei with spin $I > \frac{1}{2}$ possess not only magnetic moments but also electric moments. The lowest nonvanishing nuclear electric moment is the electric quadrupole moment, which is a measure of the departure from spherical symmetry of the charge distribution in the nucleus. The nuclear electric quadrupole moment Q interacts with electric field gradients (inhomogeneous electric fields) that exist at the nucleus due to neighboring electrical charges, such as ions or electrons external to, but associated with, the nucleus. The nuclear electric quadrupole interaction is responsible for at least four phenomena important in nuclear magnetic resonance:

(1) It may affect the nuclear spin magnetic energy level configuration.
(2) Under certain circumstances it gives rise to a set of nuclear spin energy levels—nuclear quadrupole energy levels or pure quadrupole energy levels—even when there is no DC magnetic field present.
(3) It is often responsible for the nuclear spin–lattice relaxation, to be discussed in Section 1.5.

(4) It affects the line width and line shape of the observed resonance. In nuclear acoustic resonance, further, the dynamic nuclear electric quadrupole interaction provides a means of coupling acoustic energy to the nuclear spin system.

Because of its importance in nuclear acoustic resonance, we summarize below some results of the theory of nuclear electric quadrupole interactions. In Chapters 2 and 3, a comprehensive and detailed treatment of these interactions will be given.

The nucleus can be regarded as a small, localized distribution of electric charge with a charge distribution of density $\rho(r_n)$, where r_n is the radial vector from the origin to the element of nuclear charge. If this electric charge is subjected to an electrostatic potential $V(r_n r_e)$ due to charges external to the localized distribution, the electrostatic interaction energy is given by

$$E = \int \rho(r_n) V(r_n r_e) d\tau_{r_n} , \qquad (1.12)$$

where the integral is over the volume containing the localized distribution and r_e is the radial vector from the origin to the element of electronic charge external to the nucleus. We will show in Chapter 3 that the lowest term in the orientation–dependent part of the electrostatic energy (also, see *Slichter* [1.2]) can be written in operator form as

$$\mathcal{H}_Q = \sum_{ij} Q_{ij} \left(\frac{\partial^2 V}{\partial x_i \partial x_j} \right)_{r=0} , \qquad (1.13)$$

where i, j stand for x, y or z, respectively, and the derivative is to be evaluated at the origin. The Q_{ij} are the components of the quadrupole moment tensor, which for a nuclear state of spin quantum number I has the following form:

$$Q_{ij} = \frac{eQ}{6I(2I-1)} \left[\frac{3}{2}(I_i I_j + I_j I_i) - \delta_{ij} I(I+1) \right] , \qquad (1.14)$$

where $Q = \Sigma_i \rho_i(r)(3\cos^2\theta_i - 1)r_i^2$ is a scalar quantity termed the electric quadrupole moment of the nucleus, and r_i, θ_i are the coordinates of an element of nuclear charge ρ_i with respect to a cylindrical axis. It follows that $Q = 0$ for a spherically symmetric charge distribution, that $Q > 0$ for a prolate, and that $Q < 0$ for an oblate spheroidal distribution.

It is conventional in (1.13) to define

$$V_{ij} \equiv \left(\frac{\partial^2 V}{\partial x_i \partial x_j} \right)_{r=0} . \qquad (1.15)$$

The V_{ij} are components of a symmetric, traceless, second–rank tensor, the electric field gradient (EFG) tensor. It is possible to reduce the EFG tensor to a diagonal form by transforming to the principal coordinate system, designated by x, y, z. In this coordinate system, the off–diagonal components of the tensor, $V_{xz} = V_{yz} = V_{xy} = 0$, and the only nonvanishing components are V_{xx}, V_{yy} and V_{zz}. The Hamiltonian can then be written

$$\mathcal{H}_Q = \frac{eQ}{6I(2I-1)} \sum_{ij} V_{ij} \left[\frac{3}{2}(I_i I_j + I_j I_i) - \delta_{ij} I(I+1)\right] . \qquad (1.16)$$

Even the nonvanishing components are not independent, however, since Laplace's equation is applicable, yielding $V_{xx} + V_{yy} + V_{zz} = 0$. Equation (1.16) then reduces to the form

$$\mathcal{H}_Q = \frac{eQ}{4I(2I-1)}[V_{zz}(3I_z^2 - I^2) + (V_{xx} - V_{yy})(I_x^2 - I_y^2)] . \qquad (1.17)$$

In the principal coordinate system, two parameters are thus sufficient to specify the electric field gradient tensor. The usual choice of these parameters is the following:

$$eq = V_{zz}; \quad \eta = \frac{V_{xx} - V_{yy}}{V_{zz}}, \quad 0 \leq \eta \leq 1 . \qquad (1.18)$$

The scalar quantity η is defined as the asymmetry parameter and is a measure of the departure from axial symmetry. The case $\eta = 0$ corresponds to axial symmetry. The most general type of electric field gradient is uniquely specified by the orientation of the principal axes and by the two parameters q and η.

To evaluate the contribution of the nuclear electric quadrupole interaction to the nuclear energy level configuration, as well as (as we shall see in Section 1.6) its role in coupling acoustic radiation to the nuclear spin system, we require the matrix elements of $\mathcal{H}_Q$ in the (I, m) representation. We consider, in this introductory treatment, only the case of an axially symmetric electric field gradient ($\eta = 0$) and the case in which the nuclear quadrupole interaction is small compared with the interaction energy of the nuclear magnetic moment with the external magnetic field. From conventional quantum mechanical perturbation theory, one obtains the following nonzero matrix elements [1.1,1.2]:

$$\begin{aligned}(m|\mathcal{H}_Q|m) &= \frac{A}{16}(m|[3I_z^2 - I(I+1)]V_{zz}|m) \\ &= \frac{A}{16}eq[3m^2 - I(I+1)](3\cos^2\theta - 1)/2\end{aligned} \quad (1.19a)$$

$$\begin{aligned}(m\pm 1|\mathcal{H}_Q|m) &= -\frac{A}{16}\sqrt{6}(m\pm 1|(I_\pm I_z + I_z I_\pm)V^{\mp 1}|m) \\ &= -\frac{3}{32}Aeq(2m\pm 1)[(I\pm m+1) \\ &\quad \times (I\mp m)]^{\frac{1}{2}}\sin\theta\cos\theta \, e^{\pm i\phi}\end{aligned} \quad (1.19b)$$

$$\begin{aligned}(m\pm 2|\mathcal{H}_Q|m) &= \frac{A}{16}\sqrt{6}(m\pm 2|I_\pm V^{\mp 2}|m) \\ &= \frac{3}{64}Aeq[(I\mp m)(I\mp m-1) \\ &\quad \times (I\pm m+1)(I\pm m+2)]^{\frac{1}{2}}\sin^2\theta \, e^{\pm 2i\phi} \, ,\end{aligned} \quad (1.19c)$$

where $A = e^2Q/I(2I-1)$. θ is the angle between the axis of symmetry and the direction of the DC magnetic field, and ϕ is the angle between the projection of the symmetry axis on the xy–plane and the x–axis. The operators $V^{\pm 1}$ and $V^{\pm 2}$ are given in (3.34) and (3.36).

In the high field case ($\mathcal{H}_m \gg \mathcal{H}_Q$) that we are considering, the form of the total Hamiltonian of a nuclear spin that lies in a steady magnetic field and that has a nuclear electric quadrupole coupling is

$$\mathcal{H} = \mathcal{H}_m + \mathcal{H}_Q \, , \quad (1.20)$$

where $\mathcal{H}_m$ and $\mathcal{H}_Q$ have been given previously in (1.3) and (1.17). To obtain the energies to first order in stationary perturbation theory, we use the diagonal terms in $\mathcal{H}_m$ and $\mathcal{H}_Q$ from (1.4) and (1.19a), yielding for the case $\eta = 0$

$$E_m = -\gamma_n \hbar H_0 m + \frac{e^2 qQ}{4I(2I-1)}\left(\frac{3\cos^2\theta - 1}{2}\right)[3m^2 - I(I+1)] \, . \quad (1.21)$$

As shown in Fig. 1.3(a), for a nucleus with spin $I = \frac{3}{2}$, in the absence of electric quadrupole coupling, the $2I+1$ nuclear Zeeman energy levels are equally spaced. The effect of the first–order electric quadrupole interaction is to shift these energy levels, thus splitting the single resonance line into

1.3 NUCLEAR ELECTRIC QUADRUPOLE EFFECTS

its $2I$ components, as shown in Fig. 1.3(b). Using (1.21), we obtain for the frequencies of the transitions between the new perturbed levels

$$\nu_{m \to m-1} = \frac{\gamma_n H_0}{2\pi} - (2m-1)(3\cos^2\theta - 1)\frac{3e^2qQ/h}{8I(2I-1)}, \quad (1.22)$$

where we recall that θ is the angle between the applied steady field H_0 and the principal axis of the field gradient tensor.

Inspection of (1.22) shows that the line spectrum of nuclei with odd half-integer spin quantum number contains a central ($m = \frac{1}{2} \to m = -\frac{1}{2}$) component together with $2I - 1$ satellite lines that are symmetrically disposed about the central component. The number of such components immediately provides a determination of the nuclear spin quantum number; the separation in frequency between lines yields the electric field gradient if the nuclear quadrupole moment is known. In the case of nuclei with integer spin there is no central line in the absorption spectrum. Although the position of the central line (for half-integer spin) to first order is unaffected and the center of gravity of the spectrum remains unchanged, in the nuclear quadrupole perturbation, to higher order, the central line is shifted.

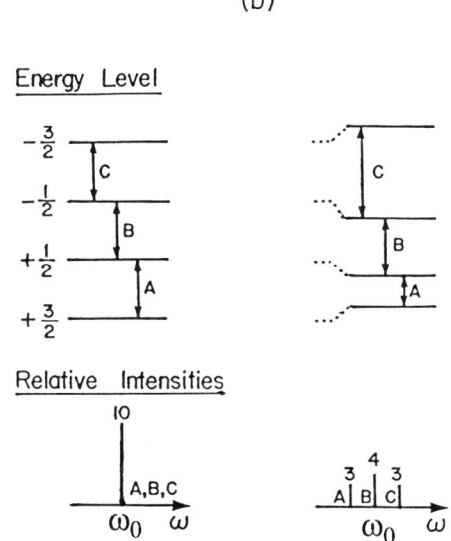

FIGURE 1.3. High field magnetic energy levels for $I = \frac{3}{2}$. (a) Magnetic field only; (b) The quadrupole interaction shifts energy levels to first order. The numerals indicate relative intensities of resonance lines.

The simplest model of the origin of the electric field gradients at the site of the nucleus is one in which the surrounding ions, considered as point charges, are responsible. It has been shown, however, that the core electrons around the nucleus, although spherically symmetric in zero order, can be polarized and can acquire a quadrupole moment by virtue of the non-central potential set up by the nucleus itself and by surrounding charges, and therefore can interact with both. The net result of this effect is to multiply the field gradient at the nucleus by a factor called $(1-\gamma_\infty)$, where γ_∞ is a constant for a given ion known as the Sternheimer antishielding factor, varying from the order of unity for light nuclei like ^{7}Li to more than 100 for ^{127}I and ^{133}Cs. A more extended discussion of electric field gradients and antishielding factors will be given in Chapter 6.

We have seen—for example, from (1.17)—that when a nucleus of spin I finds itself in an inhomogeneous electric field produced by the electric charges surrounding it, the electric quadrupole moment of the nucleus interacts with the electric field gradient to produce a set of orientation-dependent energy levels. The energy levels corresponding to the Hamiltonian of (1.17), assuming a symmetric field gradient ($\eta = 0$), are given by

$$E = \frac{e^2 qQ}{4I(2I-1)}[3m^2 - I(I+1)] \ . \tag{1.23}$$

Such energy levels, with no applied external DC magnetic field, are known as nuclear quadrupole energy levels or *pure quadrupole* energy levels. A set of levels corresponding to $I = \frac{5}{2}$ in the case of axial symmetry ($\eta = 0$) is shown in Fig. 1.4(a). In the absence of an external DC magnetic field, there are $I + \frac{1}{2}$ doubly degenerate energy levels E_m. Transitions between these levels lead to a nuclear quadrupole resonance spectrum consisting (for $I = \frac{5}{2}$) of two lines at frequencies $\nu_Q \equiv (3/20)e^2 q\, Q$ and $2\nu_Q$, as indicated in Fig. 1.4(a).

The application of a DC magnetic field removes the degeneracy, such that for $m > \frac{1}{2}$,

$$E_{\pm m} = A[3m^2 - I(I+1)] \mp m\frac{\mu}{I}H_0 \cos\theta \ . \tag{1.24}$$

The splitting of the zero-field levels and the resulting weak-field line spectrum are shown schematically in Fig. 1.4(b). The levels $|a\rangle$ and $|b\rangle$, corresponding to $m = \pm\frac{1}{2}$, are labelled so as to indicate mixing of levels at weak fields. The subject of the mixing of levels is discussed in the book by *Das and Hahn* [1.3]. The new transition frequencies for the weak field case are given by

$$\nu_{\pm m} = \frac{3A}{h}(2|m|+1) \pm \frac{\mu}{hI}H_0 \cos\theta \ . \tag{1.25}$$

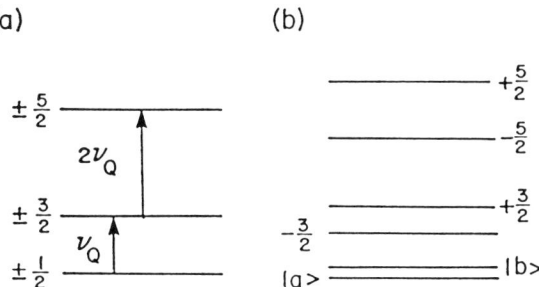

FIGURE 1.4. (a) Pure nuclear quadrupole energy levels for $I = \frac{5}{2}$. (b) Splitting of pure quadrupole energy levels on application of a weak magnetic field.

Because of the mixed levels, these are not easily shown on Fig. 1.4(b). When the field gradient tensor does not have axial symmetry (*i.e.*, $\eta \neq 0$), the expressions for the energy levels and transition frequencies are somewhat more complex. Only for $I = \frac{3}{2}$ can the solution be obtained in analytical form. For other values of I, the equations must be solved numerically for different values of η. The solutions have been tabulated by *Cohen* [1.4].

1.4 Line Width Effects

For nuclei in solids there are a number of phenomena that contribute to the width of the nuclear magnetic resonance line: lack of homogeneity of the static magnetic field; lifetime broadening due to spin–lattice relaxation processes (to be discussed in Section 1.5); coupling between magnetic dipoles; and the distribution of static electric quadrupole interactions due to the existence of strains or of impurities in the lattice. These, as well as other contributions, are discussed in the NMR literature (see, *e.g.*, [1.1,1.2]). For nuclei with spin $I = \frac{1}{2}$, the line width is determined primarily by the effect of internal magnetic fields at a given nucleus that result from the magnetic dipole coupling with surrounding nuclei. For nuclei with spin $I \geq 1$, a major additional contribution to the line width results from lattice strains due to the presence of impurities and lattice imperfections, such as dislocations and point imperfections. The distribution of strains results in a distribution of electric field gradients in both direction and magnitude. The net effect is to widen the energy levels and thereby broaden the nuclear magnetic resonance absorption line.

One can make rough estimates of the magnitudes of the above effects.

The magnetic interaction energy between identical neighboring nuclei that are a distance r apart is of the order μ^2/r^3. A typical value for the magnetic dipole field $H_{\text{loc}} \simeq \mu/r^3$ is one Oersted. The electric quadrupole interaction energy is of the order e^2Q/r^3; the ratio of electric quadrupole and magnetic dipole interaction energies, therefore, is e^2Q/μ, which for nuclei with $I > \frac{1}{2}$ falls within the range 10 to 1000.

The standard approach to the calculation of dipolar line widths in nuclear magnetic resonance is given by the method of moments, in which the second and fourth moments of the resonance line are calculated from a knowledge of the dipolar Hamiltonian and the distribution of nuclei in the crystal lattice. The calculation of quadrupolar line width effects, as related to nuclear acoustic resonance lines, is treated in some detail in Chapter 5.

In our discussion of the effect of nuclear quadrupolar coupling in the high–field case, we noted that for half integral spin, the $m = \pm\frac{1}{2}$ levels are shifted equally, so that the transition frequency between them is unaffected to first order by the quadrupole coupling. Thus the effect of a distribution of electric field gradients, due to strains in a crystal, is to shift the $m \to m \pm 1$ satellite transitions but not the $+\frac{1}{2} \leftrightarrow -\frac{1}{2}$ transition. By *satellite transition* we denote those transitions in which at least one level is not an $m = \pm\frac{1}{2}$ level. In the extreme case the satellite transitions are so broadened that they are completely wiped out and only the central $+\frac{1}{2} \leftrightarrow -\frac{1}{2}$ transition is observed. In the case of dynamic quadrupole coupling in nuclear acoustic resonance, as we shall learn in Section 1.6, the central line is not observed; thus the quadrupolar line width effects are of particular importance.

1.5 Nuclear Spin–Lattice Relaxation

The process of nuclear magnetic resonance absorption described above is predicated on the existence of more nuclear spins in a lower energy state than in a higher energy state. The effect of the externally applied RF magnetic field then is to pump up spins from a lower to a higher energy level. Were there no counterbalancing processes, this would result in an equalization of the populations of the levels and a subsequent disappearance of the absorption line. The counterbalancing process is provided by the interaction between spins and lattice. By *lattice* is meant the other internal degrees of freedom of the material in which the nuclear spins are embedded.

In a crystal the atoms move about their mean positions. Their movement is described in terms of lattice vibrations characteristic of the temperature of the crystal. Still restricting our discussion to insulating crystalline solids, we say these vibrations are made up of quasi–harmonic modes, whose excitation is described in quantum mechanical terms by the number of corresponding quasi–particles, or *phonons*, that exist in the solid. A solid

1.5 NUCLEAR SPIN—LATTICE RELAXATION

maintained at a temperature T contains a mean number of phonons per frequency mode ν, given by the Bose distribution

$$n_B = \left(e^{h\nu/k_B T} - 1\right)^{-1}. \tag{1.26}$$

The interaction between these vibration modes (phonons) and the nuclear spins produces energy exchanges between the nuclear spin system and the lattice.

The thermal lattice motion acts to keep the distribution of the nuclear spins among the states of different energy at its equilibrium (Boltzmann) value corresponding to the temperature (T) of the lattice—refer to (1.10) for the case of $I = \frac{1}{2}$. Spin–lattice relaxation is the name given to this process of maintaining thermal equilibrium in the spin system through energy exchange with the normal thermal motions of the atoms that make up the lattice. The thermal contact with the spin system results from interactions of the spin system with the random, fluctuating internal fields that derive from the thermal motions of the nuclei and electrons. We define W_+ as the probability that an upward transition in the spin system results from this nuclear spin–thermal vibration interaction; W_- is the corresponding probability of a downward transition. At equilibrium, again assuming a spin $I = \frac{1}{2}$ system,

$$(N_+ W_+)_{eq} = (N_- W_-)_{eq} \tag{1.27}$$

or, from (1.10),

$$\left(\frac{W_+}{W_-}\right)_{eq} = \left(\frac{N_-}{N_+}\right)_{eq} = e^{-2\mu H_0/k_B T}. \tag{1.28}$$

Rather than the case of the equality of transition probabilities due to a perturbing RF magnetic field, we are here considering a case in which transitions occur mutually in a coupled thermal and spin system. That there is not in this instance a complete equivalence in upward and downward transitions is demonstrated in the standard nuclear magnetic resonance texts [1.1,1.2]. Similarly, it can be shown that when the system has been perturbed from the equilibrium condition, the rate at which it returns to a thermal distribution among the spin energy states is given by

$$\frac{dn}{dt} = 2W(n_{eq} - n), \tag{1.29}$$

where $n \equiv N_+ - N_-$; $n_{eq} = (N_+ - N_-)_{eq} \cong N(\mu H_0/k_B T)$ in the conventional high–temperature regime; $N = N_+ + N_-$; and W is the average transition probability per unit time resulting from the spin–thermal phonon interaction. Solving (1.29), we obtain

$$n = n_{eq}\left(1 - \frac{n_{eq} - n_0}{n_{eq}}e^{-2Wt}\right), \tag{1.30}$$

where n_0 is the initial value of n. From (1.30) it is thus apparent that n decays exponentially from its initial value n_0 at time $t = 0$ to the equilibrium value n_{eq} at time $t = \infty$ with a characteristic time constant $1/(2W)$. We define this time constant as the spin–lattice relaxation time

$$T_1 = \frac{1}{2W}. \tag{1.31}$$

Although we have restricted our discussion to nuclei with spin $I = \frac{1}{2}$, the concepts can be generalized to the case of $I > \frac{1}{2}$ (*Slichter* [1.2]). Since the total magnetization in the z–direction (*i.e.*, direction of external magnetic field) is $n\mu$, we can write an equation analogous to (1.29) for the macroscopic magnetization:

$$\frac{dM_z}{dt} = \frac{1}{T_1}[(M_z)_{eq} - M_z]. \tag{1.32}$$

Thus T_1 can be determined in principle by observing the exponential decay of a magnetic resonance signal as a function of time after the H_1 field has been turned off. At room temperature, typical spin–lattice relaxation times for insulating crystals vary from milliseconds to seconds. These times increase markedly as the temperature is lowered.

For the lattice vibrations—or thermal phonons—to be able to transfer energy to and from the nuclear spin system, an interaction between the two systems must exist. In insulating crystals sufficiently free of paramagnetic impurities, the dominant nuclear spin–thermal phonon interaction may be that arising from the relative motion of neighboring magnetic dipoles, or it may be that arising from modulation of internal electric field gradients, which then couple to the nuclear spins via their effect on the nuclear electric quadrupole moment. Since the relaxation time T_1 is inversely proportional to the square of the interaction energy, the effect of the quadrupole interaction (for $I > \frac{1}{2}$) is $(e^2Q/\mu^2)^2 \sim 10^4$ greater than that of the magnetic dipole–dipole interaction. This was proven experimentally in a classic experiment performed by *Pound* [1.5].

Whichever internal field is being modulated by the thermal phonons, the energy exchange between the lattice and the spin system occurs primarily

1.5 NUCLEAR SPIN–LATTICE RELAXATION

through two mechanisms. In the first, a lattice vibration of energy equal to the difference between two energy levels (*e.g.*, ν_i or $2\nu_i$ of Fig. 1.3(a), where $i = A, B, C$) may be emitted or absorbed, depending upon whether a nuclear spin is excited or de-excited in the process. Because of the relatively small number of low-frequency resonant phonons at temperatures > 1 K, this direct spin–lattice interaction process is effective only at extremely low temperatures, $T \sim 10^{-3}$ K. At higher temperatures, the indirect, or Raman spin–lattice relaxation mechanism predominates. In this process, a high-energy lattice vibration is absorbed and a second high-energy lattice vibration is emitted during the interaction with the nuclear spin, the difference in energy of the two lattice vibrations being equal, again, to the difference ν_i (or $2\nu_i$ for spin $I > \frac{1}{2}$) between two spin energy levels. The indirect process is by far the more probable relaxation mechanism; it involves almost all the vibrations characteristic of the lattice at a given temperature, while the direct relaxation process involves low energy resonant phonons, which occupy only a narrow bandwidth of the normal lattice phonon spectrum.

As shown in the standard treatments of nuclear magnetic resonance [1.1,1,2], the rate of change of the population difference $n = N_+ - N_-$ of the two spin levels ($I = \frac{1}{2}$) can be expressed as the sum of the rates of change of n due to (1) the transitions induced by the RF magnetic field, and (2) the interaction of the spin system with the lattice vibrations:

$$\frac{dn}{dt} = -2W_M n + \frac{n_0 - n}{T_1}, \qquad (1.33)$$

where n_0 is the value of n when the spin system is in thermal equilibrium with the lattice, and W_M is the probability per second of inducing a spin transition between $m = \pm\frac{1}{2}$ levels by applying the RF magnetic field. T_1, the spin–lattice relaxation time, may be thought of as the time required for a spin system to attain an equilibrium distribution (*e.g.*, after a sudden change in the magnetic field).

Setting $dn/dt = 0$ in (1.33), we obtain the condition for a steady state,

$$n = \frac{n_0}{1 + 2W_M T_1}. \qquad (1.34)$$

That is, when subjected to an RF magnetic field, the nuclear spin system will continue to absorb energy from the field (*i.e.*, $n \ll n_0$) as long as $2W_M T_1 \ll 1$. When the RF magnetic field is high enough that $W_M \sim 1/(2T_1)$, the resonant absorption decreases. This saturation effect can be utilized to measure the spin–lattice relaxation time. It is also the basis of detection of acoustic saturation nuclear magnetic resonance, to be described in Section 1.7.

1.6 Interaction of Acoustic Waves with Nuclear Spins

The existence of the direct spin–lattice interaction encourages one to predict that an inverse process—the net absorption by the nuclear spin system of energy from an externally generated acoustic wave—should be observable. The acoustic waves, like the thermal phonons, should periodically modulate an internal interaction, such as the magnetic dipole–dipole interaction or the nuclear quadrupole interaction, thus exchanging energy with the nuclear spin system. Since very intense acoustic waves at ultrasonic frequencies can be introduced into the specimen, the acoustic coupling can be made quite strong even at the elevated temperatures at which the direct spin–lattice interaction is ordinarily negligible.

At least two methods capable of observing the effect of externally generated phonons on the nuclear spin system come immediately to mind. The most obvious is to monitor the effect of the acoustic phonons on a nuclear magnetic resonance line. In this case, a saturation (decrease in intensity) of the nuclear magnetic resonance line should occur because the available excess spins in the lower energy level is decreased by the combined action of RF electromagnetic and acoustic excitation. A more direct method is the observation of the resonant absorption of energy from the source of the acoustic phonons. In this method, no RF electromagnetic field is utilized; instead, the acoustic phonons alone serve the function of inducing transitions among the spin energy levels.

For most nuclei and for magnetic fields commonly available in the laboratory, the acoustic frequencies necessary to achieve nuclear spin resonance fall in the range of 1 to 100 MHz. By means of conventional piezoelectric transducers, polarized acoustic waves at these frequencies can be generated and propagated readily in solids and liquids. A brief numerical example may be helpful. The thermal lattice energy of 10-MHz phonons in a 1-kHz bandwidth (the approximate line width of nuclear spin–phonon absorption lines) at $T = 300$ K in a 1 cm^3 crystal of KI is

$$\mathcal{E}_{\text{thermal}} = \frac{8\pi}{v^3} k_B T \nu^2 d\nu \simeq 5 \times 10^{-12} \, \text{ergs/cm}^3 \,, \tag{1.35}$$

where v is the average acoustic wave velocity, and $d\nu$ is the band width at frequency ν. The energy density in the specimen corresponding to an input of 1 mW/cm^2 of coherent acoustic power at $\nu = 10$ MHz, on the other hand, is given for progressive waves by

$$\mathcal{E}_{\text{acoustic}} = \frac{P_0}{v} \simeq 3.5 \times 10^{-2} \, \text{ergs/cm}^3 \,, \tag{1.36}$$

yielding a ratio of

$$\left(\frac{\mathcal{E}_{\text{acoustic}}}{\mathcal{E}_{\text{thermal}}}\right)_{10\,\text{MHz}} \simeq 7 \times 10^9 \ . \tag{1.37}$$

P_0 is the acoustic power per unit area introduced into the specimen. Thus the energy density of coherent phonons at the resonant frequency can be made $\sim 10^{10}$ times greater than that of the incoherent thermal phonons at the same frequency.

Resonant absorption of acoustic energy by a nuclear spin system occurs when the frequency of the acoustic waves is equal to the appropriate frequency separations between the nuclear magnetic energy levels. An internal interaction capable of coupling energy to the nuclear spin system is modulated periodically by the acoustic waves, and transitions among the magnetic energy levels are induced. The greater population of the lower levels over the higher levels results in the number of transitions corresponding to absorption of energy exceeding the reverse processes. The acoustic absorption results from first–order (direct) processes involving the absorption of acoustic phonons that satisfy the resonant conditions. The calculation of the nuclear spin–acoustic phonon absorption coefficient is therefore analogous to the computation of the relaxation time T_1 of the direct thermal phonon process.

In the direct relaxation process, the nuclear spin interacts with thermal phonons in the frequency interval determined by the spin line shape factor $g(\nu)$. In the calculation of T_1, one must average over phonons of all propagation directions, polarizations and phases. The calculation of the absorption by the spin system of energy from an externally generated monochromatic acoustic wave confronts a much simpler problem: what is the probability that a single phonon will be absorbed from this wave, at the same time changing the spin state from m to m'? Analogous to the expression for nuclear magnetic resonance (1.5), the probability per second that a nucleus undergoes a transition from one magnetic level m to m', absorbing in the process one phonon, is given by

$$W_{m,m'} = \frac{1}{4\hbar^2} |\langle p', m'|\mathcal{H}_{s-p}|p, m\rangle|^2 g(\nu) \ , \tag{1.38a}$$

where $g(\nu)$ is the spin absorption line shape factor for acoustic nuclear spin transitions, and p and p' denote the initial and final phonon states.

Two mechanisms appear to be obvious candidates for the interaction of the nucleus with acoustic waves: the magnetic dipole–dipole interaction between neighboring nuclei and the nuclear electric quadrupole interaction. In the first, the relative motion of the two nuclei due to an acoustic wave produces an oscillating magnetic field component at the location of a nucleus. The oscillating magnetic field induces transitions between the Zeeman energy levels if the frequency of oscillation corresponds to the resonance condition $\nu = \Delta E/h$, where ΔE is the energy separation of the Zeeman levels. In the case of quadrupolar coupling, the acoustic wave modulates the electric field gradient at the position of the nucleus, resulting in an oscillatory nuclear quadrupole interaction that can induce transitions if the frequency of oscillation similarly obeys the resonance condition. Thus, in the case of a harmonic spin–phonon interaction via the dynamic nuclear electric quadrupole interaction, we write (1.38a) as

$$W_Q = \frac{1}{4\hbar^2}|\langle|\mathcal{H}_Q|\rangle|^2 g(\nu) \ . \tag{1.38b}$$

From (1.19b) and (1.19c) we note that there are two possible transition probabilities, corresponding respectively to $\Delta m = \pm 1$ and $\Delta m = \pm 2$. We denote these as W_{Q1} and W_{Q2}.

The probabilities for the two interactions are as follows:

(1) *Dipole–Dipole Interaction* [1.1]. An oscillatory vibration of amplitude b causes a change in the separation of the two nuclei: $d \simeq (2\pi a/\lambda)b$, where a is the equilibrium separation, and λ is the acoustic wavelength. The local magnetic field produced at one nucleus by the other is approximately $\mu/a^3 \simeq \gamma\hbar/a^3$. The effect of the acoustic wave is to produce an oscillating component of the field:

$$H_1 \simeq \frac{\gamma\hbar}{a^3}\frac{d}{a} \simeq \frac{\gamma\hbar}{a^3}\frac{2\pi b}{\lambda} \ .$$

The transition probability induced by the field H_1 is given by $W_d \sim (\gamma H_1)^2/\Delta\omega$, where $\Delta\omega$ is the line width. We evaluate

$$W_d \sim \frac{(\gamma^2\hbar 2\pi b)^2}{a^6 \lambda^2 \Delta\omega} \ ,$$

using the values

$$\gamma \simeq 2\pi \times 4.2 \times 10^7 \, \text{sec}^{-1}$$
$$a = b = 10^{-10} \, \text{m}$$
$$\lambda = 2 \times 10^{-4} \, \text{m}$$
$$\Delta\omega = 2\pi \times 10^4 \, \text{Hz} \ ,$$

to obtain $W_d \sim 10^{-5} \, \text{sec}^{-1}$.

(2) *Quadrupole Interaction.* If we assume that the electric field gradient at the nucleus is produced by a charge e (the electronic charge at each of the neighboring ions), then $\mathcal{H}_1 \sim e^2Q/\hbar a^3$, where Q is the quadrupole moment, and a is the ionic separation. With the strain again being given by $\varepsilon \sim 2\pi b/\lambda$, and we substitute into $\omega_Q \sim (e^2Q/\hbar a^3 \cdot 2\pi b/\lambda)^2/\Delta\omega$ using the values

$$Q \simeq 10^{-29} \text{m}^2$$
$$e = 1.9 \times 10^{-19} \text{ Coulombs}$$
$$a = b = 10^{-10} \text{ m}$$
$$\lambda = 2 \times 10^{-4} \text{ m}$$
$$\Delta\omega = 2\pi \times 10^4 \text{ Hz} ,$$

we obtain $W_Q \sim 10^{-2} \sec^{-1}$—substantially greater than the probability W_d for acoustically induced dipole–dipole transitions.

In the following chapters we shall find that for non-magnetic insulators and for nuclei with quadrupole moments ($I > \frac{1}{2}$) the dynamic nuclear electric quadrupole interaction is indeed the dominant means of coupling acoustic waves to nuclear spins. Together with the dynamic nuclear electric quadrupole interaction, two additional coupling mechanisms are of great importance in NAR and will be described at some length in the succeeding chapters: the dynamic Alpher–Rubin interaction, relevant to metals; and the enhanced NAR coupling, via the electron–nucleus hyperfine interaction, which is responsible for nuclear spin–phonon coupling in Van Vleck paramagnets.

1.7 Acoustic Saturation of Nuclear Magnetic Resonance

The first experiments to demonstrate acoustic coupling to nuclear spins were of the acoustic saturation type, in which acoustic coupling to the spins is observed by its effect on the nuclear magnetic resonance absorption. Specifically, the effect of the acoustic waves is determined by monitoring the magnetization of the nuclear spin system by means of a pulse technique. In this technique a short pulse of RF magnetic field H_1 at the resonant frequency is applied to a coil surrounding the specimen. Application of this pulse results in a transient pulsed signal from the spin system that is proportional in magnitude to the nuclear magnetization that existed just before the RF pulse was applied. The nuclear magnetization, as explained earlier—see (1.11) and (1.32)—is a function of the populations of various energy states. The alteration of these populations by the acoustic phonons results in a decrease in nuclear magnetization and therefore a decrease in the height of the observed transient signal.

We assume a system of N identical nuclear spins distributed among equally spaced high-field energy levels. The magnitude of the nuclear magnetization is determined by the competition between the externally applied fields—RF magnetic field and ultrasonic field—and the relaxation processes (for example quadrupolar or dipolar) which tend to restore the populations of the spin levels to thermal equilibrium with the lattice.

In a typical acoustic saturation experiment, the nuclear spin system is irradiated with ultrasonic energy at twice the NMR resonant frequency. The transition probability W_{Q2} can be made large enough so that the nuclear magnetic resonance is saturated in a time that is short compared with T_1. The populations of the equally spaced levels are monitored by measuring the magnetization M_z both before and immediately after exposure to the acoustic waves. For the case of $I = \frac{3}{2}$, the effect on the total nuclear magnetization of applying acoustic energy at frequencies corresponding to $\Delta m = \pm 2$ transitions is given by

$$\frac{A}{A_0} = \frac{M_z}{M_0} = (1 + 8W_{Q2}T_1/5)^{-1} \,, \qquad (1.39)$$

where A_0 and M_0 are, respectively, the NMR signal amplitude and equilibrium magnetization in the absence of acoustic energy, and A and M_z are the NMR signal amplitude and magnetization in the presence of acoustic energy. This relationship will be derived in Chapter 8.

The net nuclear magnetization M_z can be measured by conventional NMR pulse techniques, described in detail in standard texts [1.6,1.7]. If, in particular, a short RF pulse whose length and amplitude satisfy the condition $\gamma H_1 t_w = \pi/2$ (90° pulse, where H_1 is the amplitude, and t_w is the pulse length) is applied to an RF coil in which the specimen is placed, a transient signal will result and will be detectable in a sensitive receiver. (H_1 is perpendicular to H_0.) This transient is called a nuclear free induction decay and has an amplitude A_1 proportional to the population differences (and therefore to the net magnetization M_{z1}) at the beginning of the RF pulse. If a second 90° RF pulse is applied at some later time τ, the height A_2 of this free induction decay will measure the new magnetization M_{z2}. If (after allowing the spin system to attain thermal equilibrium) one repeats this sequence of RF pulses, but inserts during the time τ a pulse of acoustic energy whose duration is long compared to t_w, any change in the magnetization between the two cases must be attributable to the effect of the presence of resonant acoustic phonons.

A typical sequence of pulses in an acoustic saturation experiment is shown in Fig. 1.5. A_e is the amplitude of free induction decay following the initial pulse (corresponding to thermal equilibrium). A_0 is the amplitude of free induction decay following the second pulse without the intervening acoustic pulse. When the intervening acoustic pulse was applied, the pulse height A of the second free induction decay was indeed observed to depend upon the amplitude and frequency of the RF voltage at the transducer responsible for generating the acoustic pulse.

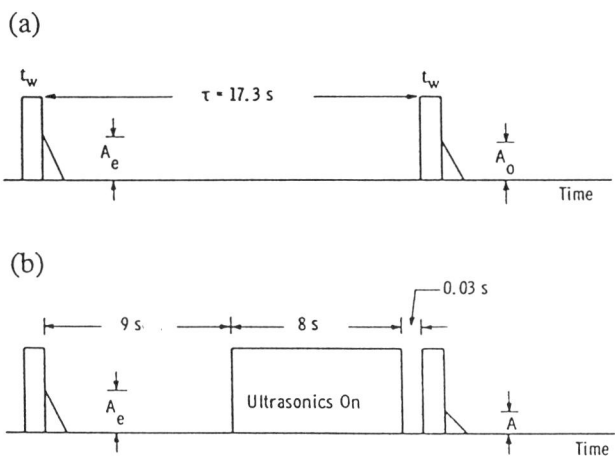

FIGURE 1.5. The sequence of events in the measurement of the effect of ultrasound in saturating magnetic energy levels. (a) Sequence of two RF pulses used to monitor magnetization at time $\tau = 17.3$ sec; $t_w = 50\,\mu$sec. (b) Ultrasonic energy is introduced into the sample during the interval $t = 9$ sec to $t = 17$ sec; the nuclear magnetization is then measured by the second RF pulse, as before. The ratio A/A_0 is a measure of the saturation due to the ultrasound. (From Ref. 1.8)

The transition probability W_{Q2} is proportional to the energy density of acoustic waves in the crystal and, hence, to the square of the peak voltage V applied to the transducer. We write k_1 for the proportionality constant, which will be discussed in greater detail in Chapter 8. Equation (1.39) can therefore be rewritten as

$$\frac{A}{A_0} = \frac{1}{1 + k_1 V^2}, \qquad (1.40)$$

where A_0 is the amplitude of the nuclear induction signal in the absence of ultrasonic excitation. A is the amplitude in the presence of ultrasonic excitation, and V the peak voltage applied to the transducer. A weakness in this technique is the necessity of independently measuring either the value of V or, alternatively, the acoustic energy density in the sample.

The acoustic saturation technique for investigating nuclear spin–phonon interactions, although seemingly less direct in its approach, is in many ways more generally applicable than the direct nuclear acoustic resonance techniques for the following reasons:

(1) It is often easier to observe an effect on a known nuclear magnetic resonance line than to search for an unknown acoustic absorption line.
(2) One can utilize very high levels of acoustic power to render an otherwise small effect observable.
(3) Since one does not rely critically on the acoustic quality factor of the specimen, one is not restricted to specimens whose background attenuation is low.

An important disadvantage of the saturation technique is that the technique is restricted because nuclear magnetic resonance cannot be observed easily in conductors. Another disadvantage is the high level of acoustic power needed, leading to the presence of non-directional, incoherent phonons. A detailed discussion of acoustic saturation NMR, as well as the related topic of double resonance, is given in Chapter 8.

1.8 Nuclear Acoustic Resonance

In the technique described in the previous section, the interaction of acoustic waves with nuclear spins is observed indirectly by the acoustic saturation of an NMR absorption line. As has been mentioned previously, NMR couples to the nuclear spins through the interaction of an RF magnetic field with the magnetic moment associated with the nuclear spin. An NMR line is observed as an absorption of energy from the source of the RF magnetic field. The closest analogy in the acoustic case is that in which *only* acoustic energy is used to couple to the nuclear spins, this coupling being observed as an absorption of energy from the source of ultrasound.

The question posed to the experimenter seeking to detect this direct acoustic coupling to nuclear spins is how to measure the very small change in acoustic power due to resonant absorption by the nuclear spin system. The frequency range lies between a few megahertz and several hundred megahertz. In this range, conventional ultrasonic pulse–echo techniques are capable of detecting changes in acoustic attenuation in solids of about 1 part in 10^3. This is sufficient sensitivity for measuring background attenuation in most solids; the background attenuation coefficient, in good acoustic materials, is of the order of 10^{-2} per centimeter. (The background attenuation is a function of type of material, state of strain, crystal imperfections, temperature, *etc.*) The attenuation coefficient due to nuclear spin–phonon coupling, however, is of the order of 10^{-7} to 10^{-8} per centimeter. Something more sensitive than the conventional acoustic techniques must therefore be utilized.

A satisfactory solution is to combine a continuous wave (CW) ultrasonic composite–resonator technique with standard nuclear magnetic resonance techniques. In practice, the sample is prepared with its two opposite faces flat and parallel, so that ultrasonic waves propagated normal to these faces reflect back and forth and potentially build up to a large amplitude. The condition for such a buildup is that the waves from successive reflections interfere constructively. For an isolated sample, an integral number of half wavelengths of ultrasound must fit within its length L: $L = n\lambda/2$ or in terms of the ultrasonic velocity, $v = \lambda\nu$. The resonant frequencies are $\omega_n/2\pi = \nu_n = nv/2L$. Such a resonator is the mechanical equivalent of a Fabry–Perot interferometer. In practice, a thin piezoelectric transducer is bonded to one face of the sample, creating a composite resonator. For such a nonisolated resonator, the impedance difference at the interface forces modification of the boundary conditions, but there still exist definite frequencies of maximum ultrasonic amplitude in one–to–one correspondence with those of the isolated sample. For accurate measurement of absolute velocity, a number of formulae have been derived to correct for the frequency shifts in composite resonators.

It is shown by *Bolef and Miller* [1.9] that the amplitude of the in–phase component of ultrasound in a mechanical resonator is

$$A_1 \sim \frac{\alpha v}{(\alpha v)^2 + (\Delta\omega)^2} \qquad (1.41)$$

for a particular mechanical resonance, provided that the departure from the resonance frequency is not too large. Here α is the attenuation in units cm^{-1}, v is the ultrasonic phase velocity, and $\Delta\omega$ is the mechanical resonance line width at the half–power points. At the center of the mechanical resonance, $A_1 \sim \alpha^{-1}$. Hence an increase in α decreases the amplitude of ultrasound. Near a resonance, the composite resonator can be represented electrically by a series RLC circuit [1.9]. A change in velocity changes the reactive elements, shifting the resonant frequency by

$$\frac{\Delta v}{v} = \frac{\Delta\nu}{\nu},$$

and a change in attenuation changes the loss elements by [1.10]

$$\frac{\Delta\alpha}{\alpha} = \frac{\Delta R}{R}.$$

Thus electrical techniques can measure the acoustic absorption and acoustic dispersion associated with nuclear spin–phonon interactions.

The frequency response of the composite resonator consists of a sharp line spectrum, such as those shown in Fig. 1.6. The spectrum is characteristic of a resonant standing wave pattern, with individual lines corresponding to mechanical resonances of the sample. For a crystal (such as an alkali halide) having low acoustic attenuation, each mechanical resonance will have a high quality factor Q, such that $Q = \nu_n/\Delta\nu$, where ν_n is a mechanical resonance frequency, and $\Delta\nu$ is the frequency difference at half–power points. Q is related to the more conventional measures of internal friction by the expression $(1/Q) = 2\alpha v/\omega = \delta/\pi$, where α is the attenuation coefficient, v is the velocity of sound, $\omega = 2\pi\nu$ and δ is the logarithmic decrement.

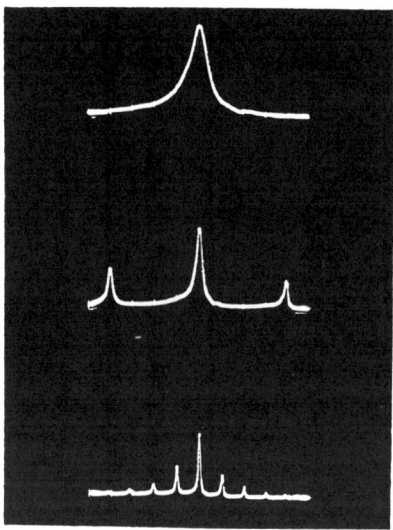

FIGURE 1.6. Typical frequency response of an ultrasonic composite resonator. The top two CRO traces show details of the complete response shown in the bottom trace.

The purely acoustic techniques are straightforward: one uses fundamental–mode or overtone piezoelectric transducers; the transducers are bonded to an oriented face of the crystal specimen; and preliminary measurements of the acoustic characteristics of the composite resonator are made on an ultrasonic pulse–echo apparatus and on an RF bridge. Special precautions in the bonding and design of the acoustic probe are necessary when measurements at low temperatures are to be made.

In one technique of NAR, the composite–resonator probe is coupled to the RF marginal oscillator of an NMR spectrometer by means of a special RF transformer and matching network. When the oscillator is adjusted to a frequency corresponding to a mechanical resonance line of high Q, it will lock in to this frequency and act effectively as a crystal–controlled oscillator. Because of the high Q of the acoustic response, with proper adjustment of the matching network, the amplitude of RF oscillations in the tank circuit of the marginal oscillator can be made to be responsive to small losses in the acoustic probe. Thus, if an additional small acoustic attenuation in the sample occurs, it will be detected as a decrease in RF oscillation level. This additional acoustic attenuation occurs when the magnetic field is swept through the value corresponding to nuclear spin–phonon resonance. A typical nuclear acoustic resonance line is shown in Fig. 1.7.

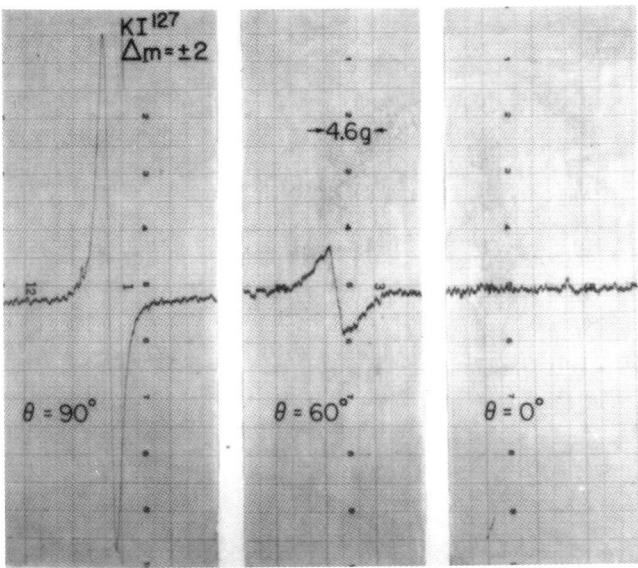

FIGURE 1.7. Nuclear acoustic resonance of ^{127}I in single–crystal KI via the dynamic nuclear electric quadrupole interaction W_{Q2}. (From Ref. 1.11)

The difficulty inherent in the acoustic saturation technique of determining the acoustic energy density in the crystal sample is overcome in part by using the direct acoustic excitation technique. The coefficient of acoustic attenuation α_n due to resonant coupling to the nuclear spin system is given by $2\alpha_n = P_n/P_0$, where P_0 is the power per unit area in the acoustic wave, and P_n is the power per unit volume absorbed by the nuclear spin system from the acoustic wave. The power density in the acoustic wave (in an isotropic solid) is given by $P_0 = \rho v^3 \varepsilon^2/2$, where ρ is the density, v the velocity of the acoustic wave, and ε the peak value of the time–varying strain due to the acoustic wave. In many of the materials investigated, the power absorbed by the nuclei depends essentially on the square of an internal interaction, such as the electric quadrupole interaction, which itself is proportional to the strain. This follows from the fact that the time–dependent part of the internal interaction is responsible for the nuclear spin–phonon coupling, and the strength of this time–dependent interaction is proportional to the strain caused by the acoustic wave. P_0 and P_n are thus both proportional to ε^2, and the acoustic attenuation coefficient α_n is therefore independent of strain.

1.9 Historical Note

The proposal that acoustic coupling to nuclear spins should be observable was first made in 1952 by *Al'tshuler* [1.12] and, independently, by *Kastler* [1.13]. Al'tshuler, in particular, predicted (in addition to the acoustic saturation of the NMR line discussed by Kastler) the observation of the direct coupling of acoustic waves to nuclear spin systems by the resonant increase in attenuation of the acoustic beam. The first detailed theory of nuclear spin–acoustic phonon interactions was given by *Al'tshuler et al.*, [1.14,1.15].

The first experiments to demonstrate acoustic coupling to nuclear spins were performed by *Proctor and Tantilla* [1.16] and by *Proctor and Robinson* [1.8] in 1955. These experiments were of the saturation type, in which the acoustic coupling was observed by its effect on decreasing the intensity of a nuclear magnetic resonance absorption. The direct observation of nuclear spin–phonon coupling as a decrease in intensity of the acoustic beam was first made by *Bolef and Menes* [1.11,1.17] in 1957. The first observation of nuclear acoustic resonance in metals (in tantalum) was by *Gregory and Bömmel* [1.18] in 1965. The theory of NAR coupling in metals via the dynamic Alpher–Rubin interaction was first suggested by *Quinn and Ying* [1.19] and was developed by *Quinn* [1.20], *Buttet* [1.21], *Fedders* [1.22] and *Müller et al.*, [1.23]. The first observation of this coupling in metals was by *Buttet, Gregory and Baily* [1.24]. SQUID–detected NAR was proposed in detail by *Mozurkewich* [1.25]. The first successful SQUID acoustomagnetic spectrometer by *Pickens* [1.26] was utilized to detect NAR in metals [1.27,1.28].

Earlier comprehensive treatments of NAR include the book by *Kessel* [1.29] and review articles by *Al'tshuler et al.*, [1.30], *Shutilov* [1.31], *Bolef* [1.32], *Burkesrode* [1.33], and *Sundfors, Bolef, and Fedders* [1.34].

CHAPTER 1 REFERENCES

1.1 A. Abragam. *The Principles of Nuclear Magnetism* (Clarendon Press, Oxford 1961).

1.2 C. P. Slichter. *Principles of Magnetic Resonance*, Third Enlarged and Updated Edition (Springer–Verlag, Berlin 1990).

1.3 T. P. Das, E. L. Hahn. *Nuclear Quadrupole Resonance Spectroscopy*, Supplement 1, *Solid State Physics*, Ed. by F. Seitz and D. Turnbull (Academic Press, New York 1958).

1.4 M. H. Cohen. "Nuclear Quadrupole Spectra in Solids." *Phys. Rev.* **96**, 1278–1284 (1954).

1.5 R. V. Pound. "Nuclear Electric Quadrupole Interactions in Crystals." *Phys. Rev.* **79**, 685–702 (1950).

1.6 N. Chandrakumar, S. Subramanian. *Modern Techniques in High-Resolution FT-NMR* (Springer–Verlag, New York 1987).

1.7 E. Fukushima, S. B. W. Roeder. *Experimental Pulse NMR* (Addison–Wesley, Reading 1981).

1.8 W. G. Proctor, W. A. Robinson. "Ultrasonic Excitation of Nuclear Magnetic Energy Levels of Na^{23} in NaCl." *Phys. Rev.* **104**, 1344–1352 (1956).

1.9 D. I. Bolef, J. G. Miller. "High–Frequency Continuous Wave Ultrasonics." Chapter 3 of *Physical Acoustics*, Vol.8, Ed. W. P. Mason and R. N. Thurston, 95–201 (Academic Press, New York 1971).

1.10 W. D. Smith, R. K. Sundfors. "An Improved Calibration Technique for CW Ultrasonic and Nuclear Resonance Spectrometers." *Rev. Sci. Instr.* **41**, 228–290 (1970).

1.11 D. I. Bolef, M. Menes. "Nuclear Magnetic Acoustic Absorption in KI and KBr." *Phys. Rev.* **114**, 1441–1451 (1959).

1.12 S. A. Al'tshuler. *Dokl. Akad. Nauk SSSR* **85**, 1235 (1952).

1.13 A. Kastler. "Quelques réflexions à propos des phénomènes de résonance magnètique dans le domaine des radiofréquences." *Experientia* **8**, 1–9 (1952).

1.14 S. A. Al'tshuler. "On the Theory of Electronic and Nuclear Paramagnetic Resonance Under the Action of Ultrasound." *Zh. ETF* **28**, 49 (1955). English Trans: *Sov. Phys. JETP* **1**, 37–44 (1955).

1.15 S. A. Al'tshuler, B. I. Kochelaev, A. M. Leushin. "Paramagnetic Absorption of Sound." *Usp. Fiz. Nauk* **75**, 1459 (1961). English trans: *Sov. Phys. Usp.* **4**, 880–903 (1962).

1.16 W. G. Proctor, W. H. Tantilla. "Influence of Ultrasonic Energy on the Relaxation of Chlorine Nuclei in Sodium Chlorate." *Phys. Rev.* **101**, 1757–1763 (1956).

1.17 M. Menes, D. I. Bolef. "Observation of Nuclear Resonance Acoustic Absorption of In^{115} in InSb." *Phys. Rev.* **109**, 218–219 (1958).

1.18 E. H. Gregory, H. E. Bömmel. "Acoustic Excitation of Nuclear Spin Resonance in Single Crystalline Tantalum." *Phys. Rev. Lett.* **15**, 404–406 (1965).

1.19 J. J. Quinn, S. C. Ying. "Acoustic Nuclear Spin Resonance in Metals". *Phys. Lett.* **23**, 61–62 (1966).

1.20 J. J. Quinn. "Direct Generation of Sound in Metals and Acoustic Nuclear Spin Resonance." *J. Phys. Chem. Solids*, **31**, 1701–1707 (1970).

1.21 J. Buttet. "Acoustic Nuclear Magnetic Resonance and Admixture of χ' and χ'' in Niobium." *Solid State Commun.* **9**, 1129–1133 (1971).

1.22 P. A. Fedders. "Acoustic Magnetic Resonance in Metals via the Alpher–Rubin Mechanism." *Phys. Rev. B* **7**, 1739–1743 (1973).

1.23 V. Müller, G. Schanz, E. Fisher, E. J. Unterhorst. "Nuclear Acoustic Resonance Absorption and Dispersion in Vanadium via Coupling to the Magnetic Dipole Moment." *Phys. Stat. Sol. (b)* **80**, 629–639 (1977).

1.24 J. Buttet, E. H. Gregory, P. K. Baily. "Nuclear Acoustic Resonance in Aluminum via Coupling to the Magnetic Dipole Moment." *Phys. Rev. B* **7**, 1739–1743 (1973).

1.25 G. Mozurkewich. "Acoustic Magnetic Resonance Investigations Utilizing Direct, Backward Wave, and SQUID Detection." Ph.D. Dissertation, Washington University, 1981 (unpublished).

1.26 K. S. Pickens. "SQUID Detection of Acoustomagnetic Effects." Ph.D. Dissertation, Washington University, 1984 (unpublished).

1.27 K. S. Pickens, G. Mozurkewich, D. I. Bolef, R. K. Sundfors, "SQUID Detection of Acousto-magnetic Effects: Nuclear Acoustic Resonance in Tantalum Metal." *Phys. Rev. Letters* **52**, 156–158 (1984).

1.28 K. S. Pickens, D. I. Bolef, M. R. Holland, R. K. Sundfors. "Superconducting Quantum Interference Device Detection of Acoustic Nuclear Quadrupole Resonance of ^{121}Sb and ^{123}Sb in Antimony Metal." *Phys. Rev. B* **30**, 3644–3648 (1984).

1.29 A. R. Kessel'. *Yadernuj Acusticheskij Resonance.* (Verlag Nauka, Moskau 1969). German trans: Akustische Kernresonanz (Akademie–Verlag, Berlin 1973).

1.30 S. A. Al'tshuler, B. I. Kochelaev, A. M. Leushin. "Paramagnetic Absorption of Sound." *Usp. Fiz. Nauk* **75**, 459–499 (1961). English trans: *Sov. Phys. Usp.* **4**, 880–903 (1962).

1.31 V. A. Shutilov. "Stimulation of Nuclear Magnetic Resonance by Ultrasound." *Akust. Zh.* **8**, 383–406 (1962). English trans: *Sov. Physics–Acoustic* **8**, 303–319 (1963).

1.32 D. I. Bolef. "Interaction of Acoustic Waves with Nuclear Spins in Solids." Chapter 3 of *Physical Acoustics*, Vol. 4A, Ed. by W. P. Mason (Academic Press, New York 1966).

1.33 W. Burkesrode. "Akustische Kernresonanz in Festkörpern." *Fortsch. der Physik* **18**, 479–526 (1970).

1.34 R. K. Sundfors, D. I. Bolef, P. A. Fedders. "Nuclear Acoustic Resonance in Metals and Alloys: A Review." *Hyperfine Interactions* **14**, 271–313 (1983).

CHAPTER 2
BASIC THEORY

2.1 Elastic Waves in Solids

2.2 Nuclear Spin Interactions

2.3 Nuclear Spin–Phonon Interactions: Fermi Golden Rule Approach

2.4 NAR Absorption and NAR Dispersion: Müller Approach

References

2. BASIC THEORY

In nuclear acoustic resonance we are concerned with the interaction of a nuclear spin (in solids and in fluids) (a) with external and internal static fields, (b) with a perturbing acoustic field introduced by means of an external source, and (c) with other nuclear spins, with electrons, with charged mobile impurities and with thermal lattice vibrations (thermal phonons). The interaction with external and internal static fields produces the set of energy levels among which spin transitions occur. This interaction thus determines the possible nuclear spin resonance spectra. The interaction with a perturbing field, produced by the externally generated acoustic wave in the case of NAR, gives rise to time–varying internal interactions, which induce specific transitions between spin energy levels. An example is the dynamic nuclear quadrupolar interaction, discussed briefly in Chapter 1. The internal interactions subsumed under (c) are responsible in large part for the width and shape of the resonance absorption lines as well as for the spin–lattice and spin–spin relaxation mechanisms. After a brief summary of the properties of elastic waves in solids in Section 2.1, a schematic and general approach to nuclear spin interactions is given in Section 2.2. (Line shape and relaxation effects will be addressed specifically in Chapter 5.)

Nuclear acoustic resonance is detected by a change of ultrasonic attenuation $\Delta\alpha$ (absorption) or ultrasonic velocity $(\Delta v/v)_\alpha$ (dispersion) when an external parameter, such as applied magnetic field or ultrasonic frequency, is swept through a nuclear resonance. Several formal approaches have been developed for calculating NAR absorption and dispersion. In Section 2.3 we present a straightforward formalism based on the Fermi Golden Rule approach, exemplified in (1.5) of Chapter 1. This calculation involves no explicit representation of the ultrasonic wave. It treats the physical problem in a formal quantum–mechanical manner. Thus, although changes in the attenuation of an ultrasonic wave are accompanied by simultaneous changes in the acoustic phase velocity or dispersion, the Fermi Golden-Rule calculation of Section 2.3 cannot give information concerning the changes in acoustic phase velocity that result from the spin–phonon interactions.

In order to obtain an expression for NAR dispersion, a term representing the ultrasound must be included in the total Hamiltonian. It is conventional to write the total Hamiltonian as

$$\mathcal{H}_{\text{total}} = \mathcal{H} + \mathcal{H}_p ,$$

where $\mathcal{H}$ is the spin Hamiltonian, and $\mathcal{H}_p$ is the phonon Hamiltonian representing the ultrasound. For the general case of coupling of acoustic modes to spin modes, *Fedders* [2.1] has obtained absorption and dispersion relationships for spin–phonon interactions in which the terms in the spin–phonon Hamiltonian are quadratic in the spin operators. The treatment of Fedders, especially appropriate to the case of dynamic nuclear electric quadrupole coupling, is presented in Chapter 3, Section 3.5, as well as in Chapter 5. A very general approach to this problem is that of Müller and his collaborators, who, building on the concepts of acoustic impedance and susceptibility, have developed an important formalism for calculating NAR absorption and dispersion. In Section 2.4 we present a summary of the Müller approach to NAR absorption and dispersion.

2.1 Elastic Waves in Solids [2.2,2.3]

The equation of motion for an element of a solid subjected to a force may be written

$$\rho \frac{\partial^2 u_i}{\partial t^2} = \frac{\partial T_{ij}}{\partial x_j} , \qquad (2.1)$$

where the right-hand side is the divergence of the stress T_{ij}, ρ is the density of the undistorted element, and u_i is a component of its displacement. We follow the convention that repeated indices are summed over. Hooke's law of elasticity further asserts that each stress component is a linear function of each strain component:

$$T_{ij} = C_{ijmn} \varepsilon_{mn} , \qquad (2.2)$$

where C_{ijmn} is the fourth-rank elastic modulus tensor. The strain tensor for small deformation is given by

$$\varepsilon_{mn} = \frac{1}{2} \left(\frac{\partial u_m}{\partial x_n} + \frac{\partial u_n}{\partial x_m} \right) . \qquad (2.3)$$

Combining (2.1), (2.2) and (2.3),

$$\rho\frac{\partial^2 u_i}{\partial t^2} = \frac{1}{2}C_{ijmn}\frac{\partial}{\partial x_j}\left(\frac{\partial u_n}{\partial x_m} + \frac{\partial u_m}{\partial x_n}\right) . \tag{2.4}$$

Since C_{ijmn} is symmetrical with respect to m and n [2.2,2.3], we may interchange m and n in the first term in brackets. For the equation of motion of an elastic body, in the absence of body–forces, this gives

$$\rho\frac{\partial^2 u_i}{\partial t^2} = C_{ijmn}\frac{\partial^2 u_m}{\partial x_j \partial x_n} . \tag{2.5}$$

It has become customary to use an abbreviated notation (due to Voigt) for the C_{ijmn} in place of the full tensor notation. Thus we have

Tensor notation:	11	22	33	23(32)	31(13)	12(21)
Voigt notation:	1	2	3	4	5	6 .

The number of independent elastic constants is reduced considerably with the degree of symmetry of the solid. In cubic crystals (*e.g.*, NaCl, GaAs), there are only three independent elastic constants: C_{11}, C_{12} and C_{44}.

We are interested in solutions of (2.5) that represent plane monochromatic elastic waves for which the displacement vector has the form

$$\boldsymbol{u} = A\boldsymbol{e}(\boldsymbol{k})\exp[i(\omega t - \boldsymbol{k}\cdot\boldsymbol{r})] , \tag{2.6}$$

where A is the amplitude of the displacement, and $\boldsymbol{k}$ is the propagation vector. The direction of $\boldsymbol{k}$ is that along which the plane wave is moving, and its magnitude $|\boldsymbol{k}| = 2\pi/\lambda$. $\boldsymbol{e}(\boldsymbol{k})$, the polarization vector, denotes the direction in which the material particles are vibrating: the wave is *longitudinal* if $\boldsymbol{e}(\boldsymbol{k})$ is parallel to $\boldsymbol{k}$, and *transverse* (or *shear*) if $\boldsymbol{e}(\boldsymbol{k})$ is perpendicular to $\boldsymbol{k}$.

In general, in an anisotropic material, the displacement is neither parallel nor perpendicular to the direction of propagation. Pure mode waves (longitudinal or transverse) are propagated only in certain crystalline directions. In standard texts [2.2,2.3] it is shown that for a cubic crystal, for example, the pure modes propagated in a [100] direction are two shear waves that propagate with the same speed ($v_{tr} = (C_{44}/\rho)^{\frac{1}{2}}$) and one longitudinal wave that propagates with the speed $v_{\text{long}} = (C_{11}/\rho)^{\frac{1}{2}}$. For propagation in the [110] direction, there are three pure modes:

$$v_1 = \left(\frac{C_{11} + C_{12} + 2C_{44}}{2\rho}\right)^{\frac{1}{2}}, \quad \text{longitudinal, displacement parallel [110];}$$

$$v_2 = (C_{44}/\rho)^{\frac{1}{2}}, \quad \text{transverse, displacement parallel [001];}$$

$$v_3 = \left(\frac{C_{11} - C_{12}}{2\rho}\right)^{\frac{1}{2}}, \quad \text{transverse, displacement parallel [1}\bar{1}\text{0].}$$

(2.7)

For relatively rigid materials with C_{ij} of the order of 10^{11} newtons/m^2, and for a typical mass density of 4×10^3 kg/m^3, the acoustic wave velocity is of order $v = 5 \times 10^3$ m/sec, and the wavelength $\lambda = 2\pi v/\omega$ is of order 5×10^{-4} m at a frequency of 10 MHz. The procedure utilized to obtain these relationships for cubic crystals can be applied equally well to other crystal symmetries, although more elastic constants are involved and more independent measurements are required to fully characterize the elastic solid.

The acoustic energy density (acoustic energy per unit volume) for an anisotropic crystal is given in [2.3] by

$$\mathcal{E} = \frac{1}{2}\sum_{ij=1}^{6} C_{ij}\varepsilon_i\varepsilon_j, \qquad (2.8)$$

where ε_i are the elastic strains ($i, j = 1, \ldots, 6$), and the C_{ij} are the elastic constants. For pure mode propagation along symmetry directions in a crystal, this reduces to

$$\mathcal{E} = \frac{1}{2}\rho v^2 \varepsilon^2, \qquad (2.9)$$

where ρ is the density of the crystal, v is the acoustic velocity, and ε is the peak strain amplitude.

Thus far we have considered only undamped, sinusoidal traveling waves, as in (2.6). If there is loss in the system, the acoustic wave has the form $e^{-\alpha x}e^{i(\omega t - kx)}$; the energy dissipation mechanisms and the scattering losses from the sound wave are expressed in terms of the attenuation factor α, given in nepers per unit length. A useful conversion formula is α(decibels/unit length) $= 8.686\alpha$(nepers/unit length). The quality of a material as a medium for wave propagation is also often described in terms of its mechanical Q, which may be defined as the number of cycles required for the wave amplitude to reduce to $e^{-\pi}$ times its initial value. The relationship between α and Q is given by

$$Q = \frac{\pi}{\alpha\lambda} \quad (\alpha \text{ in nepers/unit length}) . \tag{2.10}$$

We consider waves traveling along a rod that extends to infinity in the z-direction, the waves being maintained by a sinusoidal force acting on the available end at $z = 0$. The characteristic impedance Z_0 is defined as the complex ratio of driving force to particle velocity. (Young's modulus is generally defined to be complex. For most metals, Young's modulus is real and Z_0 is real.) For one-dimensional wave propagation, the representation of an isolated ultrasonic resonator as a section of a transmission line is often utilized. The input impedance (at $z = 0$) of a resonator of length ℓ, terminated at $z = \ell$ in some arbitrary impedance $Z(\ell)$, is given by

$$Z_{in} = Z_0 \frac{Z(\ell) + Z_0 \tanh(\theta\ell)}{Z_0 + Z(\ell) \tanh(\theta\ell)} , \tag{2.11}$$

where $Z_0 = \rho v$, $\theta = \alpha + ik$, α is the attenuation coefficient, and $k = \omega/v$ is the ultrasonic propagation constant. Equation (2.11) is the well-known equation for the transformation of impedances on either electrical or ultrasonic transmission lines.

Equation (2.11) may be applied to the case of an isolated specimen by setting $Z(\ell) = 0$. If one assumes low ultrasonic attenuation ($\alpha\ell \ll 1$), Z_{in} for the isolated specimen can be written in the approximate form

$$Z_{in} = Z_0 \frac{\alpha\ell[1 + \tan^2(k\ell)] + i\tan(k\ell)}{1 + (\alpha\ell)^2 \tan^2(k\ell)} . \tag{2.12}$$

The resonant angular frequencies ω_m are determined by the conditions for impedance minima ($k\ell = m\pi$, m an integer) in (2.12). The resonance frequencies are thus given by

$$\omega_m = m\pi v/\ell , \tag{2.13}$$

and the ultrasonic phase velocity is conveniently determined from the angular frequency separation between consecutive mechanical resonances,

$$v = (\omega_{m+1} - \omega_m)\ell/\pi . \tag{2.14}$$

For a frequency in the neighborhood of a particular mechanical resonance, $(\omega - \omega_m)\ell/v \ll 1$, and thus (2.12) can be written in the form [2.4]

$$Z_{in} = \rho v \alpha \ell + i\rho \ell(\omega^2 - \omega_m^2)/2\omega , \qquad (2.15)$$

where ρ is the density, and the substitution $Z_0 = \rho v$ is made. (Unit cross-sectional area is assumed throughout this chapter unless otherwise specified.) The impedance of a series RLC circuit is $Z = R + iL(\omega^2 - \omega_0^2)/\omega$, where ω_0 is the series resonant frequency. The impedance of the isolated specimen at the mechanical resonance can thus be related to the corresponding impedance for the electrical case if the following identifications are made:

$$R = \rho v \alpha \ell, \quad L = \rho \ell/2, \quad C = 2\ell/\pi^2 m^2 \rho v^2 . \qquad (2.16)$$

The Q of a series RLC circuit is given by $\omega_0 L/R$. Using the equivalent quantities from (2.16), one obtains for the mechanical resonance $Q = \omega_m/\Delta\omega = \omega_m/2\alpha v$ or $\omega_m/2\omega_\alpha$, where

$$\omega_\alpha \equiv \alpha v = \Delta\omega/2 , \qquad (2.17)$$

and $\Delta\omega$ is the (full) angular frequency line width of the standing wave resonance.

2.2 Nuclear Spin Interactions

The three types of spin interactions may be distinguished explicitly by writing the Hamiltonian operator of the system as

$$\mathcal{H} = \mathcal{H}_0 + \mathcal{H}_1 + \mathcal{H}'(t) . \qquad (2.18)$$

$\mathcal{H}_0$ stands for the static spin interaction (*e.g.*, Zeeman and/or static quadrupole interaction), which determines the energy level configuration. $\mathcal{H}_1$ stands for the time-dependent perturbation. In NAR this is the coupling of the ultrasonic wave to the spin system that induces transitions between energy levels and that, through absorption or dispersion, makes the nuclear spin resonance observable. In $\mathcal{H}'(t)$ are included all the remaining interactions of the nuclear spin systems; these may be static or time-dependent in nature and are primarily responsible for the shape of the observed resonance spectrum and for the relaxation effects.

An attempt at diagrammatically representing the totality of spin interactions is given in Fig. 2.1, in which are indicated eight specific types of interactions of two sets of nuclear spins in a solid medium. Among the external fields is included the perturbing Zeeman field $\boldsymbol{H}_1$, since several of the techniques to be described (for example, acoustic saturation of NMR

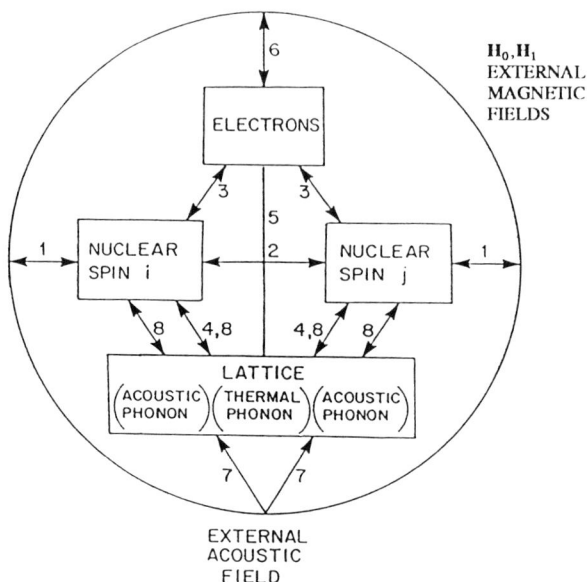

FIGURE 2.1. The eight-fold way a nuclear spin system can interact with its surroundings. 1: Zeeman interaction of spins. 2: Direct spin interaction. 3: Nuclear spin–electron interaction and indirect spin interaction. 4: Indirect (Raman) spin–lattice interaction. 3,5: Indirect spin–lattice interaction via electrons. 3,6: Shielding and polarization of nuclear spins by electrons. 7,8: Coupling of an external acoustic field to nuclear spins via acoustically induced lattice vibrations. 8: Direct spin–lattice interaction between nuclear spins and resonant thermal phonons. (Adapted from Ref. 2.6)

and double resonance techniques) utilize NMR and electron spin resonance (ESR), as well as nuclear spin–phonon interactions. The direct external nuclear spin interactions are represented by Path 1 (Zeeman fields H_0 and H_1) and by Paths 7 and 8 (external acoustic field).

The spin interactions of two different types of spins (i, j) with internal fields may be written as

$$\mathcal{H}_{\text{int}} = \mathcal{H}_{ii} + \mathcal{H}_{jj} + \mathcal{H}_{ij} + \mathcal{H}_Q + \mathcal{H}_S + \mathcal{H}_L . \qquad (2.19)$$

$\mathcal{H}_{ii}$ and $\mathcal{H}_{jj}$ represent the direct (dipolar) and the indirect interactions among i and j spins, respectively (Paths 2 and 3 in Fig. 2.1). The same paths (2 and 3) are involved in $\mathcal{H}_{ij}$, which covers the indirect and direct interactions between i spins and j spins. $\mathcal{H}_Q$ is the quadrupole Hamiltonian of the i and j spins, which has static and time–dependent terms and may contribute to all three terms [$\mathcal{H}_0$, $\mathcal{H}_1$ and $\mathcal{H}'(t)$] in (2.18).

$\mathcal{H}_S$ contains all the *shielding* Hamiltonians (*e.g.*, chemical shift and Knight shift) of the i and j spins (Paths 3 and 6 in Fig. 2.1); these interactions are important in high resolution NMR spectroscopy in solids, and are discussed in detail in the NMR literature [2.5-2.7]. Finally, $\mathcal{H}_L$ describes the spin–lattice interaction (Paths 4,8 and 3,5 in Fig. 2.1). As indicated in Fig. 2.1, the electrons may play an important role (i) in the shielding and polarization of the nuclear spins (Paths 3,6); (ii) in direct nuclear spin–electron interactions (Path 3) and indirect nuclear spin interactions, mediated by the electron spins (Path 3); (iii) in providing a spin–lattice relaxation mechanism for the nuclear spins (Paths 3,5).

A particularly elegant and general formalism for the expression of $\mathcal{H}$ in NAR is that which utilizes spherical tensor operators. In our discussion of NAR coupling mechanisms (Chapter 3), and of NAR line shapes (Chapter 5) in particular, we will find it convenient to introduce a set of irreducible tensor operators A_{lm} that fully describe the dynamics of a spin.

With respect to $\mathcal{H}_1$ of (2.18), there are two common mechanisms for the coupling of acoustic waves to nuclear spins. One is the dynamic electric quadrupole coupling, in which an acoustic wave is coupled dynamically to the electric quadrupole moments of the nuclear spins; this coupling occurs in conductors as well as non–conductors. The other mechanism, effective only in conductors, is the dynamic Alpher–Rubin coupling, in which an acoustic wave is coupled to the magnetic dipole moments of the nuclear spins. It is also theoretically possible for acoustic waves to couple to the electric hexadecapole (16–pole) moments of the nuclear spins, as well as via a dynamic dipole–dipole coupling. Since these mechanisms couple to different multipole moments of the nuclear spins, they lead to different intensities, angular dependences, line widths and line shapes of the observed resonance lines in NAR experiments. Dynamic multipole coupling is discussed in Chapter 3. NAR coupling in conductors via the dynamic Alpher–Rubin coupling is discussed in Chapter 4.

An important spin–phonon coupling mechanism is observed in Van Vleck paramagnets, in which an acoustic strain results in the modulation of the magnetic field–induced electronic magnetic moment M_e. Under these circumstances, in addition to the applied magnetic field B, there exists at the nucleus of the paramagnetic ion (*e.g.*, ^{165}Ho in HoVO$_4$) a much larger (enhanced) field B_e set up by the magnetic hyperfine interaction and proportional to M_e. The effect of an applied acoustic strain is to change the direction of M_e, resulting in a corresponding change in direction of B_e, and thus producing an oscillatory component of B_e normal to the external field B. Transitions are induced within the nuclear spin system, as usual, when the acoustic frequency $\omega = \Delta E/\hbar$. Both $\Delta m = \pm 1$ and $\Delta m = \pm 2$ transitions are allowed. A detailed discussion of this enhanced NAR in Van Vleck paramagnets is given in Chapter 9.

2.3 Nuclear Spin–Phonon Interactions: Fermi Golden Rule Approach

In nuclear acoustic resonance, the increased attenuation is due to the absorption of acoustic energy by the nuclear spin system undergoing a transition between two spin states m and m'. The power-per-unit-volume P_n absorbed by the nuclear spin system from the acoustic wave for a spin transition $h\nu = E_m - E_{m'}$ is

$$P_n = nh\nu W_{m,m'} , \qquad (2.20)$$

where n is the equilibrium population difference per unit volume between the spin energy levels, $h\nu$ is the energy exchanged between the ultrasonic wave and the spin system, and $W_{m,m'}$ is the probability per unit time that a spin makes a transition from an initial state m to a final state m'. Assuming that the absorption of energy from the acoustic wave does not cause saturation (thereby maintaining a Boltzmann distribution of spins), the population difference per unit volume is

$$n = n_m - n_{m'} = \left[\frac{Ne^{-E_m/k_BT}}{\sum_{m'=-I}^{I} e^{-E_{m'}/k_BT}} \right] \left[e^{h\nu/k_BT} - 1 \right] .$$

In the high temperature limit ($h\nu \ll k_BT$),

$$n \simeq \frac{N}{(2I+1)} \frac{h\nu}{k_BT} ,$$

where N is the number of spins per unit volume. Substituting into (2.20), we obtain

$$P_n = \frac{N}{(2I+1)} \frac{(h\nu)^2}{k_BT} \sum_m W_{m,m'} , \qquad (2.21)$$

where the summation is taken over those transitions contributing to the observed resonance line.

It is useful to define an acoustic power absorption coefficient $2\alpha_n$ as

$$2\alpha_n = \frac{P_n}{P_0} , \qquad (2.22)$$

where α_n is the amplitude attenuation coefficient due to the nuclear spin–phonon interaction, and $P_0 = \mathcal{E}v$ is the acoustic power-per-unit-area introduced into the specimen. $\mathcal{E}$, the energy density of an acoustic wave containing n_p phonons of frequency ν introduced into a specimen of volume V, is given in quantum terms by

$$\mathcal{E} = \frac{n_p h \nu}{V} . \qquad (2.23)$$

Combining (2.21), (2.22) and (2.23), we obtain for the nuclear spin–acoustic phonon absorption coefficient

$$\alpha_n = \left(\frac{N}{2I+1}\right)\left(\frac{h\nu}{k_B T}\right)\frac{V}{2n_p v}\sum_m W_{m,m'} , \qquad (2.24)$$

or, substituting from (1.38),

$$\alpha_n = \left(\frac{N}{2I+1}\right)\left(\frac{\pi\nu}{k_B T}\right)\left(\frac{V}{4\hbar n_p v}\right) g(\nu)\sum_m |\langle m', p' | \mathcal{H}_{s-p} | m, p \rangle|^2 . \qquad (2.25)$$

A related quantity is the *spin–phonon relaxation time*, or the mean time that a phonon exists in the specimen before it is absorbed by a spin:

$$\tau_n = \frac{1}{2\alpha_n v} . \qquad (2.26)$$

The calculation of the matrix elements $|\langle \mathcal{H}_{s-p} \rangle|$ of the nuclear spin–phonon interaction requires a knowledge of (1) the energy levels of the nuclear spin system, and (2) the mechanism responsible for the coupling of the acoustic wave to the nuclear spin system. The energy levels are generally known from a knowledge of the nuclear properties (spin, magnetic moment, quadrupole moment, experimentally determined static quadrupole interaction) and of the magnitude and direction of the externally applied static magnetic field. In anticipation of our discussion of specific spin–phonon interactions in Chapters 3 and 4, we here derive a general expression for the interaction of acoustic waves with nuclear spins. Following (2.6) we write for the displacement of the ith nucleus

$$\boldsymbol{u}_i = \boldsymbol{e} A \sin(\omega t - \boldsymbol{k} \cdot \boldsymbol{r}_i) , \qquad (2.27)$$

where $\boldsymbol{r}_i$ is the radius vector of the nucleus i under consideration.

The change u_{ij} in the relative separation $r_{ij} = r_i - r_j$ of nuclei i and j produced by the acoustic wave is given by

$$u_{ij} = u_j - u_i \simeq e\,A\,k\,r_{ij}\cos(\omega t - \mathbf{k}\cdot\mathbf{r}_i) \qquad (2.28)$$

and is obtained under the assumption that

$$r_{ij} \ll \lambda \quad \text{or} \quad k\,r_{ij} \ll 1 \;. \qquad (2.29)$$

In the case of standing waves, (2.28) may be written

$$u_{ij} \simeq e\,A\,k\,r_{ij}\sin\mathbf{k}\cdot\mathbf{r}_i\,\cos(\omega t) \;, \qquad (2.30)$$

subject to conditions (2.29). These conditions are generally satisfied for NAR in solids, where $r_{ij} \sim 10^{-9} - 10^{-10}$ m, and $\lambda = v/\nu \simeq 10^3$ msec$^{-1}/10^7$ Hz $\sim 10^{-4}$ m. In addition, $A \ll \lambda$, so that the relative fluctuations induced by the acoustic wave are small compared with the relative separation between nuclei,

$$u_{ij}/r_{ij} \simeq 2\pi A/\lambda \ll 1 \;. \qquad (2.31)$$

For the interaction Hamiltonian of the ith nucleus with its surroundings, which depends on the spin variable i and the relative separations r_{ij} between itself and its neighboring nuclei, we write [2.8]

$$\mathcal{H}_i = \sum_j \mathcal{H}_i(I_i, r_{ij}) \;. \qquad (2.32)$$

Under conditions of acoustic modulation, $\mathcal{H}_i$ varies periodically with time. Applying approximations (2.29) and (2.31), we may write

$$\sum_j \mathcal{H}_i(I_i, r_{ji} + u_{ji}) = \sum_j \mathcal{H}_i(I_i, r_{ji}) + \sum_j \nabla_{r_{ij}} \mathcal{H}_i(I_i, r_{ji})u_{ji} \;, \qquad (2.33)$$

retaining only the first term in the expansion in the small quantity u_{ji}. We thus arrive at

$$\mathcal{H}_i = \sum_j \nabla_{r_{ji}} \mathcal{H}_i(I_i, r_{ji})u_{ji} \qquad (2.34)$$

as the general expression for the time–dependent interaction operators of spin i with the acoustic wave. This is the nuclear spin–phonon Hamiltonian. The most important nuclear spin–phonon interactions are the dynamic nuclear quadrupole coupling, treated in Chapter 3, and the dynamic Alpher–Rubin coupling (often referred to as the dynamic nuclear dipolar coupling), treated in Chapter 4. We refer to the former as NAR1 and as NAR2, and to the latter as NAR(A-R).

2.4 NAR Absorption and NAR Dispersion: Müller Approach

Although NMR and NAR differ essentially in the applied time–varying fields and in their coupling to the nucleus, which induces nuclear spin transitions, there is no great difference in the apparatus suitable for their observation. In either case, an RF oscillator is used to produce the time–varying field—an RF magnetic field produced by a coil, in the case of NMR; an RF acoustic field produced by a piezoelectric transducer, in the case of NAR. If the electric voltage (or current) is sufficiently small, linear response may be assumed, and the device being tested can be described by an electric impedance $Z(\omega)$ which changes due to nuclear spin transitions.

A reasonable measure for the detectability of nuclear spin transitions is the relative impedance change $(\Delta Z/Z_0)$, where Z_0 is the real part of $Z(\omega)$ without spin transitions. In NMR, this ratio is usually expressed by the nuclear spin susceptibility χ_{NMR}. For a series–resonant circuit, it can be shown [2.5] that

$$\Delta Z/Z_0 = i\,\eta_e\, Q_e\, \chi_{NMR}\;, \qquad (2.35)$$

where Q_e is the quality factor of the series resonant circuit, and η_e is the filling factor of the RF coil. By following Kubo's idea of generalized susceptibility [2.9], it is possible to define a nuclear acoustic resonance spin susceptibility χ_{NAR}

$$\Delta Z/Z_0 = i\,\eta_a\, Q_a\, \chi_{NAR}\;, \qquad (2.36)$$

where $Q_a = (\omega_n \tau)$. ω_n is the frequency of an acoustic standing wave resonance, $1/\tau$ is the phonon loss rate, and η_a is the acoustic filling factor (≈ 1).

In order to relate the macroscopic observations in an NAR spin–phonon experiment to the particular spin–phonon interaction, we need an expression for the generalized NAR susceptibility χ_{NAR}. *Kubo* [2.9] uses a method of *generalized susceptibility*, correlating an external perturbation Hamiltonian $\mathcal{H}'(t)$ with relevant experimental quantities. This method is successful in NMR [2.10] and in NAR [2.11] if the perturbing Hamiltonian $\mathcal{H}'(t)$ is of the form

$$\mathcal{H}'_f(t) = W\, f(t)\;, \qquad (2.37)$$

where $f(t)$ is an external force, and W is an operator that is assumed to be independent of $f(t)$. Kubo's concept of generalized susceptibility represents one of the most powerful tools correlating an external perturbation Hamiltonian $\mathcal{H}'_f(t)$ with relevant experimental quantities. The Kubo method fails if the perturbing Hamiltonian is not of the form of (2.37). It is then impossible to express the absorbed acoustic power in terms of the Kubo susceptibility $\chi_{\mathcal{H}'_j}$, which for $\omega \neq 0$ is defined by

$$-\text{Tr}\{\rho(\omega)W\} = \chi_{\mathcal{H}'_j}(\omega) f(\omega) , \qquad (2.38)$$

where $\rho(\omega)$ is the probability density operator, and the argument ω stands for the time–Fourier transform. In the particular case of NAR(A-R), $f(t)$ is the driving acoustic wave, which results in acoustically induced electrostatic and electromagnetic fields that are, in general, phase–shifted relative to the driving $f(t)$. Therefore Kubo's treatment, whose main premise is the validity of the linear response of the density operator $\rho(t)$ to the external force $f(t)$, cannot be applied to NAR(A-R) without generalization [2.8, 2.12, 2.13].

The variation of $\rho(t)$ is given by

$$i\hbar \frac{\partial \rho}{\partial t} = [\mathcal{H}_0 + \mathcal{H}'(t), \rho(t)] . \qquad (2.39)$$

If we keep only the first–order terms in $\mathcal{H}'(t)$, we can show that the Fourier transform of (2.39) becomes

$$\rho(\omega) = \frac{1}{i\hbar}[\widetilde{\mathcal{H}}'(\omega), \rho_0] , \qquad (2.40)$$

with

$$\widetilde{\mathcal{H}}'(\omega) = \int_0^\infty dt\, e^{-i\omega t}\, e^{-(i/\hbar)\mathcal{H}_0 t}\, \mathcal{H}'(\omega)\, e^{(i/\hbar)\mathcal{H}_0 t} , \qquad (2.41)$$

where $\mathcal{H}_0$ is the Hamiltonian of the unperturbed physical system, and ρ_0 is the statistical operator at $t = -\infty$ (ρ_0 is diagonal in the energy basis). As a result of (2.40) and (2.41), the Fourier transform of the most general perturbation Hamiltonian—which is compatible with the assumption of a linear response of $\rho(t)$ to $f(t)$—must be of the form

$$\mathcal{H}'(\omega) = W(\omega) f(\omega) . \qquad (2.42)$$

By using the convolution theorem, one can write

$$\mathcal{H}'(t) = \frac{1}{2\pi} \int_{-\infty}^\infty d\tau\, W(\tau)\, f(t-\tau) . \qquad (2.43)$$

$W(\omega)$ is the Fourier transform of the response operator $W(\tau)$, which is independent of $f(\tau)$ and which, due to the causality principle, is zero for $\tau < 0$.

2.4 NAR ABSORPTION AND NAR DISPERSION: MÜLLER APPROACH

One can define a generalized susceptibility $\chi_{\mathcal{H}'}(\omega)$ by substituting $\mathcal{H}'(\omega)/f(\omega)$ for W in (2.38):

$$\chi_{\mathcal{H}'}(\omega) = \frac{\text{Tr}\{\rho(\omega)\mathcal{H}'(\omega)\}}{f(\omega)f(\omega)}. \tag{2.44}$$

It is to be noted that this defining equation for $\chi_{\mathcal{H}'}$ is not restricted to a perturbation of the type $\mathcal{H}'_f$ given in (2.37).

The concept of a generalized susceptibility may be applied to NAR when the strain tensor ε is regarded as the driving force $f(t)$. To make the generalized susceptibility a dimensionless quantity, as in the nuclear magnetic spin susceptibility, it is useful to consider $f^2(t)$ as an energy-like quantity. Since the acoustic energy is constant for a standing acoustic wave, it follows that

$$\rho_s\{v_a^2 \int d^3r[\varepsilon'(r,t)]^2 + \int d^3r[v_i(r,t)]^2\} \tag{2.45}$$

must be zero (for $\omega \neq 0$). We may therefore write

$$[f(\omega)]^2 = \rho_s v_a^2 \int_{(V_s)} d^3r[\varepsilon'(r,\omega)]^2 \tag{2.46a}$$

$$[f(\omega)]^2 = -\rho_s \int_{(V_s)} d^3r[v_i(r,\omega)]^2, \tag{2.46b}$$

where ρ_s is the mass density of the sample, V_s is the volume of the sample, v_i is the lattice displacement velocity, and v_a is the phase velocity of the acoustic wave.

We may now write the generalized NAR susceptibility as

$$\chi_{\text{NAR}}(\omega) = -\frac{\text{Tr}\{\rho(\omega)\mathcal{H}'(\omega)\}}{\rho_s v_a^2 \int_{(V_s)} d^3r[\varepsilon'(r,\omega)]^2} \tag{2.47}$$

or

$$\chi_{\text{NAR}}(\omega) = \frac{\text{Tr}\{\rho(\omega)\mathcal{H}'(\omega)\}}{\rho_s \int_{(V_s)} d^3r[v_i(r,\omega)]^2}. \tag{2.48}$$

In order to compute χ_{NAR}, we must know $f(\omega)$ and $\mathcal{H}'(\omega)$ and use (2.40) and (2.41) for $\rho(\omega)$ as follows:

$$\chi_{\text{NAR}}(\omega) = -\frac{1}{i\hbar} \frac{\text{Tr}\{[\tilde{\mathcal{H}}'(\omega), \rho_0]\mathcal{H}'(\omega)\}}{\rho_s v_a^2 \int_{(V_s)} d^3r[\varepsilon'(r,\omega)]^2} \tag{2.49a}$$

or
$$\chi_{\text{NAR}}(\omega) = \frac{1}{i\hbar} \frac{\text{Tr}\{[\tilde{\mathcal{H}}'(\omega), \rho_0]\mathcal{H}'(\omega)\}}{\rho_s \int_{(V_s)} d^3r [v_i(\boldsymbol{r},\omega)]^2} . \tag{2.49b}$$

Equations (2.49a,b) are solved in Section 4.6.1 for NAR(A-R), NAR1 and NAR2.

The relative impedance change given in (2.36) is used to define χ_{NAR} and can also be used to describe quantities measured in CW NAR experiments. In such experiments, the device being tested is the specimen with the acoustic transducer bonded to it, the whole forming a composite resonator. It can be shown (in [2.8], also see also Chapter 1) that in a CW NAR experiment in which the CW frequency ω is kept fixed at a value corresponding to a particular mechanical resonance ω_n, a change of ω_n or of the phonon loss rate $1/\tau$ causes a change in acoustic impedance for a particular acoustic component:

$$\frac{\Delta_j Z}{Z_0} = \eta_a \frac{Q_a^{(n)}}{\omega_n} [\Delta_j(1/\tau) - 2i\Delta_j \omega_n] , \tag{2.50}$$

where $Q_a^{(n)}$ is the acoustic quality factor at the mechanical resonance frequency ω_n, and Δ_j indicates a change due to the jth mechanism for acoustic absorption or dispersion. In general, $\Delta\omega_n/\omega_n = \Delta v_a/v_a$, and $\Delta(1/\tau) = 2v_a \Delta\alpha$, where α is the acoustic attenuation.

Rewriting (2.36), we find

$$\frac{\Delta_j Z}{Z_0} = i\eta_a Q_a^{(n)} \chi_{\text{NAR }j}(\omega) , \tag{2.51}$$

where $\chi_{\text{NAR }j}$ is the acoustic NAR susceptibility of the jth component. We further write

$$\chi_{\text{NAR }j} = \chi'_{\text{NAR }j} - i\chi''_{\text{NAR }j} . \tag{2.52}$$

By comparing (2.50) with (2.51) and (2.52), we may write for the change in phonon loss rate (acoustic absorption)

$$\Delta_j(1/\tau) = \omega_n \chi''_{\text{NAR }j} , \tag{2.53a}$$

and for the change in acoustic phase velocity v_a (acoustic dispersion),

$$\Delta_j \omega_n = -(\omega_n/2)\chi'_{\text{NAR }j} . \tag{2.53b}$$

In terms of acoustic quantities, (2.53) can be written as

$$\Delta_j \alpha = (\omega_n/2v_a)\chi''_{\text{NAR }j} \tag{2.54a}$$

and

$$\frac{\Delta_j v_a}{v_a} = (1/2)\chi'_{\text{NAR }j} . \tag{2.54b}$$

CHAPTER 2 REFERENCES

2.1 P. A. Fedders. "Resonant and Nonresonant Effects of Paramagnetic Spins on Acoustic Modes." *Phys. Rev. B* **12**, 2045–2048 (1975).

2.2 H. F. Pollard. *Sound Waves in Solids* (Pion Limited, London, 1977).

2.3 R. T. Beyer, S. V. Letcher. *Physical Ultrasonics* (Academic Press, New York, 1969).

2.4 R. G. Leisure, D. I. Bolef. "CW Microwave Spectrometer for Ultrasonic Paramagnetic Resonance." *Rev. Sci. Instr.* **39**, 199–205 (1968).

2.5 C. P. Slichter. *Principles of Magnetic Resonance*, Third Enlarged and Updated Edition, Second Corrected Printing (Springer–Verlag, Berlin 1990).

2.6 N. Mehring. *High Resolution NMR Spectroscopy in Solids* (Springer–Verlag, Berlin 1976).

2.7 N. Chandrakumar, S. Subramanian. *Modern Techniques in High-Resolution FT-NMR* (Springer–Verlag, New York 1987).

2.8 V. Müller, U. Bartell. "Nuclear Acoustic Resonance in Metals." *Zeitschrift für Physik B* **32**, 271–279 (1979).

2.9 R. Kubo. "Statistical Mechanical Theory of Reversible Processes. I." *J. Phys. Soc. Jap.* **12**, 510–586 (1957).

2.10 R. Kubo, K. Tomita. "A General Theory of Magnetic Resonance Absorption." *J. Phys. Soc. Jap.* **9**, 888–919 (1954).

2.11 A. R. Kessel'. *Yadernuj Acusticheskij Resonance.* (Verlag Nauka, Moskau 1969). German trans: Akustische Kernresonanz (Akademie-Verlag, Berlin 1973).

2.12 V. Müller, G. Schanz, E. Fischer, E. J. Unterhorst. "Nuclear Acoustic Resonance Absorption and Dispersion in Vanadium via Coupling to the Magnetic Dipole Moment." *Phys. Stat. Sol. (b)* **80**, 629–639 (1977).

2.13 V. Müller. "Nuclear Acoustic Resonance Absorption and Dispersion." *Phys. Lett.* **60A**, 240–242 (1977).

CHAPTER 3
DYNAMIC NUCLEAR ELECTRIC QUADRUPOLE INTERACTIONS

3.1 Multipole Expansions Using Spherical Tensors:
The A_{lm} Irreducible Tensor Operators

3.2 Nuclear Electrostatic Multipole Interactions

3.3 Dynamic Electric Quadrupole Interaction

3.4 Electric Field Gradient:
Symmetry Considerations and the S–Tensor

3.5 Fedders' Calculation of Absorption and Dispersion

3.6 Application. NAR in Noncubic Metallic Rhenium:
Pure Nuclear Electric Quadrupole Resonance
    3.6.1 Energy Levels: Low Magnetic Field Limit
    3.6.2 Dynamic Coupling
    3.6.3 Magnitude and Sign of Quadrupole Coupling Constants
    3 6.4 Other Experimental Results

3.7 Dynamic Nuclear Electric Hexadecapole Interaction

3.8 Is the Hexadecapole Interaction (HDI)
Observable by NAR Techniques?

3.9 Dynamic Magnetic Dipole–Dipole Interaction

3.10 Additional Mechanisms for Multipole Quantum Transitions
    3.10.1 Anharmonic Phonon Effects
    3.10.2 Second–Order Quadrupole–Quadrupole Transition
    3.10.3 RF-Induced Dipole–Dipole Transitions
    3.10.4 Comparison of Mechanisms

References

3. DYNAMIC NUCLEAR ELECTRIC QUADRUPOLE INTERACTION

There are two common mechanisms for the coupling of acoustic waves to nuclear spins. One is the dynamic electric quadrupole coupling, in which an acoustic wave is coupled dynamically to the electric quadrupole moments of the nuclear spins. The other, effective only in conductors, is the dynamic Alpher–Rubin coupling, in which an acoustic wave is coupled to the magnetic dipole moments of the nuclear spins. It is also theoretically possible for acoustic waves to couple to the electric hexadecapole (16–pole) moments of the nuclear spins, as well as via a dynamic dipole–dipole coupling. Since these mechanisms couple to different multipole moments of the nuclear spins, they lead to different intensities, angular dependences, line widths and line shapes of the observed resonance lines in NAR experiments. In Chapter 3 we treat dynamic multipole coupling, reserving a detailed discussion of the dynamic Alpher–Rubin coupling for Chapter 4.

In Section 3.1 of the present chapter, we introduce the A_{lm} irreducible tensor operators to fourth order ($l = 0, 1, 2, 3, 4$), summarize their properties and apply them in the derivation of the dynamic quadrupole Hamiltonians. In Section 3.2 we explicitly write out the form of the general nuclear electrostatic multipole interaction in terms of spherical harmonics and we discuss the nature of the allowed nuclear electrostatic moments. The results of these two sections are utilized in Section 3.3 to derive the detailed expressions for the static and dynamic nuclear electric quadrupole interactions in NAR. In order to elucidate the properties of the electric field gradient tensor (introduced in the previous sections), in Section 3.4 we introduce the fourth–order electrostatic field gradient–elastic strain tensor (or S–tensor), giving detailed applications to the case of cubic symmetry. (The transformation properties of the S–tensor are specified in Appendix D.) As a demonstration of the utility of the A_{lm}–tensor formalism, in Section 3.5 we derive NAR absorption and dispersion for the dynamic quadrupole interaction for the specific case of longitudinal acoustic waves propagated along a cube edge. In Section 3.6 we present the results of an NAR pure quadrupole resonance experiment in the noncubic metal rhenium. This experiment illustrates the use of much that is discussed in the previous sections of Chapter 3. The formalism of the dynamic and static nuclear electric hexadecapole interaction is given in Section 3.7. In Section 3.8 we discuss the possibility of observing the dynamic nuclear electric hexadecapole interaction. Another possible multipole interaction, acoustic wave modulation of the magnetic dipole–dipole interaction, is discussed in Section 3.9. Finally, in Section 3.10 we survey three acoustic coupling mechanisms, in addition to the electric hexadecapole, that would allow multiple quantum transitions.

3.1 Multipole Expansions Using Spherical Tensors: The A_{lm} Irreducible Tensor Operators

In dealing with multipole interactions in magnetic resonance, the usefulness of the formalism utilizing spherical tensor operators has been widely recognized. Applications of this formalism to quadrupole interactions are given

50   3. DYNAMIC NUCLEAR ELECTRIC QUADRUPOLE INTERACTION

in *Abragam* [3.1], *Slichter* [3.2] and *Poole and Farach* [3.3]; applications to dipole–dipole interactions are found in *Mehring* [3.4] and *Chandrakumar and Subramanian* [3.5]. The irreducible spherical tensor operator formalism is especially useful in NAR, as has been shown in the work of *Fedders* [3.6]. In this section we outline the formalism, referring the reader to standard treatments for details of derivations, and indicate its relevance to problems in NAR. In the following sections, the formalism will be applied to the coupling of acoustic waves to quadrupole and hexadecapole nuclear moments. The power of the technique is best demonstrated in an elegant analysis, by *Fedders* [3.6], of line shapes and relaxation in quadrupolar NAR, a topic dealt with in Chapter 5.

Experiments in NMR are often interpreted in terms of the magnetization components M_z, M_+ and M_-, corresponding to the spin operators I_z, I_+ and I_-. To fully describe the dynamics of a single spin of quantum number I, however, one requires a set of $(2I+1)^2$ operators. For $I = \frac{1}{2}$, for example, the $(2I+1)^2 = 4$ operators usually chosen, and the effect of operating with them on a given spin state $|m\rangle$, are

$$
\begin{aligned}
1 &: 1|m\rangle = |m\rangle \\
I_z &: I_z|m\rangle = m|m\rangle \\
I_\pm &: I_\pm|m\rangle = [I(I+1) - m(m\pm 1)]^{\frac{1}{2}} |m\pm 1\rangle .
\end{aligned} \quad (3.1)
$$

An infinite number of complete sets of operators may be created by making linear combinations of these. In the high temperature limit, where k_BT is much greater than any spin energy, a convenient complete set is the set of irreducible multipole operators A_{lm} [3.6].

Because the spin–dependent and coordinate–dependent parts transform as second–rank tensors, all interactions dealt with in magnetic resonance spectroscopy can be expressed in tensor notation in terms of irreducible spherical tensor operators [3.5]. The spin operators in the spherical basis and the various spherical harmonics are very similar; one may construct tensor operator equivalents by a systematic substitution of position variables by the corresponding spin components using the equivalent operator formalism. We must keep in mind, however, that while the position operators commute with each other, the spin operators need not. By using *Rose's* definition of a spherical tensor [3.7], the mth component of an lth rank tensor is given by

$$A_{lm}(I) = b_l(\mathbf{1}\cdot\nabla)^l r^l Y_l^m(r) , \quad (3.2)$$

3.1 MULTIPOLE EXPANSIONS

where the operator $(\mathbf{1} \cdot \nabla)^l$ operates on the position coordinates of the spherical harmonic Y_l^m and systematically replaces them with the appropriate spin component, and where b_l is an arbitrary constant. The spherical harmonics are given by

$$Y_l^m(\theta,\phi) = (-1)^{(m+|m|)/2} \left[\frac{2(l+1)(l-|m|)!}{4\pi(l+|m|)!} \right] P_l^m(\cos\theta) e^{im\phi}, \quad (3.3)$$

where $P_l^m(\cos\theta)$ are the associated Legendre polynomials. l takes on all values from 0 to $2I$, and m takes on all integral values $|m| \leq l$, yielding a total of $(2I+1)^2$ indices for A_{lm} for a given I. In Tables 3.1a and 3.1b, we list the spherical harmonics for $l = 2$ and $l = 4$, computed using (3.3). In Table 3.1c the A_{lm} tensor components for $l = 2$ and $l = 4$ are computed using (3.2). The constant b_l is written in terms of another constant a_l that we define below.

Table 3.1a

$l = 2$ and $l = 4$ Spherical Harmonics

		$Y_l^m(\theta,\phi)$
Y_2^0	=	$\sqrt{5/16\pi}\,(3\cos^2\theta - 1)$
$Y_2^{\pm 1}$	=	$\pm\sqrt{15/8\pi}\,\sin\theta\cos\theta e^{\pm i\phi}$
$Y_2^{\pm 2}$	=	$\sqrt{15/32\pi}\,\sin^2\theta e^{\pm 2i\phi}$
Y_4^0	=	$\sqrt{9/256\pi}\,(35\cos^4\theta - 30\cos^2\theta + 3)$
$Y_4^{\pm 1}$	=	$\mp\sqrt{45/64\pi}\,(7\cos^3\theta - 3\cos\theta)\sin\theta e^{\pm i\phi}$
$Y_4^{\pm 2}$	=	$\sqrt{45/128\pi}\,(7\cos^2\theta - 1)\sin^2\theta e^{\pm 2i\phi}$
$Y_4^{\pm 3}$	=	$\mp\sqrt{315/64\pi}\,\sin^3\theta\cos\theta e^{\pm 3i\phi}$
$Y_4^{\pm 4}$	=	$\sqrt{315/512\pi}\,\sin^4\theta e^{\pm 4i\phi}$

Table 3.1b
$l = 2$ and $l = 4$ Spherical Harmonics

$Y_l^m(r)$		
Y_2^0	=	$\sqrt{5/16\pi}\,(2z^2 - x^2 - y^2)$
$Y_2^{\pm 1}$	=	$\pm\sqrt{15/8\pi}\,z(x \pm iy)$
$Y_2^{\pm 2}$	=	$\sqrt{15/32\pi}\,(x \pm iy)^2$
Y_4^0	=	$\sqrt{9/256\pi}\,(35z^4 - 30z^2 r^2 + 3r^4)$
$Y_4^{\pm 1}$	=	$\mp\sqrt{45/64\pi}\,(7z^2 - 3r^2)z(x \pm iy)$
$Y_4^{\pm 2}$	=	$\sqrt{45/128\pi}\,(7z^2 - r^2)(x \pm iy)^2$
$Y_4^{\pm 3}$	=	$\mp\sqrt{315/64\pi}\,z(x \pm iy)^3$
$Y_4^{\pm 4}$	=	$\sqrt{315/512\pi}\,(x \pm iy)^4$

Table 3.1c
A_{lm} for $l = 2$ and $l = 4$ Based on (3.2)

A_{lm}	$b_l(\mathbf{1}\cdot\nabla)^l r^l Y_l^m$	b_l
A_{20}	$b_2[3I_z^2 - I(I+1)]$	$\sqrt{5}a_2$
A_{21}	$\mp\sqrt{3/2}\,b_2\{I_\pm, I_z\}$	$\sqrt{5}a_2$
A_{22}	$\sqrt{3/2}\,b_2 I_\pm^2$	$\sqrt{5}a_2$
A_{40}	$3b_4\{35I_z^4 - 30I(I+1)I_z^2 + 25I_z^2$ $-6I(I+1) + [3I(I+1)]^2\}$	$(1/2)a_4$
A_{41}	$\mp\sqrt{45}\,b_4\{I_\pm, [7I_z^3 - 3I(I+1)I_z - I_z]\}$	$(1/2)a_4$
A_{42}	$\sqrt{45/2}\,b_4\{I_\pm^2, [7I_z^2 - I(I+1) - 5]\}$	$(1/2)a_4$
A_{43}	$\mp\sqrt{315}\,b_4\{I_\pm^3, I_z\}$	$(1/2)a_4$
A_{44}	$\sqrt{315/2}\,b_4 I_\pm^4$	$(1/2)a_4$

In practice, it is easier to use the definition of the A_{lm} given in (3.4a) which follows from the use of a diagrammatic method by *Reiter* [3.8]. Reiter shows that the spherical harmonics can be written in terms of spin operators by use of the expression in (3.4b), where t is a dummy variable and where terms with the same t power are identified on the right- and left-hand sides of the equation in order to determine the spin operators for a specific $\mathcal{Y}_l^m$.

$$A_{lm} = \mathcal{Y}_l^m/(C_{lm})^{\frac{1}{2}}, \tag{3.4a}$$

and

$$\sum_{-l}^{l} \mathcal{Y}_l^m t^m = (-tI_+ + 2I_z + I_- t^{-1})^l. \tag{3.4b}$$

The normalization constant C_{lm} is given by

$$C_{lm} = \frac{(2I+1+l)!(2l)!(l!)^2}{(2l+1)!(2I-l)!(l-m)!(l+m)!(2I+1)}. \tag{3.4c}$$

By using the Reiter definitions, we find for

$l = 0 : A_{00} = 1$

$l = 1 : A_{10} = (3)^{\frac{1}{2}} a_1 I_z$

$$A_{1\pm 1} = \mp \left(\frac{3}{2}\right)^{\frac{1}{2}} a_1 I_\pm, \tag{3.5a}$$

where $a_1 = [I(I+1)]^{-\frac{1}{2}}$. Following [3.9], for higher values of l we obtain

$l = 2 : A_{20} = (5)^{\frac{1}{2}} a_2 [3I_z^2 - I(I+1)]$

$A_{2\pm 1} = \mp \left(\frac{15}{2}\right)^{\frac{1}{2}} a_2 \{I_\pm, I_z\}$

$$A_{2\pm 2} = \left(\frac{15}{2}\right)^{\frac{1}{2}} a_2 I_\pm^2, \tag{3.5b}$$

where $a_2 = [I(I+1)(2I-1)(2I+3)]^{-\frac{1}{2}}$, and the curly brackets $\{A, B\}$ denote the anticommutator of A and B: $\{A, B\} = AB + BA$.

$l = 3 : A_{30} = (7)^{\frac{1}{2}} a_3 [5I_z^3 - (3I^2 + 3I - 1)I_z]$

$A_{3\pm 1} = \mp \left(\frac{21}{16}\right)^{\frac{1}{2}} a_3 [5\{I_\pm, I_z^2\} - (2I^2 + 2I + 1)I_\pm]$

$A_{3\pm 2} = \left(\frac{105}{8}\right)^{\frac{1}{2}} a_3 \{I_\pm^2, I_z\}$

$$A_{3\pm 3} = \mp \left(\frac{35}{4}\right)^{\frac{1}{2}} a_3 I_\pm^3, \tag{3.5c}$$

where $a_3 = [I(I+1)(2I-1)(2I+3)(I-1)(I+2)]^{-\frac{1}{2}}$.

3. DYNAMIC NUCLEAR ELECTRIC QUADRUPOLE INTERACTION

$$l = 4 : A_{40} = \frac{3}{2}a_4\{35I_z^4 - 30I(I+1)I_z^2 + 25I_z^2 - 6I(I+1)$$
$$+ 3[I(I+1)]^2\}$$

$$A_{4\pm1} = \mp \left(\frac{45}{4}\right)^{\frac{1}{2}} a_4\{I_{\pm1}, [7I_z^3 - 3I(I+1)I_z - I_z]\}$$

$$A_{4\pm2} = \left(\frac{45}{8}\right)^{\frac{1}{2}} a_4\{I_{\pm2}^2, [7I_z^2 - I(I+1) - 5]\}$$

$$A_{4\pm3} = \mp \left(\frac{315}{4}\right)^{\frac{1}{2}} a_4\{I_{\pm3}^3, I_z\}$$

$$A_{4\pm4} = \left(\frac{315}{8}\right)^{\frac{1}{2}} a_4\{I_{\pm4}^4\} , \qquad (3.5d)$$

where $a_4 = [(2I+5)(I+2)(2I+3)(I+1)I(2I-1)(I-1)(2I-3)]^{-\frac{1}{2}}$.

Equations (3.5a–d) define a set of irreducible tensor operators, all of which are normalized so that the trace of $|A|^2$ is $2I+1$. The operator A_{00} transforms like the electric charge of the nucleus. The A_{1m} correspond to the magnetic dipole moment; the A_{2m} to the electric quadrupole moment; the A_{3m} to the magnetic octupole moment; and the A_{4m} to the electric hexadecapole moment. We list several additional important properties of the A_{lm} [3.7,3.8].

Irreducibility:

$$[I_z, A_{lm}] = m A_{lm}$$
$$[I_\pm, A_{lm}] = \sqrt{l(l+1) - m(m\pm1)}\, A_{lm\pm1} \qquad (3.6)$$
$$I_+ A_{ll} = 0, \quad I_- A_{l-l} = 0 .$$

Orthonormality:

$$\mathrm{tr}(A_\alpha A_\beta^\dagger) = \delta_{\alpha\beta}(2I+1) , \qquad (3.7)$$

where $\alpha = lm$, $\beta = l'm'$, and $A_\beta^\dagger$ is the adjoint operator. It is this property that makes the A_{lm} convenient for use only at high temperatures—that is, when $h\nu/kT \ll 1$. For nuclear spins this is not an important restriction.

Conjugate:

$$(A_{lm})^\dagger = (-1)^m A_{l-m} . \qquad (3.8)$$

3.1 MULTIPOLE EXPANSIONS

We have computed the A_{lm} from the definitions of both *Rose* [3.7] and *Reiter* [3.8], and we find consistent values for the spin operator relationships for a given lm for $l = 1$ to 4. The results of (3.5), Table 3.1a and Table 3.1b generally agree with other similar, published works, such as *Ambler, Eisenstein, and Schooley* [3.10].

Utilizing the A_{lm} tensor operators, the spin Hamiltonian, in the absence of spin–lattice interactions, can be written to include (i) terms linear in the spin operators, which represent the interaction of a spin with spatially uniform and time–independent external fields, and (ii) terms bilinear in the spin operators at different sites, which represent the interaction between spins at different sites [3.11]. The most general Hamiltonian that one can write under these circumstances is

$$\mathcal{H}_0 = \sum_l \mathcal{H}_l \qquad (3.9)$$

$$\mathcal{H}_l = -\sum \hbar\omega_{lm} A_{lm}(i) + \sum \hbar\Omega^{(l)}_{m,m'}(i,j) A_{lm}(i) A_{lm'}(j) , \qquad (3.10)$$

where $A_{lm}(i)$ and $A_{lm'}(j)$ are the spin operators at sites i and j, respectively. The first summation in (3.10) is over all sites i and all allowed m; the second summation in (3.10) is over all lattice sites i and j (with $j \neq i$) and over all allowed m and m'. $\hbar\omega_{lm}$ and $\hbar\Omega_{lm}$ are coupling constants; ω_{lm} and Ω_{lm} are expressed in frequency units. The summation in (3.9) is over all the integral values of l for $1 \leq l \leq 2I$, where I refers to the nuclear spins in question. It is to be noted that a bilinear Hamiltonian cannot mix operators of different l.

The first term on the right–hand side of (3.10) for $l = 1$ is the Zeeman term. If the external magnetic field H is in the z–direction, then $\omega_{10} = [\frac{1}{3}I(I+1)]^{\frac{1}{2}} \gamma H$, and $\omega_{11} = \omega_{1-1} = 0$. The second term on the right–hand side of (3.10) for $l = 1$ includes all spin–spin interactions via the dipole or vector spin operators—for example, the dipolar and the isotropic and anisotropic exchange interaction terms. For $l = 2$ the first term on the right–hand side of (3.10) is the quadrupolar term that describes the interaction of a spin with static electric field gradients. The second term on the right–hand side of (3.10) for $l = 2$ describes the general interaction between a pair of spins via their quadrupole ($l = 2$) moments. Higher values of l similarly result in terms representing interactions of spins via higher–order nuclear multipole moment operators: $l = 3$, nuclear magnetic octupole moment operators; $l = 4$, nuclear electric hexadecapole moment operators.

Although the concept of spin–spin interactions in a magnetic system is most often thought of in terms of the mutual interactions of spins via their dipole ($l = 1$) moments, the spherical tensor operator formalism indicates that the spins in such a system may also interact via their higher–order ($l > 1$) multipole moments. The second term on the right–hand side of (3.10) thus represents the most general bilinear spin–spin interaction between pairs of spins (or, for that matter, between pairs of any physical quantities that are representable in terms of spin or angular momentum operators).

As another example of application of the use of the A_{lm} tensors, we consider a set of noninteracting nuclear spins in an external magnetic field $\boldsymbol{H}_0$, which makes spherical polar angles θ and ϕ with the coordinate system of the lattice. In addition, we assume that the spins are coupled to time–varying electric field gradients by the nuclear electric quadrupole interaction. The correct form of the Hamiltonian in this case can be shown to be [3.9]

$$\mathcal{H}_Q = \sum_{i,j;m} \left[\frac{eQ}{2I(2I-1)} \right] V_{ij} d(i,j;m) A_{2m} , \qquad (3.11)$$

where V_{ij} can be related to time–varying elastic strain and is expressed in the coordinate system of the lattice. $d(i,j;m)$ is a transformation tensor which changes the reference system of V_{ij} to the reference system of the magnetic field $\boldsymbol{H}_0$. (Other methods for such a transformation are given by *Taylor* [3.12] and *Mieher* [3.13,3.14]). For illustrating the expansion of (3.11), we take $\theta = 0$ and $\phi = 0$. The non–zero components of V_{ij} for $m = 0$ are V_{xx}, V_{yy}, and V_{zz}. For $m = \pm 1$, the non–zero components are $V_{xz} = V_{zx}$ and $V_{yz} = V_{zy}$; for $m = \pm 2$, the non–zero components are V_{xx}, V_{yy} and $V_{xy} = V_{yx}$.

The transformation tensor $d(i,j;m)$ has the following relationships:

$$d(i,j;m) = d(j,i;m) = (-1)^m d^*(i,j;-m) . \qquad (3.12)$$

With $\boldsymbol{H}_0$ along the z–direction, the $d(i,j;m)$ values associated with the

3.2 NUCLEAR ELECTROSTATIC MULTIPOLE INTERACTIONS

nonzero V_{ij} values in (3.11) are

$$d(z,z;0) = \frac{2}{\sqrt{180a_2}} \qquad d(x,x;0) = d(y,y;0) = \frac{-1}{\sqrt{180a_2}}$$

$$d(x,z;1) = \frac{-1}{\sqrt{120a_2}} \qquad d(x,z;-1) = \frac{1}{\sqrt{120a_2}}$$

$$d(y,z;1) = \frac{i}{\sqrt{120a_2}} \qquad d(y,z;-1) = \frac{i}{\sqrt{120a_2}}$$

$$d(x,x;2) = \frac{1}{\sqrt{120a_2}} \qquad d(x,x;-2) = \frac{1}{\sqrt{120a_2}}$$

$$d(y,y;2) = \frac{-1}{\sqrt{120a_2}} \qquad d(y,y;-2) = \frac{-1}{\sqrt{120a_2}}$$

$$d(x,y;2) = \frac{-i}{\sqrt{120a_2}} \qquad d(x,y;-2) = \frac{i}{\sqrt{120a_2}} \; . \qquad (3.13)$$

With the use of (3.11–3.13), we write the elements of the sum of (3.11) as

$$\mathcal{H}_Q(m=0) = \frac{eQ}{2I(2I-1)} \frac{1}{\sqrt{180a_2}} A_{20}(2V_{zz} - V_{xx} - V_{yy})$$

$$= \frac{eQ}{4I(2I-1)} V_{zz}[3I_z^2 - I(I+1)]$$

$$\mathcal{H}_Q(m=\pm 1) = \frac{eQ}{2I(2I-1)} 2[(-V_{xz} + iV_{yz})A_{21}$$

$$+ (V_{xz} + iV_{yz})A_{2-1}]\frac{1}{\sqrt{120a_2}}$$

$$= \mp \frac{eQ}{4I(2I-1)}[V^{-1}\{I_+, I_z\} + V^{+1}\{I_-, I_z\}]$$

$$\mathcal{H}_Q(m=\pm 2) = \frac{eQ}{2I(2I-1)}[(V_{xx} - V_{yy} - 2iV_{xy})A_{22}$$

$$+ (V_{xx} - V_{yy} + 2iV_{xy})A_{2-2}\frac{1}{\sqrt{120a_2}}$$

$$= \frac{eQ}{4I(2I-1)}(V^{+2}I_+^2 + V^{-2}I_-^2) \; , \qquad (3.14)$$

where

$$V^{\pm 1} = V_{xz} \pm iV_{yz}$$
$$V^{\pm 2} = (1/2)(V_{xx} - V_{yy}) \pm iV_{xy} \; . \qquad (3.15)$$

The results of (3.14) will produce the matrix elements of (3.39) if we recognize the difference between the definition of $V^{\pm 1}$ and $V^{\pm 2}$ between (3.15) and (3.34), (3.36).

3.2 Nuclear Electrostatic Multipole Interactions

The conventional treatment of the decomposition of the potential field due to a given charge distribution into multipolar contributions leads naturally to the introduction of the spherical multipole tensor operators discussed in the preceding section. The Hamiltonian for the general electrostatic interaction between an element of nuclear charge and an electron charge distribution external to the nucleus is given by

$$\mathcal{H}_{el} = \int_{\tau_e}\int_{\tau_n} \frac{\rho_e(\mathbf{r}_e)\rho_n(\mathbf{r}_n)d\tau_e d\tau_n}{r}, \qquad (3.16)$$

where $\rho_e(\mathbf{r}_e)$ is the charge density of the electrons in the volume element $d\tau_e$ at position $\mathbf{r}_e$ relative to the origin of the nuclear charge distribution. $\rho_n(\mathbf{r}_n)$ is the nuclear charge density in the volume element $d\tau_n$ at position $\mathbf{r}_n$ relative to the origin, and r is the magnitude of the radius vector $\mathbf{r}$ joining $d\tau_e$ and $d\tau_n$. (See Fig. 3.1.) (A factor $1/4\pi\varepsilon_o$ multiplies the right-hand side of (3.15) when the *mks* system of units is used; here we are assuming the *cgs* system.)

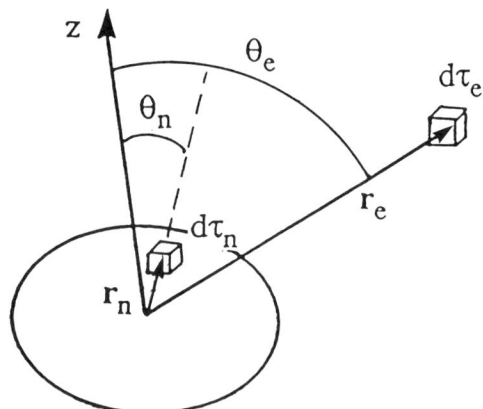

FIGURE 3.1. Interaction between nuclear charge density $(\rho_n, d\tau_n)$ and electron charge density $(\rho_e, d\tau_e)$.

The term $r \equiv |\mathbf{r}_e - \mathbf{r}_n|$ may be expanded in terms of the normalized spherical harmonics

$$\frac{1}{r} = 4\pi \sum_{l=0}^{\infty} \sum_{m=-l}^{m=l} \left(\frac{1}{2l+1}\right) \frac{r_n^l}{r_e^{l+1}} Y_l^m(\theta_e, \phi_e) Y_l^{-m}(\theta_n, \phi_n) , \qquad (3.17)$$

where l takes the integer values $0, 1, 2 \ldots$; θ_e, ϕ_e and θ_n, ϕ_n are the coordinate angles of r_e and r_n (Fig. 3.1). In the expansion (3.17), r_n has been assumed to be smaller than r_e, which is generally true, except for s electrons. The s electrons, however, are distributed in spherical symmetry and therefore do not contribute to the multipolar interactions (quadrupole, hexadecapole) we shall be considering. The resulting expression for $\mathcal{H}_{el}$ is

$$\mathcal{H}_{el} = \sum_{l=0}^{\infty} \mathcal{H}_{el}^l \qquad (3.18a)$$

$$\mathcal{H}_{el}^l = \frac{4\pi}{2l+1} \sum_{m=-l}^{m=l} \int_{\tau_e} \int_{\tau_n} (-1)^m \rho_e(r_e) \rho_n(r_n) \frac{r_n^l}{r_e^{l+1}} Y_l^m(\theta_e, \phi_e)$$
$$\times Y_l^m(\theta_n, \phi_n) d\tau_e d\tau_n \qquad (3.18b)$$

$$= \sum_m (-1)^m A_l^m B_l^m = \sum_m A_l^m B_l^{-m} , \qquad (3.18c)$$

where

$$A_l^m = \left(\frac{4\pi}{2l+1}\right)^{\frac{1}{2}} \int_{\tau_n} \rho_n(r_n) r_n^l Y_l^m(\theta_n, \phi_n) \, d\tau_n \qquad (3.18d)$$

and

$$B_l^m = \left(\frac{4\pi}{2l+1}\right)^{\frac{1}{2}} \int \rho_e(r_e) r_e^{-(l+1)} Y_l^m(\theta_e, \phi_e) d\tau_e . \qquad (3.18e)$$

The B_l^m correspond to the spherical harmonic expansion of a potential inside a sphere and are produced by a charge distribution external to the sphere. As we shall see below, the B_l^m are related to the derivatives of the overall potential produced by charges external to the nucleus. In this connection it is useful to recall the property of spherical harmonics that any derivative of a spherical harmonic is also a spherical harmonic; i.e., $\partial^\alpha/\partial x^\alpha \cdot \partial^\beta/\partial y^\beta \cdot \partial^\gamma/\partial z^\gamma V_n$ is a harmonic of degree $[n - (\alpha + \beta + \gamma)]$.

The terms A_l^m are related to the components of the electrostatic multipole. The electrostatic multipole moment is given by

$$\sqrt{\frac{4\pi}{(2l+1)!}} \int_{\tau_n} r_n^l Y_l^m(\theta_n, \phi_n) \rho_n(r_n) d\tau_n . \qquad (3.19)$$

Assuming that the nucleus has an angular spin quantum number I, and that ψ_n^* and ψ_n are the eigenfunctions corresponding to the angular momentum $I\hbar$, the multipole moment of order 2^l is proportional to

$$\int_{\tau_n} r_n^l \, \psi_n^* \, Y_l \, \psi_n \, d\tau_n \;, \tag{3.20}$$

where Y_l is an eigenfunction corresponding to the angular momentum $I\hbar$. Since r_n^l has no angular dependence, $r_n^l \psi_n^*$ is an eigenfunction corresponding to the angular momentum I. Thus the multipole moment (3.20) is the product of three angular momentum eigenfunctions $Y_l^{-m_1}$, $Y_l^{-m_2}$ and Y_l^m, which can be shown to be proportional to $(\delta_{m_1, I, m_2, m})$, where

$$\delta = 0, \quad |m_2| > |m + m_1|$$
$$\delta = 1, \quad |m_2| \leq |m + m_1| \;.$$

Since m_1 and m can take on the values $+I \cdots -I$, and m_2 takes the values $-l \cdots +l$, $\delta = 0$ unless $l \leq 2I$. This places an upper limit on the order of the electrostatic multipole that can exist; *i.e.*,

$$I = \frac{1}{2}, \quad \text{no quadrupole moment};$$
$$I = \frac{3}{2}, \quad \text{quadrupole but no hexadecapole};$$
$$I = \frac{5}{2}, \quad \text{hexadecapole but no higher moment}.$$

A further restriction on the existence of electrostatic multipoles results from considerations of parity, if we assume that parity is conserved for nuclear forces. A quantum mechanical system is said to have even parity if the sign of its wave function is unaltered on reflection, and it is said to have odd parity if its sign is reversed on reflection. Since a change of sign in the electrostatic charge distribution results in a different physical description of the nucleus, only charge distributions with even parity can be present. The parity of a spherical harmonic Y_n is $(-1)^n$, so that for the electrostatic charge distribution only even order electric multipoles exist. Conversely, only odd order magnetic multipoles exist.

If the state of the nucleus is described by the wave function $\psi_n(R_1 \ldots R_i \ldots R_n)$, where R_i is the coordinate of the ith nucleon, the nuclear charge density at r_n is $\rho_n(r_n) = (\psi_n | \sum_{i=1}^{A_N} e_i \delta(r_n - R_i) | \psi_n)$, where

$$e_i = e \quad \text{(the electronic charge) for a proton},$$
$$e_i = 0 \quad \text{for a neutron},$$

and A_N is the atomic mass number of the nucleus. Thus we can write A_l^m as the expectation value of the operator $\boldsymbol{A}_l^m$ defined by

$$\boldsymbol{A}_l^m = \left(\frac{4\pi}{2l+1}\right)^{\frac{1}{2}} \sum_{i=1}^{N} e_i\, R_i^l\, Y_l^m(\theta_i, \phi_i)\,, \tag{3.21}$$

and, similarly, B_l^m as the expectation value of the operator $\boldsymbol{B}_l^m$ defined by

$$\boldsymbol{B}_l^m = e\left(\frac{4\pi}{2l+1}\right) \sum_{k=1}^{N} r_k^{-(l+1)} Y_l^m(\theta_k, \phi_k)\,, \tag{3.22}$$

where r_k, θ_k and ϕ_k are the coordinates of the N electrons. The energy of the electrostatic interaction is the expectation value of the Hamiltonian

$$\mathcal{H}_e = \sum_l \sum_m \boldsymbol{A}_l^m\, \boldsymbol{B}_l^{-m}\,. \tag{3.23}$$

Since the operators $\boldsymbol{A}_l^m$ and $\boldsymbol{B}_l^m$ transform under rotation of coordinate axes in the same way as spherical harmonics $Y_l^m(\theta, \phi)$ of order l, they constitute spherical tensor operators of order l.

The electrostatic energy E_0 of the nucleus regarded as a point charge (electric monopole term, $l = 0$) is $(Ze)^2/r_e$, where Z is the atomic number of the nucleus, e is the electronic charge, and r_e is a typical atomic dimension ($\sim 10^{-10}$ m). This energy is approximately equal to 10^{-13} Joules or to a photon of wavelength 10^{-7} m. Since, from the form of the operators (3.21) and (3.22), the interaction energy varies as $(R/r)^l \equiv (r_n/r_e)^l$, and since the nuclear radius is of the order of 10^{-14} m, the quadrupole ($l = 2$) energy corresponds approximately to $10^{-8} E_0$ or 10 MHz, comparable to the magnetic dipole interaction discussed in Chapter 1. The hexadecapole term ($l = 4$) then corresponds to a frequency of the order of 1 Hz. Although the hexadecapole interaction is usually thought to be negligible compared with the quadrupole interaction, we will find that considerations presented in Section 3.8 will give us reason to treat the hexadecapole interaction as a serious candidate for coupling energy from an acoustic wave to a nuclear spin system.

In the actual computation of the multipole moments [3.1,3.2], the Wigner–Eckart theorem [3.7] with the Clebsch–Gordon coefficients allow the replacement of the $\boldsymbol{A}_l^m$ operator sum of the coordinates of each nucleon, expressed in Cartesian coordinates, by the related spin operators for the whole nucleus. The expectation value of $\boldsymbol{A}_l^m$ between appropriate spin states can be used to define the multipole moment (quadrupole or hexadecapole), as illustrated in the next section.

3.3 Dynamic Electric Quadrupole Interaction

The most effective mechanism for coupling acoustic waves to nuclear spin systems with $I > \frac{1}{2}$ in nonconductors (and in some conductors) is the dynamic quadrupolar interaction, which, except for the trivial $l = 0$ monopole term representing a point charge, is the first allowed term in the expansion (3.23). In deriving the detailed expressions for the nuclear electric quadrupole interactions in NAR, following the pioneer work of *Pound* [3.15], we express the components of the tensor operators in Cartesian coordinates. The quadrupolar Hamiltonian is written

$$\mathcal{H}_Q = \sum_{m=-2}^{2} Q_2^m V_2^{-m} , \qquad (3.24)$$

in which Q_2^m is termed the nuclear quadrupole moment operator and is equal to A_2^m in (3.21). V_2^{-m} is the electric field gradient (EFG) operator and is equal to B_2^{-m} in (3.22). In further use of the operators Q_2^m, V_2^m and their components, we shall omit the subscript. ($l = 2$ is assumed, by definition of the quadrupole operator.) In Cartesian coordinates, the components of Q^m can be written in irreducible form as

$$Q^0 = \frac{1}{2} \sum_i e_i (3z_i^2 - r_i^2)$$

$$Q^{\pm 1} = \frac{\sqrt{6}}{2} \sum_i e_i z_i (x_i - iy_i) \qquad (3.25)$$

$$Q^{\pm 2} = \frac{\sqrt{6}}{4} \sum_i e_i (x_i \pm iy_i)^2 .$$

The Wigner–Eckart theorem [3.7] now permits the replacement of the Cartesian coordinates by a constant multiple of the appropriate components of the nuclear angular momentum operator I. Since the components of I in irreducible form are

$$I_\pm = (I_x \pm iI_y)$$
$$I_z = I_z , \qquad (3.26)$$

we obtain for the quantum mechanical operator components of the quadrupole moment

$$Q^0 = \frac{1}{2} A [3I_z^2 - I(I+1)]$$

$$Q^{\pm 1} = \frac{\sqrt{6}}{4} A [I_z I_\pm + I_\pm I_z] \qquad (3.27)$$

$$Q^{\pm 2} = \frac{\sqrt{6}}{4} A (I_\pm)^2 ,$$

where we note that in writing $Q^{\pm 1}$, the angular momentum operators do not necessarily commute. The constant A is defined by requiring that the classical quadrupole moment (eQ) be equal to the expectation value of Q^0,

$$eQ = A(II|3I_z^2 - I^2|II)$$
$$= A[3I^2 - I(I+1)] , \quad (3.28)$$

where $(II|$ is the state in which $I_z = I$, and

$$A = \frac{eQ}{I(2I-1)} . \quad (3.29)$$

Q is a scalar quantity usually termed the *electric quadrupole moment* of the nucleus. The quantity measured by magnetic resonance—NAR or NMR—is the spectroscopic nuclear quadrupole moment Q, which is related to the intrinsic nuclear quadrupole moment Q_o measured in nuclear physics by the relation [3.16]

$$Q = \frac{I(2I-1)}{(2I+3)(I+1)} Q_o . \quad (3.30)$$

Thus for nuclear spins $I = \frac{3}{2}, \frac{5}{2}$ and $\frac{7}{2}$, Q is respectively $\frac{1}{5}$, $\frac{5}{14}$ and $\frac{7}{15}$ the size of Q_o.

The component V^m can also be written in Cartesian coordinates or, alternatively, in terms of the derivatives of the potential $V(r_i)$ resulting from all the charge e_i external to the nucleus under consideration. Thus,

$$V^0 = \frac{1}{2} \sum_{i=1}^{N} \frac{e_i(3z_i^2 - r_i^2)}{r_i^5} \quad (3.31)$$

$$= -\frac{1}{2}\left(\frac{\partial^2 V}{\partial z^2}\right)_{r=0} \equiv -\frac{1}{2} V_{zz} , \quad (3.32)$$

where the origin is taken at the nucleus, and $V(x, y, z)$ is the potential produced at the point x, y, z by the electrons. V_{zz} is an operator, and the observable V_{zz} is its expectation value over the electronic wave function. The other components of V_2^m are obtained in a similar fashion:

$$V^{\pm 1} = \pm \frac{\sqrt{6}}{2} \sum_i \frac{z_i(x_i \pm iy_i)}{r_i^5} \quad (3.33)$$

$$= \mp \frac{\sqrt{6}}{6}(V_{xz} \pm iV_{yz}) \quad (3.34)$$

$$V^{\pm 2} = \frac{\sqrt{6}}{4} \sum_i (x_i^2 - y_i^2 \pm 2ix_iy_i) \quad (3.35)$$

$$= \frac{\sqrt{6}}{12}(V_{xx} - V_{yy} \pm 2iV_{xy}) . \quad (3.36)$$

3. DYNAMIC NUCLEAR ELECTRIC QUADRUPOLE INTERACTION

In summary, we have found that the quadrupole Hamiltonian can be written in terms of two traceless operators,

$$\mathcal{H}_Q = \sum_{j,k} \left(\frac{\partial^2 V}{\partial x_j \partial x_k}\right)_{r=0} Q_{jk} , \qquad (3.37)$$

where

$$Q_{jk} = \frac{eQ}{6I(2I-1)} \left\{\frac{3}{2}(I_j I_k + I_k I_j) - \delta_{jk} I(I+1)\right\} , \qquad (3.38)$$

and j and k take on the values x, y and z.

At high magnetic fields such that the quadrupole interaction is weak in comparison with the separation of the magnetic energy levels, the nonzero matrix elements of the electric quadrupole tensor in an I, m representation (which involve only the well-known elements of the angular momentum [3.1–3.5]), are

$$(m|\mathcal{H}_Q|m) = \frac{A}{2}[3m^2 - I(I+1)]V^0 \qquad (3.39a)$$

$$(m|\mathcal{H}_Q|m \pm 1) = \mp \left(\frac{6^{\frac{1}{2}}}{4}\right) A(2m \pm 1)[(I \pm m + 1)(I \mp m)]^{\frac{1}{2}} V^{\pm 1} \qquad (3.39b)$$

$$(m|\mathcal{H}_Q|m \pm 2) = \left(\frac{6^{\frac{1}{2}}}{4}\right) A[(I \mp m)(I \mp m - 1)(I \pm m + 1)$$
$$\times (I \pm m + 2)]^{\frac{1}{2}} V^{\pm 2} . \qquad (3.39c)$$

For comparison, we give the nonzero matrix elements for the case of a dipolar interaction between an RF magnetic field and the magnetic dipole moment of the nucleus:

$$(m|\mathcal{H}_D|m) = -\gamma\hbar(m|I_z|m)H'_z = -\gamma\hbar m H'_z \qquad (3.39d)$$

$$(m|\mathcal{H}_D|m \pm 1) = -\gamma\hbar(m|I_\pm|m \pm 1)(H'_x \pm iH'_y)$$
$$= \gamma\hbar[(I \pm m)(I \pm m + 1)]^{\frac{1}{2}}(H'_x \pm iH'_y) , \qquad (3.39e)$$

where (3.39d) indicates the (slight) change in the high field energy levels $-\gamma\hbar m E_0$ due to the z-component of the perturbing magnetic field. (3.39e) establishes that dipolar transitions are allowed only between adjacent energy levels ($\Delta m = \pm 1$).

3.3 DYNAMIC ELECTRIC QUADRUPOLE INTERACTION

For the case of axial symmetry, the principal axes x', y', z' of the symmetric tensor V_{ij} are chosen as the axes x, y and z in (3.31)–(3.36). In that case, $V_{x'z'} = V_{y'z'} = V_{x'y'} = 0$. We also choose the axes such that $|V_{z'z'}| \geq |V_{x'x'}| \geq |V_{y'y'}|$.

We define

$$eq = V_{z'z'}$$
$$\eta = \frac{V_{x'x'} - V_{y'y'}}{V_{z'z'}} \; . \tag{3.40a}$$

Since the trace is zero ($V_{x'x'} + V_{y'y'} + V_{z'z'} = 0$ from Laplace's equation), the field gradient tensor is completely described by these two quantities.

The symmetry of the field gradient at a nuclear site is largely determined by the lattice symmetry at the site. For a site having a four-fold rotation axis passing through it, for example, a 90° rotation about this axis leaves the crystal unchanged. Hence, the field gradient also must be unchanged. Choosing the axis of rotation as the z-axis, the effect of the 90° rotation is to change $x \to y$ and $y \to -x$. For the field gradient to remain unchanged, $V_{xx} = V_{yy}$, $V_{xy} = -V_{yx}$, $V_{xz} = V_{yz}$, and $V_{yz} = -V_{xy}$. For a nuclear site having a four-fold axis of symmetry, therefore, $\eta = 0$. Similarly, any n-fold rotation axis with $n > 2$ requires $\eta = 0$. Any site with two or more n-fold rotation axes with $n > 2$, furthermore, must have zero-field gradient. Thus in a perfect cubic crystal, in which every site has three four-fold axes, the field gradient must vanish.

For axial symmetry the components of the field gradient tensor, from (3.31), (3.33) and (3.35), may be written

$$V^0 = \frac{1}{4} eq (3 \cos^2 \theta - 1) \tag{3.40b}$$

$$V^{\pm 1} = \pm \frac{1}{4} (6^{\frac{1}{2}}) eq \sin \theta \cos \theta \, e^{\pm i\phi} \tag{3.40c}$$

$$V^{\pm 2} = \frac{1}{8} (6^{\frac{1}{2}}) eq \sin^2 \theta \, e^{\pm 2i\phi} \; , \tag{3.40d}$$

where θ is the angle between the axis of symmetry and the z-direction (usually chosen as the direction of quantization—e.g., the direction of the external DC magnetic field $\boldsymbol{H}_0$). ϕ is the angle between the projection of the symmetry axis on the xy-plane and the x-axis.

The scalar quantity eq is a measure of the magnitude of the axially symmetric electric field gradient due to all charges e_j external to the nucleus: $eq = \partial E_z / \partial z = \Sigma_j e_j (3 \cos^2 \theta_j - 1) r_j^{-3}$. r_j and θ_j are the coordinates (with center of nucleus as origin) of the element of charge e_j. If the distribution of external charge is spherically symmetric, then $q = 0$.

3. DYNAMIC NUCLEAR ELECTRIC QUADRUPOLE INTERACTION

In the principal axes system, using (3.24), (3.27), (3.32), (3.34), (3.36) and (3.40a), the quadrupolar Hamiltonian becomes

$$\mathcal{H}_Q = \frac{e^2qQ}{4I(2I-1)}\{3I_{z'}^2 - I(I+1) + \frac{1}{2}\eta(I_+^2 + I_-^2)\} . \tag{3.41}$$

It is this quadrupolar Hamiltonian that alone accounts for the pure quadrupole energy levels and that combined with the Zeeman Hamiltonian $\mathcal{H}_m$, accounts for the quadrupole–split high–field energy levels. Both these cases have been discussed briefly in Chapter 1. Further examples of these energy level configurations will occur in this and the ensuing chapters.

From (1.38) in Chapter 1, we recall that the probability per second that a nucleus will undergo a transition from one magnetic level m to another level m' and will absorb in the process one acoustic phonon, is proportional to the square of the interaction matrix elements. For the quadrupolar interaction we are considering the most general statement of the transition probabilities, using (3.39b) and (3.39c), is

$$W_{Q1} \equiv W_{m,m\pm1} = 6\left(\frac{A}{4\hbar}\right)^2 \xi_\pm^2 g_{Q1}(\nu)|V^{\pm1}|^2 \tag{3.42a}$$

$$W_{Q2} \equiv W_{m,m\pm2} = 6\left(\frac{A}{4\hbar}\right)^2 \eta_\pm^2 g_{Q2}(\nu)|V^{\pm2}|^2 , \tag{3.42b}$$

where $g_{Q1}(\nu)$ and $g_{Q2}(\nu)$ are the absorption line shape factors for $\Delta m = \pm 1$ and $\Delta m = \pm 2$ transitions. For comparison, the transition probability for magnetic dipole transitions in NMR is given as

$$W_M \equiv \frac{1}{4}\gamma_n^2 H_1^2 g_o(\nu)\zeta_\pm^2 \sin^2\theta , \tag{3.42c}$$

where $2H_1$ is the peak amplitude of the RF magnetic field. $\gamma_n = g\mu_n/\hbar$, and θ is the angle that $\mathbf{H}_1$ makes with the static field $\mathbf{H}_0$. We define

$$\xi_\pm = (2m \pm 1)[(I \pm m + 1)(I \mp m)]^{\frac{1}{2}} \tag{3.43a}$$

$$\eta_\pm = [(I \mp m)(I \mp m - 1)(I \pm m + 1)(I \pm m + 2)]^{\frac{1}{2}} \tag{3.43b}$$

$$\zeta_\pm = [(I \mp m)(I \pm m + 1)]^{\frac{1}{2}} . \tag{3.43c}$$

As indicated by the notation adopted, W_{Q1} denotes dynamic quadrupolar transitions in which $\Delta m = \pm 1$, and W_{Q2} denotes transitions in which $\Delta m = \pm 2$. The transitions $m = -\frac{1}{2} \leftrightarrow m = +\frac{1}{2}$ are forbidden for the

3.3 DYNAMIC ELECTRIC QUADRUPOLE INTERACTION

NAR dynamic quadrupolar interaction. The allowed transitions between the energy levels of a nuclear spin $I = \frac{7}{2}$ system are indicated in Fig. 3.2, which shows (a) the magnetic high field (Zeeman) energy levels, with equal spacing $\omega_m = E_0/\hbar = \gamma H_0$; (b) the Zeeman energy levels split by a static quadrupole interaction, with transitions induced by an RF magnetic field as in NMR and also in dipolar [dynamic Alpher–Rubin] NAR, to be discussed in Chapter 4; (c) transitions induced by W_{Q1} and (d) by W_{Q2}.

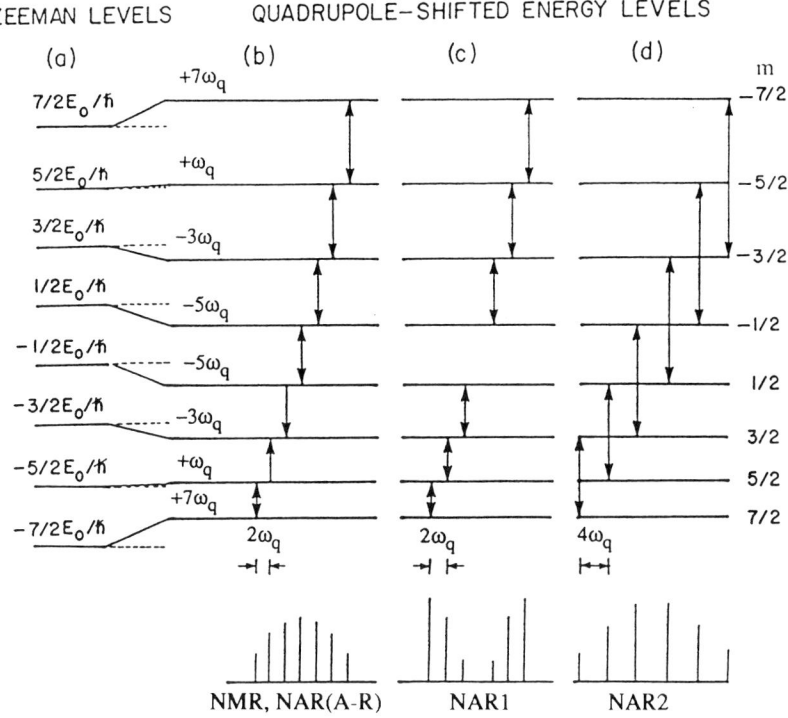

FIGURE 3.2. Quadrupole–shifted Zeeman energy levels and resulting spectra for spin $I = \frac{7}{2}$.

The quadrupole-split energy levels are given by

$$E = E_m + (m|\mathcal{H}_Q|m) \tag{3.44a}$$

$$= -m\hbar\omega_0 + \hbar\omega_q[m^2 - \frac{1}{3}I(I+1)] . \tag{3.44b}$$

If we assume, for simplicity, an axially symmetric EFG, then

$$\omega_q = \frac{e^2 qQ}{\hbar \cdot 2I(2I-1)} \cdot \frac{3}{4}(3\cos^2\theta - 1) , \tag{3.44c}$$

where θ is the angle between $\boldsymbol{H}_0$ and the axis of symmetry of q.

The signal strengths of the NAR absorption lines are obtained, using (3.42), from the application of (2.22):

$$\alpha_{Q1} = \frac{3\pi^2}{8} \cdot \frac{Ne^2Q^2}{\rho v^3 k_B T} \nu^2 g_{Q1}(\nu) \cdot f_Q(I) \cdot \frac{|V^{\pm 1}|^2}{\varepsilon^2} \quad (3.45a)$$

$$\alpha_{Q2} = \frac{3\pi^2}{8} \cdot \frac{Ne^2Q^2}{\rho v^3 k_B T} \nu^2 g_{Q2}(\nu) \cdot f_Q(I) \cdot \frac{|V^{\pm 2}|^2}{\varepsilon^2} \quad (3.45b)$$

$$\alpha_{\text{A-R}} = \frac{\pi^2}{3} \cdot \frac{Nh^2}{\rho v^3 k_B T} \nu^4 g_D(\nu) \cdot f_{\text{A-R}}(I) , \quad (3.45c)$$

where $\alpha_{\text{A-R}}$, the dynamic Alpher–Rubin absorption (treated in Chapter 4), is given for completeness. The expressions for $f_Q(I)$ and $f_{\text{A-R}}(I)$ are given in Table 3.2. In (3.45c) the specific case of $\beta = 0$ and observation of $\chi''(\nu)$ alone is given (see Chapter 4). The relative signal intensities within a particular type of transition ($\alpha_{\text{A-R}}$, α_{Q1} or α_{Q2}) are also given in Table 3.2 and are indicated schematically in Fig. 3.2. A summary of transition frequencies and relative transition probabilities (within a given category) for $I = \frac{3}{2}, \frac{5}{2}, \frac{7}{2}$ and $\frac{9}{2}$ is given in Table 3.3, where

$$\Delta = \frac{e^2 qQ}{h} \frac{(3\cos^2\theta - 1)}{8I(2I-1)} . \quad (3.45d)$$

Table 3.2

Relative Intensities of NAR Absorption Lines

I	3/2	5/2	7/2	9/2
$\sum_{m=-I}^{I} \xi^2(m) = \sum_{m=-I}^{I} \eta^2(m)$	24	224	1008	3168
$f_Q(I) = \dfrac{\sum_{m=-I}^{I} \xi^2(m)}{(2I)^2 (2I-1)^2 (2I+1)}$	1/6	7/75	1/14	11/180
$f_{\text{A-R}}(I) = I(I+1)$	15/4	35/4	63/4	99/4

3.3 DYNAMIC ELECTRIC QUADRUPOLE INTERACTION

Table 3.3

High–Field Transition Frequencies and Probabilities in NMR and NAR

	NMR		NAR 1		NAR 2		
Trans.	Freq.	Prob.	Freq.	Prob.	Trans.	Freq.	Prob.
$I = \frac{3}{2}$							
$+\frac{1}{2} \leftrightarrow +\frac{3}{2}$	$\omega_0 + 6\Delta$	3	$\omega_0 + 6\Delta$	3	$-\frac{1}{2} \leftrightarrow +\frac{3}{2}$	$2\omega_0 + 6\Delta$	3
$-\frac{1}{2} \leftrightarrow +\frac{1}{2}$	ω_0	4					
$-\frac{3}{2} \leftrightarrow -\frac{1}{2}$	$\omega_0 - 6\Delta$	3	$\omega_0 - 6\Delta$	3	$-\frac{3}{2} \leftrightarrow +\frac{1}{2}$	$2\omega_0 - 6\Delta$	3
$I = \frac{5}{2}$							
$+\frac{3}{2} \leftrightarrow +\frac{5}{2}$	$\omega_0 + 12\Delta$	5	$\omega_0 + 12\Delta$	5	$+\frac{1}{2} \leftrightarrow +\frac{5}{2}$	$2\omega_0 + 18\Delta$	5
$+\frac{1}{2} \leftrightarrow +\frac{3}{2}$	$\omega_0 + 6\Delta$	8	$\omega_0 + 6\Delta$	2	$-\frac{1}{2} \leftrightarrow +\frac{3}{2}$	$2\omega_0 + 6\Delta$	9
$-\frac{1}{2} \leftrightarrow +\frac{1}{2}$	ω_0	9					
$-\frac{3}{2} \leftrightarrow -\frac{1}{2}$	$\omega_0 - 6\Delta$	8	$\omega_0 - 6\Delta$	2	$-\frac{3}{2} \leftrightarrow +\frac{1}{2}$	$2\omega_0 - 6\Delta$	9
$-\frac{5}{2} \leftrightarrow -\frac{3}{2}$	$\omega_0 - 12\Delta$	5	$\omega_0 - 12\Delta$	5	$-\frac{5}{2} \leftrightarrow -\frac{1}{2}$	$2\omega_0 - 18\Delta$	5
$I = \frac{7}{2}$							
$+\frac{5}{2} \leftrightarrow +\frac{7}{2}$	$\omega_0 + 18\Delta$	7	$\omega_0 + 18\Delta$	21	$+\frac{3}{2} \leftrightarrow +\frac{7}{2}$	$2\omega_0 + 30\Delta$	7
$+\frac{3}{2} \leftrightarrow +\frac{5}{2}$	$\omega_0 + 12\Delta$	12	$\omega_0 + 12\Delta$	16	$+\frac{1}{2} \leftrightarrow +\frac{5}{2}$	$2\omega_0 + 18\Delta$	15
$+\frac{1}{2} \leftrightarrow +\frac{3}{2}$	$\omega_0 + 6\Delta$	15	$\omega_0 + 6\Delta$	5	$-\frac{1}{2} \leftrightarrow +\frac{3}{2}$	$2\omega_0 + 6\Delta$	20
$-\frac{1}{2} \leftrightarrow +\frac{1}{2}$	ω_0	16					
$-\frac{3}{2} \leftrightarrow -\frac{1}{2}$	$\omega_0 - 6\Delta$	15	$\omega_0 - 6\Delta$	5	$-\frac{3}{2} \leftrightarrow +\frac{1}{2}$	$2\omega_0 - 6\Delta$	20
$-\frac{5}{2} \leftrightarrow -\frac{3}{2}$	$\omega_0 - 12\Delta$	12	$\omega_0 - 12\Delta$	16	$-\frac{5}{2} \leftrightarrow -\frac{1}{2}$	$2\omega_0 - 18\Delta$	15
$-\frac{7}{2} \leftrightarrow -\frac{5}{2}$	$\omega_0 - 18\Delta$	7	$\omega_0 - 18\Delta$	21	$-\frac{7}{2} \leftrightarrow -\frac{3}{2}$	$2\omega_0 - 30\Delta$	7
$I = \frac{9}{2}$							
$+\frac{7}{2} \leftrightarrow +\frac{9}{2}$	$\omega_0 + 24\Delta$	9	$\omega_0 + 24\Delta$	12	$+\frac{5}{2} \leftrightarrow +\frac{9}{2}$	$2\omega_0 + 42\Delta$	6
$+\frac{5}{2} \leftrightarrow +\frac{7}{2}$	$\omega_0 + 18\Delta$	16	$\omega_0 + 18\Delta$	12	$+\frac{3}{2} \leftrightarrow +\frac{7}{2}$	$2\omega_0 + 30\Delta$	14
$+\frac{3}{2} \leftrightarrow +\frac{5}{2}$	$\omega_0 + 12\Delta$	21	$\omega_0 + 12\Delta$	7	$+\frac{1}{2} \leftrightarrow +\frac{5}{2}$	$2\omega_0 + 18\Delta$	21
$+\frac{1}{2} \leftrightarrow +\frac{3}{2}$	$\omega_0 + 6\Delta$	24	$\omega_0 + 6\Delta$	2	$-\frac{1}{2} \leftrightarrow +\frac{3}{2}$	$2\omega_0 + 6\Delta$	25
$-\frac{1}{2} \leftrightarrow +\frac{1}{2}$	ω_0	25					
$-\frac{3}{2} \leftrightarrow -\frac{1}{2}$	$\omega_0 - 6\Delta$	24	$\omega_0 - 6\Delta$	2	$-\frac{3}{2} \leftrightarrow +\frac{1}{2}$	$2\omega_0 - 6\Delta$	25
$-\frac{5}{2} \leftrightarrow -\frac{3}{2}$	$\omega_0 - 12\Delta$	21	$\omega_0 - 12\Delta$	7	$-\frac{5}{2} \leftrightarrow -\frac{1}{2}$	$2\omega_0 - 18\Delta$	21
$-\frac{7}{2} \leftrightarrow -\frac{5}{2}$	$\omega_0 - 18\Delta$	16	$\omega_0 - 18\Delta$	12	$-\frac{7}{2} \leftrightarrow -\frac{3}{2}$	$2\omega_0 - 30\Delta$	14
$-\frac{9}{2} \leftrightarrow -\frac{7}{2}$	$\omega_0 - 24\Delta$	9	$\omega_0 - 24\Delta$	12	$-\frac{9}{2} \leftrightarrow -\frac{5}{2}$	$2\omega_0 - 42\Delta$	6

3. DYNAMIC NUCLEAR ELECTRIC QUADRUPOLE INTERACTION

The coupling of acoustic energy to the nuclear spin system in the case of the dynamic nuclear electric quadrupole interaction is accomplished by means of the time–varying electric field gradients produced by the acoustic wave. The effect of the acoustic wave is to create time–varying strains in the locality of the nucleus. Treating the electric field gradient as a function of strain, we may expand the field gradient in a Taylor series in the strain components. If, for simplicity, we restrict our discussion to the case of longitudinal waves propagated either in an isotropic medium or along the [100] axis of a cubic crystal, the electric field gradient resulting from the acoustically induced strain has axial symmetry along the direction of propagation. The Taylor expansion of the field gradient in terms of the strain may be written explicitly as

$$q = q_0 + q_1\varepsilon + \frac{1}{2}q_2\varepsilon^2 + \cdots, \qquad (3.46)$$

where ε is the peak strain caused by the longitudinal acoustic wave, and where q_1 and q_2 are, respectively, the first and second derivatives of q with respect to the strain ε. The constant terms correspond to the diagonal quadrupolar matrix element (1.19a). In isotropic specimens and in cubic crystals these terms vanish (except for line width effects) because of symmetry.

The linear terms of (3.46) are of particular interest to us, since they relate the time–varying (or dynamic) strain caused by the impressed acoustic wave to the time–varying electric field gradient. They enter, similarly, into the direct or the single–phonon relaxation process. In the latter case, it is the thermal lattice phonons that interact with the spins. The linear terms also enter into the broadening of the nuclear magnetic resonance and the nuclear acoustic resonance lines by static strains caused by imperfections and impurities in the crystal lattice. The quadratic terms of (3.46) enter into the indirect or Raman–type relaxation process.

From (3.45a) and (3.45b), we see that the NAR signal strength for both single–frequency (α_{Q1}) and double–frequency (α_{Q2}) absorption lines is proportional to the square of the nuclear quadrupole moment. A useful graph of nuclear quadrupole moments as a function of the number of odd nucleons in the nucleus is given by *Segre* [3.16] and is reproduced here in Fig. 3.3. The quantity $Q/(ZR^2)$, in which Z is the atomic number and R the mean nuclear radius, is a measure of the nuclear deformation independent of the size of the nucleus. Values of the nuclear quadrupole moment Q are listed in Appendix B.

3.3 DYNAMIC ELECTRIC QUADRUPOLE INTERACTION

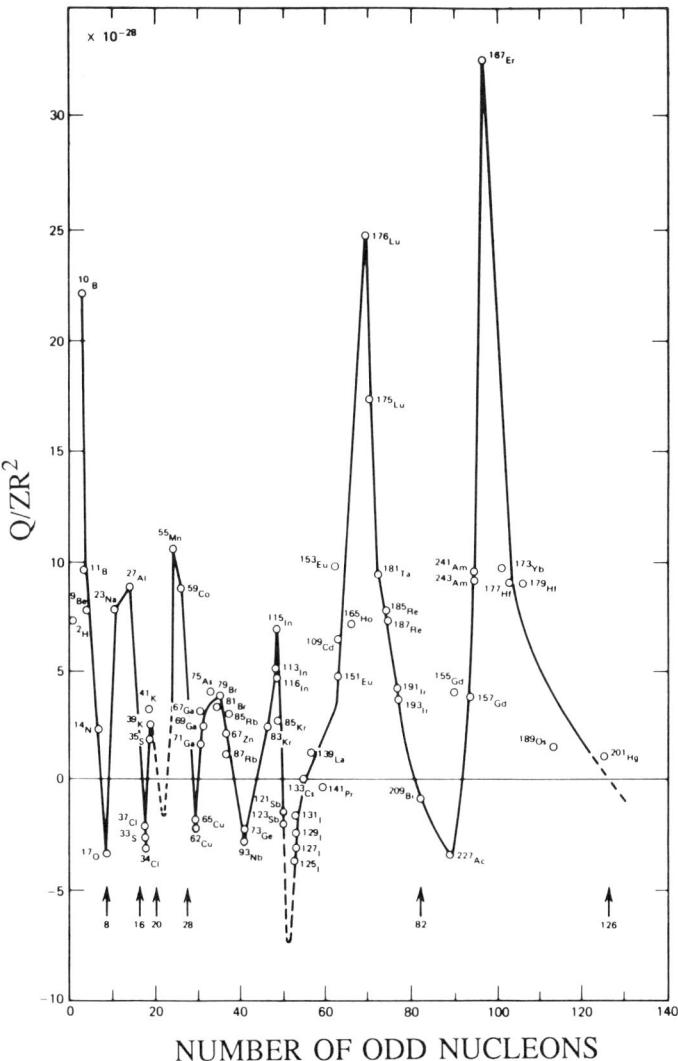

FIGURE 3.3. Reduced nuclear quadrupole moments as a function of the number of odd nucleons. The quantity (Q/ZR^2) gives a measure of the nuclear deformation independent of the size of the nucleus. (From Ref. 3.16)

The nuclear shell model has been particularly successful in explaining nuclear features for nuclei composed of a closed shell plus one or a few additional nucleons. Closed shells are indicated by the arrows and by the magic numbers in Fig. 3.3. In the closed–shell configuration, the nucleus is spherical. The addition of one or more nucleons produces only small deformations. Midway between closed shells, however, the nuclei depart appreciably from the spherical form, and collective motions involving many nucleons become important. Most notable from Fig. 3.3 are the exceptionally large positive quadrupole moments for odd A nuclei in the range $155 < A < 185$, for $A > 225$, for nuclei in the s–d shell $19 \leq A \leq 25$, and for p–shell nuclei $9 \leq A \leq 14$. Further discussion of the relationship of quadrupole moments to nuclear structure models may be found in standard texts on nuclear physics [3.16,3.17].

As indicated above, there are two NAR dynamic coupling interactions present in non–magnetic conductors and in highly degenerate semiconductors: the dynamic nuclear electric quadrupole interaction and the dynamic Alpher–Rubin interaction. Since the dynamic Alpher–Rubin interaction requires the presence of free electrons, one expects that only the dynamic nuclear electric quadrupole interaction will be present in semiconductors and ionic compounds. If an NAR $\Delta m = \pm 2$ transition is found experimentally for small elastic–strain amplitudes (strains in the linear region), one expects the coupling to the the nuclear spin system to occur by the dynamic nuclear electric quadrupole interaction. The possibility that $\Delta m = \pm 1$ and $\Delta m = \pm 2$ are due to the dynamic quadrupole interaction is further confirmed if they obey the theoretical angular dependence of the angle between the propagation direction and the external magnetic field direction. In general, this angular dependence will be different among NAR1, NAR2 and NAR(A-R). A third way to determine the dynamic quadrupole interaction is to detect when isotopic pairs are present in the same crystal. As an example, in KBr single crystal the experimental ratio of observed NAR attenuation for ^{79}Br and ^{81}Br is

$$\frac{\alpha_{2x}(^{79}\text{Br})}{\alpha_{2x}(^{81}\text{Br})} = 1.2 ,$$

where $2x$ indicates $\Delta m \pm 2$ transitions and longitudinal elastic wave propagation. Theoretically, from (3.45b), the ratio is expected to be

$$\frac{N_{79}}{N_{81}} \frac{Q_{79}}{Q_{81}} \frac{f_Q(3/2)}{f_Q(3/2)} \frac{g_{79}}{g_{81}} = 1.3 .$$

For NAR(A-R), the ratio for $\Delta m = \pm 1$ transitions is given by (3.45c) as

$$\frac{N_{79}}{N_{81}} \frac{g_{79}}{g_{81}} \frac{f_D(3/2)}{f_D(3/2)} = 1.0 .$$

3.4 ELECTRIC FIELD GRADIENT: SYMMETRY CONSIDERATIONS

For NMR the theoretical ratio would be 0.88. The conclusion one must reach is that the observed NAR attenuation ratio for KBr is dynamic nuclear electric quadrupole coupling and not NAR(A-R) (because there are no free electrons) or NMR (because no signal is caused by stray RF waves).

3.4 Electric Field Gradient: Symmetry Considerations and the S–Tensor

The off–diagonal matrix elements of the nuclear electric quadrupole Hamiltonian (3.39b–c) describe the mechanism that induces transitions between the spin energy levels under the effect of an acoustic wave. The coupling of acoustic energy to the nuclear spin system is accomplished by the time–varying electric field gradients resulting from the action of the externally generated acoustic wave. The effect of the acoustic wave is to create a time–varying strain in the locality of the nucleus. We recall, for example, that in an ideal cubic crystal, every element V_{ij} of the electric field gradient tensor V vanishes at every point of cubic symmetry. In the presence of elastic waves, however, the crystal is deformed, and the field gradient elements have values that depend upon the elastic strain components at the lattice point being considered. Like the components of the electric field gradient, the elastic strain components ε_{kl} are those of a second–rank tensor. If the strains produced by the elastic waves are sufficiently small, as is always the case in NAR experiments (but not necessarily in ASNMR experiments or in external stress experiments), then the part of V_{ij} that is proportional to strain is also proportional to the strain tensor components,

$$V_{ij}^S = \sum_{kl} S_{ijkl}\varepsilon_{kl} \qquad i, j, k, l = x, y, z , \qquad (3.47)$$

where x, y, and z are the coordinate axes. The S_{ijkl} are the components of a fourth–rank tensor termed the electrostatic field gradient elastic strain tensor, or S–tensor [3.18]. As in the case of C_{ijmn} (Section 2.1), the components S_{ijkl} of the S–tensor are usually given in the Voigt notation $S_{\alpha\beta}$, where the product $\alpha\beta$ takes on the values of 1 to 6, as follows:

$\alpha\beta$:	xx	yy	zz	$yz(zy)$	$zx(xz)$	$xy(yx)$
Voigt Notation	1	2	3	4	5	6

In this notation, (3.47) can be rewritten as

$$V_j^S = \sum_{i=1}^{6} S_{ji}\varepsilon_i . \qquad (3.48)$$

3. DYNAMIC NUCLEAR ELECTRIC QUADRUPOLE INTERACTION

The six equations that relate the strain components to the electric field gradients may be displayed in symbolic representation:

$$
\begin{array}{c|cccccc}
 & \varepsilon_{xx} & \varepsilon_{yy} & \varepsilon_{zz} & \varepsilon_{yz} & \varepsilon_{zx} & \varepsilon_{xy} \\
V^S_{xx} & S_{11} & S_{12} & S_{13} & S_{14} & S_{15} & S_{16} \\
V^S_{yy} & S_{21} & S_{22} & S_{23} & S_{24} & S_{25} & S_{26} \\
V^S_{zz} & S_{31} & S_{32} & S_{33} & S_{34} & S_{35} & S_{36} \\
V^S_{yz} & S_{41} & S_{42} & S_{43} & S_{44} & S_{45} & S_{46} \\
V^S_{zx} & S_{51} & S_{52} & S_{53} & S_{54} & S_{55} & S_{56} \\
V^S_{xy} & S_{61} & S_{62} & S_{63} & S_{64} & S_{65} & S_{66}
\end{array}
\quad (3.49)
$$

The relationship between the field gradient tensor and the strain produced by the acoustic waves is analogous to the relationship between the components of the strain tensor and the stress tensor in the theory of elastic constants of single crystals [3.19,3.20]. Both the elastic constant tensor C_{ijmn} and the elastic field gradient tensor S are symmetric, fourth-rank tensors in cubic crystals. In noncubic crystals, S may have antisymmetric forms [3.21].

The predominant number of investigations of the $S_{\alpha\beta}$ has been made in cubic crystals. We will show that in cubic crystals with coordinate axes along the cubic axes, the S-tensor has only two distinct elements at points of the unstrained lattice that have cubic symmetry.

Since the three cubic axes are equivalent,

$$S_{11} = S_{22} = S_{33}$$
$$S_{44} = S_{55} = S_{66}$$
$$S_{12} = S_{21} = S_{13} = S_{31} = S_{23} = S_{32} \ .$$

For a simple compression along the x-axis (only $\varepsilon_{xx} \neq 0$), V^S must be an even function of x, of y and of z about any point that has cubic symmetry in the unstrained lattice. Thus $\partial V^S/\partial x$ is even with respect to y and z, $\partial V^S/\partial y$ is even with respect to z and x, and $\partial V^S/\partial z$ is even with respect to x and y, so that

$$\frac{\partial^2 V^S}{\partial y \partial x} = \frac{\partial^2 V^S}{\partial z \partial x} = \frac{\partial^2 V^S}{\partial y \partial z} = 0 \ ,$$

or

$$S_{54} = S_{64} = 0 \ .$$

Repeating with shear in the zx-plane and the xy-plane:

$$S_{45} = S_{65} = S_{46} = S_{56} = 0 \ .$$

3.4 ELECTRIC FIELD GRADIENT: SYMMETRY CONSIDERATIONS

By symmetry, the value of V_{xx} should not be altered by reversing the direction in which y or z is measured. But $\varepsilon_{zx} = -\varepsilon_{-z,x}$ and so forth, so that
$$S_{14} = S_{15} = S_{16} = 0 .$$
Similarly for V_{yy}^S and V_{zz}^S,
$$S_{24} = S_{25} = S_{26} = S_{34} + S_{35} = S_{36} = 0 .$$

The only nonzero elements remaining in the S-tensor occupy a three-by-three block in the upper left corner and the diagonal in the lower right corner. The field gradients under consideration in the experiment to be described are those due to electrons and ions external to a nucleus that is at a point of cubic symmetry in the unstrained crystal. The only electronic wave functions that have appreciable probability at the nucleus are the s functions, and they are spherically symmetric so they contribute nothing to the field gradients. The field gradients at the nucleus that are due to charges external to the nucleus follow Laplace's equation:
$$V_{xx}^S + V_{yy}^S + V_{zz}^S = 0 .$$
Substituting into this equation the equivalent expressions involving the strain,
$$(S_{11} + S_{21} + S_{31})\varepsilon_{xx} + (S_{12} + S_{22} + S_{32})\varepsilon_{yy} + (S_{13} + S_{23} + S_{33})\varepsilon_{zz} = 0 .$$
This expression must hold for all ε_{xx}, ε_{yy} and ε_{zz}. Using the S-equalities proved earlier,
$$S_{12} = S_{21} = S_{32} = S_{23} = S_{31} = S_{13} = -\frac{S_{11}}{2} .$$

Thus for a cubic crystal, the S-tensor becomes

$$S_{\alpha\beta} = \begin{pmatrix} S_{11} & -\frac{1}{2}S_{11} & -\frac{1}{2}S_{11} & 0 & 0 & 0 \\ -\frac{1}{2}S_{11} & S_{11} & -\frac{1}{2}S_{11} & 0 & 0 & 0 \\ -\frac{1}{2}S_{11} & -\frac{1}{2}S_{11} & S_{11} & 0 & 0 & 0 \\ 0 & 0 & 0 & S_{44} & 0 & 0 \\ 0 & 0 & 0 & 0 & S_{44} & 0 \\ 0 & 0 & 0 & 0 & 0 & S_{44} \end{pmatrix} . \quad (3.50)$$

3. DYNAMIC NUCLEAR ELECTRIC QUADRUPOLE INTERACTION

The assumptions made above in deriving (3.50) are as follows:

(1) A linear relation exists between the electric field gradient and the elastic strain components.
(2) The coordinate axes are taken along the axes of cubic symmetry.
(3) The above tensor is valid only at points which have cubic symmetry in the unstrained crystal.

For a more detailed discussion of trace relations and the S-tensor, reference should be made to *Harrison and Sagalyn* [3.19].

Returning to the general expression (3.49), the symmetry condition for S_{ijkl} and C_{ijkl} immediately reduces the number of their components from 81 to 36. In addition, because of the conservation of energy requirements the most general elastic constant tensor has 21 independent components. These are further reduced by using the symmetry of the crystal. Because conservation of energy does not apply to the components of the elastic field gradient tensor S_{ijkl}, only the symmetry relationships are available. Since the electric field gradient satisfies the Laplace equation at the nucleus ($\nabla^2 V = 0$), the trace of the field gradient tensor is zero:

$$\sum_{\mu} V_{\mu\mu}^S = \sum_{\mu} \sum_{kl} S_{\mu\mu kl} \varepsilon_{kl} = 0 \ . \tag{3.51}$$

Since the direction of strain is arbitrary,

$$\sum_{\mu} S_{\mu\mu kl} = 0 \ , \tag{3.52}$$

or, in Voigt notation

$$\sum_{\alpha} S_{\alpha\beta} = 0 \quad \text{for all } \beta \ , \tag{3.53}$$

where α has the values 1, 2 or 3, and β has integral values 1-6. The relationships (3.51) reduce the number of independent components of S_{jk} from 36 to 30. In Appendix C are given the forms of $S_{\alpha\beta}$ for a number of crystal structures under the assumption of (3.51)—that the trace of the electric field gradient tensor is zero at the nuclear position. The number of independent $S_{\alpha\beta}$ components is as follows:

3.4 ELECTRIC FIELD GRADIENT: SYMMETRY CONSIDERATIONS

Structure	$S_{\alpha\beta}$ Components
Triclinic	30
Monoclinic	16
Orthorhombic	9
Trigonal, (C_{3v}, D_3, D_{3d})	6
Tetragonal, (C_{4v}, V_d, D_4, D_{4h})	6
Hexagonal, (D_{3h}, D_6, C_{6v}, D_{6h})	4
Cubic	2
Isotropic	1

The expression (3.53) for $S_{\alpha\beta}$ determines the tensor relationship

$$V_i^S = \sum S_{ij}\varepsilon_j \quad , \quad i,j = 1,2,\cdots,6 \tag{3.54}$$

in the coordinate system fixed with respect to the crystalline axes, and, in the case of a cubic crystal, fixed with respect to the cubic axes of the crystal. In practice it is necessary to use coordinate systems whose axes do not coincide with the crystal axes. Thus, in (3.53) the z'-axis was chosen to be the direction of the external magnetic field $\boldsymbol{H}_0$. It is often convenient, indeed, to use a coordinate system in which the DC magnetic field lies along the z'-axis and the direction of propagation of the acoustic waves is along the x'-axis. It is then necessary to find the form of (3.54) in such a rotated coordinate system. The matrix formalism for finding the proper expressions for S_{kl} and V_{ij} in rotated coordinate systems has been given explicitly by *Taylor* [3.12]; a summary of the procedure is given in Appendix D. A more concise and elegant approach utilizing the transformation matrix for second–order spherical harmonics has been described by *Mieher* [3.13,3.14]. As discussed in Section 3.1, *Fedders* [3.9] has demonstrated how yet another tensor relationship can be used to transform V_{ij}.

Applying the rotation procedure of Appendix D, we list the results for cubic crystals for the rotated axes most often used* in NAR experiments. For convenience in applying them to theory and in evaluating experimental results, we list the results in terms of $W_{Q1}(\equiv W_{m,m\pm 1})$ given in (3.42a) and of $W_{Q2}(\equiv W_{m,m\pm 2})$ given in (3.42b). We also repeat the expressions for A, $\xi_\pm$ and $\eta_\pm$:

$$\begin{aligned} A &= \frac{eQ}{I(2I-1)} \\ \xi_\pm^2 &= (2m \pm 1)^2 (I \pm m + 1)(I \mp m) \\ \eta_\pm^2 &= (I \mp m)(I \mp m - 1)(I \pm m + 1)(I \pm m + 2) \end{aligned} \tag{3.55}$$

---

*The [111] axis in cubic crystals is seldom used.

78   3. DYNAMIC NUCLEAR ELECTRIC QUADRUPOLE INTERACTION

Case I: Propagation Along [100] Crystalline Direction

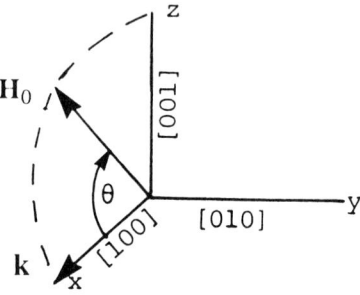

FIGURE 3.4. θ-variation

a) Longitudinal wave

$$W_{m,m\pm1} = \frac{9}{64}g_1(\nu)\left(\frac{A}{2\hbar}\right)^2 \zeta_\pm^2 [S_{11} - S_{12}]^2 \varepsilon^2 \sin^2\theta \cos^2\theta$$

$$W_{m,m\pm2} = \frac{9}{256}g_2(\nu)\left(\frac{A}{2\hbar}\right)^2 \eta_\pm^2 [S_{11} - S_{12}]^2 \varepsilon^2 \sin^4\theta$$

b) Shear wave polarized along [010]

$$W_{m,m\pm1} = \frac{1}{16}g_1(\nu)\left(\frac{A}{2\hbar}\right)^2 \zeta_\pm^2 S_{44}^2 \varepsilon^2 \cos^2\theta$$

$$W_{m,m\pm2} = \frac{1}{16}g_2(\nu)\left(\frac{A}{2\hbar}\right)^2 \eta_\pm^2 S_{44}^2 \varepsilon^2 \sin^2\theta$$

c) Shear wave polarized along [001]

$$W_{m,m\pm1} = \frac{1}{16}g_1(\nu)\left(\frac{A}{2\hbar}\right)^2 \zeta_\pm^2 S_{44}^2 \varepsilon^2 [\sin^2\theta - \cos^2\theta]^2$$

$$W_{m,m\pm2} = \frac{1}{16}g_2(\nu)\left(\frac{A}{2\hbar}\right)^2 \eta_\pm^2 S_{44}^2 \varepsilon^2 \sin^2\theta \cos^2\theta$$

3.4 ELECTRIC FIELD GRADIENT: SYMMETRY CONSIDERATIONS

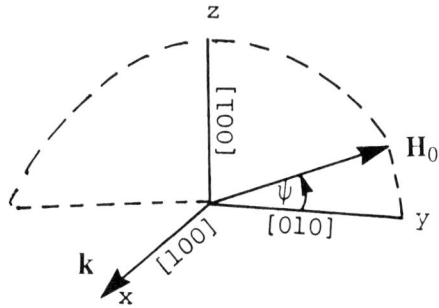

FIGURE 3.5. ψ-variation

a) Longitudinal wave

$$W_{m,m\pm1} = 0$$

$$W_{m,m\pm2} = \frac{9}{256} g_2(\nu) \left(\frac{A}{2\hbar}\right)^2 \eta_\pm^2 [S_{11} - S_{12}]^2 \varepsilon^2$$

b) Shear wave polarized along [010]

$$W_{m,m\pm1} = \frac{1}{16} g_1(\nu) \left(\frac{A}{2\hbar}\right)^2 \zeta_\pm^2 S_{44}^2 \varepsilon^2 \cos^2\psi$$

$$W_{m,m\pm2} = \frac{1}{16} g_2(\nu) \left(\frac{A}{2\hbar}\right)^2 \eta_\pm^2 S_{44}^2 \varepsilon^2 \sin^2\psi$$

c) Shear wave polarized along [001]

$$W_{m,m\pm1} = \frac{1}{16} g_1(\nu) \left(\frac{A}{2\hbar}\right)^2 \zeta_\pm^2 S_{44}^2 \varepsilon^2 \sin^2\psi$$

$$W_{m,m\pm2} = \frac{1}{16} g_2(\nu) \left(\frac{A}{2\hbar}\right)^2 \eta_\pm^2 S_{44}^2 \varepsilon^2 \cos^2\psi$$

80   3. DYNAMIC NUCLEAR ELECTRIC QUADRUPOLE INTERACTION

Case II: Propagation Along [110] Crystallographic Direction

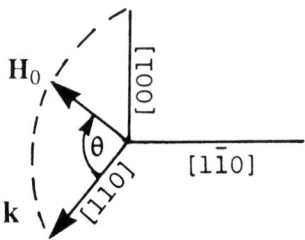

FIGURE 3.6.   Rotation in $(\bar{1}10)$ plane (θ)

a) Longitudinal wave

$$W_{m,m\pm1} = \frac{9}{64}g_1(\nu)\left(\frac{A}{2\hbar}\right)^2 \zeta_\pm^2 \left[\frac{1}{2}(S_{11}-S_{12})+S_{44}\right]^2 \varepsilon^2 \sin^2\theta \cos^2\theta$$

$$W_{m,m\pm2} = \frac{9}{256}g_2(\nu)\left(\frac{A}{2\hbar}\right)^2 \eta_\pm^2 \left[S_{44}(\sin^2\theta+1)-\frac{1}{2}(S_{11}-S_{12})\cos^2\theta\right]^2 \varepsilon^2$$

b) Shear wave polarized along [001]

$$W_{m,m\pm1} = \frac{1}{16}g_1(\nu)\left(\frac{A}{2\hbar}\right)^2 \zeta_\pm^2 S_{44}^2 \varepsilon^2 [\sin^2\theta - \cos^2\theta]^2$$

$$W_{m,m\pm2} = \frac{1}{16}g_2(\nu)\left(\frac{A}{2\hbar}\right)^2 \eta_\pm^2 S_{44}^2 \varepsilon^2 \sin^2\theta \cos^2\theta$$

c) Shear wave polarized along $[\bar{1}10]$

$$W_{m,m\pm1} = \frac{1}{64}g_1(\nu)\left(\frac{A}{2\hbar}\right)^2 \zeta_\pm^2 [S_{11}-S_{12}]^2 \varepsilon^2 \cos^2\theta$$

$$W_{m,m\pm2} = \frac{1}{64}g_2(\nu)\left(\frac{A}{2\hbar}\right)^2 \eta_\pm^2 [S_{11}-S_{12}]^2 \varepsilon^2 \sin^2\theta$$

3.4 ELECTRIC FIELD GRADIENT: SYMMETRY CONSIDERATIONS

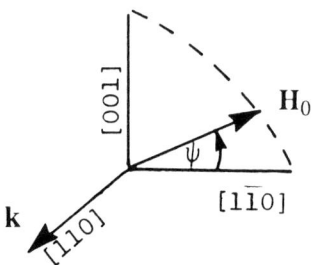

FIGURE 3.7. Rotation in (110) plane (ψ)

a) Longitudinal wave

$$W_{m,m\pm1} = \frac{9}{64}g_1(\nu)\left(\frac{A}{2\hbar}\right)^2 \zeta_\pm^2 \left[\frac{1}{2}(S_{11}-S_{12})-S_{44}\right]^2 \varepsilon^2 \sin^2\psi \cos^2\psi$$

$$W_{m,m\pm2} = \frac{9}{256}g_2(\nu)\left(\frac{A}{2\hbar}\right)^2 \eta_\pm^2 \left[S_{44}(\sin^2\psi+1)+\frac{1}{2}(S_{11}-S_{12})\cos^2\psi\right]^2 \varepsilon^2$$

b) Shear wave polarized along [001]

$$W_{m,m\pm1} = \frac{1}{16}g_1(\nu)\left(\frac{A}{2\hbar}\right)^2 \zeta_\pm^2 S_{44}^2 \varepsilon^2 \sin^2\psi$$

$$W_{m,m\pm2} = \frac{1}{16}g_2(\nu)\left(\frac{A}{2\hbar}\right)^2 \eta_\pm^2 S_{44}^2 \varepsilon^2 \cos^2\psi$$

c) Shear wave polarized along [$\bar{1}$10]

$$W_{m,m\pm1} = \frac{1}{64}g_1(\nu)\left(\frac{A}{2\hbar}\right)^2 \zeta_\pm^2 [S_{11}-S_{12}]^2 \varepsilon^2 \cos^2\psi$$

$$W_{m,m\pm2} = \frac{1}{64}g_2(\nu)\left(\frac{A}{2\hbar}\right)^2 \eta_\pm^2 [S_{11}-S_{12}]^2 \varepsilon^2 \sin^2\psi$$

3. DYNAMIC NUCLEAR ELECTRIC QUADRUPOLE INTERACTION

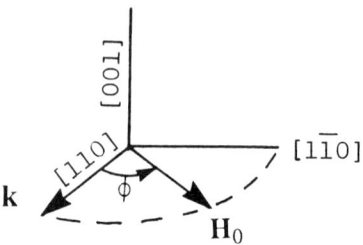

FIGURE 3.8. Rotation in (001) plane (ϕ)

a) Longitudinal wave

$$W_{m,m\pm 1} = \frac{9}{16} g_1(\nu) \left(\frac{A}{2\hbar}\right)^2 \zeta_\pm^2 \varepsilon^2 S_{44}^2 \sin^2 \phi \cos^2 \phi$$

$$W_{m,m\pm 2} = \frac{9}{256} g_2(\nu) \left(\frac{A}{2\hbar}\right)^2 \eta_\pm^2 \varepsilon^2 \left[S_{44}(\sin^2 \phi - \cos^2 \phi) + \frac{1}{2}(S_{11} - S_{12})\right]^2$$

b) Shear wave polarized along [001]

$$W_{m,m\pm 1} = \frac{1}{16} g_1(\nu) \left(\frac{A}{2\hbar}\right)^2 \zeta_\pm^2 S_{44}^2 \varepsilon^2 \cos^2 \phi$$

$$W_{m,m\pm 2} = \frac{1}{16} g_2(\nu) \left(\frac{A}{2\hbar}\right)^2 \eta_\pm^2 S_{44}^2 \varepsilon^2 \sin^2 \phi$$

c) Shear wave polarized along [$\bar{1}$10]

$$W_{m,m\pm 1} = \frac{1}{64} g_1(\nu) \left(\frac{A}{2\hbar}\right)^2 \zeta_\pm^2 [S_{11} - S_{12}]^2 \varepsilon^2 (\sin^2 \phi - \cos^2 \phi)$$

$$W_{m,m\pm 2} = \frac{1}{64} g_2(\nu) \left(\frac{A}{2\hbar}\right)^2 \eta_\pm^2 [S_{11} - S_{12}]^2 \varepsilon^2 \sin^2 \phi \cos^2 \phi$$

3.5 Fedders' Calculation of Absorption and Dispersion

We give Fedders' treatment of absorption and dispersion for the specific case of the excitation of the nuclear spin system of a cubic crystal by longitudinal acoustic waves propagated along a cube edge. The general treatment is given in [3.22]. The Hamiltonian appropriate to this case is

$$\mathcal{H}_{s-p} = \left(\frac{3eQ}{4I(2I-1)}\right)(S_{11} - S_{12})\left[I_x^2 - \frac{1}{3}I(I+1)\right]\varepsilon_{xx} .$$

This Hamiltonian can be rewritten in terms of the irreducible multipole operators A_{2m}:

$$\mathcal{H}_{s-p} = \sum_{m=0,1,2} f_{2m}(x,x)\varepsilon_{xx} A_{2m} . \qquad (3.56)$$

This form of the Hamiltonian emphasizes the coupling of the acoustic mode ε_{xx} to the spin modes A_{2m}. Here $f_{2m}(x,x)$ is the transformation coefficient between the quadrupole operators in the crystal lattice coordinate system and the irreducible spin operators in the coordinate system where $\mathbf{H}$ is along z. We wish the results of this section to correspond to the $\mathbf{H}$-plane rotation of Fig. 3.4. This is done below by setting $\phi = 0$ and replacing θ by $\pi/2 - \theta$ in the $d(x,x;m)$ expressions taken from [3.22]. We can express this by writing

$$f_{2m}(x,x) = d(x,x;m)\left(\frac{3eQ}{2I(2I-1)}\right)\left(\frac{S_{11} - S_{12}}{2}\right) . \qquad (3.57)$$

The A_{2m} and $d(x,x;m)$ needed for our problem are given in [3.22].

$m = \pm 2$:

$$A_{2\pm 2} = \left(\frac{15}{2}\right)^{\frac{1}{2}} a_2 I_\pm^2 \qquad (3.58)$$

$$d(x,x;\pm 2) = \sin^2\theta/(\sqrt{120}a_2) .$$

$m = \pm 1$:

$$A_{2\pm 1} = \mp\left(\frac{15}{2}\right)^{\frac{1}{2}} a_2\{I_\pm, I_z\} \qquad (3.59)$$

$$d(x,x;\pm 1) = 2\sin\theta\cos\theta/(\sqrt{120}a_2) .$$

$m = 0$:

$$A_{20} = (45)^{\frac{1}{2}} a_2\left[I_z^2 - \frac{1}{3}I(I+1)\right] \qquad (3.60)$$

$$d(x,x;0) = (3\cos^2\theta - 1)/(\sqrt{180}a_2) .$$

In (3.58)–(3.60), $a_2 = [I(I+1)(2I-1)(2I+3)]^{-\frac{1}{2}}$, and $\{A, B\}$ is an anticommutator.

These irreducible multipole operators form a convenient and complete set in the limit $kT \gg \hbar\omega$ because their expectation values satisfy a set of semi–phenomenological Bloch equations. In the presence of an external magnetic field $H\hat{z}$, of a driving torque due to $\mathcal{H}_{s-p}$ and of some intrinsic decay mechanism (relaxation), a set of spins interacting with the lattice satisfy the following effective Bloch equations [3.22]:

$$\left(\frac{d}{dt} + im\omega_o\right) \langle A_{2m}(t) \rangle$$
$$= \frac{\langle [A_{2m}(t), \mathcal{H}_{s-p}] \rangle}{i\hbar} - \Gamma_{2m}[\langle A_{2m}(t) \rangle - \langle A_{2m}(t) \rangle_o] ,$$
(3.61)

where $\langle A_{2m}(t) \rangle$ is the average value of $A_{2m}(t)$, $\langle A_{2m}(t) \rangle_o$ is the instantaneous equilibrium value of $\langle A_{2m}(t) \rangle$, and Γ_{2m} is the relaxation decay rate of the $2m$ mode. All terms except the one proportional to Γ_{2m} arise rigorously from taking the expectation value of the Heisenberg equation of motion for $A_{2m}(t)$. The last term includes the spin–lattice relaxation. This relaxation term is added phenomenologically by assuming that the $\langle A_{2m}(t) \rangle$ are good normal modes of the spin system that exponentially relax with a decay rate Γ_{2m}.

We assume an $e^{i(\mathbf{k}\cdot\mathbf{r}-\omega t)}$ space–time dependence for the acoustic wave and the spin operators. In the high temperature limit, (3.61) reduces to

$$(-i\omega + im\omega_o + \Gamma_{2m})\langle A_{2m}(t) \rangle = (im\omega_o + \Gamma_{2m})$$
$$\times \left(-\beta \sum_m f^*_{2m}(x,x)\varepsilon_{xx}\right) ,$$
(3.62)

where $\beta \equiv (1/k_B T)$, and $f_{lm} = f^*_{lm}$. This can be written in the more convenient form

$$\langle A_{2m}(t) \rangle = \frac{m\omega_o - \Gamma_{2m}}{m\omega_o - \omega - i\Gamma_{2m}} \left(-\beta \sum_m f^*_{2m}(x,x)\varepsilon_{xx}\right) .$$
(3.63)

The equation of motion for the lattice displacement u_x is

$$\rho\left(\frac{d^2 u_x}{dt^2}\right) = f_x ,$$
(3.64)

where $f_x = \rho v_o^2 (d^2 u_x/dx^2)$ and v_0 is the acoustic velocity in the absence of spins. The effects of the spin–phonon interactions can be derived from an additional energy density in the equation of motion for u_x. The spin–phonon Hamiltonian can be put into the form of an energy density in the elastic continuum limit. We multiply $\mathcal{H}_{s-p}$ (3.56) by the number of the spins per unit volume N and replace $A_{2m}(t)$ with $\langle A_{2m}(t)\rangle$. The additional energy density U_{s-p} associated with the spins is

$$U_{s-p} = \sum_m N f_{2m}(x,x)\varepsilon_{xx}\langle A_{2m}(t)\rangle . \tag{3.65}$$

This energy density gives rise to a force Δf_x,

$$\Delta f_x = \sum_m N f_{2m}(x,x)\left(\frac{d\langle A_{2m}(t)\rangle}{dx}\right) . \tag{3.66}$$

The equation of motion for the lattice displacement $u = u_x \hat{x}$ is now

$$\rho\left(\frac{d^2 u_x}{dt^2}\right) = f_x + \Delta f_x . \tag{3.67}$$

For the space–time dependence stated earlier, we have

$$(\omega^2 - v_o^2 k^2) u_x = -\frac{\beta N}{\rho}\sum_m k u_x f_{2m}(x,x)\langle A_{2m}(t)\rangle , \tag{3.68}$$

and

$$(\omega^2 - v_o^2 k^2) = -\frac{\beta N}{\rho} k \sum_m f_{2m}(x,x)\langle A_{2m}(t)\rangle . \tag{3.69}$$

Using the expression for $\langle A_{2m}(t)\rangle$ in (3.63), we have

$$(\omega^2 - v_o^2 k^2) = -\frac{\beta N}{\rho} k \sum_m f_{2m}(x,x) f_{2m}^*(x,x) F_{2m}(\omega) , \tag{3.70}$$

where

$$F_{2m}(\omega) = \frac{m\omega_o - \Gamma_{2m}}{m\omega_o - \omega - i\Gamma_{2m}} \tag{3.71}$$

and Γ_{2m} is the inverse lifetime $(1/\tau_{2m})$ of the $(l = 2m)$ spin mode.

One can write $k = k_o + i\Delta\alpha$ and $\omega = k_o(v_o + \Delta v)$, where $\Delta\alpha$ and Δv are the changes in attenuation and velocity, respectively. Assuming small $\Delta\alpha$ and Δv, we can write

$$\omega^2 - v_o^2 k^2 = 2 v_o^2 k_o^2 \left(\frac{\Delta v}{v_o} - \frac{i\Delta\alpha}{k_o}\right) . \tag{3.72}$$

The terms on the right-hand side of (3.70) are small compared to changes in k. Therefore, one can write

$$\left(\frac{\Delta v}{v_o} - i\frac{\Delta \alpha}{k_o}\right) = -\frac{\beta N}{2\rho v_o^2} \sum_m f_{2m}(x,x) f_{2m}^*(x,x) F_{2m}(\omega), \qquad (3.73)$$

where $\Delta v/v_o$ is the velocity change, and $-i\Delta\alpha/k_o$ is the attenuation change. Using the explicit representation of $f_{2m}(x,x)$ for $m = 0, 1, 2$, we have

$$\left(\frac{\Delta v}{v_o} - i\frac{\Delta \alpha}{k_o}\right) = -\frac{a}{2}\frac{N}{k_B T \rho v_o^2}[\sin^4\theta(F_{22}(\omega) + F_{2-2}(\omega))$$
$$+ 4\sin^2\theta \cos^2\theta(F_{21}(\omega) + F_{2-1}(\omega))$$
$$+ \frac{2}{3}(3\cos^2\theta - 1)^2 F_{20}(\omega)], \qquad (3.74)$$

where $a = 1/120[\frac{3}{2}eQ/2I(2I-1)(S_{11} - S_{12})]^2 I(I+1)(2I-1)(2I+3)$.

The terms in (3.74) that are proportional to $F_{2m}(\omega)$ with $m \neq 0$ describe the resonant spin–phonon interactions $\Delta m = 1$ and $\Delta m = 2$. The term in (3.74) that is proportional to $F_{20}(\omega)$ describes the coupling of the acoustic wave to the nonresonant $\Delta m = 0$ mode. In the frequency regime far below the resonant frequencies ($\omega \ll \omega_o$), the $F_{2m}(\omega)$ for $m \neq 0$ approach 1.

Considering the $\Delta m = 1$ and $\Delta m = 2$ modes separately, we obtain

$$\left[\frac{\Delta v}{v_o} - i\frac{\Delta \alpha}{k_o}\right]_{\Delta m=1}$$
$$= -\left(\frac{Na}{2\rho v_o^2 k_B T}\right) 4\sin^2\theta \cos^2\theta$$
$$\times \left[\frac{(\omega_o - \omega)(\omega_o - \Gamma_{21}) + i\Gamma_{21}(\omega_o - \Gamma_{21})}{(\omega_o - \omega)^2 + \Gamma_{21}^2}\right]. \qquad (3.75)$$

The imaginary part of this equation gives an expression for changes in attenuation. This expression contains a Lorentz line shape function. If one normalizes this line shape function and writes it as $g(\nu)$, then the change in attenuation is identical with the results of Section 3.3. The real part gives an expression for the dispersion of the acoustic wave for the $\Delta m = 1$ mode:

$$\left(\frac{\Delta v}{v_o}\right)_{\Delta m=1} = -\frac{2Na}{\rho v_o^2 k_B T}\frac{\omega_o(\omega_o - \omega)}{(\omega_o - \omega)^2 + \Gamma_1^2} \sin^2\theta \cos^2\theta, \qquad (3.76)$$

where $\Gamma_{21} \equiv \Gamma_1 \equiv 1/\tau_1$, with τ_1 denoting the spin–lattice relaxation time for the $\Delta m = 1$ mode, and $\Gamma_1 \ll \omega_o$. Similarly, for the NAR $\Delta m = 2$ mode, we obtain the dispersion

$$\left(\frac{\Delta v}{v_o}\right)_{\Delta m=2} = \frac{Na}{2\rho v_o^2 k_B T}\frac{\omega_o(\omega_o - \omega)}{(\omega_o - \omega)^2 + \Gamma_2^2} \sin^4\theta, \qquad (3.77)$$

where $\Gamma_{22} \equiv \Gamma_2 \equiv 1/\tau_2$. τ_2 is the spin–lattice relaxation time for the $\Delta m = 2$ mode, and $\Gamma_2 \ll \omega_o$.

At high enough frequencies, all of the $F_{lm}(\omega)$ are vanishingly small and the dispersion $\Delta v/v$ is zero. In this limit the spins are decoupled from the lattice since they cannot respond to such a high frequency. This situation corresponds to the isothermal limit: the spins remain in thermal equilibrium because they are unaffected by the lattice. In the opposite limit of $\omega \to 0$, all of the $F_{lm}(\omega)$ are equal to one. This corresponds to the adiabatic limit: the driving frequency is so low that the spins adiabatically follow the lattice, remaining in local instantaneous equilibrium instead of thermal equilibrium. In this limit there is no attenuation, and $\Gamma_2 = a(8/3)$, independent of angle.

3.6 Application. NAR in Noncubic Metallic Rhenium: Pure Nuclear Electric Quadrupole Resonance

For a nucleus possessing a quadrupole moment, magnetic resonance experiments generally are performed under one of two limiting conditions: (a) a strong magnetic field, so that the Larmor frequency ν_L is much larger than the zero–field or pure quadrupole frequency ν_Q; (b) zero or small external magnetic field, such that $\nu_L < \nu_Q$. For a quadrupole nucleus, a zero-field splitting of energy levels is available, of course, only in the presence of a static electric field gradient, which is present in most noncubic structures. To observe transitions among the magnetic field–quadrupole levels, metallic Re is especially well suited because of its relatively large quadrupole moment and because there are two stable isotopes, ^{185}Re and ^{187}Re, with large natural abundances of 37.40% and 62.60%, respectively. Rhenium metal has a close–packed hexagonal structure with the c–axis along the [0001] crystallographic direction. The nuclear spin of ^{185}Re and ^{187}Re is $\frac{5}{2}$, the magnetic dipole moments are 3.1437 and 3.1760 nuclear magnetons, and the quadrupole moments are 2.36 and 2.24 barn, respectively.

3.6.1 Energy Levels: Low Magnetic Field Limit

In contrast to the high–field case given in (3.44), in the present instance we are concerned with a noncubic crystal with large zero–field energy level splittings due to the static nuclear electric quadrupole interaction, perturbed by a small applied magnetic field. In the case of an axially symmetric electric field gradient, which exists in Re, and for a nucleus with half integral spin, there are $I + \frac{1}{2}$ doubly degenerate levels, E_m, at $H = 0$. The application of a small magnetic field H at an angle θ with respect to the symmetry axis removes the degeneracy. The resultant energy levels for $m > \frac{1}{2}$ are given by

$$E_{\pm m} = B[3m^2 - I(I+1)] \mp m\hbar\gamma H \cos\theta \;, \tag{3.78}$$

3. DYNAMIC NUCLEAR ELECTRIC QUADRUPOLE INTERACTION

where
$$B = \frac{eQ}{4} A \equiv \frac{e^2 qQ}{4I(2I-1)},$$
and eq, as before, is the zz-component of the electric field gradient tensor.

In the case of the $m = \frac{1}{2}$ energy level, originally degenerate, the application of a small magnetic field H results in energy states (denoted ψ_+ and ψ_-) that are mixtures of the $\psi_{+\frac{1}{2}}$ and the $\psi_{-\frac{1}{2}}$ states. The energies associated with ψ_+ and ψ_- are

$$E_\pm = B\left[\frac{3}{4} - I(I+1)\right] \mp \frac{f}{2}\hbar\gamma H \cos\theta, \qquad (3.79)$$

where
$$f\cos\theta = \left[\cos^2\theta + \left(I+\frac{1}{2}\right)^2 \sin^2\theta\right]^{\frac{1}{2}}.$$

The states themselves are given by
$$\psi_+ = \psi_{+\frac{1}{2}} \sin\alpha + \psi_{-\frac{1}{2}} \cos\alpha \qquad (3.80a)$$
$$\psi_- = \psi_{-\frac{1}{2}} \sin\alpha - \psi_{+\frac{1}{2}} \cos\alpha, \qquad (3.80b)$$

where $\tan\alpha = [(f+1)/(f-1)]^{\frac{1}{2}}$.

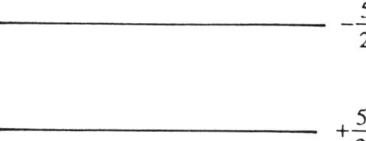

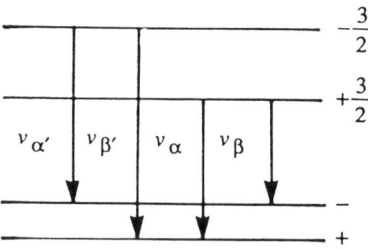

FIGURE 3.9. Schematic energy levels at a given small magnetic field for a nucleus with $I = \frac{5}{2}$ and an axially symmetric static quadrupole interaction. B is positive. Transitions observed in the case of magnetic field split pure quadrupole NAR are shown with downward arrows.

3.6 APPLICATION: NAR IN METALLIC RHENIUM

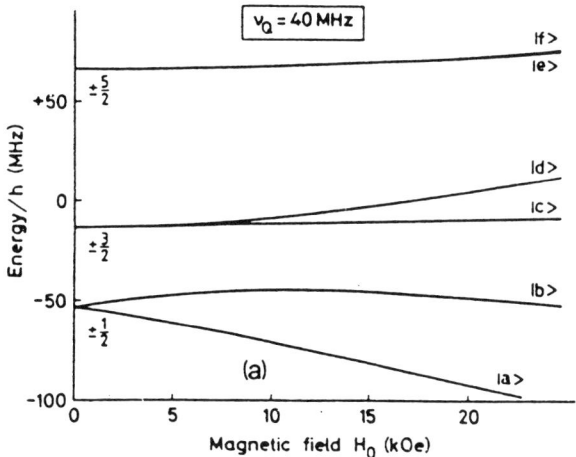

FIGURE 3.10. Energy levels of ^{187}Re for $\theta = \pi/2$. A value of $\nu_Q = 40$ MHz is assumed on the ordinate scale. (From Ref. 3.23)

For spin $I = \frac{5}{2}$, therefore, application of an external field results in splitting each of the three degenerate energy levels ($H = 0$) into two levels, as shown schematically in Fig. 3.9. The behavior of the energy levels of ^{187}Re as a function of increasing magnetic field is given in Fig. 3.10 for an assumed value of $\nu_Q = 40$ MHz. Several investigations [3.23,3.24] of the magnetic resonance of ^{185,187}Re in rhenium metal have been made utilizing NAR in the intermediate field range, where the energy level splittings due to the Zeeman effect are approximately equal to those due to the nuclear electric quadrupole interaction. A drawback, in the intermediate field case, is that in order to determine the nuclear spin energy levels, one must solve a secular equation that involves a sum of both quadrupole and Zeeman Hamiltonians. In the present treatment, we restrict ourselves to the low–field approximation, in which the Zeeman energy can be treated as a perturbation on the zero-field levels. Applying (3.78), one obtains for the frequencies of the transitions shown with arrows in Fig. 3.9

$$\nu_\alpha = \frac{6B}{h} - \frac{3-f}{2}\frac{\gamma}{2\pi}H\cos\theta \tag{3.81a}$$

$$\nu_\beta = \frac{6B}{h} - \frac{3+f}{2}\frac{\gamma}{2\pi}H\cos\theta \tag{3.81b}$$

$$\nu_{\alpha'} = \frac{6B}{h} + \frac{3-f}{2}\frac{\gamma}{2\pi}H\cos\theta \tag{3.81c}$$

90   3. DYNAMIC NUCLEAR ELECTRIC QUADRUPOLE INTERACTION

and
$$\nu_{\beta\prime} = \frac{6B}{h} + \frac{3+f}{2}\frac{\gamma}{2\pi}H\cos\theta . \qquad (3.81d)$$

3.6.2 Dynamic Coupling

By applying (3.47) to the case (for Re) of hexagonal close–packed symmetry (6/mmm), the non–zero components of the fourth-rank polar tensor can be determined using the methods of *Fumi and Ripamonti* [3.25] and taking advantage of the fact that the S-tensor is symmetric in i and j and also in k and l. In Voigt notation, the S-tensor in the present case is

$$\begin{pmatrix} S_{11} & S_{12} & S_{13} & 0 & 0 & 0 \\ S_{12} & S_{11} & S_{13} & 0 & 0 & 0 \\ S_{31} & S_{31} & S_{33} & 0 & 0 & 0 \\ 0 & 0 & 0 & S_{44} & 0 & 0 \\ 0 & 0 & 0 & 0 & S_{44} & 0 \\ 0 & 0 & 0 & 0 & 0 & \tfrac{1}{2}(S_{11}-S_{12}) \end{pmatrix} . \qquad (3.82)$$

The S-tensor for this crystal symmetry, unlike those discussed in Section 3.4, is not symmetric ($S_{13} \neq S_{31}$). We choose x to be along the $[11\bar{2}0]$ direction, y along $[\bar{1}100]$ and z along $[0001]$ as shown in Fig. 3.11. In the plane defined by the $[11\bar{2}0]$ and $[0001]$ directions, the angle θ is measured from the $[0001]$ axis to the direction of the external magnetic field $\boldsymbol{H}$. For the particular case of acoustic shear mode propagation along the z-direction, the appropriate electric field gradients are computed following the procedures outlined in Appendix D:

$$V_{\zeta[x]\pm 1} = \frac{1}{\sqrt{6}}(\cos^2\theta - \sin^2\theta)S_{44}\varepsilon_{zx} \qquad (3.83a)$$

$$V_{\zeta[x]\pm 2} = -\frac{1}{\sqrt{6}}\sin\theta\cos\theta S_{44}\varepsilon_{zx} \qquad (3.82b)$$

$$V_{\zeta[y]\pm 1} = \frac{1}{\sqrt{6}}\cos\theta S_{44}\varepsilon_{zy} \qquad (3.83c)$$

$$V_{\zeta[y]\pm 2} = -\frac{1}{\sqrt{6}}\sin\theta S_{44}\varepsilon_{zy} \qquad , \qquad (3.83d)$$

where ζ is the polarization direction of the shear mode. Following the procedures described in Section 3.2 above, the dynamic quadrupole transition probabilities for $\Delta m \pm 1$ and $\Delta m = \pm 2$ in the present instance are, respectively,

$$W_{m,m\pm 1} = D\xi^2 V_{\zeta\pm 1}^2 \tag{3.84a}$$
$$W_{m,m\pm 2} = D\eta^2 V_{\zeta\pm 2}^2 , \tag{3.84b}$$

where $D = (B/\hbar)^2 g(\nu)/4$,

$$\xi^2 = (2m \pm 1)^2 (I \pm m + 1)(I \mp m)$$
$$\eta^2 = (I \mp m)(I \mp m - 1)(I \pm m + 1)(I \pm m + 2) .$$

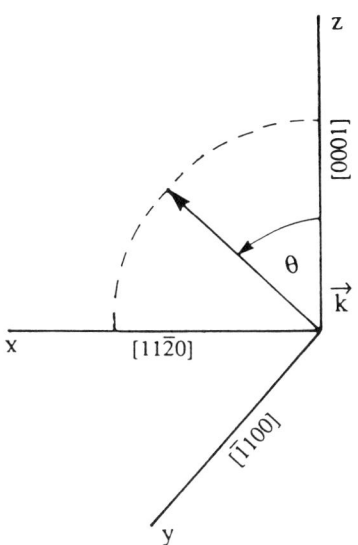

FIGURE 3.11. Definition of x-, y-, z-axes and the angle θ in terms of the hexagonal crystal axes.

For a nucleus with spin $I = \frac{5}{2}$ undergoing transitions between the $\pm\frac{3}{2}$ levels and the $\pm$ mixed levels (Fig. 3.9), $\xi^2 = 32$, and $\eta^2 = 72$. Both $\Delta m = \pm 1$ and $\Delta m = \pm 2$, depending upon the angle θ, may contribute to the four transitions indicated in Fig. 3.9, the frequencies of which are given in (3.81a–d).

As described in some detail in Section 7.7, the ^{185,187}Re NAR quadrupole resonances were studied by recording the first derivatives of the dispersion signals at magnetic field angles θ of 0°, 43.3° and 90° at 77 K and 4.2 K, using both polarization directions of the shear transducer associated with the strains in (3.81a–d). For $\theta = 0°$, $f = 1$, and $\alpha = 90°$, one can show that $\psi_+ = \psi_{\frac{1}{2}}$ and $\psi_- = \psi_{-\frac{1}{2}}$. Thus ν_α and $\nu_{\alpha'}$ correspond to $\Delta m = \pm 1$ transitions; ν_β and $\nu_{\beta'}$ correspond to $\Delta m = \pm 2$ transitions. Only the $\Delta m = \pm 1$ transitions ν_α and $\nu_{\alpha'}$ are observed. The transition probabilities for dynamic quadrupole coupling (3.84) and for the dynamic Alpher–Rubin coupling (Chapter 4) predict this result. For $\theta = 90°$, $\alpha = 45°$; now the $m = \pm\frac{1}{2}$ levels are mixed, with equal contributions from $\psi_{+\frac{1}{2}}$ and $\psi_{-\frac{1}{2}}$, while the $m = \pm\frac{3}{2}$ levels coincide. Both the dynamic quadrupole coupling field gradient element in (3.83c) and the theory of the Alpher–Rubin coupling predict in this case that $\Delta m = \pm 1$ transitions are forbidden. On the other hand, $\Delta m = \pm 2$ transitions are allowed by the dynamic quadrupole coupling (3.83d), and, indeed, a resonance narrower than that observed at $\theta = 0°$ is found—thus justifying the conclusion that the dynamic coupling in ^{185,187}Re is quadrupolar.

3.6.3 Magnitude and Sign of Quadrupole Coupling Constants

Applying the experimental techniques described in Section 7.7, the quadrupole coupling constants for ^{185,187}Re at 4.2 K and 77.8 K (given in Table 3.4) are obtained. The sign of the quadrupole coupling constant may be determined from the fortuitous circumstances that ψ_+ and ψ_- are mixed states of $\psi_{\frac{1}{2}}$ and $\psi_{-\frac{1}{2}}$, and that dynamic quadrupole coupling allows both $\Delta m = \pm 1$ and $\Delta m = \pm 2$ transitions. The theoretical argument for the sign determination, as well as the experimental determination of the sign of e^2qQ for ^{185,187}Re as positive, has been given by *Sundfors* [3.26].

Table 3.4

Magnitudes of the Measured Nuclear Electric Quadrupole Coupling Constants for ^{185}Re and ^{187}Re*

Temperature (K)	Nucleus	e^2qQ/h (MHz)
77.8	^{187}Re	255.920 ± 0.013
77.8	^{185}Re	270.966 ± 0.013
4.2	^{187}Re	255.478 ± 0.013
4.2	^{185}Re	270.430 ± 0.013

*(From Ref. 3.26)

3.6.4 Other Experimental Results

Studies have been made of both the line broadening and the temperature dependence of the quadrupole coupling constants in ^{185,187}Re in rhenium metal. Details of experimental line widths and possible sources of the line broadening are given in Chapter 5. Anticipating that discussion, we shall now say the dominant contribution to line broadening in Re is attributed to indirect exchange and to static nuclear electric quadrupole effects. A theory of bonding in metals as it influences the temperature dependence of e^2qQ is given in Chapter 6. The conclusion, to be justified in Chapter 5, is that for ^{185,187}Re in metallic rhenium in the temperature range $2K \leq T \leq 150K$, the response of the electron density distribution, which is mainly responsible for the total electric field gradient, is almost completely determined by the temperature dependence of the rigid, nonvibrating lattice.

3.7 Dynamic Nuclear Electric Hexadecapole Interaction

After the nuclear quadrupole moment, the next-higher order nuclear electric moment, the nuclear hexadecapole moment, corresponds to $l = 4$, and can be defined from (3.19). We write an expression for the hexadecapole moment from [3.27] as

$$eM^{(4)} = \sum_{(p)} \langle r_p^4 (35 \cos^4 \theta_p - 30 \cos^2 \theta_p + 3) \rangle_{m_I = I} , \quad (3.85)$$

where r_p is the distance of a proton from the center of the nucleus, and the angle θ_p is measured with respect to the z-axis (direction of quantization). The expectation value in (3.85) is taken over the nuclear wave function with magnetic quantum number $m_I = I$, and the sum extends over all protons in the nucleus. Alternatively, using the notation of Table 3.1b,

$$eM^{(4)} = \langle II| \sum_{(p)} (35z_i^4 - 30 r_i^2 z_i^2 + 3 r_i^4) |II \rangle , \quad (3.86)$$

where r_i is the distance to the ith proton, and z_i its projection on the z-axis.

Referring to the general expression (3.23) for the electrostatic multipole interaction and to (3.5d) for the A_{4m} tensor operators (also see Table 3.1), we write the A_4^m, or hexadecapole moment operators, as

$$A_4^0 = A^{(4)} [35 I_z^4 - 30 I(I+1) I_z^2 + 25 I_z^2 \\ - 6I(I+1) + 3[I(I+1)]^2]_{op} \quad (3.87a)$$

$$A_4^{\pm 1} = 5^{\frac{1}{2}} A^{(4)} [(7 I_z^3 - 3I(I+1) I_z - I_z) I_\pm + I_\pm \\ \times (7 I_z^3 - 3I(I+1) I_z - I_z)]_{op} \quad (3.87b)$$

$$A_4^{\pm 2} = \frac{1}{2} (10)^{\frac{1}{2}} A^{(4)} [(7 I_z^2 - I(I+1) - 5)(I_\pm)^2 \\ + (I_\pm)^2 (7 I_z^2 - I(I+1) - 5)]_{op} \quad (3.87c)$$

94   3. DYNAMIC NUCLEAR ELECTRIC QUADRUPOLE INTERACTION

$$A_4^{\pm 3} = (35)^{\frac{1}{2}} A^{(4)} [I_z(I_\pm)^3 + I(\pm)^3 I_z]_{op} \tag{3.87d}$$

$$A_4^{\pm 4} = \frac{1}{2}(70)^{\frac{1}{2}} A^{(4)} [(I_\pm)^4]_{op} , \tag{3.87e}$$

where $A^{(4)} = eM^{(4)}/16I(I-1)(2I-1)(2I-3)$, and $I_\pm = I_x \pm iI_y$. Proceeding along similar lines, we express the electronic terms for $l = 4$ in (3.23) using the notation of reference [3.28] and the values of $r^l Y_l^m(\theta, \phi)$ from Table 3.1a. The $B_l^{\pm m}$ are written in terms of the components of the fourth derivative of the electric potential at a nuclear position V_{ijkl} by comparing particular Cartesian coordinate V_{ijkl} with the particular $B_l^{\pm m}$ in (3.88a–e). The fourth-order potential operators are defined in (3.89a–e).

$$B_4^0 = 1 V_4^0 \tag{3.88a}$$

$$B_4^{\pm 1} = \frac{1}{5}(5^{\frac{1}{2}}) V_4^{\pm 1} \tag{3.88b}$$

$$B_4^{\pm 2} = \frac{1}{5}(10)^{\frac{1}{2}} V_4^{\pm 2} \tag{3.88c}$$

$$B_4^{\pm 3} = \frac{1}{35}(35)^{\frac{1}{2}} V_4^{\pm 3} \tag{3.88d}$$

$$B_4^{\pm 4} = \frac{1}{35}(70)^{\frac{1}{2}} V_4^{\pm 4} , \tag{3.88e}$$

where

$$V_4^0 = \frac{1}{24} V_{zzzz} \tag{3.89a}$$

$$V_4^{\pm 1} = \frac{1}{12}(V_{zzzx} \pm iV_{zzzy}) \tag{3.89b}$$

$$V_4^{\pm 2} = \frac{1}{24}(V_{zzxx} - V_{zzyy} \pm 2iV_{zzxy}) \tag{3.89c}$$

$$V_4^{\pm 3} = \frac{1}{12}[(V_{zxxx} - 3V_{zxyy} \pm i(3V_{zxxy} - V_{zyyy})] \tag{3.89d}$$

$$V_4^{\pm 4} = \frac{1}{48}[V_{xxxx} + V_{yyyy} - 6V_{xxyy} \pm 4i(V_{xxxy} - V_{xyyy})] . \tag{3.89e}$$

In the expressions for $V_4^{\pm m}$, $V_{xyzx} \equiv [(\partial^4/\partial x \partial y \partial z \partial x)V]_0$, where V is the electrostatic potential, and $[\ldots]_0$ signifies the evaluation of the quantity in the bracket at the nuclear site. Combining (3.87a–e) and (3.89a–e) we can write the hexadecapole Hamiltonian as

$$\mathcal{H}_4 = \sum_{m=-4}^{m=+4} A_4^m B_4^{m*} = \sum_{m=-4}^{m=+4} A_4^m B_4^{-m}$$

$$= (A_4^0 B_4^0 + A_4^{+1} B_4^{-1} + \cdots + A_4^{+4} B_4^{-4} + A_4^{-4} B_4^{+4}) . \tag{3.90}$$

3.7 DYNAMIC NUCLEAR ELECTRIC HEXADECAPOLE INTERACTION

The first term in (3.90) gives the energy levels for the hexadecapole interaction, corresponding to a small shift in the nuclear magnetic energy levels:

$$\langle |\mathcal{H}_4^0| \rangle \equiv E_4$$
$$= \frac{ehM^{(4)}[35m^4 - 30m^2 I(I+1) + 25m^2 + 3I^2(I+1)^2 - 6I(I+1)]}{384 I(I-1)(2I-1)(2I-3)},$$

(3.91)

where

$$h \equiv V_4^0 = \int_{\tau_e} \frac{\rho_e(r_e)(35\cos^4\theta_e - 30\cos^2\theta_e + 3)d\tau_e}{r_e^5}.$$

(3.92)

We treat the effect of the hexadecapole interaction on the energy levels for certain cases. We note that the splitting due to $E_4 (\equiv \langle |\mathcal{H}_4^0| \rangle)$ is degenerate in m, just as in the quadrupole case. For the case of a perfect cubic crystal in a DC magnetic field—in which case the static quadrupole interaction vanishes—we have

$$\mathcal{H} = \mathcal{H}_m + \mathcal{H}_4^0,$$

(3.93)

where $\mathcal{H}_m = -mh\nu_L$. ν_L is the Larmor frequency. Due to $\mathcal{H}_m$, the energy levels are split into $2I+1$ equally spaced levels, on which the hexadecapole interaction can be regarded as a very small perturbation. Since the hexadecapolar splitting is degenerate in m, as in the quadrupolar case, we obtain symmetric hexadecapolar energy level shifts.

Table 3.5
Hexadecapole Energy Level Shifts*

m	$I = 3/2$	$I = 7/2$	$I = 9/2$
1/2	+2/192	+9/1344	+18/3456
3/2	−3/192	−3/1344	+3/3456
5/2	+1/192	−13/1344	−17/3456
7/2		+7/1344	−22/3456
9/2			+18/3456

*Units of $e^2 M^{(4)} h$

96   3. DYNAMIC NUCLEAR ELECTRIC QUADRUPOLE INTERACTION

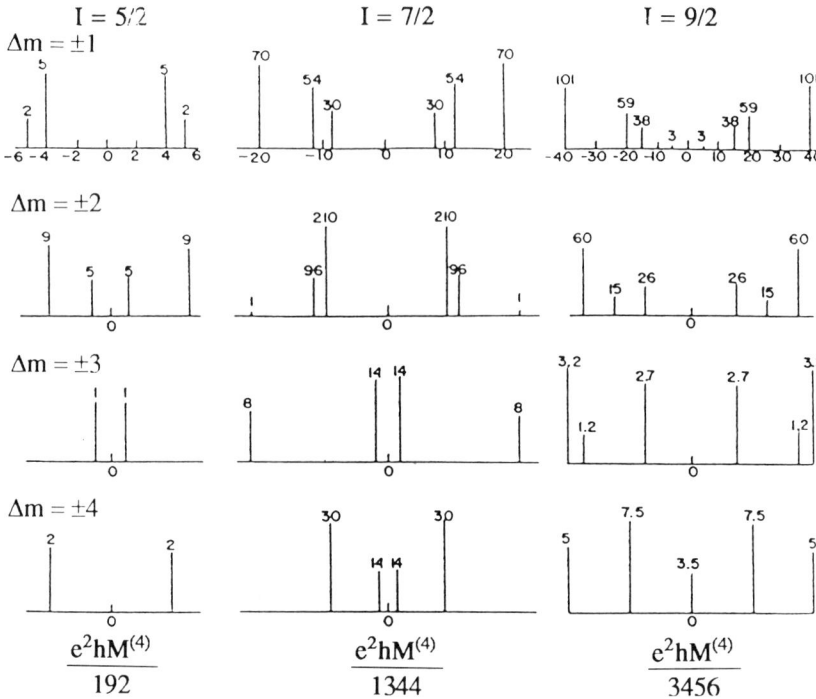

FIGURE 3.12. Hexadecapolar–split high–field Zeeman lines for nuclei at sites of cubic symmetry. Relative intensities for a given spin value are given. Horizontal scales for a given spin value are the same for $\Delta m = \pm 1, \pm 2, \pm 3, \pm 4$.

Zeeman line spectra subjected to these shifts for $I = \frac{5}{2}, \frac{7}{2}$ and $\frac{9}{2}$, and for $\Delta m = 1, 2, 3, 4$ are shown in Fig. 3.12. The relative intensities of the hexadecapole–split lines are indicated by the numbers labelling each line. Comparisons are valid only among multiplets ($\Delta m = \pm 1, \pm 2, \pm 3, \pm 4$) for a given spin (*e.g.*, $I = \frac{7}{2}$). The intensities are calculated from the expressions for $|A_4^{\pm n}(m)|^2$ given below in (3.96a–d). For the case of noncubic solids, the hexadecapolar energy level shift is added to the quadrupolar shift, which is given in (1.21). In the case of zero–field or pure quadrupole spectra the quadrupolar energy levels are also modified by the hexadecapolar shifts. Restricting ourselves to systems in which there exists cylindrical symmetry for the two coordinate frames of reference (*i.e.*, the nuclear spin system and the electron charge system) we can apply (1.23) and (3.91), obtaining the levels for spin $I = \frac{5}{2}, \frac{7}{2}, \frac{9}{2}$ given in Fig. 3.13. For convenience, the

3.7 DYNAMIC NUCLEAR ELECTRIC HEXADECAPOLE INTERACTION

hexadecapole energy level shifts for this case are listed in Table 3.5. The remaining terms in (3.90) describe the interaction of phonons with the nuclear spin system via the hexadecapole interaction, corresponding to one phonon $\Delta m = \pm 1, \pm 2, \pm 3$ and ± 4 nuclear spin transitions.

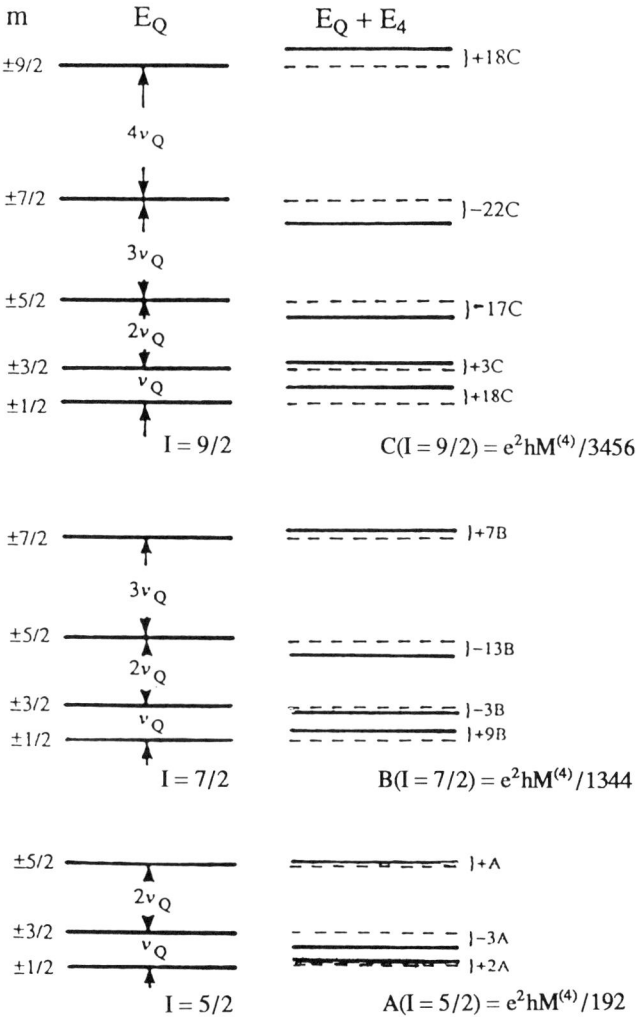

FIGURE 3.13. Hexadecapolar–shifted pure nuclear quadrupole energy levels.

3. DYNAMIC NUCLEAR ELECTRIC QUADRUPOLE INTERACTION

Analogously to our treatment of the dynamic quadrupolar interaction, we introduce phonon operators into the expression for the interaction energy by expanding the terms in (3.90) in a Taylor's series about the equilibrium position of the nucleus:

$$\mathcal{H}_{mm}(n) = A_4^{\pm n} B_4^{\pm n} = K^n A^{\pm n} V^{\mp n}$$
$$= [K^n A^{\pm n} V^{\mp n}]_0 + K^n A^{\pm n} \sum_{i,\alpha} \left[\frac{\partial}{\partial \xi_i^\alpha} V^{\mp n} \right]_0 \xi_i^\alpha$$
$$+ \frac{1}{2!} K^n A^{\pm n} \sum_{\substack{i,\alpha \\ j,\beta}} \left[\frac{\partial^2}{\partial \xi_i^\alpha \partial \xi_j^\beta} V^{\mp n} \right]_0 \xi_i^\alpha \xi_j^\beta + \cdots, \quad (3.94)$$

where ξ_i^α is the ith component of the displacement of the αth particle. $n > 0$, and Σ_α denotes a lattice sum. The first term in the expansion is the static term; the second term corresponds to a one–phonon hexadecapole interaction; the third term corresponds to a two–phonon hexadecapole interaction. Thus for the linear term of interest in NAR, a phonon is absorbed (or emitted), and a corresponding energy conserving nuclear spin transition occurs. The allowed transitions are those for $n = 1, 2, 3$ and 4, corresponding to $\Delta m = \pm 1, \pm 2, \pm 3$ and ± 4 respectively.

We may write the transition probabilities for the dynamic hexadecapolar interaction as

$$W_{\text{HDI}}^{\pm n} = (B_4^{\pm n})^2 |\langle m' | A_4^{\pm n} | m \rangle|^2 . \quad (3.95)$$

The matrix elements are:

$$|A_4^{\pm 1}(m)|^2 = \frac{5e^2 (M^{(4)})^2}{[16 I(I-1)(2I-1)(2I-3)]^2}$$
$$\times \{(m \pm 1)[7(m \pm 1)^2 - \{3I(I+1) + 1\}]$$
$$+ m[7m^2 - \{3I(I+1) + 1\}]\} [(I \mp m)$$
$$\times (I \pm m + 1)] \quad (3.96a)$$

$$|A_4^{\pm 2}(m)|^2 = \frac{5e^2 (M^{(4)})^2}{2[16 I(I-1)(2I-1)(2I-3)]^2}$$
$$\times \{7[(m \pm 2)^2 + m^2] - 2[I(I+1) + 5]\}^2$$
$$\times [(I \mp m)(I \pm m + 1)][(I \mp m - 1)$$
$$\times (I \pm m + 2)] \quad (3.96b)$$

3.7 DYNAMIC NUCLEAR ELECTRIC HEXADECAPOLE INTERACTION

$$|A_4^{\pm 3}(m)|^2 = \frac{35e^2(M^{(4)})^2}{16I(I-1)(2I-1)(2I-3)]^2}$$
$$\times (2m \pm 3)^2[(I \mp m)(I \mp m - 1)$$
$$\times (I \mp m - 2)(I \pm m + 1)(I \pm m + 2)$$
$$\times (I \pm m + 3)] \qquad (3.96c)$$

$$|A_4^{\pm 4}(m)|^2 = \frac{35}{2} \frac{e^2(M^{(4)})^2}{[16I(I-1)(2I-1)(2I-3)]^2}$$
$$\times [(I \mp m)I \mp m - 1)(I \mp m - 2)$$
$$\times (I \mp m - 3)(I \pm m + 1)(I \pm m + 2)$$
$$\times (I \pm m + 3)(I \pm m + 4)] . \qquad (3.96d)$$

The spin factors for the dynamic hexadecapole interaction, analogous to those for the dynamic quadrupolar interaction—ξ^2 and η^2 in (3.43)—are given by

$$\frac{1}{[16I(I-1)(2I-1)(2I-3)]^2} \sum_m |\xi(I,m)|^2 ,$$

where the sum is taken over all the allowed transitions. These may be tabulated as follows.

Table 3.6

Hexadecapole Spin Factors

Δm	$I = 5/2$		$I = 7/2$		$I = 9/2$
±1	0.0875	(2.2)	0.0393	(0.42)	0.0276
	(0.5)		(0.5)		(0.5)
±2	0.1750	(2.2)	0.0780	(0.42)	0.0550
	(14)		(14)		(14)
±3	0.0125	(2.2)	0.0056	(0.42)	0.0039
	(0.5)		(0.5)		(0.5)
±4	0.0250	(2.2)	0.0112	(0.42)	0.0079

Note that the ratios (shown in parentheses) between spin factors in adjacent rows and adjacent columns are the same. To obtain explicit expressions for the $V_4^{\pm n}$, including dependence on angle (between the direction of acoustic propagation and H_0), corresponding to imposed strains (acoustic waves

100   3. DYNAMIC NUCLEAR ELECTRIC QUADRUPOLE INTERACTION

of given polarization and direction of propagation), we expand the fourth derivative of the electric potential evaluated at a nuclear position as a Taylor series in the strains. We write V_{ijkl} for the term in the expansion that is linear in strain and introduce a sixth-rank polar tensor, the T-tensor, to relate the strain ε_{mn} to V_{ijkl}^s:

$$V_{ijkl}^s = T_{ijklmn}\varepsilon_{mn} , \qquad (3.97)$$

where ε_{mn} are the components of the elastic strain tensor. From the definitions of V_{ijkl}^s and ε_{mn}, the $ijkl$ indices permute, and the mn indices permute. Following a procedure similar to that of Taylor–Bloembergen for the case of dynamic quadrupolar coupling (see Appendix D), *Sundfors* [3.29] has performed the calculations to obtain the expressions for V_{ijkl}, in terms of the appropriate components of the T-tensor in systems rotated with respect to the principal axis system. A summary of these calculations, for the case of cubic $\bar{4}3m$ symmetry is given in Appendix E.

The hexadecapole interaction in solids is an extremely small effect. However, as in the case of the interaction of the nuclear electric quadrupole moment with the electric field gradient, the interaction of the nuclear hexadecapole moment with the fourth-order derivative terms of the potential due to the ionic lattice may be considerably amplified by antishielding effects, by charge–overlap effects and by conduction electron effects. The hexadecapole antishielding factors η_∞ have been calculated by *Sternheimer* [3.30] for several nuclei; for the same nucleus, they are generally considerably larger than the corresponding quadrupolar antishielding factors, γ_∞. For atoms with filled or partially filled f or d shells, η_∞ may approximate 10^4 or 10^5; for Hg^{2+}, for example, Sternheimer computes $\eta_\infty = -63,000$ [3.30]. A more complete discussion of hexadecapole antishielding factors is given in Chapter 6.

3.8 Is the Hexadecapolar Interaction (HDI) Observable by NAR Techniques?

It is widely recognized that electromagnetic moments of nuclei provide sensitive tests for theories of nuclear structure. Values of nuclear electric hexadecapole moments, in particular, are valuable adjuncts to nuclear electric quadrupole values in the theoretical understanding of shape collective phenomena in nuclear structure.

Although quadrupole moments Q_2 and quadrupole deformation parameters β_2 have been measured accurately for many nuclei, the same is not true for the corresponding hexadecapole moments M_4 and the hexadecapole deformation parameters β_4. Measurements of hexadecapole moments have

3.8 IS THE HDI OBSERVABLE BY NAR TECHNIQUES?

been concentrated (i) in and near the rare earth region and (ii) in the region of the actinides. A wide variety of high energy nuclear techniques have been utilized in obtaining this data on hexadecapole moments and β_4 deformation parameters; the experiments appear to be relatively difficult, and some of the results have been contradictory. Concomitantly, significant and on–going theoretical efforts have been made both to predict $E4$ (E being the nuclear form factor) effects for certain nuclei and to incorporate the experimentally obtained hexadecapole data. Characteristic discussions are given in references [3.31,3.32].

Precision measurements of nuclear quadrupole moments and of hexadecapole moments have been made in recent years by observing the hyperfine structure of muonic X–rays. The technique has been used primarily to measure ground–state quadrupole moments in rare earth atoms and in nuclear excited states. A review of the early work has been given by *Powers* [3.33]. It is also possible to extract the intrinsic $E4$ static hexadecapole moment from the muonic $3d_{\frac{5}{2}}$ state. In their work on the actinides, for example, the Los Alamos muonic atom group has demonstrated that the hexadecapole moments of heavy nuclei can be determined with a precision of ±5% or better [3.34].

In addition to the nuclear techniques discussed above, spectroscopic methods involving nuclei in atomic and molecular beams and in solids have been used to search for hexadecapole interactions. *Wang* [3.35] credits at least three observations [3.28,3.36–3.37], including his own, of hexadecapole interactions in solids. *Doering and Waugh* [3.38], on the other hand, point out that the effects attributed to hexadecapole moment interactions typically have been very close to the stated experimental errors. Doering and Waugh themselves report that they were unsuccessful in their attempt using very sophisticated nuclear magnetic resonance techniques to observe the splitting pattern of ^{181}Ta in KTaO$_3$. A dismissal of all claims to having observed hexadecapole interactions in solids was made by *Segal* [3.39] after a thorough experimental study using nuclear quadrupole resonance techniques of a number of isotopes (^{185,187}Re, ^{127}I, ^{209}Bi, ^{93}Nb, ^{121,123}Sb). A similar conclusion with respect to atomic beams was drawn by *Hammerle and Zorn* [3.40].

As we have seen in Chapter 1 and in earlier sections of the present chapter, the electric quadrupole interaction of nuclei with spin $I \geq \frac{3}{2}$ affects—or even determines—the hyperfine energy levels among which transitions may be observed. Thus one also expects that the effect of a nuclear electric hexadecapole interaction could be detected if the NMR or NQR transition frequencies in a solid or in a molecular or atomic beam containing a nucleus with $I \geq 2$ were measured with high precision. This is precisely the basis

3. DYNAMIC NUCLEAR ELECTRIC QUADRUPOLE INTERACTION

of most claims to having observed hexadecapole interactions in solids. For the case of cylindrical symmetry for the two coordinate frames of reference (*i.e.*, the nuclear spin system and the electron charge system), it is the shift in energy level—see (3.91)—that most often has been claimed to have been observed [3.36, 3.37] or claimed *not* to have been observed [3.38–3.40]. Among this group of experiments, *Wang* [3.36] and *Gotou* [3.37] both measured the NQR frequencies for ^{121}Sb ($I = \frac{5}{2}$) and ^{123}Sb ($I = \frac{7}{2}$), with values of $eM_{16}m_{16}$ between 20 kHz and 36 kHz, roughly $3 \times 10^{-4}\, eqQ$. *Hewitt and Williams* ([3.41], page 1191) obtain $eM_{16}m_{16} \approx 15$ kHz for the same isotopes in Sb metal, but conclude that because of irreproducibility of these results on annealing the metal, the presence of a hexadecapole interaction "has not been demonstrated".

If the nucleus being investigated by the spectroscopic technique is located on a site of perfect cubic symmetry, quadrupole effects—which for $I \geq 2$ in noncubic or imperfectly cubic crystals are orders of magnitude more effective than hexadecapole effects in contributing to energy shifts, line widths, and spin–lattice relaxation—are eliminated. This is the approach taken by *Doering and Waugh* [3.38] (NMR of ^{181}Ta in KTaO$_3$) and *Pfeifer* [3.42] (^{73}Ge Mössbauer resonance); assuming that the line width of the observed resonance is due to HDI, an upper limit on $em_{16}M_{16}$ for Ge was set at 700 kHz and at 400 kHz for Ta by these experimenters. As pointed out by Doering and Waugh, however, deviation from cubic symmetry at the nuclear site, due to crystal imperfections, could account for a quadrupolar contribution to the observed line width.

Mahler et al. [3.43] utilized acoustic saturation of NMR (ASNMR) in an attempt to observe the dynamic coupling of energy from an externally generated acoustic wave to the nuclear hexadecapole moment as a means of inducing transitions among the Zeeman levels characteristic of the spin system (^{115}In in InAs). We recall that in the case of NMR, the only allowed selection rule for transitions among the Zeeman levels, corresponding to RF coupling to nuclear magnetic dipole moments, is $\Delta m = \pm 1$; for nuclear acoustic resonance or ASNMR, dynamic coupling occurs either via the electric quadrupole moment, with allowed transitions $\Delta m = \pm 1$ and ± 2, or via the electric hexadecapole moment, with allowed transitions $\Delta m = \pm 1, \pm 2, \pm 3$ and ± 4. Thus the observation of $\Delta m = \pm 3$ (which is also allowed for magnetic octupole transitions) and $\Delta m = \pm 4$ transitions is a strong indication that the HDI may be involved. At least two caveats to this interpretation, however, must be made: (1) a competing mechanism, the second-order quadrupole–quadrupole interaction, for many nuclei may be considerably larger than the HDI [3.44]; (2) at high acoustic intensities, such as those used in ASNMR by *Mahler et al.* it has been demonstrated [3.45] that fractional harmonic phonons are generated in solids.

Mahon, et al., [3.45] repeated the experiment of Mahler and showed, indeed, that fractional phonons were generated that could induce the $\Delta m = \pm 1$ and $\Delta m = \pm 2$ transitions allowed by the dynamic quadrupole interactions. These, as well as other competing mechanisms, will be discused in Section 3.9.

An extended and detailed search for the dynamic nuclear electric hexadecapole interaction, utilizing the most sensitive NAR techniques, has been made by *Sundfors* [3.29]. Even though the theoretically predicted effect is unobservably low, the search was predicated on the possibility—perhaps even likelihood—of the existence of large, nonionic contributions to the components of the T-tensor that enter into the dynamic hexadecapole interaction. A description of the experimental search in Re metal, ^{115}InP, In^{115}As and Ta metal is given in Section 7.8.

3.9 Dynamic Magnetic Dipole–Dipole Interaction

The magnetic interaction between two nuclear spins I_1 and I_2 takes the form

$$\mathcal{H}_d = \frac{g^2 \beta_N^2}{r^3}\left[I_1 \cdot I_2 - \frac{3(I_1 \cdot r)(I_2 \cdot r)}{r^2}\right] , \tag{3.98}$$

where β_N is the nuclear magneton, and r is the spin–spin distance. At equilibrium, r becomes a, the near neighbor distance. For purposes of treating the modulation of this interaction by phonons, it is convenient to rewrite it as [3.46]

$$\mathcal{H}_d = g^2 \beta_N^2 \sum_{i,j} I_{1i} I_{2j} f_{ij}(r) , \tag{3.99}$$

where $f_{ij}(r) = (1/r^3)(\delta_{ij} - 3r_i r_j / r^2)$, and $f_{ij}(r)$ and (i, j) run over the Cartesian indices (x, y, z).

To take into account the effect of introducing small amplitude acoustic waves, which modulate the interaction, we expand $\mathcal{H}_d$ in a Taylor's series

$$\mathcal{H}_d = (\mathcal{H}_d)|_{r=a} + (\nabla \mathcal{H}_d)|_{r=a} u + \cdots , \tag{3.100}$$

where u is the small time–dependent strain due to the acoustic wave. The second term in (3.100) may be interpreted as a time–dependent dipole–dipole interaction, capable of inducing transitions between spin states. Explicitly, the time–dependent interaction takes the form

$$\mathcal{H}_d(t) = \frac{3g^2 \beta_N^2}{a^4} u_k(t) I_{1i} I_{2j} [5\hat{r}_i \hat{r}_j \hat{r}_k - (\hat{r}_i \delta_{jk} + \hat{r}_j \delta_{ik} + \hat{r}_k \delta_{ij})] , \tag{3.101}$$

where sums on repeated indices are assumed, and $\hat{r}_i$ is the ith Cartesian component of a unit vector in the r direction.

3. DYNAMIC NUCLEAR ELECTRIC QUADRUPOLE INTERACTION

To simplify the mathematics, we assume a simple cubic lattice and evaluate (3.101) in a coordinate system whose Cartesian axes coincide with the cubic crystalline axes. We assume, further, that the spin–spin axis is along the positive z-axis of this coordinate system. This assumption is made without loss of generality, since a cyclic permutation of x, y, z gives the form of the interaction for all other nearest neighbors. Substituting $\hat{r}_i = \delta_{iz}$ and assuming that the direction of phonon propagation is along the spin–spin (or z-) axis ($u_i = \delta_{iz} u_z$), (3.101) simplifies to

$$\mathcal{H}_d(t) = \frac{3g^2\beta_N^2}{a^4} u_z(t)[3I_{1z}I_{2z} - \mathbf{I}_1 \cdot \mathbf{I}_2] \ . \tag{3.102}$$

Were the propagation direction to be chosen differently, the interaction would take a slightly different form, but would not change the properties for inducing spin transitions.

The nuclear spins are quantized along the external field $\mathbf{H}_0$, while the direction of phonon propagation is along a cube axis. Since the spin components in (3.101) and (3.102) are expressed in the cubic system, to take the matrix elements of (3.102) between spin states, the spins must be transformed to the magnetic field system. The details of such a transformation are discused in Appendix D. When such a transformation is made and the magnetic field is chosen to be in the xz-plane of the crystal system ($\phi = 0$), the interaction takes the form [3.46]

$$\begin{aligned}\mathcal{H}_d(t) =& \frac{3g^2\beta_N^2 u_z(t)}{4a^4}[3(1 - \cos 2\theta)(I_{1+}I_{2+} + I_{1-}I_{2-}) \\ &+ 2(1 + 3\cos 2\theta)(2I_{10}I_{20} - I_{1+}I_{2-} - I_{1-}I_{2+}) \\ &- 12\sqrt{2}\sin 2\theta(I_{1+}I_{20} + I_{10}I_{2+} + I_{1-}I_{20} + I_{10}I_{2-})] \ ,\end{aligned} \tag{3.103}$$

where $I_\pm = (1/\sqrt{2})(I_{xh} \pm iI_{yh})$, and $I_0 = I_{zh}$ in the coordinate system in which the z-axis coincides with the direction of $\mathbf{H}_0$ (Appendix D).

Utilizing the irreducible spherical tensor formalism of Section 3.1, the time–dependent dipole–dipole Hamiltonian between two spins I_i and I_j can be written in *Abragam's* [3.1] notation:

$$\hbar\mathcal{H}_d^m(t) = \sum_{m=-2}^{+2} C^{(m)} F^{(m)} A^{(m)} \ . \tag{3.104}$$

The $A^{(m)}$ are spin operators, $F^{(m)}(t)$ are time–dependent functions of the position coordinates, and the $C^{(m)}$ are constants. They have the properties $A^{(m)} = A^{(-m)\dagger}$, $F^{(m)} = F^{(-m)*}$ and $C^{(m)} = C^{(-m)}$, where the dagger

3.9 DYNAMIC MAGNETIC DIPOLE—DIPOLE INTERACTION

and asterisk denote the Hermitian conjugate and the complex conjugate, respectively. The irreducible components of the $F^{(m)}$ tensor are

$$F^{(0)} = Y_2^0/2r_{ij}^3 = (1 - 3\cos^2\theta_{ij})r_{ij}^{-3}$$
$$F^{(1)} = (8\pi/15)^{\frac{1}{2}} Y_2^1/r_{ij}^3 = \sin\theta_{ij}\cos\theta_{ij}e^{-i\phi_{ij}}r_{ij}^{-3} = F^{(-1)*}$$
$$F^{(2)} = 2(8\pi/15)^{\frac{1}{2}} Y_2^2/r_{ij}^3 = \sin^2\theta_{ij}e^{-2i\phi_{ij}}r_{ij}^{-3} = F^{(-2)*} ,$$
(3.105)

where r_{ij}, θ_{ij} and ϕ_{ij} are the spherical coordinates of the radial vector $\mathbf{r}_{ij}$ between the two nuclei. Except for constants, the components of the $A^{(m)}$ tensor are given by corresponding operators from Column 3 of Table 3.1b:

$$A^{(0)} = [I_{iz}I_{jz} - (I_{i+}I_{j-} + I_{i-}I_{j+})/4]$$
$$A^{(1)} = I_{iz}I_{j+} + I_{i+}I_{jz} = A^{(-1)*} \qquad (3.106)$$
$$A^{(2)} = I_{i+}I_{j+} = A^{(-2)*} ,$$

and

$$C^{(0)} = \gamma_i\gamma_j\hbar^2 , \quad C^{(1)} = -3\gamma_i\gamma_j\hbar^2/2 , \quad C^{(2)} = -3\gamma_i\gamma_j\hbar^2/4 , \quad (3.107)$$

where γ_i and γ_j are the gyromagnetic ratios of nuclei i and j, respectively. Terms in (3.104) with $m = 0$ are responsible for static broadening and for opposite spin flips, $\Delta(m_i+m_j)=0$. Terms with $m=\pm 1$ and $m=\pm 2$ result in spin transitions with $\Delta(m_i+m_j)=\pm 1$ and $\Delta(m_i+m_j)=\pm 2$, respectively. The components with $A^{(1)}$ flip one spin only, while the components with $A^{(2)}$ flip both spins up and down. The corresponding resonance frequencies for the case of high-field Zeeman energy levels, are $\pm(\omega_0^i - \omega_0^j)$, $\pm\omega_0^i$, $\pm\omega_0^j$ and $\pm(\omega_0^i + \omega_0^j)$, where ω_0^i and ω_0^j are the Larmor frequencies for spins i and j, respectively.

We recall that the raising and lowering operators yield the following relations when they operate on a function $|I, m\rangle$:

$$I^+|I, m\rangle = [I(I + 1) - m(m + 1)]^{\frac{1}{2}}|I, m + 1\rangle$$
$$I^-|I, m\rangle = [I(I + 1) - m(m - 1)]^{\frac{1}{2}}|I, m - 1\rangle ;$$
(3.108)

also

$$\langle I, m|I^+I^-|I, m\rangle = (I - m)(I - m + 1)\langle I, m|I, m\rangle .$$

Stated in words, the matrix element $(m'|I^+|m)$ vanishes unless $m' = m+1$, and $(m'|I^-|m)$ vanishes unless $m' = m - 1$. The effect of operating with

106   3. DYNAMIC NUCLEAR ELECTRIC QUADRUPOLE INTERACTION

I^+I^- is to leave the state unchanged. For the transition corresponding to $\Delta m = -1$, for example,

$$C^{(1)}F^{(-1)}A^{(-1)} = -\frac{3}{2}\gamma_i\gamma_j\hbar^2[I_{jz}I_{i-} + I_{j-}I_{iz}] \\ \times \sin\theta_{ij}\cos\theta_{ij}e^{i\phi_{ij}}/r_{ij}^3 \;. \tag{3.109}$$

The corresponding transition probability per unit time that the spin I_i will interact with the spin I_j is [3.47]

$$P_{m_i \to m'_i} = \frac{9}{4}\gamma_i^2\gamma_j^2\hbar^2|(m_jm_i|I_{jz}I_{i-}|m'_im'_j) \\ + (m_jm_i|I_{j-}I_{iz}|m'_im'_j)|^2|F^{-1}|^2g(\nu)\;, \tag{3.110}$$

which for a downward transition of the spin I_i reduces to

$$P_{m_i \to m_i-1} = \frac{9}{4}\gamma_i^2\gamma_j^2\hbar^2 m_j^2(I_i+m_i)(I_i-m_i+1)|F^{-1}|^2g(\nu)\;. \tag{3.111}$$

Since the summation is to be taken over all neighbors j, the factor m_j^2 can be replaced by its mean value $\frac{1}{3}I_j(I_j+1)$, giving

$$P_{m_i \to m_j-1} = \frac{3}{4}\gamma_i^2\gamma_j^2\hbar^2 \\ \times I_j(I_j+1)(I_i+m_i)(I_i-m_i+1)|F^{(-1)}|^2g(\nu)\delta\;, \tag{3.112}$$

where δ is the number of nearest neighbors I_j of I_i.

The acoustic vibrations produce a time variation in $F^{(\pm m)}$ ($m = 1, 2$) and result in transitions between the high–field energy levels corresponding to $\Delta m = \pm 1$ and $\Delta m = \pm 2$. For longitudinal waves propagated parallel to r_{ij}, only changes in r_{ij} need be taken into account, in which case the linear variation in $F^{(-1)}$ may be represented in the Taylor expansion by $(\partial F^{(-1)}/\partial r)\Delta r$. The relative displacement of the two nuclei is $\Delta r_{ij} = Aka\cos\omega t$ if $ka \ll 1$, where A, k, and ω are the amplitude, wave number ($k = 2\pi/\lambda$) and angular frequency of the acoustic wave. If r_{ij} is not parallel to the direction of propagation of the acoustic wave, the relative displacement is reduced by a factor $\cos\theta'\cos\phi'$, where θ' and ϕ' are the spherical polar angles with respect to the direction of the acoustic wave. In the more general case, therefore,

$$\Delta r_{ij} = Aka\cos\theta'\cos\phi'\cos\omega t\;. \tag{3.113}$$

Carrying out the differentiation of $F^{(-1)}$, multiplying by (3.113), and squaring, we obtain

$$|F^{(-1)}|^2 \simeq \left|\frac{\partial F^{(-1)}}{\partial r}\Delta r_{ij}\right|^2$$
$$= \frac{1}{r_{ij}^6}(9\sin^2\theta_{ij}\cos^2\theta_{ij}\cos^2\theta'\cos^2\phi')A^2k^2\left(\frac{\pi}{2}\right), \quad (3.114)$$

where $\pi/2$ is the time average of $\cos^2\omega t$. Substituting into (3.112),

$$P_{m_i \to m_i-1} =$$
$$\frac{3}{4}\gamma_i^2\gamma_j^2\hbar^2 I_j(I_j+1)(I_i+m_i)(I_i-m_i+1)$$
$$\times \frac{A^2k^2}{r_{ij}^6}\left(\frac{2}{\pi}\right)^2 \{9\sin^2\theta_{ij}\cos^2\theta_{ij}\cos^2\theta'\cos^2\phi'\}g(\nu)\delta. \quad (3.115)$$

Of particular importance is the case in which one of the gyromagnetic ratios γ_j results from the unpaired electron of a paramagnetic impurity ion. The probability per unit time that the nucleus i will undergo a transition, with the impurity ion state remaining unchanged, is given by [3.47]

$$P_{m_i \to m_i-1} =$$
$$\frac{3}{4}\gamma_i^2(\mu\beta)^2(I_i+m_i)(I_i-m_i+1)$$
$$\times \frac{A^2k^2}{R_{ij}^6}\left(\frac{2}{\pi}\right)^2 \{9\sin^2\theta_{ij}\cos^2\theta_{ij}\cos^2\theta'\cos^2\phi'\}g(\nu)N\delta. \quad (3.116)$$

$\gamma_j^2\hbar^2 I_j(I_j+1)$ of (3.115) has been replaced by $(\mu\beta)^2$. μ the effective magnetic moment of the impurity ion, and β is the Bohr magneton. R_{ij} is the distance between nucleus i and the impurity ion, and N is the impurity ion concentration in the sample. Modifications in (3.116) due to the effects of spin diffusion and impurity line broadening are discussed by *Luukkala* [3.47]. Although dynamic dipole-dipole interactions have not been directly observed, they contribute to the experimentally observed acoustic-solid effect to be described in Chapter 8.

3.10 Additional Mechanisms for Multipole Quantum Transitions

Compared to the dynamic nuclear quadrupole interaction, the dynamic hexadecapole and dynamic dipole–dipole interactions are extremely weak.

108   3. DYNAMIC NUCLEAR ELECTRIC QUADRUPOLE INTERACTION

As discussed above, among their distinguishing features are the occurrence of allowed multiple–quantum transitions: $\Delta m = 3, 4$ for hexadecapolar; $\Delta m = 2$ for dipole–dipole. It is necessary, therefore, to tabulate those mechanisms in NAR and ASNMR that also can lead to the observance of multiple quantum ($\Delta m = 2, 3, 4$) transitions. In the present section, we address several of the more obvious nuclear spin–phonon effects in this category.

3.10.1   Anharmonic Phonon Effects

Incoherent fractional phonons may be generated via lattice anharmonicities in the solid sample or in the transducer that is utilized to generate the ultrasound [3.45]. Fractional phonons at the appropriate frequencies $e.g., (\frac{1}{3}\nu_{\pm 3}$ and $\frac{2}{3}\nu_{\pm 3}$, $\frac{1}{2}\nu_{\pm 4}$ and $\frac{1}{4}\nu_{\pm 4}$) can induce allowed quadrupolar transitions at $\Delta m = \pm 1$ and $\Delta m = \pm 2$, resulting in a spurious hexadecapolar absorption at the frequencies (corresponding to $\Delta m = \pm 3$ or ± 4) of the unfractionated phonons. This likelihood may be minimized by (a) utilizing several acoustic modes, in particular the slow shear mode, which is relatively unaffected by anharmonic lattice effects [3.48]; and (b) utilizing a level of ultrasound corresponding to very weak strains ($\sim 10^{-6} - 10^{-8}$). A disturbing effect, even for low–strain ultrasonics, however, is the heightened likelihood in the NAR technique of coherent fractional phonon generation within the high–Q mechanical resonance maintained during the nuclear spin–resonance experiment [3.49].

3.10.2   Second–Order Quadrupole–Quadrupole Transition

The nuclear electric quadrupole interaction taken to second order (NQ^2I) can contribute pseudo–hexadecapole effects, including higher–order transitions corresponding to $\Delta m = \pm 3$ and $\Delta m = \pm 4$. This effect has been investigated by *Sternheimer* [3.44], who (effectively) calculates the ratio of the transition probabilities:

$$(\rho_{\text{ion}})^2 = W_{NQ^2I}/W_{\text{HDI}} ,$$

for several ions. He obtains, for example:

$$[\rho_{\text{ion}}^{Ag^+}]^2 \simeq 0.02 \frac{Q^4}{M_4^2}$$

and

$$[\rho_{\text{ion}}^{Hg^{++}}]^2 \simeq 0.01 \frac{Q^4}{M_4^2} .$$

For the case of ^{121}Sb, for example, using a known value of $Q = 0.53b$ and an estimated value of M_4 ($\sim 0.1b^2$), we obtain

$$\left[\rho_{\text{ion}}^{^{121}\text{Sb}}\right]^2 \simeq 0.16 \ .$$

3.10.3 RF–Induced Dipole–Dipole Transitions

Nuclear acoustic resonance treats the effect of acoustically induced nuclear spin–phonon interactions; ASNMR and double resonance effects, as discussed briefly in Chapter 1, consider the combined effects of acoustic and electromagnetic RF induced interactions. However, even in NAR the RF responsible for driving the acoustic transducer may, under certain circumstances, interact directly with nuclei of the sample under study. It has been demonstrated [3.50,3.51] that at low magnetic fields and at low temperatures, transitions are driven by the dipole–dipole interactions, which are modulated by the RF field, rather than directly by the applied RF field. Resonance lines corresponding to this RF–modulated dipole–dipole interaction are expected to occur at fields such that H_{loc} is comparable to the external field, where H_{loc} is a measure of the dipole–dipole fields. Under such circumstances, $\Delta m = \pm 1$ and $\Delta m = \pm 2$ spectra are of comparable magnitude, and $\Delta m = \pm 3$ is much weaker.

3.10.4 Comparison of Mechanisms

The distinctions among the mechanisms responsible for the experimentally observed nuclear spin–phonon spectra are often obvious. For nuclei with spins $I \geq \frac{3}{2}$, the dynamic quadrupole interaction is by far the most likely, except for transitions between $m = -\frac{1}{2}$ and $m = \frac{1}{2}$ levels, which can be induced only by RF–induced dipolar interaction or by ultrasonic- or RF–induced dipole–dipole interaction. For spins $I = \frac{1}{2}$, only the non-quadrupolar interactions are effective. Doubts as to the origin of $\Delta m = \pm 2$ transitions (for $I \geq 1$), unlikely as they are to arise, may be resolved by a direct comparison of (i) the dependence of transition probabilities on the magnitudes of dipole moments and quadrupole moments, and (ii) the relative angular dependencies.

Identifying observed $\Delta m = \pm 3$ and $\Delta m = \pm 4$ transitions (for nuclei with $I \geq 2$) is more difficult. Of the 3 possible mechanisms mentioned in this chapter (HDI, NQ^2I, Fractional phonon $+NQI$), particular circumstances alone may determine which is predominant. The two most obvious methods of distinction are (1) quadrupole moment ratios and (2) angular dependences.

(1) Quadrupole–Moment Ratios

110   3. DYNAMIC NUCLEAR ELECTRIC QUADRUPOLE INTERACTION

(a) In the case of paired isotopes, such as ^{121}Sb and ^{123}Sb, both of which have nuclear spin $I \geq 2$ and are therefore characterized by quadrupole and hexadecapole moments, one can utilize the ratio of the known Q–moments to distinguish among the three mechanisms. The ratios of the observed resonance absorptions for the two isotopes A and B should go as follows:

(i) Second–order quadrupolar, NQ^2I,

$$(Q_A/Q_B)^4 \ ;$$

(ii) Fractional phonon (for NQI spin–phonon coupling),

$$(Q_A/Q_B)^2 \ ;$$

(iii) Hexadecapole, HDI,

$$(M_A/M_B)^2 \ .$$

The effect of the static quadrupole interaction on the shape of the observed resonance line is not taken into account in the above ratios.

(b) In the case of different species nuclei in the same specimen (e.g., ^{69,71}Ga and ^{121,123}Sb), in which one species (^{69,71}Ga) has $I < 2$ and the other (^{121,123}Sb) $I \geq 2$, one can distinguish among

(i) NQ^2I: as in (a) ;

(ii) fractional phonons: both species should show a $(Q_A/Q_B)^2$ behavior;

(iii) HDI: no observed effect for species with $I < 2$.

(2) Angular Dependence

The angular dependencies of the three mechanisms are quite distinct. We cite for illustration the angular dependencies (angle θ between the crystal axis along which the ultrasound is propagated and the direction of magnetic field H_0) for longitudinal acoustic waves propagating along the [100] axis of a cubic crystal (e.g., GaSb), with H_0 in the (001) plane as follows:

(i) Second–order quadrupole:

$$W_{NQ^2I} \propto \sin^6\theta \cos^2\theta, \quad \text{for} \quad \Delta m = \pm 3 ,$$
$$W_{NQ^2I} \propto \sin^8\theta, \quad \text{for} \quad \Delta m = \pm 4 .$$

(ii) Fractional phonon–induced $\Delta m = \pm 1, \pm 2$ transitions:

$$W_{\pm 1} \propto \sin^2 \theta \cos^2 \theta$$
$$W_{\pm 2} \propto \sin^4 \theta \quad .$$

(We assume here that the fractional phonons are coherent.)

(iii) Hexadecapole:

$$W_{\text{HDI}} \propto \{[(t_1-t_4) - 6(t_3-t_2)] \sin \theta \cos \theta (\sin^2 \theta + 1)\}, \text{ for } \Delta m = \pm 3$$

$$W_{\text{HDI}} \propto \{[(t_1-t_4) - 6(t_3-t_2)](\frac{1}{2}\sin^4 \theta + \sin^2 \theta - 1)\}, \text{ for } \Delta m = \pm 4.$$

In W_{HDI}, the components t_i are four of the six independent components of the sixth–rank tensor T, which relates the dynamic fourth derivative of the electric potential to the dynamic elastic strain (Appendix E).

CHAPTER 3 REFERENCES

3.1 A. Abragam. *The Principles of Nuclear Magnetism* (Clarendon Press, Oxford 1961).

3.2 C. P. Slichter. *Principles of Magnetic Resonance*, Second Revised and Expanded Edition, Second Corrected Printing (Springer–Verlag, Berlin 1980).

3.3 C. P. Poole, Jr., H. A. Farach. *The Theory of Magnetic Resonance*, Second Edition (Wiley, New York 1987).

3.4 M. Mehring. *High Resolution NMR Spectroscopy in Solids* (Springer–Verlag, Berlin 1976).

3.5 N. Chandrakumar, S. Subramanian. *Modern Techniques in High Resolution FT-NMR* (Springer–Verlag, New York 1987).

3.6 P. A. Fedders. "Spin–Resonance Line–Shape Changes Induced by Intraspin Cross Relaxation." *Phys. Rev.* **B11**, 995–1000 (1975).

3.7 M. E. Rose. *Elementary Theory of Angular Momentum* (Wiley, New York 1957).

3.8 C. F. Reiter. "Spin Fluctuations in Heisenberg Paramagnets. I. Diagrammatic Expansion for the Moments of the Spectral Density at Finite Temperature." *Phys. Rev.* **B5** 222–235 (1972).

3.9 P. A. Fedders. "Theory of Dynamic Quadrupole Spin Relaxation—Application to Ta Nuclei with Mobile Hydrogen Impurities," *Phys. Rev.* **B10**, 4510–4514 (1974).

3.10 E. Ambler, J. C. Eisenstein, J. F. Schooley. "Traces of Products of Angular Momentum Matrices." *J. Math. Phys.* **3**, 118–130 (1962).

3.11 P. A. Fedders. "Some Quadrupolar Effects on $T_1(H)$ for Nuclear Spins." *Phys. Rev.* **B13**, 4678–4681 (1974).

3.12 E. F. Taylor. "Nuclear Spin Saturation by Ultrasonics in Sodium Chloride." Ph. D. Dissertation, Harvard University, 1958 (unpublished).

3.13 R. L. Mieher. "Quadrupole Nuclear Relaxation in the III–V Compounds." *Phys. Rev.* **125**, 1537–1551 (1962).

3.14 R. L. Mieher. *Semiconductors and Semimetals, Vol. 2*, Ed. by R. K. Willardson and A. C. Beer (Academic Press, New York 1966).

3.15 R. V. Pound. "Nuclear Electric Quadrupole Interactions in Crystals." *Phys. Rev.* **79**, 685–702 (1950).

3.16 E. Segre. *Nuclei and Particles*, Second Edition, Completely Revised, Reset, Enlarged (W. A. Benjamin, Reading 1977).

3.17 A. de Shalit, H. Feshbach. *Theoretical Physics, Volume I: Nuclear Structure* (John Wiley, New York 1974).

3.18 E. F. Taylor, N. Bloembergen. "Nuclear Spin Saturation by Ultrasonics in Sodium Chloride." *Phys. Rev.* **113**, 431–438 (1959).

3.19 R. J. Harrison, P. L. Sagalyn. "Trace Relations Relating Electric Fields and Elastic Strains to Nuclear Quadrupole Effects." *Phys. Rev.* **128**, 1630–1631 (1962).

3.20 R. G. Shulman, B. J. Wyluda, P. W. Anderson. "Nuclear Magnetic Resonance in Semiconductors. II. Quadrupole Broadening of Nuclear Magnetic Resonance Line by Elastic Axial Deformation." *Phys. Rev.* **107**, 953–958 (1957).

3.21 B. Strobel, V. Müller. "Nuclear Acoustic Resonance Determination of the Strain Electric–Field Gradient Tensor S and its Temperature Dependence in Ta Single Crystals." *Phys. Rev.* **B24**, 6292–6303 (1981).

3.22 P. A. Fedders. "Resonant and Nonresonant Effects of Paramagnetic Spins on Acoustic Modes." *Phys. Rev.* **B12**, 2045–2048 (1975).

3.23 J. Buttet, P. K. Baily. "Knight Shift and Zero–Field Splitting in Rhenium Determined by Nuclear Acoustic Resonance." *Phys. Rev. Letters* **24**, 1220–1223 (1970).

3.24 M. Stachel, H. E. Bömmel. "Temperature Dependence of the Nuclear Quadrupole Interaction and the Knight Shift in Rhenium Metal," *Appl. Phys.* **A30**, 27–32 (1983).

3.25 F. G. Fumi, C. Ripamonti. "Tensor Properties and Rotational Symmetry of Crystals: A New Method for Group 3(3z) and Its Application to General Tensors up to Rank 8." *Acta Crystallogr.* **A36**, 535–551 (1980).

3.26 R. K. Sundfors. "Determination of the Magnitude and Sign of the [185,187]Re Nuclear Electric Quadrupole Coupling Constants Using Nuclear Acoustic Resonance." *Phys. Rev.* **B42**, 1922–1928 (1990).

3.27 R. M. Sternheimer. "Antishielding of Nuclear Electric Hexadecapole Moments." *Phys. Rev. Lett.* **6**, 190–192 (1961).
3.28 R. J. Mahler. "Nuclear Hexadecapole Interactions." *Phys. Rev.* **152**, 325–330 (1966).
3.29 R. K. Sundfors, unpublished calculations.
3.30 R. M. Sternheimer. "Antishielding of Nuclear Electric Hexadecapole Moments." *Phys. Rev.* **123**, 870–872 (1961).
3.31 B. H. Wildenthal, B. A. Brown, I. Sick. "Electric Hexadecapole Transition Strength in ^{32}S and Shell Model Predictions for E4 Systematics in the sd Shell." *Phys. Rev.* **C32**, 2185–2188 (1985).
3.32 F. K. McGowan, E. E. Bemis, Jr., J. L. C. Ford, Jr., W. T. Milner, R. L. Robinson, P. H. Stelson. "Equilibrium Quadrupole and Hexadecapole Deformations in ^{232}Th and ^{239,240,242}Pu." *Phys. Rev. Lett.* **27**, 1741–1744 (1971).
3.33 R. J. Powers. "The Measurement of E2 and E4 Moments Using Exotic Atoms." *Hyperfine Interactions IV*, 1977, Ed. by R. S. Raghavan, D. E. Murnick, pp. 123–142 (North-Holland, New York, 1977)
3.34 J. D. Zumbro, R. A. Naumann, M. V. Hoehn, W. Rueter, E. B. Shera, C. E. Bemis, Jr., Y. Tanaka. "E2 and E4 Deformations in ^{232}Th and ^{239,240,242}Pu." *Phys. Lett.* **167B**, 383–387 (1986).
3.35 T.-J. Wang. "Nuclear Electric Hexadecapole Interactions in Solids." *J. Magn. Res.* **64**, 194–198 (1985).
3.36 T.-C. Wang. "Pure Nuclear Quadrupole Spectra of Chlorine and Antimony Isotopes in Solids." *Phys. Rev.* **99** 566–577 (1955). [Note: T.-C. Wang is now known as T.-J. Wang.]
3.37 Hiroshi Gotou. "Nuclear Electric Hexadecapole Coupling Constant in Antimony Trichloride." *J. Magn. Res.* **54** 36–45 (1983).
3.38 E. B. Doering, J. S. Waugh. "Search for Hexadecapole Interaction in $KTaO_3$ by ^{181}Ta NMR." *J. Chem. Phys.* **85**, 1753–1758 (1986).
3.39 S. L. Segal. "Nuclear Electric Hexadecapole Interactions in Solids." *J. Chem. Phys.* **69**, 2434–2438 (1978).
3.40 R. H. Hammerle, J. C. Zorn. "Search for the Nuclear Electric hexadecapole Moment of ^{115}In by Molecular-Beam Electric-Resonance Spectroscopy." *Phys. Rev.* **C7**, 1591–1596 (1973).
3.41 R. R. Hewitt, B. F. Williams. "Nuclear Quadrupole Interaction of ^{121}Sb and ^{123}Sb in Antimony Metal." *Phys. Rev.* **129**, 1188–1192 (1963).
3.42 L. Pfeiffer. "Measurement of large E2 Dispersive Interference in the High-Resolution ^{73}Ge Mössbauer Transition at Natural Linewidth." *Phys. Rev. Lett.* **38**, 862–865 (1977).
3.43 R. J. Mahler, L. W. James, W. H. Tantilla. "Possible Observation

of In^{115} Nuclear Hexadecapole Transitions." *Phys. Rev. Letters* **16**, 259–261 (1966).

3.44 R. M. Sternheimer. "Second–Order Quadrupole Effect for the Nuclear Hexadecapole Coupling in Ions." *Phys. Rev.* **127**, 812–816 (1962).

3.45 H. Mahon, E. Brun, M. Luukkala, W. G. Proctor. "Excitation of Fractional Harmonic Phonons in Solids." *Phys. Rev. Lett.* **19**, 430–432 (1967).

3.46 C. W. Myles. "Higher Order Acoustic Paramagnetic Resonance Transitions of Magnetic Impurities in Dielectrics." Ph.D. Dissertation, Washington University, 1973 (unpublished).

3.47 M. Luukkala. "Acoustically Induced Nuclear Magnetic Dipole Transitions in $NaClO_3$ at 4.2° K." *Ann. Acad. Sci. Fenn. AVI* **193**, 1–39 (1965).

3.48 M. Lax, V. Narayanamurti, P. Hu, W. Weber. "Lifetimes of High Frequency Phonons." *Journal de Physique* **42**, *Colloque* **C6**, 161–163 (1981).

3.49 V. A. Golenishchev–Kutuzov, I. I. Sadykov, E. P. Khaimovich. "Investigation of the Behavior of Incoherent Phonons in Crystals by the Nuclear Acoustic Resonance Method." *Fiz. Tverd. Tela (Leningrad)* **21**, 802–807 (1979). English trans: *Sov. Phys. Solid State* **21**, 469–472 (1979).

3.50 G. P. Jones, J. T. Daycock. "Subsidiary Proton Resonances in NMR." *Phys. Lett.* **24A**, 302–303 (1967).

3.51 S. Clough, A. J. Horsewill, P. J. McDonald, F. O. Zelaya. "Molecular Tunneling Measured by Dipole–Dipole—Driven Nuclear Magnetic Resonance." *Phys. Rev. Lett.* **55**, 1794–1796 (1985).

CHAPTER 4
DYNAMIC ALPHER–RUBIN (DIPOLAR) INTERACTION

4.1 Nature of the Coupling

4.2 Classical Derivation of Dipolar NAR Absorption and Dispersion: I. Method of Quinn–Buttet–Fedders

4.3 Classical Derivation of Dipolar NAR Absorption and Dispersion: II. Method of Müller

4.4 Experimental Verification in Metals

    4.4.1 The Line Shape
    4.4.2 The Case of Vanadium

4.5 The 'Low Temperature' Limit

    4.5.1 Theory of Quinn and Buttet
    4.5.2 Experimental Verification: Copper and Lead

4.6 NAR Absorption and Dispersion: Quantum Mechanical Approach

    4.6.1 Method of Müller and Bartell
    4.6.2 NAR Alpher–Rubin Susceptibility
    4.6.3 NAR Quadrupole Susceptibility
    4.6.4 NAR Interference Term

4.7 Influence of Electronic Bandstructure

References

4. DYNAMIC ALPHER–RUBIN (DIPOLAR) INTERACTION
4.1 Nature of the Coupling

The second common mechanism for coupling acoustic energy to nuclear spins is the dynamic Alpher–Rubin coupling (often also termed *dynamic dipolar* coupling), which depends on the electromagnetic interaction of the conduction electrons with the applied acoustic wave in the presence of an externally applied static magnetic field. This mechanism exists only in conductors. The coupling is to the magnetic dipole moments of the nuclear spins; thus the mechanism is operative for all values I of the nuclear spin, including $I = \frac{1}{2}$. In terms of the multipole operators, the dynamic Alpher–Rubin interaction couples to the A_{lm} with $l = 1$. These are the same operators used for coupling in NMR experiments.

In NMR, energy is channeled into the nuclear spin system via the interaction of the magnetic dipole moment μ and the applied RF magnetic field H_1. The same coupling applies to dynamic dipolar NAR, except that the oscillating magnetic field H_1 is internally generated as a consequence of the detailed electrodynamics of the conducting medium in which the ultrasonic wave is propagating. Thus dynamic dipolar NAR can be considered to be NMR by devious means.

It is essential that the sample be a conductor, because local internal magnetic fields arise from microscopic currents set up by the ultrasonic wave. The electrons must be able to move somewhat independently of the (positive) ion cores. In practice, the collision processes that account for finite resistivity introduce strong correlations between the electronic and ionic motions. Because the electrons follow the ions very closely under most circumstances, even in good conductors, the ionic and electronic microscopic currents nearly cancel, and any magnetic fields set up are exceedingly small.

When an acoustic wave is launched in a conductor in the presence of a large static magnetic field B_0, however, the conduction electrons do not instantaneously follow the motion of the charged ion cores. The resulting electronic and ionic currents, therefore, do not cancel, and thus they act as source or driving terms in Maxwell's equations. The direction of the Lorentz force on the electrons is opposite to that on the ions, and the two types of charge are forced to follow different paths and set up small but important currents transverse to both B_0 and the ultrasonic dispacement ξ. These transverse currents, in turn, induce electric and magnetic fields in accordance with Maxwell's equations. The oscillating magnetic fields can couple to the magnetic dipole of the nucleus and, if the frequency of the oscillating fields corresponds to a resonance frequency, excite magnetic resonance. The currents also are dissipated by Joule heating, leading to a nonresonant attenuation (the *Alpher–Rubin effect*), which does not depend on the spins. The interplay of the resonant and nonresonant effects is intricate and will be discussed in some detail in Section 4.2.

The nuclear spins enter the picture in two, distinct ways. In the first and more straightforward way, the dynamic magnetic fields generated by the currents couple to the dipole moments of the nuclear spins and lead directly to the absorption of power from the acoustic wave. This process is analogous to the absorption of power in an NMR experiment and always yields a positive contribution to the attenuation of the acoustic wave. The second way in which the nuclear spins enter the picture is somewhat more subtle. The transfer of energy between the acoustic wave and the conduction electrons is described by Maxwell's equations, which involve both B and H and, thus, also the magnetic susceptibility. One contribution to the susceptibility is the resonant part of the nuclear susceptibility. In this context, the resonant nuclear susceptibility affects the amount of energy transferred from the acoustic wave to the conduction electrons, but this energy is not absorbed by the nuclear spins. Curiously enough, this resonant contribution can lead to either an increase or a decrease in the amount of energy lost by the acoustic wave and can thus appear to be a negative resonant attenuation. However, the total attenuation, including the background Alpher–Rubin attenuation, is always positive.

Alpher and Rubin [4.1] developed a phenomenological theory that accounts for the electromagnetic fields generated by the ultrasonic wave and the resulting nonresonant effects on ultrasonic velocity and attenuation. The dynamical Alpher–Rubin coupling to nuclear spins was first suggested by *Quinn and Ying* [4.2,4.3].

Buttet [4.4] presented the first detailed theory of the resonant ultrasonic attenuation due to this mechanism for the special case of $B_0 \parallel q$, where q is the ultrasonic wavevector. That theory was extended to arbitrary orientation of B_0 by *Fedders* [4.5]. The theory was extended to the low temperature regime by *Buttet* [4.6]. *Müller, et al.* [4.7–4.10] have presented both an alternative classical and a quantum mechanical derivation of NAR absorption and dispersion in metals.

4.2 Classical Derivation of Dipolar NAR Absorption and Dispersion: I. Method of Quinn–Buttet–Fedders

The effect of an external field on the acoustic attenuation and velocity dispersion in a metal, first suggested by *Alpher and Rubin* [4.1], has been treated extensively in the literature [4.11–4.15]. The treatment has evolved into a generalized high temperature and low temperature theory involving both resonant and nonresonant interactions. In this section we address the calculation of the high temperature interaction, and we follow the method of Quinn, Buttet and Fedders. The most comprehensive treatment, given by *Fedders* [4.5], is outlined. Although, in principle, the calculation is very general, the final results are specialized to a free electron cubic paramagnetic metal in the long wavelength limit. The theory treats all the fields and currents self–consistently to obtain a dispersion relation for ultrasonic propagation in a metal, in an applied B_0 and in the presence of a nonzero, frequency–dependent magnetic susceptibility. The real and imaginary parts of the dispersion relation are then separated, giving the ultrasonic dispersion and attenuation.

We consider the propagation of an acoustic plane wave of frequency ω and wave vector q whose displacement ξ is

$$\xi = \xi_0 \, e^{-i\omega t + i q \cdot r} \ .$$

4. DYNAMIC DIPOLAR INTERACTION

The equation for the motion of an ion of mass M is obtained from Newton's second law,

$$M\frac{\partial^2 \boldsymbol{\xi}}{\partial t^2} = C_l \nabla(\nabla \cdot \boldsymbol{\xi}) - C_t \nabla \times (\nabla \times \boldsymbol{\xi}) + \boldsymbol{F}_L + \boldsymbol{F}_c \,, \qquad (4.1)$$

where $\boldsymbol{\xi}$ is the ionic displacement, C_l and C_t are the longitudinal and transverse elastic constants respectively, $\boldsymbol{F}_L$ is the Lorentz force on the ion, and $\boldsymbol{F}_c$ is the force on the ion due to collisions with electrons. The Lorentz force on the ion is

$$\boldsymbol{F}_L = Ze\boldsymbol{E} + \frac{Ze}{c}\frac{\partial \boldsymbol{\xi}}{\partial t} \times \boldsymbol{B} \,, \qquad (4.2)$$

where $\boldsymbol{E}$ is the Alpher–Rubin induced electric field, and $\boldsymbol{B}$ is the sum of the external magnetic field and the induced magnetic field, $\boldsymbol{B} = \boldsymbol{B}_0 + \boldsymbol{b}$. Ze is the charge on an ion, $-e$ is the electronic charge, and c is the speed of light. It is necessary to express $\boldsymbol{F}_c$ and $\boldsymbol{E}$ in terms of $\boldsymbol{\xi}$. Rodriguez [4.12] has related the collision force $\boldsymbol{F}_c$ to a phenomenological collision time τ by using the equation

$$\boldsymbol{F}_c = \frac{mZ}{\tau}\left(\langle \boldsymbol{v} \rangle - \frac{\partial \boldsymbol{\xi}}{\partial t}\right) = \frac{mZ}{n_0 e \tau}\left[n_0 e \langle \boldsymbol{v} \rangle - n_0 e \frac{\partial \boldsymbol{\xi}}{\partial t}\right] \,, \qquad (4.3)$$

where m is the electron mass, $\langle \boldsymbol{v} \rangle$ is the average electron velocity, and n_0 is the number of conduction electrons per unit volume. The terms in brackets can be written as current densities:

$$\boldsymbol{F}_c = \frac{mZ}{n_0 e \tau}[-\boldsymbol{j}_e - \boldsymbol{j}_{\text{ion}}] \,, \qquad (4.4\text{a})$$

where $\boldsymbol{j}_e$ is the electronic current density, and $\boldsymbol{j}_{\text{ion}}$ is the ionic current density. The two current densities can be combined to give a total current density

$$\boldsymbol{j} = \boldsymbol{j}_e + n_0 e \frac{\partial \boldsymbol{\xi}}{\partial t} \,. \qquad (4.4\text{b})$$

A more general form of the collision force, useful in the following analysis, can be written. We write the collision force as [4.5]

$$\boldsymbol{F}_c = \boldsymbol{A} \cdot \left(\langle \boldsymbol{v} \rangle - \frac{d\boldsymbol{\xi}}{dt}\right) \,, \qquad (4.5\text{a})$$

where $\boldsymbol{A}$ is a tensor of proportionality. To determine this tensor, we write

$$-\boldsymbol{j}_e = n_0 e \langle \boldsymbol{v} \rangle = \boldsymbol{\sigma}_0 \cdot \boldsymbol{E} \,, \qquad (4.5\text{b})$$

where σ_0 is the conductivity tensor. In the limit that $B = 0$, the ions are at rest, and there can be no force on the ions. Thus the collisional force must cancel the ZeE of the Lorentz force (4.2):

$$F_c = A \cdot \langle v \rangle = A \cdot \left(-\frac{\sigma_0}{n_0 e} \cdot E\right) = ZeE \ . \tag{4.5c}$$

Equation (4.5c) determines

$$A = \frac{Ze^2 n_0}{\sigma_0} \ . \tag{4.5d}$$

In general, the collision force can now be written

$$F_c = -\left(\frac{eZ}{\sigma_0}\right) \cdot j = -(eZR_0) \cdot j \ , \tag{4.5e}$$

where $R_0 = \sigma_0^{-1}$ is the static resistivity tensor in zero magnetic field. If the current is due to holes rather than to electrons, the two terms cannot be combined into a total current density. The σ_0 tensor is introduced to account for anisotropy in the material. In case the conductivity is isotropic the static resistivity tensor is $R_0 = (1/\sigma_0)$. In order to get a solution to (4.1), it is necessary to relate j, E and B to the velocity $u = \partial \xi/\partial t$. The dynamic fields are all taken to vary as $\exp(-i\omega t - iq \cdot r)$.

The total current density can be expressed in terms of the induced electric field E. The Alpher–Rubin oscillating fields can be expressed as

$$E = E_0 e^{-i\omega t + iq \cdot r} \tag{4.6a}$$

and

$$b = b_0 e^{-i\omega t + iq \cdot r} \ . \tag{4.6b}$$

These expressions are used in Maxwell's two curl equations,

$$\frac{4\pi}{c} j = \nabla \times \mu^{-1} \cdot B - \frac{1}{c} \frac{\partial E}{\partial t} \tag{4.7a}$$

and

$$\nabla \times E + \frac{\mu}{c} \cdot \frac{\partial H}{\partial t} = 0 \ , \tag{4.7b}$$

to obtain

$$j = \frac{i\omega}{4\pi} E + \frac{ic^2}{4\pi\omega} q \times (\mu^{-1} \cdot (q \times E)) \ , \tag{4.8a}$$

where μ is the magnetic permeability tensor. We can now write

$$j = \Gamma \cdot E \quad \text{or} \quad J_i = \Gamma_{ni} E_n , \qquad (4.8b)$$

where Γ is a tensor defined by

$$\Gamma_{ni} = \left(\frac{i\omega}{4\pi} \delta_{ni} + \frac{ic^2}{4\pi\omega} \epsilon_{ijk} \epsilon_{lmn} q_j q_m \mu_{kl} \right) , \qquad (4.9)$$

and where ϵ_{ijk} is the Levi–Civita density. It should be noted that at this point the nuclear spin system enters through the permeability μ, since

$$\mu = 1 + 4\pi\chi , \qquad (4.10)$$

with χ being the total complex susceptibility tensor and $\mathbf{1}$ the unit tensor. The susceptibility includes nuclear and electronic contributions, as well as the static susceptibility. With the use of (4.5) and (4.8), the collision force F_c may be expressed in terms of E. It now remains to express E in terms of the displacement ξ. Using

$$j = j_e + n_0 \frac{\partial \xi}{\partial t} = \Gamma \cdot E$$

$$j_e = \sigma \cdot \left(E - n_0 \, e \, R_0 \cdot \frac{\partial \xi}{\partial t} \right) , \qquad (4.11)$$

we find that

$$\Gamma \cdot E = \sigma \cdot E - \sigma \cdot n_0 \, e \, R_0 \cdot \frac{\partial \xi}{\partial t} + n_0 \, e \frac{\partial \xi}{\partial t}$$
$$[\Gamma - \sigma] \cdot E = -i\omega n_0 \, e(-\sigma \cdot R_0 + 1) \cdot \xi$$
$$(R \cdot \Gamma - 1) \cdot E = i\omega n_0 \, e(R_0 - R) \cdot \xi \qquad (4.12)$$

or

$$E = -i\omega n_0 \, e(1 - R \cdot \Gamma)^{-1}(R_0 - R) \cdot \xi . \qquad (4.13)$$

σ and R are, respectively, the magnetoconductivity and the magnetoresistivity tensors. A complete solution to (4.1) can now be found if μ, R and R_0 are known. If (4.1) through (4.8a–b) are combined, replacing B with B_0 (the static external field) and retaining terms to first order in ξ, we arrive at the following equation:

$$M\omega^2 \xi - (C_l - C_t) qq \cdot \xi - C_t q^2 \xi$$
$$= -Ze[1 - R_0 \cdot \Gamma] \cdot E + \frac{Zei\omega}{c} \xi \times B_0 . \qquad (4.14)$$

The above two equations yield the dispersion relationship

$$\left\{\omega^2 \mathbf{1} - \left[\frac{(C_l - C_t)\mathbf{qq} + C_t q^2 \mathbf{1}}{M}\right]\right\} \cdot \boldsymbol{\xi}$$
$$= \frac{iZe^2 n_0 \omega}{M}(\mathbf{1} - \mathbf{R}\cdot\boldsymbol{\Gamma})^{-1}$$
$$\cdot (\mathbf{R}_0 - \mathbf{R}) \cdot (\mathbf{1} - \mathbf{R}_0 \cdot \boldsymbol{\Gamma}) \cdot \boldsymbol{\xi} + \frac{Zei\omega}{Mc}\boldsymbol{\xi} \times \mathbf{B}_0 \ . \tag{4.15}$$

Expressions for the acoustic velocity and attenuation are obtained from the real and imaginary parts of this equation.

Following the standard notation, we define

$$\omega_c = \frac{eB_0}{mc}$$

$$\beta = \frac{\omega c^2}{4\pi\sigma_0 v^2}$$

$$\omega = vq \ ,$$

where v is the velocity of sound, and ω_c is the cyclotron frequency. We restrict ourselves to the limit $\omega_c\tau$, $\omega\tau$ and $ql \ll 1$, where τ is the electron collision time, ω_c is the cyclotron frequency, ω is the frequency of the ultrasonic wave, and $l = v_F\tau$, v_F is the Fermi velocity. Indeed, the microscopic theory of *Rodriguez* [4.11,4.12] demonstrates that the assumption that the Lorentz force acts directly on the lattice is justified if $\omega\tau$, ql and $\omega_c\tau \ll 1$. These conditions require that the conduction electrons attain momentum equilibrium with the lattice in times and distances that are short compared to the period and wavelength, respectively, of the ultrasonic wave. At 300 K in pure metals, $\tau \sim 10^{-14}$ sec, $l \sim 10^{-5} - 10^{-6}$ cm, and $\omega_c \sim 10^{11}$ Hz for 100 kOe. The restrictions thus apply to the case of pure metals at 300 K for frequencies below 300 MHz and fields under 100 kOe.

We assume the three acoustic modes—two transverse and one longitudinal, represented by ξ_x, ξ_y and ξ_z respectively—to be independent. The acoustic wave is defined to propagate in the $\hat{z}$-direction: $\mathbf{q} = q\hat{z}$. We choose the Rodriguez coordinate system [4.12], shown in Fig. 4.1, wherein the magnetic field $\mathbf{B}_0$ lies in the yz-plane and is given by $\mathbf{B}_0 = B_0(0, \sin\theta, \cos\theta)$. We let the permeability tensor $\boldsymbol{\mu} = \mu_0\mathbf{1} + \delta\boldsymbol{\mu}$. $\mathbf{R}$ and $\delta\boldsymbol{\mu}$ are expressed in this coordinate system.

The resistivity tensor $\mathbf{R}$ in (4.15) may be expanded in powers of $\omega_c\tau$. Also, $\boldsymbol{\Gamma}$, a function of the permeability tensor $\boldsymbol{\mu}$, may be expanded in powers of $\delta\boldsymbol{\mu}$. Let $\mathbf{R} = \mathbf{R}_0 + \delta\mathbf{R}$, where the static (zero-field) resistivity tensor is

$$\mathbf{R}_0 = \sigma_0^{-1}\mathbf{1} \equiv \frac{1}{\sigma_0}\begin{pmatrix} 1 & 0 & 0 \\ 0 & 1 & 0 \\ 0 & 0 & 1 \end{pmatrix} \ .$$

4. DYNAMIC DIPOLAR INTERACTION

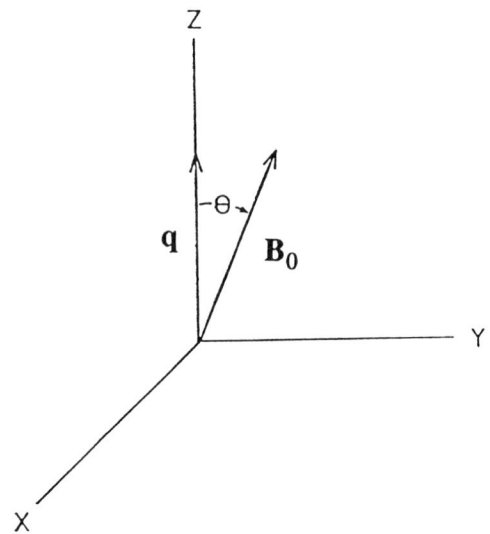

FIGURE 4.1. The Rodriguez coordinate system.

Then

$$\delta R = R - R_0 = (R\sigma_0 - R_0\sigma_0)\sigma_0^{-1} = -\sigma_0^{-1}(1 - R\sigma_0) . \qquad (4.16)$$

In the coordinate system of Fig. 4.1, the magnetoresistivity tensor is given by

$$R \equiv \sigma^{-1} = \sigma_0^{-1} \begin{pmatrix} 1 & \omega_c\tau \cos\theta & -\omega_c\tau \sin\theta \\ -\omega_c\tau \cos\theta & 1 & 0 \\ \omega_c\tau \sin\theta & 0 & 1 \end{pmatrix} . \qquad (4.17)$$

Therefore, from (4.16) and (4.17),

$$\delta R = \sigma_0^{-1} \begin{pmatrix} 0 & \omega_c\tau \cos\theta & -\omega_c\tau \sin\theta \\ -\omega_c\tau \cos\theta & 0 & 0 \\ \omega_c\tau \sin\theta & 0 & 0 \end{pmatrix} . \qquad (4.18)$$

We further let $\Gamma = \Gamma_0 + \delta\Gamma$, where

$$\Gamma_0 = -i\beta\sigma_0 \begin{pmatrix} 1 & 0 & 0 \\ 0 & 1 & 0 \\ 0 & 0 & 0 \end{pmatrix} \qquad (4.19a)$$

or, in summation notation,

$$(\Gamma_0)_{ij} = -i\beta\sigma_0 \delta_{ij}(1 - \delta_{iz}) . \qquad (4.19b)$$

$\delta\Gamma$ contains all of the resonant changes in the dispersion relation, since it depends on the susceptibility tensor $\delta\mu$. The contributing components for the cubic coordinate system we have chosen are

$$\delta\mu_{xx} = 4\pi(\chi_+ + \chi_-)$$
$$\delta\mu_{yy} = 4\pi[(\chi_+ + \chi_-)\cos^2\theta + \chi_0 \sin^2\theta]$$
$$\delta\mu_{xy} = -\delta\mu_{yx} = 4\pi i\cos\theta(\chi_+ - \chi_-) , \qquad (4.20)$$

yielding

$$\delta\Gamma = i\beta\sigma_0 \begin{pmatrix} \delta\mu_{xx} & \delta\mu_{xy} & 0 \\ -\delta\mu_{xy} & -\delta\mu_{yy} & 0 \\ 0 & 0 & 0 \end{pmatrix} \qquad (4.21a)$$

or

$$(\delta\Gamma)_{ij} = -i\beta\sigma_0(\delta\mu)_{ji}(1 - 2\delta_{ij})(1 - \delta_{iz})(1 - \delta_{jz}) . \qquad (4.21b)$$

The yz- and xz-components do not contribute to $\delta\Gamma$, as will be seen shortly. χ_+ and χ_- are the polar components of the nuclear spin susceptibility given by

$$\chi_\pm = \frac{1}{2}(\chi_{xx} \pm i\chi_{xy}) ,$$

where

$$\chi_{ij} = \chi'_{ij} + i\chi''_{ij} . \qquad (4.22)$$

The subscript notation xy, for example, refers to a magnetization in the x-direction induced by a field in the y-direction. χ'' and χ' are the usual Bloch susceptibilities. For a Lorentzian line shape, they are,

$$\chi'' = \frac{1}{2}\omega\chi_0 T_2 \left[\frac{1}{1 + [(H_e - H_0)/\delta]^2}\right]$$

$$\chi' = \frac{1}{2}\omega\chi_0 T_2 \left[\frac{[(H_e - H_0)/\delta]}{1 + [(H_e - H_0)/\delta]^2}\right] . \qquad (4.23)$$

χ_0 is the static susceptibility. H_0 is assumed fixed for a fixed $\omega_0 = \gamma H_0$, while H_e is externally varied. δ is the magnetic field half-width at half maximum of the absorption line shape χ'' and is equal to $(1/\gamma T_2)$.

124   4. DYNAMIC DIPOLAR INTERACTION

Combining (4.19), (4.20), (4.21) and (4.22), we obtain

$$\Gamma = -\beta\sigma_0 \begin{pmatrix} (1 - 4\pi\chi_{xx}) & -i4\pi\chi_{xy} & 0 \\ 4\pi\chi_{xy} & (1 - 4\pi\chi_{xx}) & 0 \\ 0 & 0 & 0 \end{pmatrix} . \quad (4.24)$$

This result can also be obtained from (4.9) and the expansion $\mu = \mu_0 + \delta\mu$. Explicitly, in (4.9) we substitute

$$\mu^{-1} = [\mu_0(1 + \mu_0^{-1}\delta\mu)]^{-1} = \mu_0^{-1}(1 - \mu_0^{-1}\delta\mu) = \mu_0^{-1}\mathbf{1} - \mu_0^{-2}\delta\mu .$$

We now proceed to expand (4.15) to second order in $\omega_c\tau$ (since first-order contributions vanish) and to first order in $\delta\mu$. For convenience, in (4.15) we define the tensor Φ as

$$\Phi \equiv (1 - R_0\Gamma)(1 - R\cdot\Gamma)^{-1}(R_0 - R) \quad (4.25)$$

and define a tensor A as

$$A = (1 - R_0\Gamma) = \begin{pmatrix} 1+i\beta & 0 & 0 \\ 0 & 1+i\beta & 0 \\ 0 & 0 & 1 \end{pmatrix} \quad (4.26a)$$

or

$$(A)_{ij} = \delta_{ij}[1 + i\beta(1 - \delta_{iz})] \quad (4.26b)$$

and

$$A^{-1} = \begin{pmatrix} (1+i\beta)^{-1} & 0 & 0 \\ 0 & (1+i\beta)^{-1} & 0 \\ 0 & 0 & 1 \end{pmatrix} . \quad (4.26c)$$

Defining $A_0 = 1 - R_0\Gamma_0$, it follows that

$$A = (A_0 - R_0\delta\Gamma) = (1 - R_0\delta\Gamma A_0^{-1})A_0$$

and

$$A^{-1} = A_0^{-1}(1 + R_0\delta\Gamma A_0^{-1}) .$$

To calculate Φ, we first expand $R = R_0 + \delta R$, leaving Γ unaltered:

$$\begin{aligned}\Phi &= (1 - R_0\Gamma)[1 - (R_0 + \delta R)\Gamma]^{-1}(-\delta R) \\ &= A(A - \delta R\Gamma)^{-1}(-\delta R) \\ &= A[(1 - \delta R\Gamma A^{-1})A]^{-1}(-\delta R) \\ &= AA^{-1}(1 - \delta R\Gamma A^{-1})^{-1}(-\delta R) \\ \Phi &\simeq (1 + \delta R\Gamma A^{-1})(-\delta R) .\end{aligned} \quad (4.27)$$

We expand $\Gamma = \Gamma_0 + \delta\Gamma$ in (4.24), dropping fourth-order differentials:

$$\Phi = [1 + \delta R(\Gamma_0 + \delta\Gamma)A_0^{-1}(1 + R_0\delta\Gamma A_0^{-1})](-\delta R)$$
$$= (1 + \delta R\Gamma_0 A_0^{-1} + \delta R\delta\Gamma A_0^{-1} + \delta R\Gamma_0 A_0^{-1} R_0\delta\Gamma A_0^{-1})(-\delta R)$$
$$\Phi = (\Phi_1 + \Phi_2 + \Phi_3 + \Phi_4)(-\delta R) . \qquad (4.28)$$

We now rewrite (4.15) in terms of Φ:

$$\{\omega^2\boldsymbol{\xi} - [v_l\boldsymbol{qq}\cdot\boldsymbol{\xi} - v_t\boldsymbol{q}\times(\boldsymbol{q}\times\boldsymbol{\xi})]\}$$
$$= \frac{Zn_0e^2\omega i}{M}\boldsymbol{\Phi}\cdot\boldsymbol{\xi} + \frac{Zei\omega}{Mc}(\boldsymbol{\xi}\times\boldsymbol{B}_0) , \qquad (4.29)$$

where $v_l = \sqrt{c_l/M}$ and $v_t = \sqrt{c_t/M}$. The Φ_1 term exactly cancels the $\boldsymbol{\xi} \times \boldsymbol{B}_0$ term:

(i)

$$\frac{Zei\omega}{Mc}(\boldsymbol{\xi}\times\boldsymbol{B}_0) = \frac{Zei\omega}{Mc}\begin{vmatrix}\hat{i} & \hat{j} & \hat{k} \\ \xi_x & \xi_y & \xi_z \\ 0 & \sin\theta & \cos\theta\end{vmatrix}B_0$$

$$= \frac{Zei\omega}{Mc}B_0\begin{pmatrix}\xi_y\cos\theta - \xi_z\sin\theta \\ -\xi_x\cos\theta \\ \xi_x\sin\theta\end{pmatrix}$$

and
(ii)

$$\frac{Zn_0e^2\omega i}{M}\delta\boldsymbol{R}\cdot\boldsymbol{\xi} = -\frac{cZn_0e^2\omega}{M}\sigma_0^{-1}\omega_c\tau\begin{pmatrix}\xi_y\cos\theta - \xi_z\sin\theta \\ -\xi_x\cos\theta \\ \xi_x\sin\theta\end{pmatrix}$$

$$= -\frac{Zei\omega}{Mc}B_0\begin{pmatrix}\xi_y\cos\theta - \xi_z\sin\theta \\ -\xi_x\cos\theta \\ \xi_x\sin\theta\end{pmatrix} .$$

Franz [4.16] has shown clearly that this cancellation occurs, within the framework of Fedders' calculation, only in the free-electron case. The term $(\boldsymbol{\xi}\times\boldsymbol{B}_0)$ comes from the direct effect of the magnetic field on the moving ions. The term $(\delta\boldsymbol{R}\cdot\boldsymbol{\xi})$ comes from the electron-ion interaction. The fact that the two terms cancel indicates that in the free-electron case, the electrons completely shield the ions from direct interaction with the magnetic field.

The second term in (4.28) corresponds to the nonresonant Alpher–Rubin interaction:

$$\Phi_2 = -\delta R \Gamma_0 A_0^{-1} \delta R$$

$$= +\frac{i\beta}{1+i\beta} \begin{pmatrix} \cos^2\theta & 0 & 0 \\ 0 & \cos^2\theta & -\sin\theta\cos\theta \\ 0 & -\sin\theta\cos\theta & \sin^2\theta \end{pmatrix} \frac{(\omega_c\tau)^2}{\sigma_0}$$

or, from (4.15), the nonresonant dispersion relationship is

$$(\omega^2 - v_s^2 q^2)\xi = \left(\frac{Z n_0 e^2 \omega}{\sigma_0 M}\right)$$

$$\times \begin{pmatrix} \cos^2\theta & 0 & 0 \\ 0 & \cos^2\theta & -\sin\theta\cos\theta \\ 0 & -\sin\theta\cos\theta & \sin^2\theta \end{pmatrix} \xi \frac{i\beta}{1+i\beta}(\omega_c\tau)^2 , \quad (4.30)$$

where v_s is the appropriate acoustic velocity. Since the dispersion relation is complex, it is customary to designate a complex wave number:

$$q = k + i\Delta\alpha .$$

With the displacement given by $\xi = \xi_0 \exp(-i\omega t + i q \cdot r)$, a complex $q = q\hat{z}$ implies a damped wave along z:

$$\xi = \xi_0 e^{-\Delta\alpha z} e^{(-i\omega t + ikz)} . \quad (4.31)$$

α is the attenuation in cm^{-1}. Typically $k^2 \gg (\Delta\alpha)^2$ for solids at frequencies above 1 MHz. We also use the relationship for small changes in acoustic velocity: $v^2 - v_0^2 = \Delta(v^2) \simeq 2v\Delta v$. Using this approximation and setting $v = \omega/k$, we find from the real and imaginary parts of (4.30), respectively, the change in acoustic velocity and the attenuation:

$$\left(\frac{\Delta v}{v_0}\right)_{AR} = \frac{B_0^2}{8\pi\rho v^2(1+\beta^2)} f(\theta) \quad (4.32)$$

$$\alpha_{AR} = \frac{\omega\beta B_0^2}{8\pi\rho v^3(1+\beta^2)} f(\theta) = \frac{\omega\beta}{v}\left(\frac{\Delta v}{v_0}\right)_{AR} , \quad (4.33)$$

where $\rho = M n_0/z$ is the density, θ is the angle between B_0 and q, and $f(\theta) = \sin^2\theta$ for longitudinal waves and $\cos^2\theta$ for transverse waves. This nonresonant behavior, called the Alpher–Rubin effect [4.1], and its usefulness in NAR, is discussed in more detail in Chapter 7. The Alpher–Rubin attenuation originates in the dissipation of the transverse currents set up by the Lorentz force. The Alpher–Rubin velocity shift arises from a phase lag between the oscillating self–consistent current j and ionic displacements ξ. The tangent of that phase lag is β, the *phase shift parameter*:

$$\beta = \frac{q^2 c^2}{4\pi\omega\sigma_0} = \frac{\omega c^2}{4\pi\sigma_0 v^2} = 2\pi^2 \left(\frac{\delta}{\lambda}\right)^2 , \qquad (4.34)$$

where c is the speed of light, σ_0 the DC conductivity, λ the ultrasonic wavelength, and δ the classical skin depth.

The resonant Alpher–Rubin dispersion arises from the $\delta\mu$ contribution contained in the third and fourth terms of (4.28). We first evaluate

$$\Phi_3 = -\delta R \delta \Gamma A_0^{-1} \delta R = -\frac{i\beta(\omega_c \tau)^2}{(1+i\beta)\sigma_0}$$

$$\times \begin{pmatrix} -\delta\mu_{xx}\cos^2\theta & -\delta\mu_{xy}\cos^2\theta & -\delta\mu_{xy}\sin\theta\cos\theta \\ -\delta\mu_{yx}\cos^2\theta & -\delta\mu_{yy}\cos^2\theta & \delta\mu_{yy}\sin\theta\cos\theta \\ \delta\mu_{yx}\sin^2\theta\cos\theta & \delta\mu_{yy}\sin\theta\cos\theta & -\delta\mu_{yy}\sin^2\theta \end{pmatrix} .$$

$$(4.35)$$

Since, to lowest order, off–diagonal terms do not contribute, we obtain

$$(\Phi_3)_{zz} = \left(\frac{i\beta}{1+i\beta}\right)\frac{(\omega_c\tau)^2}{\sigma_0}(\delta\mu_{yy})\sin^2\theta$$

$$(\Phi_3)_{\bar{i}\bar{j}} = \left(\frac{i\beta}{1+i\beta}\right)\frac{(\omega_c\tau)^2}{\sigma_0}(\delta\mu_{\bar{i}\bar{j}})\cos^2\theta , \qquad (4.36)$$

where $\bar{i}$ and $\bar{j}$ run over x and y. Similarly,

$$\Phi_4 = -\delta R \Gamma_0 A_0^{-1} R_0 \delta \Gamma A_0^{-1} \delta R = -\frac{\beta^2(\omega_c\tau)^2}{(1+i\beta)^2}\frac{1}{\sigma_0}$$

$$\times \begin{pmatrix} -\delta\mu_{xx}\cos^2\theta & -\delta\mu_{xy}\cos^2\theta & \delta\mu_{xy}\sin\theta\cos\theta \\ -\delta\mu_{yx}\cos^2\theta & -\delta\mu_{yy}\cos^2\theta & \delta\mu_{yy}\sin\theta\cos\theta \\ \delta\mu_{yx}\sin^2\theta\cos\theta & \delta\mu_{yy}\sin\theta\cos\theta & -\delta\mu_{yy}\sin^2\theta \end{pmatrix} \quad (4.37)$$

or, following the same procedure as for Φ_3 above,

$$(\Phi_4)_{zz} = \frac{\beta^2}{(1+i\beta)^2}\frac{(\omega_c\tau)^2}{\sigma_0}(\delta\mu_{yy})\sin^2\theta$$

$$(\Phi_4)_{\bar{i}\bar{j}} = \frac{\beta^2}{(1+i\beta)^2}\frac{(\omega_c\tau)^2}{\sigma_0}(\delta\mu_{\bar{i}\bar{j}})\cos^2\theta . \qquad (4.38)$$

Combining (4.36) and (4.35) and using the relations (4.20) and (4.22), the resonant change in attenuation is obtained from the imaginary part of the dispersion relation of the acoustic wave, and the velocity shift $\Delta v/v$ is obtained from the real part as follows:

$$\Delta\alpha_{\text{res}} = \omega \frac{B_0^2 \cos^2\theta \sin^2\phi}{2\rho v^3 (1+\beta^2)^2}[(1-\beta^2)\chi_+'' - 2\beta\chi_+'] \tag{4.39}$$

$$\left(\frac{\Delta v}{v}\right)_{\text{res}} = \frac{B_0^2 \cos^2\theta \sin^2\phi}{2\rho v^2 (1+\beta^2)^2}[(\beta^2-1)\chi_+' - 2\beta\chi_+''], \tag{4.40}$$

where ϕ is the angle between $\boldsymbol{B}_0$ and $\boldsymbol{\xi}$. A similar self-consistent calculation by *Müller et al.* [4.17], to be sketched out in Section 4.3, obtains the same result. Their compact expression for the angular dependence, which is the same for all three ultrasonic modes, is used here.

It should be noted that $\Delta\alpha_{\text{res}}$ is not positive definite. As explained by *Fedders* [4.5], this does not mean that the ultrasonic wave can be amplified. Rather, the presence of the dynamic susceptibility can cause a decrease of the nonresonant attenuation α_{AR}, while leaving the total attenuation greater than zero.

The mixture of the real and imaginary parts of the susceptibility bears some relation to NMR line shapes in metals [4.18]. In NMR the appropriate scale is the smallest sample dimension, while in NAR it is the ultrasonic wavelength. When the skin depth $\delta \simeq \lambda$, then $\beta \simeq 1$, and the phase lag between the fields in the sample causes strong mixing of χ' and χ''. For either case, $\delta \gg \lambda$ or $\lambda \gg \delta$, the observed signal is essentially pure χ''.

For later reference, we give the peak resonant attenuation at the center of a Lorentzian line. From (4.23), the susceptibility can be written

$$\chi_+'' = \frac{\chi_0 \omega_0}{2} \frac{i\Gamma_2}{(\omega_0 - \omega)^2 + \Gamma_2^2}, \tag{4.41}$$

where Γ_2 is the spin–spin relaxation rate, assumed to be much less than $\omega_0 \simeq \omega$. The static susceptibility is

$$\chi_0 = \frac{N}{V} \cdot \frac{\gamma^2 \hbar^2 I(I+1)}{3 k_B T}, \tag{4.42}$$

where N/V is the number density of spins of the appropriate isotope, γ is the gyromagnetic ratio, $\omega_0 = 2\pi\nu_0 = \gamma B_0$, T is the absolute temperature, and k_B is the Boltzmann constant. Thus at $\omega = \omega_0$,

$$\Delta\alpha_{\text{res}} = \frac{4\pi^4 \hbar^2}{3 k_B} \cdot \frac{N}{V} \cdot \frac{I(I+1)}{\rho \Gamma_2} \cdot \frac{\nu_0^4 \cos^2\theta \sin^2\phi}{v^3 T} \cdot \frac{1-\beta^2}{(1+\beta^2)^2}. \tag{4.43}$$

For $\beta \ll 1$ (low temperature, good conductor), $\Delta\alpha_{\text{res}} \propto v^{-3}B_0^4$; $\Delta\alpha_{res}$ is larger for transverse waves and large magnetic fields. For $\beta \gg 1$ (high temperature, poor conductor), $\Delta\alpha_{\text{res}} \propto vB_0^2$; $\Delta\alpha_{res}$ is larger for longitudinal waves.

It is of interest to calculate the magnetic fields in the sample. From (4.13) the self-consistent electric field is

$$\boldsymbol{E} = -i\omega n_0 e(1 - \boldsymbol{R}\cdot\boldsymbol{\Gamma})^{-1}(\boldsymbol{R}_0 - \boldsymbol{R})\cdot\boldsymbol{\xi} \ . \tag{4.44}$$

Expanding in powers of $\omega_c\tau$ with $\boldsymbol{A} = 1 - \boldsymbol{R}_0\cdot\boldsymbol{\Gamma}$,

$$\boldsymbol{E} = i\omega n_o e \boldsymbol{A}^{-1}[1 + \boldsymbol{A}^{-1}\delta\boldsymbol{R}\boldsymbol{\Gamma}_0 + \boldsymbol{A}^{-1}\delta\boldsymbol{R}\delta\boldsymbol{\Gamma}]\delta\boldsymbol{R}\cdot\boldsymbol{\xi} \ . \tag{4.45}$$

The first term contains one power of $\omega_c\tau = (\sigma_0/n_0 ce)B_0$, while the other two contain $(\omega_c\tau)^2$. The nuclear susceptibility enters through $\delta\Gamma$. The susceptibility is very small and does not change $\boldsymbol{E}$ very much, although synchronous detection and signal averaging may allow detection of the small effect on the attenuation. The oscillating magnetic field $\boldsymbol{h}$ can be expressed in terms of $\boldsymbol{E}$ and the magnetization $\boldsymbol{M}$ by using one of Maxwell's equations,

$$\boldsymbol{\nabla}\times\boldsymbol{E} + \frac{1}{c}\frac{\partial}{\partial t}(\boldsymbol{h} + 4\pi\boldsymbol{M}) = 0 \ , \tag{4.46}$$

or by assuming harmonic dependences:

$$\boldsymbol{h} + 4\pi\boldsymbol{M} = \frac{c}{\omega}\boldsymbol{q}\times\boldsymbol{E} \ . \tag{4.47}$$

The magnetization changes $\boldsymbol{h}$ very little and will be ignored beyond this point. The part of $\boldsymbol{h}$ that arises from the first term in $\boldsymbol{E}$ is proportional to B_0 through $\omega_c\tau$; the other terms in $\boldsymbol{E}$ make a negligible contribution to $\boldsymbol{h}$ because $\omega_c\tau \ll 1$. Assuming $\boldsymbol{q} = (\omega/v)\hat{\boldsymbol{z}}$ and $\boldsymbol{B}_0 = B_0(\cos\theta\hat{\boldsymbol{z}} + \sin\theta\hat{\boldsymbol{y}})$ and taking the cross product, the leading term in $\boldsymbol{h}$ is

$$\begin{pmatrix} h_x \\ h_y \\ h_z \end{pmatrix} = \frac{iB_0}{1+i\beta}\begin{pmatrix} \varepsilon_{xz}\cos\theta \\ \varepsilon_{yz}\cos\theta - \varepsilon_{zz}\sin\theta \\ 0 \end{pmatrix} \ , \tag{4.48}$$

using $\varepsilon_{iz} = d\xi_i/dz + iq\xi_i$. NMR signals are proportional to $H_1^2 B_0^2$, but here $H_1 \propto h \propto B_0$ for small β, which explains the dynamic dipolar NAR B_0^4 dependence. However, for large β, h is independent of B_0, and a B_0^2 dependence is regained. The power absorbed by the nuclear spins is $\langle P\rangle = 2\omega\chi''|H_1^2|$ per unit volume, where $2\boldsymbol{H}_1 = \boldsymbol{h}^\perp$ is the component of $\boldsymbol{h}$

perpendicular to $\boldsymbol{B}_0$: $h_x^\perp = h_x \sin\theta$, $h_y^\perp = h_y \cos\theta$. Thus assuming that only one ultrasonic mode is excited in a given experiment,

$$\langle P \rangle = \frac{1}{2}\omega\chi'' \frac{B_0^2 S(\theta)|\varepsilon|^2}{(1+\beta^2)^2} \qquad (4.49)$$

and

$$\Delta\alpha_{\text{spins}} = \frac{\langle P \rangle}{\rho v^3 |\varepsilon|^2} = \frac{\omega B_0^2 S(\theta)\chi''}{2\rho v^3(1+\beta^2)^2}. \qquad (4.50)$$

$S(\theta) = \cos^2\theta$ for ε_{xz}, $\cos^4\theta$ for ε_{yz} and $\sin^2\theta\cos^2\theta$ for ε_{zz}, which can be written concisely as $\cos^2\theta \sin^2\phi$ for all three modes. The term is labeled *spins* because it gives that part of the total attenuation that is due to energy absorbed by the nuclear spin system. The difference between (4.40) and (4.50) is

$$\Delta\alpha_{\text{res}} - \Delta\alpha_{\text{spins}} = \frac{\omega B_0^2 \cos^2\theta \sin^2\phi}{2\rho v^3(1+\beta^2)^2} \cdot [-2\beta^2 \chi'' - 2\beta\chi']. \qquad (4.51)$$

Thus, as remarked earlier, the overall attenuation cannot be calculated simply by determining the power absorbed by the spins.

4.3 Classical Derivation of Dipolar NAR Absorption and Dispersion: II. Method of Müller

The existence of a comprehensive theory [4.5] of the dynamic Alpher–Rubin coupling with resultant expressions for the NAR absorption and dispersion in metals (4.39,4.40) does not preclude our presenting an alternative classical derivation, given by *Müller et al.* [4.17], which utilizes the generalized Kubo susceptibility approach given in Chapter 2 and which also elucidates clearly the physical origin of the terms causing the asymmetry of the NAR lines. In Section 4.5 we will compare their experimental results in vanadium with theoretical predictions. In presenting this model we follow closely the classical paper of *Müller et al.* [4.17] and we refer, where necessary, to the earlier discussion of acoustic susceptibility (Section 2.2) and NAR susceptibility (Section 2.4.3).

As in the earlier derivation, a model is chosen in which the real metal is replaced by a gas of conduction electrons embedded in a background of positive ions; the interaction between the charged particles is replaced by the interaction with a self-consistent electromagnetic field derived from Maxwell's equations [4.15]. When an acoustic wave propagates in the metal, the ions experience oscillatory motions that are screened by the conduction electrons. In the presence of an external magnetic field, however, the screening is not quite complete so there is a small resultant current density j. Hence, in addition to the ordinary elastic force density, the Lorentz force density

$$f_L = j \times B \tag{4.52}$$

acts on the charge–carrier system, where $B = B_0 + b$, and b is the acoustically induced RF magnetic field.

In the framework of the original Alpher–Rubin phenomenological theory [4.1] and in linear approximation to the time derivative of the ion displacement u, the acoustically induced current density is assumed to be

$$j = \sigma \left(e + \frac{du}{dt} \times B_0 \right) , \tag{4.53}$$

where e is the ultrasonically induced electric field, and σ is the DC conductivity of the material. We recall that the microscopic theory of *Rodriguez* [4.11] demonstrates that (4.53) is justified if $\omega\tau$, ql and $\omega_c\tau \ll 1$. Here τ is the mean scattering time for the electrons, l the electron mean free path, ω_c the cyclotron angular frequency in the external magnetic field B_0, q the propagation vector, and ω the angular frequency of the ultrasonic wave.

If the acoustic displacement field is sufficiently small, a linear theory is applicable to the acoustically induced quantities. Keeping only first–order terms in the time derivative of the acoustic displacement field, (4.52) takes the form

$$f_L = j \times B_0 . \tag{4.54}$$

Since stationarity can be realized only if the Lorentz force density is compensated by an external force density produced by the transducer, the external power fed into the sample is

$$P(t) = \int_{(V_s)} d^3r [j \times B_0] \frac{d\xi}{dt} , \tag{4.55}$$

where V_s is the volume of the sample, and $d\xi/dt$ the velocity of an infinitesimal element of the sample at the position r. If v_e is the electron velocity averaged over the Fermi surface, $d\xi/dt$ is given by

$$\rho_s \frac{d\xi}{dt} = \rho_e v_e + \rho_i \frac{du}{dt} , \tag{4.56}$$

where ρ_e is the effective mass density of the electrons, ρ_i the mass density of the ions, and $\rho_s = \rho_e + \rho_i$. Since the electric current density is defined by

$$j = \eta_e v_e + \eta_i \frac{du}{dt}, \qquad (4.57)$$

where the charge density of the electrons is designated by η_e, and that of the ions by η_i, it can be shown, by aid of simple algebraic operations and $\eta_e + \eta_i = 0$, that (4.55) is equivalent to

$$P(t) = -\int_{(V_s)} d^3r [j(r,t) \times B_0] \frac{du(r,t)}{dt}. \qquad (4.58)$$

As j can be replaced by $\eta_e v_e$ without altering the value of $P(t)$, (4.58) demonstrates that this portion of the external energy is transferred to the ions (forming the lattice) via the electron–ion interaction.

To calculate the acoustic power fed into the nuclear spin system, one has to take into account that the nuclei not only absorb energy when resonating, but also behave like an electromagnetic source whose field is superimposed on the acoustically induced one, thus changing the current density in the sample. If the correlation of j with the electromagnetic field is known, the contribution $j_N(r,t)$ of the nuclei to the current density can be calculated by solving Maxwell's equations,

$$\nabla \times b = \mu_0 j + \mu_0 \nabla \times m + \mu_0 \frac{\partial d}{\partial t} \qquad (4.59a)$$

$$\nabla \times e = -\frac{\partial b}{\partial t} \qquad (4.59b)$$

$$\nabla \cdot b = 0 \qquad (4.59c)$$

$$\nabla \cdot d = \eta_e + \eta_i, \qquad (4.59d)$$

where m is the nuclear spin magnetization, and d the electric displacement. After j_N is computed, the power P_N absorbed by the nuclei can be deduced from (4.58), yielding

$$P_N(t) = -\int_{(V_s)} d^3r [j_N(r,t) \times B_0] \frac{du(r,t)}{dt}. \qquad (4.60)$$

In Section 2.2 we have shown that the relevant experimental quantity in NAR is the relative change $\Delta_N Z / Z_0$ of the electric impedance, where the real part describes the NAR absorption, and the imaginary part describes the NAR dispersion. Z_0 is the electric impedance without spin transitions and is a real quantity at an acoustic standing wave resonance frequency $\omega_n/2\pi$. Using the identities $Z_0 = \langle P_0(t)/I^2(t) \rangle$ and $\Delta_N Z(\omega) = [P_N(t)/I(t)]_\omega/I(\omega)$, the ratio $\Delta_N Z/Z_0$ may be written

$$\frac{\Delta_N Z(\omega)}{Z_0(\omega)} = \frac{[P_N(t)/I(t)]_\omega/I(\omega)}{\langle P_0(t)/I^2(t) \rangle} , \qquad (4.61a)$$

where the index ω symbolizes the Fourier transform with respect to the parameter t. $I(\omega)$ is the Fourier transform of the electric current, and the $\langle \, \rangle$ quantities are time–averaged values. In CW acoustic resonance experiments with composite resonators, the acoustic displacement field of the ion has the form

$$\boldsymbol{u}(\boldsymbol{r},t) = \hat{\boldsymbol{u}}(\boldsymbol{r})f_u(t) , \qquad (4.61b)$$

where $f_u(t)$ is a dimensionless harmonic function of time. By restricting our considerations to the vicinity of an acoustic standing wave resonance at ω_n and to piezoelectric transducers, where $I(t)$ is found to be proportional to $df_u(t)/dt = \dot{f}_u(t)$, and by using the identities

$$\langle [\dot{f}(t)]^2 \rangle_f = \omega_n^2/2 ,$$
$$\dot{f}_u(\omega) = i\omega f_u(\omega) ,$$
$$\langle P_0(t) \rangle = \omega_n E_a/Q_a^{(n)} ,$$

we obtain from (4.61)

$$\frac{\Delta_N Z(\omega)}{Z_0} = -iQ_a^{(n)} \frac{[P_N(t)/\dot{f}_u(t)]_\omega}{2E_a f_u(\omega)} , \qquad (4.62)$$

where $Q_a^{(n)}$ is the quality factor of the composite resonator at $\omega = \omega_n$. E_a is the total acoustic energy in the sample with transducer defined by

$$E_a = \frac{\omega_n^2}{2} \int_{(V)} d^3r \rho(\boldsymbol{r})[\hat{\boldsymbol{u}}(\boldsymbol{r})]^2 . \qquad (4.63)$$

$\rho(\boldsymbol{r})$ is the mass density, and V is the volume of the composite resonator.

By inserting (4.60), (4.61b) and (4.63) into (4.62), we obtain, using Parseval's identity,

$$\frac{\Delta_N Z(\omega)}{Z_0} = i\eta_a Q_a^{(n)} \frac{\int_{-\infty}^{\infty} d^3k [\boldsymbol{j}_N(\boldsymbol{k},\omega) \times \boldsymbol{B}_0] \boldsymbol{u}(-\boldsymbol{k},\omega)}{\rho_s \omega_n^2 \int_{-\infty}^{\infty} d^3k \, \boldsymbol{u}(\boldsymbol{k},\omega) \boldsymbol{u}(-\boldsymbol{k},\omega)} , \qquad (4.64)$$

where the argument $\boldsymbol{k}$ stands for the Fourier transform with respect to the position variable $\boldsymbol{r}$. η_a is the filling factor defined by

$$\eta_a = \rho_s \frac{\int_{(V_s)} d^3r [\hat{\boldsymbol{u}}(\boldsymbol{r})]^2}{\int_{(V)} d^3r \rho(\boldsymbol{r})[\hat{\boldsymbol{u}}(\boldsymbol{r})]^2} .$$

The mass density ρ_s of the sample is assumed to be constant. If the mass of the transducer plus the bond is small compared to $\rho_s V_s$, it is evident that η_a is of order 1. Equation (4.64) correlates the relative change of the electric impedance to the Fourier transform of j_N and can be compared with (2.50) and (2.53), yielding for the resonant attenuation and velocity shift

$$\Delta_N \alpha_{res} = \frac{\omega_n}{2v_a} \chi''_{\text{NAR}_{(D)}}(\omega) \tag{4.65a}$$

$$\left(\frac{\Delta_N v_a}{v_a}\right)_{res} = -\frac{1}{2}\chi'_{\text{NAR}_{(D)}}(\omega), \tag{4.65b}$$

where

$$\chi_{\text{NAR}_{(D)}} = \chi'_{\text{NAR}_{(D)}} - i\chi''_{\text{NAR}_{(D)}} \tag{4.66}$$

and

$$\chi_{\text{NAR}_{(D)}}(\omega) = \frac{\int_{-\infty}^{\infty} d^3k [j_N(k,\omega) \times B_0] u(-k,\omega)}{\rho_s \omega_n^2 \int_{-\infty}^{\infty} d^3k\, u(k,\omega) u(-k,\omega)}. \tag{4.67}$$

We call $\chi_{\text{NAR}_{(D)}}$ the NAR dipole susceptibility. The relations (4.65a–b) are quite general and are not restricted to ql, $\omega\tau$ and $\omega_c\tau \ll 1$. Following [4.17], to calculate the NAR dipole susceptibility, we apply the method of Fourier transforms for solving Maxwell's equations, and we split e, b and j into $e = e_0 + e_N$, $b = b_0 + b_N$ and $j = j_0 + j_N$, where the index 0 characterizes that part of e, b and j that does not depend on the nuclear spin magnetization. If $\omega\tau$, ql, and $\omega_0\tau \ll 1$ the electric current density is given by (4.3). By combining (4.3) and (4.59) and by neglecting the displacement current in (4.59a), we obtain with $v_a^2 = \omega^2/k^2$ the self-consistent expressions

$$e_0(k,\omega) = \frac{j_0(k,\omega)}{\sigma} - i\omega[u(k,\omega) \times B_0] \tag{4.68a}$$

$$b_0(k,\omega) = \frac{1}{1-i\beta(\omega)} \{ik \times [u(k,\omega) \times B_0]\} \tag{4.68b}$$

$$j_0(k,\omega) = \frac{1}{\mu_0} \frac{1}{1-i\beta(\omega)} \{ik \times [ik \times (u(k,\omega) \times B_0)]\} \tag{4.68c}$$

and

$$e_N(k,\omega) = \frac{j_n(k,\omega)}{\sigma} \tag{4.69a}$$

$$b_N(k,\omega) = -\frac{\mu_0}{k^2} \frac{i\beta(\omega)}{1-i\beta(\omega)} \{ik \times [ik \times m(k,\omega)]\} \tag{4.69b}$$

$$j_N(k,\omega) = -\frac{1}{1-i\beta(\omega)} [ik \times m(k,\omega)], \tag{4.69c}$$

where

$$\beta(\omega) = \frac{\omega}{\mu_0 \sigma v_a^2} \ . \tag{4.70}$$

The neglect of the displacement current is justified for the frequencies applied in NAR, since the vacuum wavelength λ_0 of an electromagnetic wave is large compared to the skin depth δ, and since $v_a \ll c$. Furthermore, it should be mentioned that the neglect of the displacement current does not alter the expression (4.69) if $v_a \ll c$.

We are now in the position to express $\chi_{\mathrm{NAR_{(D)}}}(\omega)$ by the nuclear spin magnetization. Since stationarity demands

$$\frac{d[m(r,t) B_0]}{dt} = 0 \ ,$$

only the component $m_\perp$ of the nuclear spin magnetization perpendicular to the applied magnetic field B_0 can be time–dependent; that is, $m(k,\omega) B_0 = 0$, whereas $m_\perp(k,\omega) \neq 0]$. With (4.66), (4.67), (4.69c) and (4.18b), we obtain the expression

$$\chi_{\mathrm{NAR_{(D)}}}(\omega) = \frac{\int_{-\infty}^{\infty} d^3 k\, m_\perp(k,\omega) b_{0\perp}(-k,\omega)}{\rho_s \omega_n^2 \int_{-\infty}^{\infty} d^3 k\, u(k,\omega) u(-k,\omega)} \ , \tag{4.71}$$

which, by aid of Parseval's identity and the conventional nuclear magnetic spin susceptibility χ defined by

$$\chi(\omega) = \frac{\int_{(V_s)} d^3 r\, m_\perp(r,\omega) b_{0\perp}(r,\omega)}{\int_{(V_s)} d^3 r\, b_{0\perp}(r,\omega) b_{0\perp}(r,\omega)/\mu_0} \ , \tag{4.72}$$

leads to

$$\chi_{\mathrm{NAR_{(D)}}}(\omega) = \frac{\int_{(V_s)} d^3 r\, b_{0\perp}(r,\omega) b_{0\perp}(r,\omega)/\mu_0}{\rho_s \omega_n^2 \int_{(V_s)} d^3 r\, u(r,\omega) u(r,\omega)} \chi(\omega) \ . \tag{4.73}$$

To give (4.73) a more convenient form, we consider a crystal carrying an acoustic standing wave with the wave vector $\pm q$. In this case, (2.18) and (4.68a) yield

$$b_{0\perp}(r,\omega) = \hat{b}_{0\perp}(r,\omega) e^{i\phi(\omega)} f_u(\omega) \ , \tag{4.74}$$

where

$$\phi(\omega) = \arctan \beta(\omega) \tag{4.75}$$

and

$$\hat{b}_{0\perp}(r,\omega) = \frac{q B_0 \cos\theta \sin\Phi}{\sqrt{1 + [\beta(\omega)]^2}} \hat{u}\left(r + \frac{\pi}{2q} \frac{q}{q}\right) \ . \tag{4.76}$$

In these formulas, θ is the angle between q and B_0, Φ is the angle between u and B_0, and $\hat{b}_{0\perp}(r,\omega)$ is a real quantity. Inserting (4.74) into (4.73), we obtain

$$\chi_{\text{NAR}_{(D)}}(\omega) = \frac{E_{m\perp}^{(s)}}{E_a^{(s)}} e^{2i\phi(\omega)} \chi(\omega) , \qquad (4.77)$$

where

$$E_{m\perp}^{(s)} = \int_{(V_s)} d^3 r \frac{[\hat{b}_{0\perp}(r,\omega)]^2}{2\mu_0} \qquad (4.78\text{a})$$

and

$$E_a^{(s)} = \rho_s \omega_n^2 \int_{(V_s)} d^3 r \frac{[\hat{u}(r)]^2}{2} . \qquad (4.78\text{b})$$

In these formulas, $E_{m\perp}^{(s)}$ is the contribution of the component $b_\perp$ to the acoustically induced electromagnetic field energy, and $E_a^{(s)}$ is the total acoustic energy stored in the specimen (without transducer).

There are some interesting features contained in (4.74) through (4.77). The first is the magnitude of the NAR dipole susceptibility compared to that of the nuclear spin susceptibility χ that is detected in conventional NMR. Since both quantities differ by the factor $(E_{m\perp}^{(s)}/E_a^{(s)})$, which for vanadium is of the order 10^{-5} at $B_0 = 20\,\text{kG}$, the detection of dipole NAR seems to be impossible at first sight. However, in NAR, as well as in NMR, the detectability is given by the relative change of electric impedance $\Delta Z/Z_0$ of the device under test, and not by the corresponding susceptibilities only. In NAR the ratio $\Delta Z/Z_0$ is given by (4.64), which, combined with (4.67), yields (2.57), reproduced here:

$$\left(\frac{\Delta Z}{Z_0}\right)_{\text{NAR}} = -i Q_a^{(n)} \eta_a \chi_{\text{NAR}} , \qquad (4.79)$$

which is very similar to the expression (2.56) $[(\Delta Z/Z_0)_{\text{NMR}} = -iQ_e\eta_e\chi]$, often used in NMR for a parallel–resonant circuit with the quality factor Q_e and the filling factor η_e. Since $\eta_e Q_e$ is of the order 1 when NMR is applied to investigate typical single crystals, $\eta_a Q_a$ should be about 10^5 if roughly the same detectability is desired in NAR as in NMR. To obtain such high quality factors, optical flatness and parallelism of the end faces of the specimen forming the acoustic resonator are required.

The second interesting aspect of (4.74)–(4.77) is the temporal phase shift ϕ of the acoustically induced magnetic field with respect to the acoustic wave. According to (4.75) and (4.77), this shift generally leads to a contribution of χ' and χ'' to both the NAR absorption and the NAR dispersion if $\beta \neq 0$. A similar situation exists for NMR in bulk metals, where the external RF magnetic field, and the RF magnetic field seen by the nuclei within the skin depth are not in phase. To become more familiar with the consequences of the phase shift, we consider the cases $\beta = 0$ and $\beta = 1$. The case $\beta > 1$ has been discused in Section 4.2. With $\chi = \chi' - i\chi''$, and with (4.66) and (4.77), we obtain for $\beta = 0$ (that is, $\phi = 0$):

$$\chi'_{NAR_{(D)}} = \frac{E_{m\perp}^{(s)}}{E_a^{(s)}}\chi', \qquad \chi''_{NAR_{(D)}} = \frac{E_{m\perp}^{(s)}}{E_a^{(s)}}\chi''$$

and for $\beta = 1$ (that is, $\phi = \pi/4$)

$$\chi'_{NAR_{(D)}} = \frac{E_{m\perp}^{(s)}}{E_a^{(s)}}\chi'', \qquad \chi''_{NAR_{(D)}} = -\frac{E_{m\perp}^{(s)}}{E_a^{(s)}}\chi'.$$

Hence, the time–averaged value of the acoustic power absorbed by the nuclei $\langle P_N \rangle = 2v_a E_a^{(s)} \Delta_N \alpha = E_a^{(s)} \omega_n \chi''_{NAR_{(D)}}$ is equal to the electromagnetic power $\langle P_N^{(m)} \rangle = E_{m\perp}^{(s)} \omega_n \chi''$ if the ionic displacement field and the ultrasonically induced magnetic field are in phase ($\beta = 0$). If, on the other hand, $\beta = 1$ and the acoustically induced magnetic field is phase shifted by the angle $\phi = 45°$, the time–averaged value of the acoustic power absorbed by the nuclei is zero, since at the center of a nuclear resonance $\chi' = 0$. Thus the electromagnetic power absorbed by the nuclei must be fed back coherently into the acoustic displacement field of the ions. Finally, we give the results for $\chi'_{NAR_{(D)}}$ and $\chi''_{NAR_{(D)}}$, which, by combining (4.74) to (4.78) and by using the identity

$$e^{2i\phi} = \frac{1}{1+\beta^2}[(1-\beta^2) + 2i\beta], \qquad (4.80)$$

are found to be

$$\chi'_{NAR_{(D)}} = \frac{B_0^2 \cos^2\theta \sin^2\Phi}{\mu_0 \rho_s v_a^2} \left(\frac{1}{1+\beta^2}\right)^2$$
$$\times [(1-\beta^2)\chi' + 2\beta\chi''] \qquad (4.81a)$$

$$\chi''_{NAR_{(D)}} = \frac{B_0^2 \cos^2\theta \sin^2\Phi}{\mu_0 \rho_s v_a^2} \left(\frac{1}{1+\beta^2}\right)^2$$
$$\times [(1-\beta^2)\chi'' - 2\beta\chi'], \qquad (4.81b)$$

where $\chi_{NAR_{(D)}} = \chi'_{NAR_{(D)}} - i\chi''_{NAR_{(D)}}$. $\chi = \chi' - i\chi''$, θ is the angle between $\boldsymbol{q}$ and $\boldsymbol{B}_0$, and Φ is the angle between $\boldsymbol{u}$ and $\boldsymbol{B}_0$. As expected, (4.81a) is related to (4.81b) via the Kramers–Kronig relations.

Inserting (4.81) into (4.65), we find for the acoustic attenuation and velocity shift the same expressions as given by the Fedders' derivation, (4.39) and (4.40). Müller et al. [4.17] point out, however, that if χ in (4.81) is interpreted as the conventional nuclear spin susceptibility defined in (4.72), there does exist a discrepancy between the two sets of results. Using (4.72), the discrepancy occurs because the time–averaged value of the electromagnetic power $\langle P_N^{(m)} \rangle = - \int d^3 r\, m_\perp(\boldsymbol{r},t) \partial b_\perp(\boldsymbol{r},t)/\partial t$ absorbed by the nuclei is related to χ by $\langle P_N^{(m)} \rangle = E_{m\perp}^{(s)} \text{Re}(i\omega\chi)$. As $\langle P_N^{(m)} \rangle$ is a positive quantity, $\Re(i\omega\chi)$ also must be positive, which requires $\omega\chi''$ to be positive if $\chi = \chi' - i\chi''$, and $\omega\chi''$ to be negative if $\chi = \chi' + i\chi''$. Hence, *the actual formulas of Section 4.2 (with $\chi = \chi' + i\chi''$) lead to a negative attenuation shift for $\beta = 0$, which violates physical laws, since the Alpher-Rubin attenuation is zero for $\beta = 0$.*

4.4 Experimental Verification in Metals

4.4.1 The Line Shape

The most interesting feature of $\Delta\alpha_{res}$ and $(\Delta v/v)_{res}$ is their dependence on an admixture of absorption χ'' and dispersion χ' through the phase shift parameter β. This feature is similar to the line shape effects seen in metal foils [4.18]. For sensitivity only to acoustic attenuation at fixed frequency, a unique admixture is observed in the signal. Since the χ' coefficient can have large or small positive or negative values, the line shape can appear either absorptive or dispersive, but in general is asymmetric. As seen in (4.81), the amount of mixing of χ'' and χ' in the signal is dependent on β. Experimentally, this mixing can be varied by changing the frequency of the acoustic wave, by changing the velocity through the use of shear or longitudinal waves, or by changing the conductivity σ_0. The conductivity can be easily changed over a wide range by changing the temperature. This allows a wide range of mixing between χ'' and χ' to be studied. Figures 4.2 and 4.3 show line shapes corresponding, respectively, to a $\sim 25\%$ dispersive admixture $(\chi'' + 0.3\chi')$ and a predominantly dispersive admixture $(\chi'' + 3.1\chi')$ [4.19]. Also shown are the first–derivative line shapes (which are typically observed experimentally). The peak–to–peak line width can be defined in general as the field spacing between the two most up–field extrema of the first derivative signal as the external field is swept from low values to high values. This is true for pure absorption where there are only two extrema and for pure dispersion with three extrema. The general admixture has three extrema; it can be shown that as the admixture varies from purely dispersive to purely absorptive, the low–field extremum moves toward large $-x$ and vanishes leaving the two high–field extrema for line width definition. The half–widths at half–maximum, δ, for a pure Lorentzian absorption or dispersion can be obtained from the peak–to–peak values defined above:

$$\delta = \frac{\sqrt{3}}{2}\Delta H_{p-p} \quad \text{(absorption)}$$

$$\delta = \frac{1}{\sqrt{3}}\Delta H_{p-p} \quad \text{(dispersion)} .$$

δ can be obtained for any Lorentzian admixture $(\chi'' + a\chi')$ from the general result [4.20]

$$\delta = \frac{a}{2\sqrt{3}}\left(\frac{\cos\phi}{\sin\phi/3}\right)\Delta H_{p-p} , \qquad (4.82)$$

where $\cos\phi = 1/\sqrt{1+a^2}$. This result is valid for any $a \neq 0$. For $a = 0$, we have, of course, pure absorption, and the δ shown earlier is correct. δ (admixture) approaches the correct pure absorption value for $a \ll 1$ and the correct pure dispersion values for $a \gg 1$. An extensive discussion of line shapes can be found in *Poole* [4.21].

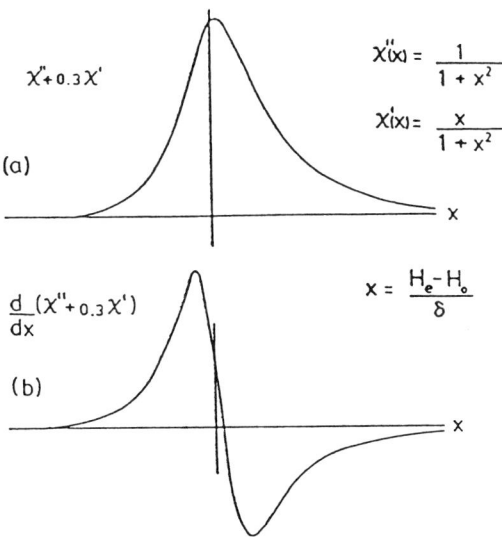

FIGURE 4.2. NAR Lorentzian line shape (a) and derivative line shape (b) for a predominantly absorptive admixture $(\chi'' + 0.3\chi')$. δ is the half-width at half maximum of the absorption line. H_e is varied from low to high values, while H_0 is kept fixed. (From Ref. 4.20)

140   4. DYNAMIC DIPOLAR INTERACTION

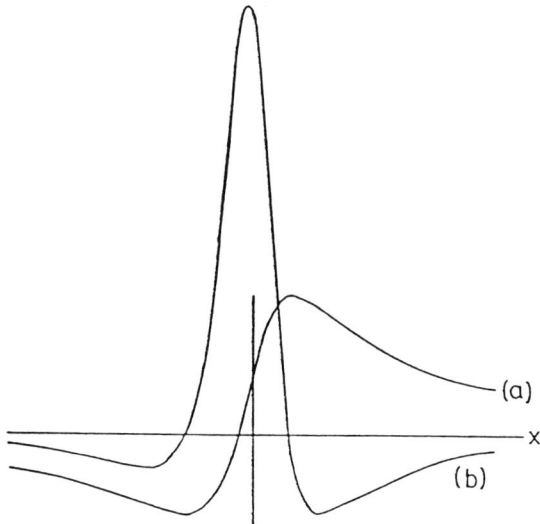

FIGURE 4.3. NAR Lorentzian line shape (a) and double scale derivative line shape (b) for a predominantly dispersive admixture $(\chi'' + 3.1\chi')$. (From Ref. 4.20)

4.4.2 The Case of Vanadium

The assumptions made in deriving (4.39) and (4.40) are valid for the frequencies, magnetic fields and temperatures used in the NAR study of vanadium by *Müller et al.* [4.17], since at 5 K the electron mean free path l is of the order 10^{-5} cm, τ is less than 10^{-13} sec, and the value of the effective mass is of the order 2 in units of the free–electron mass. Studying both the NAR absorption and NAR dispersion signals in a single crystal of vanadium and using longitudinal acoustic waves propagating along the [100] axis, the authors are able to demonstrate that in the temperature range of 5 to 294 K, the observed NAR line shapes in single crystals of vanadium are in good agreement with the formula (4.81).

The 99.94%–pure V single crystal with growth axis along a [100] crystal axis is a cylindrical specimen of about 1 cm in length and 1.2 cm in diameter. After grinding and polishing the end faces of the sample are flat and parallel to within 1 μm and are measured to be within $\pm 1°$ of the (100) crystal plane. In order to remove interstitial hydrogen and to reduce internal strains caused by crystal cutting and grinding, the specimen is carefully annealed in vacuum of 4×10^{-8} Torr for 36 hours at 1320 K. After heating, the sample is allowed to cool very slowly (35 K/h) to room temperature.

The longitudinal acoustic waves are generated by means of a gold–plated X–cut transducer (one side plated) bonded to the sample with polysulfide liquid polymers (Thiokol LP33). The transducer is 0.8 cm in diameter with a fundamental frequency of 19 MHz. The quality factor of the composite resonator is determined by means of a CW acoustic bridge spectrometer and is found to be about 4×10^4 at 17 kG and to have a working frequency $\omega_n/2\pi \approx 19\,\text{MHz}$.

The experimental NAR absorption line $\partial \chi''_{\text{NAR}_{(D)}}/\partial B_0$, obtained at 5 K for $\boldsymbol{B}_0$ parallel to the [110] direction, is given in Fig. 4.4. As predicted by (4.81) the observed NAR line width (about 14 G peak–to–peak) and the NAR line shape are in good agreement with NMR measurements [4.22] of $\partial \chi''/\partial B_0$ in single–crystal vanadium, if β is assumed to be negligibly small. The marked difference between the experimental curve and the Gaussian fit may be attributed to the indirect exchange interaction [4.18].

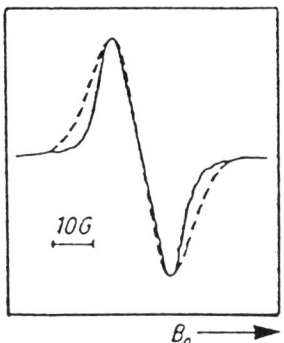

FIGURE 4.4. $\partial \chi''_{\text{NAR}_{(D)}}/\partial B_0$ as a function of magnetic field B_0 at $T = 5$ K. The 19–MHz longitudinal wave is propagated along the [100] direction, $B_0 = 16.9$ kG, time constant $\tau = 3$ sec, modulation field is 0.4 G peak–to–peak. The dashed line gives a Gaussian line fit. (From Ref. 4.17)

Figure 4.5 shows both the NAR absorption line $\partial \chi''_{NAR_{(D)}}/\partial B_0$ and the NAR dispersion line $\partial \chi'_{\text{NAR}_{(D)}}/\partial B_0$ at 5 K [4.17]. The solid lines are the experimental results, and the open circles represent a computer fit to the data. The computer fit to the NAR absorption line is from a model of a sum of three empirical Gaussian line shape functions, while the open circles, fitting the dispersion line, were obtained from the absorption line fit by applying the Kramers–Kronig relations and by using the expressions given by *Pake and Purcell* [4.23]. The observed NAR lines are in good

142   4. DYNAMIC DIPOLAR INTERACTION

agreement (within experimental error) with the Kramers–Kronig relations and with formulas (4.81a–b).

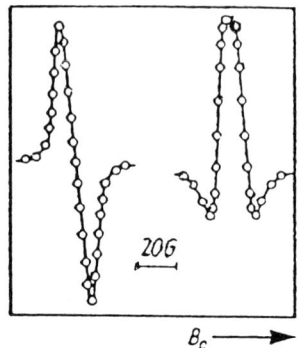

FIGURE 4.5. Derivative of the NAR absorption and dispersion line at $T = 5$ K. The parameters are the same as in Fig. 4.4. The open circles give a computer fit. (From Ref. 4.17)

Equations (4.81a–b) predict that β is the important parameter for the contribution of χ' and χ'' to the NAR dipole signal. To study the signal asymmetry as a function of β, the parameter β is varied via the temperature dependence of the DC conductivity. Figure 4.6a shows the experimental absorption signals observed in the temperature range of 5 to 294 K. The change of β with temperature, calculated from published values of the impurity–independent DC conductivity [4.24] and acoustic velocity [4.25], is given by the dashed line in Fig. 4.6b. As shown in Fig. 4.6a and as expected from formula (4.81b) and Fig. 4.6b, the shape of the NAR absorption signal at room temperature is essentially altered as compared to that at 5 K. With the use of (4.81b) and of the empirical $\partial \chi''/\partial B_0$ curve applied to fit the NAR absorption and NAR dispersion line at 5 K, a computer is programmed to find the appropriate admixtures of $\partial \chi'/\partial B_0$ and $\partial \chi''/\partial B_0$ fitting the experimental curves at $T > 5$ K. Taking into account the temperature dependence of the field modulation amplitude and of χ', the width and height of the computed curves are adjusted to match the experimental data. The β-values are calculated at each temperature. The solid dots in Fig. 4.7 represent the predicted line shape with the assumption of a Gaussian line shape function in the computer fit. Although there is some uncertainty in the analysis, the observed asymmetries of the NAR absorption lines are in agreement with Formula (4.81b) and the temperature dependence of β.

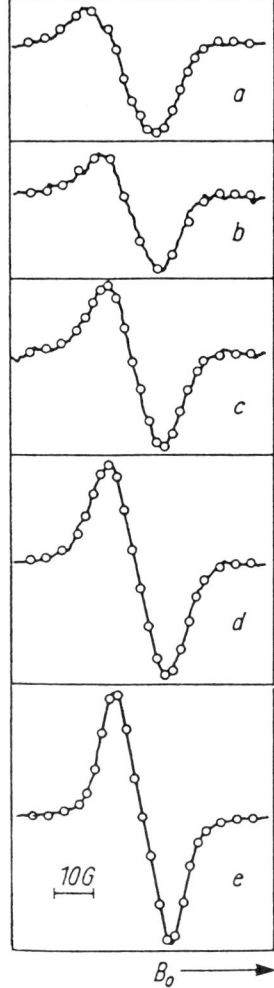

FIGURE 4.6A. $\partial \chi''_{NAR_{(D)}}/\partial B_0$ at various temperatures. The time constant increases with increasing temperature from 3 to 30 sec and the modulation field increases from 0.4 to 10 G peak-to-peak. The open circles give a computer fit taking into account the admixture of χ' and χ''. Because of different adjustment of the receiver gain, the peak–to–peak height of the NAR signals cannot be taken as a true–to–scale measure of the signal intensity. (a) $T = 294$ K, $\beta = 0.5$; (b) $T = 221$ K, $\beta = 0.3$; (c) $T = 152$ K, $\beta = 0.2$; (d) $T = 61$ K, $\beta = 0.1$; (e) $T = 5$ K, $\beta = 0$. (From Ref. 4.17)

Figure 4.6C shows the angular dependence of $\chi''_{NAR_{(D)}}$ (measured by the peak-to-peak height of $\partial \chi''_{NAR_{(D)}}/\partial B_0$) at 61 K for the DC magnetic field rotated in the (001) plane. For longitudinal acoustic waves ($\theta = \Phi$), the angular dependence of the signal intensity is expected to be proportional to $\cos^2\theta \sin^2\theta$—see (4.81). In this geometry, the $|\Delta m| = 1$ signal due to the electric quadrupole coupling has the same angular dependence. In order to check the contribution of the $|\Delta m| = 1$ quadrupole coupled signal to the observed NAR line, an unsuccessful search is made for the $|\Delta m| = 2$ transition. The lack of success indicates that the interaction is indeed via the nuclear magnetic dipole moment.

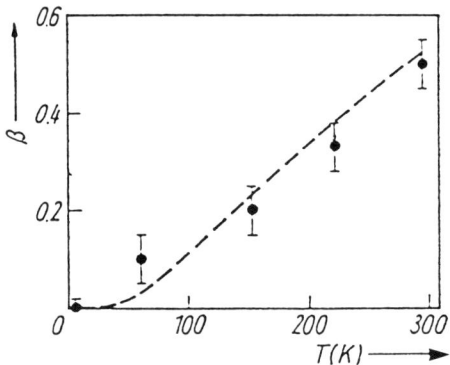

FIGURE 4.6B. Temperature dependence of β. The dashed line represents the calculated values. The solid dots are determined from the computer fit in Fig. 4.6A. (From Ref. 4.17)

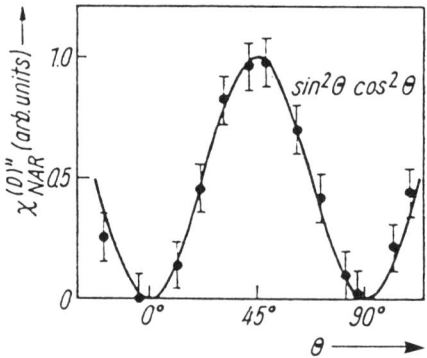

FIGURE 4.6C. Angular dependence of $\chi''_{NAR_{(D)}}$ on the angle θ. (From Ref. 4.17)

4.5 The 'Low Temperature' Limit

Thus far we have discussed dipole–coupled NAR in metals in the high temperature limit, corresponding to the conditions $\omega_c\tau \ll 1$ and $ql \ll 1$. The NAR absorption and dispersion signals in this limit have been found to be asymmetric, which can be explained in terms of a mixture of the real and imaginary parts of the complex nuclear susceptibility χ. In this section we summarize the calculation of the NAR attenuation $\Delta\alpha_{res}$ and velocity shift $(\Delta v_a/v_a)_{res}$ in the low temperature limit, and we compare experimental results taken under conditions in which $\omega_c\tau$ and ql are no longer much smaller than unity.

4.5.1 Theory of Quinn and Buttet

It is convenient to limit the discussion to the case of circularly polarized shear waves $\xi_\pm = \xi_x \pm i\xi_y$, where $\xi = \xi_0 e^{i\omega t + iqz}$ and $q||B_0$. Following the model of *Quinn and Ying* [4.2], one may write the equation of the motion of a gas of free electrons embedded in an isotropic lattice of positive ions, for the limit $\omega_c\tau \gg 1 \gg \omega\tau$, and $\omega_c\tau \gg ql$ [4.6] as follows:

$$\omega^2 - (v_a^2 q^2) \pm \Omega_c\omega - \frac{Zmi\omega}{M\tau}\frac{(1-i\beta\nu_\pm)(\sigma_0 R_\pm - 1)}{(1 - i\beta\sigma_0\nu_\pm R_\pm)}\xi_\pm = 0 \;, \qquad (4.83)$$

where v_a is the velocity of the acoustic wave in the absence of the DC magnetic field. Z and M are, respectively, the charge and the mass of the ions, and $\Omega_c = ZeB_0/Mc$ is the cyclotron frequency of the ions. $R_\pm = \sigma_\pm^{-1}$, with $\sigma_\pm = \sigma_{xx} \mp \sigma_{xy}$, where $\sigma_{\alpha\beta}$ is the conductivity tensor. σ_0 is the DC conductivity, and $\nu_\pm = 1 - 4\pi\chi_\pm$ where $\chi_\pm$ are the spherical components of the nuclear spin susceptibility. As in preceding sections, $\beta = q^2c^2/4\pi\omega\sigma_0$ is the parameter giving the mixture of χ' and χ'' in the high temperature limit. At low temperature, β is usually small; we assume $\beta \ll 1$ and $\beta(\omega_c\tau) \ll 1$, conditions which are satisfied in pure metals at low temperature and for moderate DC magnetic fields.

Following a procedure similar to that of Fedders (detailed above in Section 4.2), the real and imaginary parts of the eigenvalue equation are calculated. After neglecting higher–order terms and collecting others, one obtains

$$\Delta\alpha_{res} = \frac{1}{2}\frac{\omega B_0^2}{\rho v_a^3}(-A\chi'' - B\chi') \qquad (4.84a)$$

$$\left(\frac{\Delta v_a}{v_a}\right)_{res} = \frac{1}{2}\frac{B_0^2}{\rho v_a^2}(-B\chi'' + A\chi') \;, \qquad (4.84b)$$

where A and B are, respectively, the real and imaginary part of $(\sigma_0 R_\pm - 1)^2/(\omega_c\tau)^2$. Equations (4.84a-b) are similar in form to (4.39),

(4.40) and (4.81), obtained in the high temperature case. The evaluation of the coefficients, however, is more difficult. An important parameter in the evaluation of the coefficients is $\gamma = \omega_c \tau/ql$. In NAR, where B_0 and the frequency are related by the Larmor condition, γ is independent of frequency and relaxation time and may be written as

$$\gamma = \frac{v_s\, e}{v_F m^* c \gamma_I}, \qquad (4.85)$$

where v_F is the Fermi velocity, m^* is the effective mass, and γ_I is the gyromagnetic ratio of the nucleus. In aluminum, for $q||[110]$ and $\xi||[1\bar{1}0]$, assuming $m^* = m$, $\gamma \simeq 3.6$.

In the high temperature limit, where $ql \ll 1$, $A \to -1$ and $B = 0$. (4.84a–b) then agree with Fedders' results for $\beta = 0$. For arbitrary ql, a computer program is used to obtain A and B as functions of ql [4.6]. The theory predicts a measurable mixture of χ' and χ'' for $\gamma < 2$ and enhancement of the NAR signal for $\gamma < 0.5$. For $\gamma > 3$, $-A \simeq 1$ and $B \simeq 0$ for all values of ql.

4.5.2 Experimental Verification: Copper and Lead

The NAR(A-R) absorption of ^{63}Cu in single crystal copper has been studied over the temperature range from 4 to 300 K [4.26, 4.27] and has been found to follow the expected angular dependence of (4.39) for magnetic dipolar coupling. The temperature–dependent study conducted by *Leisure et al.* [4.27] from 4 to 250 K encompassed the low, intermediate and high temperature conditions: they found, for ^{63}Cu, $\omega_c \tau \simeq 1$ for $T \approx 55$ K and $ql \approx 1$ for $T \simeq 45$ K. Despite the fact that (4.39) and (4.40) were derived from the high temperature conditions, a good fit is obtained for line shapes taken at all temperatures, even those in regions where $\omega_c \tau \approx 1$ and $\omega_c \tau \gg 1$. It should be noted, of course, that for copper, the low temperature theory of Buttet makes essentially the same prediction as the high temperature theory.

The absorption curves shown in Fig. 4.7 correspond to a range of $\beta \ll 1$ to $\beta > 1$. Since for NAR(A-R) the percentage of χ'' is given by $|1-\beta^2|/(|1-\beta^2|+|2\beta|)$, for $\beta \ll 1$ (high temperatures) the line shape is essentially 100% χ''. As β increases, χ' is admixed. At $\beta = 1$, the line shape is 100% χ'; mixing occurs as the temperature is lowered, and β becomes greater than 1. For $\beta \gg 1$, the line shape is again 100% χ''.

The intensity of the NAR signal is in qualitative agreement with that predicted by theory. Since the resonant absorption $\Delta\alpha \propto [(1-\beta^2)\chi'' - 2\beta\chi']/(1+\beta^2)^2$, where χ' and χ'' are inversely proportional to temperature, the signal strength is seen to decrease as the temperature is increased. This

4.5 THE 'LOW TEMPERATURE' LIMIT

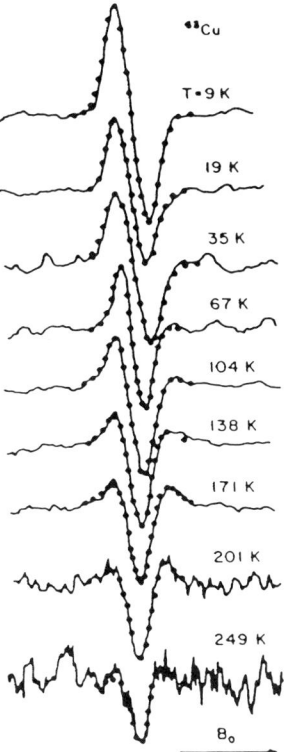

FIGURE 4.7. Temperature dependence of the NAR absorption signal line shape of ^{63}Cu in single-crystal copper. 64 MHz acoustic waves, polarized along [$\bar{1}$10], are propagated along the [110] direction, which is also the magnetic field direction. The solid dots represent a theoretical Gaussian fit to the experimental curves. (From Ref. 4.27)

represents an evident decrease in χ' and χ'' (due to $1/T$ dependence) and a $1/\beta^2$ dependence as β becomes larger. The line width, fairly independent of temperature, is 7.4 Oe.

Comparison of the observed NAR(A-R) dispersion signal in Cu at 76 K with theory (4.40) is given in Fig. 4.8. The observed NAR(A-R) absorption line width in ^{65}Cu is 5.6 Oe at 76 K; the intensity of the line is less than that of ^{63}Cu in the ratio of their isotopic abundances. The NAR line widths and Knight shifts of both isotopes agree well with those measured by NMR techniques.

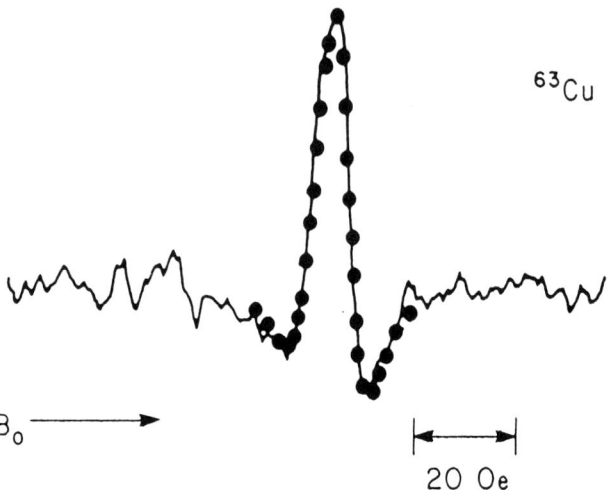

FIGURE 4.8. NAR dispersion signal of ^{63}Cu in copper at a temperature of 67 K. Other conditions are the same as in Fig. 4.7. (From Ref. 4.27)

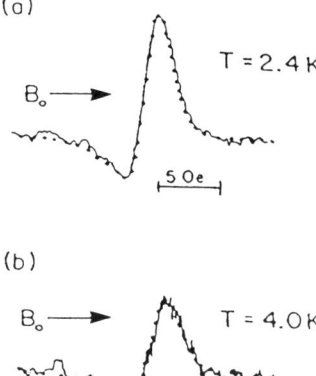

FIGURE 4.9. NAR absorption signal in Pb. The solid lines are the experimental curves, and the dots represent a theoretical Lorentzian fit to the data. 52.73-MHz shear waves were propagated along the [110] direction, which was also the magnetic field direction. (From Ref. 4.28)

The Alpher–Rubin dipolar absorption and dispersion signal in single-crystal lead has been observed in the temperature range of 2.4 to 17.6 K [4.28]. Lead, like tungsten, is cubic, and the 21% magnetically active nucleus, ^{207}Pb, has spin $\frac{1}{2}$; therefore neither dynamic nor static electric quadrupole effects are possible. The NAR(A-R) absorption signals in Pb at 2.4 K and 4.0 K are shown in Fig. 4.9. At these temperatures, for the frequency and magnetic field employed, both $\omega_c\tau$ and ql are much greater than unity; hence we are in the low temperature limit. A comparison of the experimental data with the Buttet theory appropriate for this limit (4.84a–b) gives good agreement for signal shape, signal intensity and their temperature dependence. The symmetrized absorption line width $H_{pp} = 2.5 \pm 0.1$ G at 2.4 K; this is, at least in part, attributed to the inhomogeneity of the applied magnetic field. The attenuation coefficient $\Delta\alpha_n = 4 \times 10^{-5} \text{cm}^{-1}$ at 2.4 K and 5.85 T. The signal intensity varies inversely with the temperature.

4.6 NAR Absorption and Dispersion: Quantum Mechanical Approach

Depending primarily upon the nuclei involved, either of the two main interactions—dynamic electric quadrupole or dynamic Alpher-Rubin—can be responsible for the observed NAR absorption and dispersion. For nuclei with spin $\frac{1}{2}$ (*e.g.*, Pb, W), only the dynamic Alpher–Rubin is available; for nuclei with large quadrupole moments (*e.g.*, Ta, Re), the quadrupole interaction is generally responsible. In a number of metals, such as V and Nb, there is competition between the two coupling mechanisms at high magnetic fields.

Thus far, the two NAR interactions have been treated separately. The formal approaches used in Chapter 3 did not allow for consideration of the dynamic Alpher–Rubin coupling; the approaches considered thus far in the present chapter for the most part utilize classical electrodynamics and neglect nuclear quadrupole effects. The possibility of interactions between the coupling mechanisms (in cases in which both are allowed) and consequently of interference terms in the expressions for NAR absorption and dispersion has not been taken into account.

A comprehensive quantum mechanical treatment of NAR in metals that takes both coupling mechanisms into account has been given by *Müller and Bartell* [4.8]. This treatment combines the use of a general acoustically induced perturbation Hamiltonian with the generalized susceptibility approach described in Chapter 2; application to a metal with nuclei of spin $> \frac{1}{2}$ results in the appearance of interference terms that enables one (in theory) to determine the signs of the gradient–elastic tensor components, in addition to their absolute value.

In this section, we first outline the Müller and Bartell formalism, emphasizing its application to combined dynamic Alpher–Rubin and quadrupolar NAR absorption and dispersion. In Section 4.7, we discuss briefly the influence of electronic band structure on NAR Alpher–Rubin absorption in metals.

4.6.1 Method of Müller and Bartell

The nuclear spin–phonon Hamiltonian of an acoustic wave in a metal is written

$$\mathcal{H}'(t) = \mathcal{H}'_D(t) + \mathcal{H}'_Q(t) , \qquad (4.86\text{a})$$

where

$$\mathcal{H}'_D(t) = -\int d^3r \, \boldsymbol{m}(\boldsymbol{r}) \cdot \boldsymbol{b}(\boldsymbol{r}, t) \qquad (4.86\text{b})$$

and

$$\mathcal{H}'_Q(t) = \int d^3r \, \boldsymbol{Q}(\boldsymbol{r}) : \boldsymbol{V}(\boldsymbol{r}, t) . \qquad (4.86\text{c})$$

$\mathcal{H}'_D(t)$, the dynamic Alpher–Rubin term, represents the coupling of nuclear spins to the acoustically induced magnetic field $\boldsymbol{b}$. $\mathcal{H}'_Q(t)$, the dynamic electric quadrupole term, represents the coupling of nuclear spins to the acoustically induced electric field gradient. $\boldsymbol{m}(\boldsymbol{r})$ is the nuclear spin magnetization operator. $\boldsymbol{V}(\boldsymbol{r}, t)$ is the dynamic electric field gradient tensor, and $\boldsymbol{Q}(\boldsymbol{r})$ is the tensor operator of the nuclear electric quadrupole moment density.

We utilize the expressions for generalized NAR susceptibility derived in Section 2.4. Specifically, we recall (2.49a)

$$\chi_{\text{NAR}}(\omega) = -\frac{1}{i\hbar} \frac{\text{Tr}\{[\widetilde{\mathcal{H}'}(\omega), \rho_0]\mathcal{H}'(\omega)\}}{\rho_s v_a^2 \int_{(V_s)} d^3r [\varepsilon'(\boldsymbol{r}, \omega)]^2} , \qquad (2.49\text{a})$$

in which we substitute for $\mathcal{H}'(\omega)$ from (4.86a) above, yielding for the numerator

$$\text{Tr}\{[\widetilde{\mathcal{H}'}(\omega), \rho_0]\mathcal{H}'(\omega)\} = \sum_{i,j = D \text{ or } Q} \text{Tr}\{\widetilde{\mathcal{H}}'_i(\omega), \rho_0]\mathcal{H}'_j(\omega)\} . \qquad (4.87)$$

Further, we write

$$
\begin{aligned}
&\text{Tr}\{[\widetilde{\mathcal{H}}'_i(\omega), \rho_0]\mathcal{H}'_j(\omega)\} \\
&= \sum_m \langle m|[\widetilde{\mathcal{H}}'_i(\omega), \rho_0]\mathcal{H}'_j(\omega)|m\rangle \\
&= \sum_m \langle m|\widetilde{\mathcal{H}}'_i(\omega)\rho_0\mathcal{H}'_j(\omega) - \rho_0\widetilde{\mathcal{H}}'_i(\omega)\mathcal{H}'_j(\omega)|m\rangle \\
&= \sum_{mm'm''} \{\langle m|\widetilde{\mathcal{H}}'_i(\omega)|m'\rangle\langle m'|\rho_0|m''\rangle\langle m''|\mathcal{H}'_j(\omega)|m\rangle \\
&\quad - \langle m|\rho_0|m'\rangle\langle m'|\widetilde{\mathcal{H}}'_i(\omega)|m''\rangle\langle m''|\mathcal{H}'_j(\omega)|m\rangle\} \\
&= \sum_m \langle m|\rho_0|m\rangle\langle m|\mathcal{H}'_j(\omega), \widetilde{\mathcal{H}}'_i(\omega)|m\rangle \\
&= \text{Tr}\{\rho_0[\mathcal{H}'_j(\omega), \widetilde{\mathcal{H}}'_i(\omega)]\} \equiv \langle[\mathcal{H}'_j(\omega), \widetilde{\mathcal{H}}'_i(\omega)]\rangle_0 \,. \quad (4.88)
\end{aligned}
$$

We substitute (4.88) into (4.87) and obtain

$$
\begin{aligned}
\text{Tr}\{[\widetilde{\mathcal{H}}'(\omega), \rho_0]\mathcal{H}'(\omega)\} &= \sum_{ij}\text{Tr}\{[\widetilde{\mathcal{H}}'_i(\omega), \rho_0]\mathcal{H}'_j(\omega)\} \\
&= \sum_{ij}\langle[\mathcal{H}'_j(\omega), \widetilde{\mathcal{H}}'_i(\omega)]\rangle_0 \,. \quad (4.89)
\end{aligned}
$$

With this result for the numerator, we can write for the generalized NAR susceptibility in (2.49a) (for the case in which dynamic Alpher–Rubin and quadrupole coupling exist simultaneously)

$$
\begin{aligned}
\chi_{\text{NAR}} = -\frac{1}{i\hbar}\frac{1}{\rho_s v_a^2 \int d^3r[\varepsilon'(\mathbf{r},\omega)]^2} &\times \Big(\langle[\mathcal{H}'_D(\omega), \widetilde{\mathcal{H}}'_D(\omega)]\rangle_0 \\
+ \langle[\mathcal{H}'_D(\omega), \widetilde{\mathcal{H}}'_Q(\omega)]\rangle_0 &+ \langle[\mathcal{H}'_Q(\omega), \widetilde{\mathcal{H}}'_D(\omega)]\rangle_0 \\
+ \langle[\mathcal{H}'_Q(\omega), \widetilde{\mathcal{H}}'_Q(\omega)]\rangle_0\Big) &\,, \quad (4.90)
\end{aligned}
$$

where the expectation value is to be evaluated in the equilibrium spin ensemble. The first term of (4.90) corresponds to a pure dynamic Alpher–Rubin (or NAR dipole) susceptibility; the fourth term, to a pure dynamic quadrupole susceptibility; and the two middle terms, to an interference susceptibility between dynamic Alpher–Rubin and dynamic quadrupole, or

$$
\chi_{\text{NAR}}(\omega) = \chi_{\text{NAR}_{(D)}}(\omega) + \chi_{\text{NAR}_{(\text{int})}}(\omega) + \chi_{\text{NAR}_{(Q)}}(\omega) \,, \quad (4.91)
$$

where

4. DYNAMIC DIPOLAR INTERACTION

$$\chi_{NAR_{(D)}}(\omega) = -\frac{1}{i\hbar} \frac{\langle [\mathcal{H}'_D(\omega), \widetilde{\mathcal{H}}'_D(\omega)] \rangle_0}{\rho_s v_a^2 \int_{(V_s)} d^3 r [\varepsilon'(r,\omega)]^2} \quad (4.92a)$$

$$\chi_{NAR_{(Q)}}(\omega) = -\frac{1}{i\hbar} \frac{\langle [\mathcal{H}'_Q(\omega), \widetilde{\mathcal{H}}'_Q(\omega)] \rangle_0}{\rho_s v_a^2 \int_{(V_s)} d^3 r [\varepsilon'(r,\omega)]^2} \quad (4.92b)$$

and

$$\chi_{NAR_{(int)}}(\omega)$$
$$= -\frac{1}{i\hbar} \frac{\langle [\mathcal{H}'_D(\omega), \widetilde{\mathcal{H}}'_Q(\omega)] + [\mathcal{H}'_Q(\omega), \widetilde{\mathcal{H}}'_D(\omega)] \rangle_0}{\rho_s v_a^2 \int_{(V_s)} d^3 r [\varepsilon'(r,\omega)]^2} . \quad (4.92c)$$

The sign of $\chi_{NAR_{(int)}}(\omega)$ depends on the sign both of the gyromagnetic ratio and of the dynamic electric field gradient tensor components. Therefore the NAR $\Delta m = \pm 1$ lines should differ, depending on whether the electric field gradient terms are positive or negative.

The metal is assumed to consist of N identical particles of nuclear spin $I > \frac{1}{2}$. The effective one particle static Hamiltonian of the nuclear spin system in an external magnetic field is written as

$$\mathcal{H}'_0 = \mathcal{H}'^s_z + \mathcal{H}'^s_D + \mathcal{H}'^s_Q$$
$$= \gamma \hbar B_0 I_z + \hbar a_D I_z + \hbar a_Q \left(\frac{I_z^2 - I(I+1)}{3} \right) , \quad (4.93)$$

where the magnetic dipole interaction Hamiltonian $\mathcal{H}'^s_D$ describes the homogeneous line broadening. $a_D = -\gamma B_{Dz}$, with B_{Dz} being the static local magnetic fields produced by neighboring spins. a_Q, the quadrupole frequency, is given by $\hbar a_Q = 3eQV^s_{zz}/4I(2I-1)$, where V^s_{zz} is the component of the static electric field gradient in the direction of the external B_0. The distribution of a_D and a_Q in the sample contribute to the line shape functions $g_D(a_D)$ and $g_Q(a_Q)$ introduced below.

To obtain the general NAR susceptibility for this system under the influence of the time–dependent spin–phonon interaction $\mathcal{H}'(t)$ (4.86a), we rewrite (4.88), incorporating the explicit dependence of $\mathcal{H}'(\omega)$, $\widetilde{\mathcal{H}}'(\omega)$ on the position of ρ_0:

$$\text{Tr}\{\rho_0[\mathcal{H}'_j(\omega), \widetilde{\mathcal{H}}'_i(\omega)]\}$$
$$= \frac{N}{V_s} \int_{(V_s)} \text{Tr}\{\rho_0(r)[\mathcal{H}'_j(r,\omega), \widetilde{\mathcal{H}}'_i(r,\omega)]\} d^3 r . \quad (4.94)$$

4.6 QUANTUM MECHANICAL APPROACH

Since the variation in ρ_0, the density operator at equilibrium, can be accounted for by the distribution functions $g_D(a_D)$ and $g_Q(a_Q)$, (4.94) can be written as

$$\text{Tr}\{\rho_0[\mathcal{H}'_j(\omega), \widetilde{\mathcal{H}}'_i(\omega)]\}$$
$$= \frac{N}{V_s} \int_{(V_s)} d^3 r \int_{-\infty}^{\infty} da_D \int_{-\infty}^{\infty} da_Q\, g_Q(a_Q) g_D(a_D)$$
$$\times \text{Tr}\{\rho_0[\mathcal{H}'_j(r,\omega), \widetilde{\mathcal{H}}'_i(r,\omega)]\}\ . \tag{4.95}$$

We expand:

$$\text{Tr}\{\rho_0[\mathcal{H}'_j(r,\omega), \widetilde{\mathcal{H}}'_i(r,\omega)]\}$$
$$= \sum_{mm''} [\langle m|\rho_0|m\rangle(\langle m|\mathcal{H}'_j(r,\omega)|m'\rangle$$
$$\times \langle m'|\widetilde{\mathcal{H}}'_i(r,\omega)|m\rangle$$
$$- \langle m|\widetilde{\mathcal{H}}'_i(r,\omega)|m'\rangle\langle m'|\mathcal{H}'_j(r,\omega)|m\rangle)]\ , \tag{4.96}$$

where

$$\langle m|\widetilde{\mathcal{H}}'(r,\omega)|m'\rangle$$
$$= \langle m|\int_0^\infty dt\, e^{-i\omega t} e^{-i(\mathcal{H}'_0/\hbar)t} \mathcal{H}'(r,\omega) e^{i(\mathcal{H}'_0/\hbar)t}|m\rangle$$
$$= \int_0^\infty e^{-i(\omega+\omega_m-\omega_{m'})t}\langle m|\mathcal{H}'(r\omega)|m'\rangle dt\ , \tag{4.97}$$

and where $\omega_m = E_m/\hbar$. Since

$$\mathcal{H}'_0|m\rangle = \left\{-\gamma\hbar B_0 I_z + \hbar a_D I_z\right.$$
$$\left. + \hbar a_Q\left[I_z^2 - \frac{I(I+1)}{3}\right]\right\}|m\rangle$$
$$= \left\{(-\gamma\hbar B_0 m + \hbar a_D m\right.$$
$$\left. + \hbar a_Q\left[m^2 - \frac{I(I+1)}{3}\right]\right\}|m\rangle$$
$$= E_m|m\rangle\ ,$$

4. DYNAMIC DIPOLAR INTERACTION

we can define

$$\omega_{m'm} = \frac{1}{\hbar}(E_{m'} - E_m)$$
$$= (m' - m)[(-\gamma B_0 + a_D) + (m' + m)a_Q] .$$

By substituting (4.97) into (4.96), with

$$\langle m|\rho_0|m\rangle = \frac{e^{-E_m/k_B T}}{\sum_m e^{-E_m/k_B T}} = \frac{N_m}{N} ,$$

we find

$$\mathrm{Tr}\{\rho_0[\mathcal{H}'_j(r,\omega), \widetilde{\mathcal{H}}'_i(r,\omega)]\}$$
$$= \sum_{mm'} \left\{ \frac{N_m - N_{m'}}{N} \right.$$
$$\times \langle m|\widetilde{\mathcal{H}}'_i(r,\omega)|m'\rangle$$
$$\left. \times \langle m'|\mathcal{H}'_j(r,\omega)|m\rangle \int_0^\infty e^{-i(\omega - \omega_{mm'})t} dt \right\} . \qquad (4.98)$$

Combining (4.95) and (4.98) with (4.90), we finally obtain the generalized NAR susceptibility:

$$\chi_{\mathrm{NAR}_{ij}}(\omega) = \frac{\pi}{\hbar \rho_s v_a^2 \int_{(V_s)} d^3 r [\varepsilon'(r,\omega)]^2} \sum_{mm'} \frac{N_m - N_{m'}}{N_m}$$
$$\times \int_{(V_s)} d^3 r \langle m|\widetilde{\mathcal{H}}'_i(r,\omega)|m'\rangle$$
$$\times \langle m'|\mathcal{H}'_j(r,\omega)|m\rangle g(\omega - \omega_{m,m'}) , \qquad (4.99)$$

where the normalized line shape function is

$$g(\omega - \omega_{mm'})$$
$$= \frac{1}{i\pi} \int_0^\infty dt \int_{-\infty}^\infty da_D \int_{-\infty}^\infty da_Q e^{-i(\omega - \omega_{mm'})t} g_D(a_D) g_Q(a_Q) . \qquad (4.100)$$

In the Müller and Bartell formalism, each of the three terms in the generalized NAR susceptibility is treated separately.

4.6.2 NAR Alpher–Rubin Susceptibility

For the NAR Alpher–Rubin susceptibility, we can utilize the procedure described in Section 4.3 that begins with (4.71). We write the expression for $\chi_{\text{NAR}_{(D)}}$ as

$$\chi_{\text{NAR}_{(D)}}(\omega) = \frac{\int_{-\infty}^{\infty} d^3k\, b_\perp(k\omega)b_\perp(-k,\omega)}{\rho_s v_a^2 \mu_0 \int_{-\infty}^{\infty} d^3k\, \varepsilon'(k,\omega)\varepsilon'(-k,\omega)} \chi(\omega) , \qquad (4.101a)$$

where $b_\perp$ is the projection of the acoustically induced RF magnetic field $b(r,t)$ on the plane perpendicular to B_0, $\chi(\omega)$ is the conventional NMR nuclear magnetic spin susceptibility, μ_0 is the vacuum permeability, and the arguments ω and k stand for the time- and space-Fourier transforms. Since the elastic strain ε' is perpendicular to v_i the velocity of the metal ions [see (2.43)], (4.101a) can be written alternatively as

$$\chi_{\text{NAR}_{(D)}}(\omega) = -\frac{\int_{-\infty}^{\infty} d^3k\, b_\perp(k,\omega)b_\perp(-k,\omega)}{\mu_0 \rho_s \int_{-\infty}^{\infty} d^3k\, v_i(k,\omega)v_i(-k,\omega)} \chi(\omega) . \qquad (4.101b)$$

In the limit in which band structure effects are negligibly small, the resultant expressions for absorption and dispersion agree with the classical result, (4.81a–b). The quantum treatment of Alpher–Rubin NAR in conductors is not restricted by the classical Alpher–Rubin limits, however, so that band structure effects may be incorporated readily (see below).

4.6.3 NAR Quadrupole Susceptibility

For the NAR quadrupole susceptibility, the procedure incorporates the quantum treatment for the dynamic electric quadrupole interaction already presented in Section 3.3. Because both $\Delta m = \pm 1$ and $\Delta m = \pm 2$ transitions are allowed in this case, we express $\chi_{\text{NAR}_{(Q)}}$ as

$$\chi_{\text{NAR}_{(Q)}} = \chi_{\text{NAR1}} + \chi_{\text{NAR2}} , \qquad (4.102)$$

where

$$\chi_{\text{NAR1}}(\omega) = D_1(\omega) \frac{\int_{(V_s)} d^3r\, V_{+1}(r,\omega) V_{-1}(r,\omega)}{\int_{(V_s)} d^3r\, [\varepsilon'(r,\omega)]^2} ,$$

where

$$D_1(\omega) = \frac{\pi A^2}{16\hbar \rho_s v_a^2} \sum_{m=-I}^{I-1} \frac{N_m - N_{m+1}}{V_s}(2m+1)^2$$

$$\times f_I^2(m)[g(\omega - \omega_{m+1,m}) - g(\omega + \omega_{m+1,m})]$$

(4.103a)

4. DYNAMIC DIPOLAR INTERACTION

and

$$\chi_{NAR2}(\omega) = D_2(\omega) \frac{\int_{(V_s)} d^3r\, V_{+2}(r,\omega) V_{-2}(r,\omega)}{\int_{(V_s)} d^3r\, [\varepsilon'(r,\omega)]^2},$$

with

$$D_2(\omega) = \frac{\pi A^2}{16\hbar \rho_s v_a^2} \sum_{m=-I}^{I-2} \frac{N_m - N_{m+2}}{V_s} f_I^2(m) f_I^2(I+m)$$
$$\times [g(\omega - \omega_{m+2,m}) - g(\omega + \omega_{m+2,m})] \quad (4.103b)$$

and $f_I = [I(I+1) - m(m+1)]^{\frac{1}{2}}$. We treat the case of longitudinal acoustic waves propagated along the [110] axis of a cubic crystal, with the static magnetic field rotated in different planes. The angle between B_0 and the [001] axis in the (110) plane is ψ, that between the acoustic wave vector and B_0 in the $(1\bar{1}0)$ plane is θ, and that between B_0 and the propagation direction in the (001) plane is ϕ. The resulting expressions may be compared with those in Figs. 3.6–3.8.

a. B_0 Rotated in the (110) Plane:

$$\chi_{NAR1}(\omega) = \frac{1}{4} D_1(\omega) [(S_{11} - S_{12}) - 2S_{44}]^2 \sin^2 \psi \cos^2 \psi \quad (4.104a)$$

$$\chi_{NAR2}(\omega) = \frac{1}{16} D_2(\omega)$$
$$\times [(S_{11} - S_{12}) \cos^2 \psi + 2S_{44}(1 + \sin^2 \psi)]^2. \quad (4.104b)$$

b. B_0 Rotated in the $(\bar{1}10)$ Plane:

$$\chi_{NAR1}(\omega) = \frac{1}{4} D_1(\omega) [(S_{11} - S_{12}) + 2S_{44}]^2 \sin^2 \theta \cos^2 \theta \quad (4.104c)$$

$$\chi_{NAR2}(\omega) = \frac{1}{16} D_2(\omega)$$
$$\times [(S_{11} - S_{12}) \cos^2 \theta - 2S_{44}(1 + \sin^2 \theta)]^2. \quad (4.104d)$$

c. B_0 Rotated in the (001) Plane:

$$\chi_{NAR1}(\omega) = \frac{1}{4} D_1(\omega)(4S_{44})^2 \sin^2 \phi \cos^2 \phi \quad (4.104e)$$

$$\chi_{NAR2}(\omega) = \frac{1}{16} D_2(\omega)$$
$$\times [(S_{11} - S_{12}) + 2S_{44}(1 - 2\cos^2 \phi)]^2. \quad (4.104f)$$

Equations (4.104a–e) serve to remind us that as we saw earlier in Chapter 3 the NAR quadrupole susceptibilities are bilinear in the appropriate S-tensor components, so that, in addition to their absolute values, only the relative sign of $S_{11} - S_{12}$ and S_{44} can be determined from $\chi_{\mathrm{NAR}_{(\mathrm{Q})}}$.

4.6.4 NAR Interference Term

For the interference NAR susceptibility, the Müller–Bartell formalism yields

$$\chi_{\mathrm{NAR}_{(\mathrm{int})}}(\omega) = D_{11}(\omega) \frac{\int d^3k [b_x(\mathbf{k},\omega)V_{xz}(-\mathbf{k},\omega) + b_y(\mathbf{k},\omega)V_{yz}(-\mathbf{k},\omega)]}{\int d^3k \varepsilon'(\mathbf{k},\omega)\varepsilon'(-\mathbf{k},\omega)},$$

where

$$D_{11}(\omega) = -\frac{\pi \gamma A}{4\rho_s v_a^2} \sum_{m=-I}^{I-1} \frac{N_m - N_{m+1}}{V_s}(2m+1)f_I^2(m) \\ \times [g(\omega - \omega_{m+1,m}) - g(\omega + \omega_{m+1,m})]. \quad (4.105)$$

The argument $\mathbf{k}$ stands for the Fourier transform, b_i is the ith component of the acoustically induced magnetic field, and V_{ij}^s is the ijth component of the strain–dependent electric field gradient tensor. From (4.105) it is clear that the interference term contributes to the $\Delta m = \pm 1$ NAR lines only and, further, that $\chi_{\mathrm{NAR}_{(\mathrm{int})}}$ vanishes if either χ_{NAR1} or $\chi_{\mathrm{NAR}_{(\mathrm{D})}}$ is zero.

For the same experimental circumstances as were treated in Section 4.6.3 for longitudinal waves propagating along the [110] axis of a cubic crystal, we obtain the following:

a. $\mathbf{B}_0$ Rotated in the (110) plane,

$$\chi_{\mathrm{NAR}_{(\mathrm{int})}} = 0. \quad (4.106)$$

b. $\mathbf{B}_0$ Rotated in the $(\bar{1}10)$ plane,

$$\chi_{\mathrm{NAR}_{(\mathrm{int})}} = \frac{D_{11}(\omega)}{2} \frac{B_0[(S_{11}-S_{12}) + 2S_{44}]\sin^2\theta \cos^2\theta}{(1-i\beta)}. \quad (4.106\mathrm{b})$$

c. for $\mathbf{B}_0$ Rotated in the (001) plane,

$$\chi_{\mathrm{NAR}_{(\mathrm{int})}} = D_{11}(\omega) \frac{2B_0 S_{44} \sin^2\phi \cos^2\phi}{(1-i\beta)}. \quad (4.106\mathrm{c})$$

From (4.106) it is seen that $X_{\text{NAR}_{(\text{int})}}$ is linear in the appropriate S-tensor components. Hence, provided that the relative sign of γ and Q is known, NAR observation of the interference term may yield the sign of $(S_{11} - S_{12})$ and S_{44}.

4.7 Influence of Electronic Band Structure

From the discussion in Section 4.1, we know that in conductors the same electron–phonon interaction that gives rise to magneto–acoustic effects via the acoustically induced electric and magnetic fields, $e(r,t)$ and $b(r,t)$, also serve to couple the acoustic wave to the magnetic dipole moments of the nuclear spin system. It is well known [4.29] that the magneto–acoustic effects strongly depend on the shape of the Fermi surface. Thus a similar dependence on electron band–structure is expected for NAR in metals.

In his classical derivation (Section 4.2) *Fedders* does state that his calculation is intended to include non–free-electron effects in the conductivity [4.5]; it is necessary only to use the correct magnetoresistivity tensor. A brief discussion of this point has been given by *Franz* [4.16]. The influence of electronic band structure on dipolar NAR [NAR(A-R)] can also be discussed by further treatment of the quantum mechanical approach of the previous section. Referring to (4.101), one sees that the electronic band structure enters into X_{NAR} mainly through the intrinsic magnetic field components $b_\perp$; v_i is an external quantity, and X is a measure of the response of the nuclear spin magnetization to $b_\perp$.

Using either the classical approach [4.6] or the quantum mechanical formalism [4.9, 4.10], one obtains explicit expressions for the NAR absorption $\Delta_N \alpha$ in terms of the dynamic magnetoresistivity tensor components ρ_{ij}. For the cases relevant to NAR(A-R), in which the acoustic wave vector q is either along or perpendicular to the applied DC magnetic field B_0, the results for cubic metals using the Müller–Bartell approach are the following:

a. $q \perp B_0$,

$$\Delta_N \alpha = \frac{\omega \eta_e^2 \cos^2 \Phi}{2\mu_0 \rho_s v_a^3} \frac{m^2}{\langle m^* \rangle^2} \left(\frac{\Delta \rho_{zz}}{1+\beta_\|^2} \right)^2 [(1-\beta_\|^2)\chi'' - 2\beta_\| \chi'] . \quad (4.107a)$$

b. $q \parallel B_0$,

$$\Delta_N \alpha = \frac{\omega \eta_e^2 \sin^2 \Phi}{2\mu_0 \rho_s v_a^3} \frac{m^2}{\langle m^* \rangle^2} \frac{(\Delta \rho_{xx})^2 + (\Delta \rho_{xy})^2}{(1+\beta_\perp^2)^2}$$
$$\times [(1-\beta_\perp^2)\chi'' - 2\beta_\perp \chi'] , \quad (4.107b)$$

where η_e is the conduction electron charge density, Φ is the angle between the velocity of the acoustic displacement field and B_0, ρ_{xy} is the dynamic Hall resistivity. $\Delta\rho_{zz} = \rho_{zz} - \rho_0$ and $\Delta\rho_{xx} = \rho_{xx} - \rho_0$ are, respectively, the dynamic longitudinal and transverse magnetoresistivity. ρ_0 is the DC resistivity, $\beta_\parallel = \beta(\rho_{zz}/\rho_0)$, $\beta_\perp = \beta(\rho_{xx}/\rho_0)$, $\beta = \omega\rho_0/\mu_0 v_a^2$, m is the free-electron mass, and $\langle m^* \rangle$ is the averaged effective mass. Equation (4.107b) holds only for q along an axis of four-fold crystal symmetry.

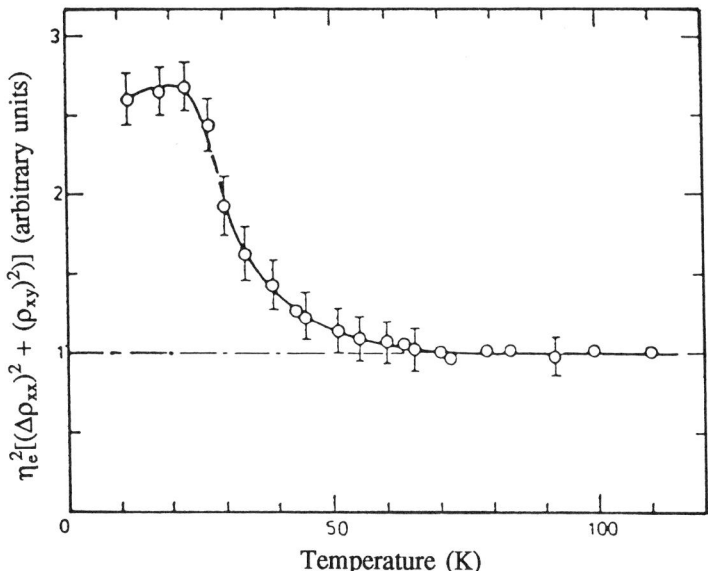

FIGURE 4.10. $\eta_e^2[(\Delta\rho_{xx})^2 + (\rho_{xy})^2]$ as a function of temperature in ^{27}Al. The open circles are obtained by quantitative evaluation of the NAR signal intensity. The chain curve, obtained from ultrasonic velocity measurements, gives the temperature dependence of $\eta_e^2(\rho_{xy})^2$ if $\langle m/m^* \rangle = $ const. Note that at low temperatures the enhancement expected from (4.107b) should be proportional to $\eta_e^2[(\Delta\rho_{xx})^2 + (\rho_{xy})^2]$. (From Ref. 4.10)

Experimental verification [4.10] of (4.107b) is shown is Fig. 4.10, in which the $\Delta m = \pm 1$ NAR signal intensity in single crystal aluminum for 18.5–MHz shear waves propagated along the [100] direction (parallel to the magnetic field) is plotted as a function of temperature. The signal intensity is evaluated by taking into account the temperature dependence of Q (the acoustic quality factor), the magnetic field modulation, χ and

$\eta_e^2 \langle m/m^* \rangle \rho_{xy}^2$ (determined independently by ultrasonic velocity measurements). The observed variation with temperature of $\eta_e^2[(\Delta\rho_{xx})^2 + (\rho_{xy})^2]$ (shown in Fig 4.12) agrees well with that found in conventional and ultrasonic resistivity measurements [4.10]. At low temperatures, it is seen that the NAR signal intensity in Al is caused mainly by $\eta_e^2(\Delta\rho_{xx})^2$, demonstrating the important role of the dynamic transverse magnetoresistivity in NAR(A-R). Even more dramatic is the emergence at temperatures below 25 K of an NAR(A-R) signal from the background noise for case (a) for $q \perp B_0$ and v_i along B_0 above (4.107a). At 15 K the measured ρ_{zz} is found to be equal to $\rho_{xx}/2$, in agreement with (4.107a-b). For $q \perp B_0$ and v_i perpendicular to B_0, the NAR signal vanishes, in agreement with (4.107b).

CHAPTER 4 REFERENCES

4.1 R. A. Alpher, R. J. Rubin. "Magnetic Dispersion and Attenuation of Sound in Conducting Fluids and Solids." *J. Acoust. Soc. Am.* **26**, 452–453 (1954).

4.2 J. J. Quinn, S. C. Ying. "Acoustic Nuclear Spin Resonance in Metals." *Phys. Lett.* **23**, 61–62 (1966).

4.3 J. J. Quinn. "Direct Generation of Sound in Metals and Acoustic Nuclear Spin Resonance." *J. Phys. Chem. Solids* **31**, 1701–1707 (1970).

4.4 J. Buttet. "Acoustic Nuclear Magnetic Resonance and Admixture of χ' and χ'' in Niobium." *Solid State Commun.* **9**, 1129–1133 (1971).

4.5 P. A. Fedders. "Acoustic Magnetic Resonance in Metals via the Alpher–Rubin Mechanism." *Phys. Rev.* **B7**, 1739–1743 (1973). [Note: There are a number of typographical errors in this paper.]

4.6 J. Buttet. "Nuclear Acoustic Resonance in Metals at Low Temperatures." *Solid State Commun.* **16**, 397–399 (1975).

4.7 V. Müller. "Nuclear Acoustic Resonance and Dispersion." *Phys. Lett.* **60A**, 240–242 (1977).

4.8 V. Müller, U. Bartell. "Nuclear Acoustic Resonance in Metals." *Z. Physik* **B32**, 271–279 (1979).

4.9 V. Müller, E. J. Unterhorst, D. Maurer, G. Schanz. "Influence of Electron Band Structure on Dipole–Nuclear Acoustic Resonance." *Phys. Lett.* **69A**, 139–141 (1978).

4.10 V. Müller, E. J. Unterhorst, D. Maurer, G. Schanz. "Influence of Longitudinal and Transverse Magnetoresistivity on Nuclear Acoustic Resonance in Aluminum." *J. Phys. F* **9**, L11–L14 (1979).

4.11 S. Rodriguez. "Velocity of Sound in Metals in a Uniform Magnetic Field." *Phys. Lett.* **2**, 271–272 (1962).

4.12 S. Rodriguez. "Modification of the Velocity of Sound in Metals by an Applied Magnetic Field." *Phys. Rev.* **130**, 1778–1783 (1963).

4.13 J. J. Quinn, S. Rodriguez. "Helicon–Phonon Interaction in Metals." *Phys. Rev.* **133**, A1589–1594 (1964).

4.14 Y. Shapira. "Acoustic Wave Propagation in High Magnetic Fields." Chapter 1, pp. 1–58. In *Physical Acoustics*, ed. by W. P. Mason (Academic Press, New York, 1968) Vol. V.

4.15 M. H. Cohen, M. J. Harrison, W. H. Harrison. "Magnetic–Field Dependence of the Ultrasonic Attenuation in Metals." *Phys. Rev.* **117**, 937–952 (1960).

4.16 J. R. Franz. "Discussion of Anomalous Nuclear–Acoustic–Resonance Signals in Aluminum." *Phys. Rev.* **B14**, 5110–5113 (1976).

4.17 V. Müller, G. Schanz, E. Fischer, E. J. Unterhorst. "Nuclear Acoustic Resonance Absorption and Dispersion in Vanadium via Coupling to the Magnetic Dipole Moment." *Phys. Stat. Sol. (b)* **80**, 629–639 (1977).

4.18 J. Winter. *Magnetic Resonance in Metals* (Oxford Press, London 1971).

4.19 H. I. Ringermacher, R. E. Norberg, R. K. Sundfors, D. G. Westlake. "Observation of the Nuclear Acoustic Resonance of Hydrogen in $NbH_{0.026}$." *Phys. Rev. Lett.* **42**, 910–912 (1979).

4.20 H. I. Ringermacher. "Nuclear Acoustic Resonance of ^1H in $NbH_{0.026}$." Ph. D. Thesis, Washington University, 1980 (unpublished).

4.21 C. P. Poole, Jr. *Electron Spin Resonance* (John Wiley, New York, 1967).

4.22 D. Hechtfischer, R. Karcher, K. Lüders. "A Nuclear Magnetic Resonance Spectrometer Using a Superconducting Magnet and Digital Data Processing: Application to Single Crystal Vanadium." *Appl. Spectroscopy* **26**, 549–552 (1972); D. Hechtfischer. Thesis, Freie Universität Berlin, Berlin (West) 1974).

4.23 G. E. Pake, E. M. Purcell. "Line Shapes in Nuclear Paramagnetism." *Phys. Rev.* **74**, 1184–1188 (1948).

4.24 *American Institute of Physics Handbook*, Ed. by D. E. Gray, (McGraw-Hill Publ. Co., New York 1972).

4.25 D. I. Bolef, R. E. Smith, J. G. Miller. "Elastic Properties of Vanadium. I. Temperature Dependence of the Elastic Constants and the Thermal Expansion." *Phys. Rev.* **B3**, 4100–4108 (1971).

4.26 A. A. Manuel, J. Buttet. "Nuclear Acoustic Resonance in Copper via Coupling to the Magnetic Dipole." *Sol. St. Commun.* **13**, 289–291 (1973).

4.27 R. G. Leisure, D. K. Hsu, E. J. Ozimek. "Nuclear Acoustic Resonance in Single Crystal Copper over the Temperature Range 4–250 K." *Sol. St. Commun.* **20**, 489–491 (1976).

4.28 R. G. Leisure, D. K. Hsu, E. J. Ozimek. "Observation of Nuclear Acoustic Resonance in Lead." *Sol. St. Commun.* **17**, 67–70 (1975).

4.29 A. B. Pippard. "Theory of Ultrasonic Attenuation in Metals and Magneto–Acoustic Oscillations." *Proc. Roy. Soc.* **A257**, 165–193 (1960).

CHAPTER 5
LINE BROADENING AND RELAXATION EFFECTS

5.1 Introduction

5.2 The Method of Moments Applied to NAR1 and NAR2
    5.2.1 Method of Moments Applied to Cubic Symmetry
    5.2.2 Second Moment Analysis in Rhenium Metal

5.3 Generalized Bloch Equations

5.4 Quadrupole–Split High Field Case: Hamiltonian, Energy Levels, Transitions

5.5 Line Shape of the Quadrupole–Split High Field Line: Intraspin Cross Relaxation

5.6 Spin Relaxation by Dynamic Quadrupole Interaction
    5.6.1 General Theory of Relaxation
    5.6.2 Application of Relaxation Theory to the Tantalum–Hydrogen System
    5.6.3 Explanation of the Anomalous Tantalum NAR1 and NAR2 Line Shapes in the Tantalum–Hydrogen System

5.7 Dipolar and Quadrupolar Exchange Contributions to NAR Line Shapes

5.8 Relaxation by the Intrinsic Direct Process in Van Vleck Paramagnets

References

5. LINE BROADENING AND RELAXATION EFFECTS
5.1 Introduction

When the Zeeman interaction is much larger than the static quadrupole interaction, resonance line shapes and line widths are more sensitive to the effects of static and dynamic electric field gradients and nuclear and electron exchange when the coupling is dynamic quadrupolar (NAR1, NAR2) than when the coupling is via NAR(A-R) or NMR. This enhanced resonance line shape sensitivity for NAR1 derives from two facts: first, although the $m = -\frac{1}{2}$ to $m = +\frac{1}{2}$ transition dominates the NAR(A-R) and NMR line shapes, in the case of NAR1, the $m = -\frac{1}{2}$ to $m = +\frac{1}{2}$ transition is missing; second, although in NMR and NAR(A-R) the probability for transitions between smaller values of $|m|$ (as seen in Table 3.3) is greater than for larger values of $|m|$, the exact opposite is true for NAR1. Line widths and line shapes for NAR2 may have the same contributions as for NAR(A-R) and NMR, but the magnitude of these contributions will differ because of the different coupling mechanisms for NAR2 and for NMR and NAR(A-R).

In NMR and NAR(A-R) experiments, the analysis of a given resonance line width by the method of moments allows assessment of the contributions to the line broadening [5.1, 5.2]. For NAR1 and NAR2, the method of moments also serves as a useful method of analysis of line widths when the homogeneous interactions (dipole–dipole, pseudo–dipole and pseudo–exchange) dominate the effects of time–varying or static electric field gradients. When the effects of time–varying or static electric field gradients are of the same order of magnitude as, or larger than, the homogeneous interactions, however, the NAR1 and NAR2 line shapes often become anomalous. The generalized Bloch equation solutions of *Fedders* [5.3] allow an understanding of these anomalous NAR line shapes by providing a method of determining the contributions of various decay rates. In the particular case in which homogeneous interactions dominate the effects of electric field gradients, the generalized Bloch Equation solutions of Fedders allow both line width and line shape analysis.

In the present chapter, we first discuss the method of moments as applied to NAR1 and NAR2. We then show an application of the method of moments to the analysis of the line widths and line shapes of ^{185,187}Re in rhenium single crystal. The generalized Bloch Equations are introduced and written in terms of the multipole operator expectation values $\langle A_{lm} \rangle$. The Bloch Equations for the high–field quadrupole Hamiltonian are solved for the energy levels and the allowed transitions. We interpret the high–field quadrupole coupling line shapes in terms of the Bloch Equation solutions for decay rates and intraspin cross–relaxation (the interference between different decay rates). A more powerful method of determining decay rates and spectral shape functions, utilizing a correlation function of the quadrupole electric field, is then discussed. As an application, the Fedders relaxation theory is applied to the narrowing of the ^{181}Ta NAR1 and NAR2 line widths with increasing temperature in a tantalum single crystal with a small hydrogen impurity concentration. In a second application, the theory of intraspin cross relaxation is used to explain the anomalous behavior of the ^{181}Ta NAR1 and NAR2 line shapes. In the final section, we address the problem of understanding the anomalous behaviour of the observed isotropic, temperature–independent NAR1, NAR2 and NAR(A-R) resonances of ^{181}Ta in hydrogen-free tantalum metal. Agreement between theory and experiment is obtained by invoking a novel interaction, pseudo–quadrupolar nuclear exchange, as the major contributor to the line width, with smaller contributions from (i) isotropic interactions due to static charge defect–originated electric field gradients, and (ii) an anisotropic spin–spin interaction.

5.2 The Method of Moments Applied to NAR1 and NAR2

One of the first methods of analysis of NAR line shapes and line widths, provided that the homogeneous line broadening mechanisms are much greater than the inhomogeneous, is usually the method of moments because of its physical insight and simplicity of calculation. Most NAR measurements have been made in crystals with cubic symmetry; therefore, we first give an illustration of the calculation of the method of moments with this symmetry, in Section 5.2.1. In Section 5.2.2, an application of the method of moments is given in a different symmetry: the hexagonal close–packed structure.

5.2.1 Method of Moments Applied to Cubic Symmetry

We consider a system of N nuclear spins per unit volume in a single–crystal environment with an external magnetic field H directed along the z–direction. By specifying the angular momentum labels $\hbar I$ and $\hbar S$, we allow the possibility of two different nuclear spin systems. We assume, however, that the resonance frequencies of the two nuclear spin systems do not overlap. The assumption is also made that the spacing between the Zeeman split levels is large compared to the energy splitting produced by dipole–dipole, exchange or quadrupole interactions. Finally, we assume a particular model for the source of electric field gradients.

The second moment expression can be written [5.2] as

$$\langle \hbar^2 \omega^2 \rangle_{av} = -\frac{Tr[\mathcal{H}, X]^2}{Tr(X^2)} . \qquad (5.1)$$

The system Hamiltonian $\mathcal{H}$ is defined by

$$\mathcal{H} = \mathcal{H}_{0I} + \mathcal{H}_{0S} + \mathcal{H}_{DI} + \mathcal{H}_{DS} + \mathcal{H}_{DIS} + \mathcal{H}_{EI} + \mathcal{H}_{ES} + \mathcal{H}_{EIS} + \mathcal{H}_Q , \qquad (5.2)$$

where $\mathcal{H}_{0I}$ is the Zeeman term for the resonant spins I, $\mathcal{H}_{0S}$ is the Zeeman term for the spins S, $\mathcal{H}_{DI}$ is the dipole–dipole term for like spins I, $\mathcal{H}_{DS}$ is the dipole–dipole term for like spins S, and $\mathcal{H}_{DIS}$ is the dipole–dipole term for unlike spins I and S.

$$\mathcal{H}_{EI} = \sum_{j<k} a_{jk} \boldsymbol{I}_j \cdot \boldsymbol{I}_k \,, \tag{5.3a}$$

$$\mathcal{H}_{ES} = \sum_{j<k} \tilde{a}'_{jk} \boldsymbol{S}_j \cdot \boldsymbol{S}_k \,, \tag{5.3b}$$

$$\mathcal{H}_{EIS} = \sum_{j,k} \tilde{a}_{jk} \boldsymbol{I}_j \cdot \boldsymbol{S}_k \tag{5.3c}$$

and

$$\mathcal{H}_Q = \sum_j E_j [3I_{zj}^2 - I_j(I_j+1)] \,. \tag{5.3d}$$

In the above equations, a_{jk} is the exchange term for the like spins I, $\tilde{a}'_{jk}$ is the exchange term for like spins S, $\tilde{a}_{jk}$ is the exchange term for the unlike spins I and S, and we choose

$$E_j = \frac{A}{4}\left(\frac{\partial^2 V}{\partial z^2}\right)_j = \frac{A}{4} V_{zz_j} \,. \tag{5.3e}$$

From (3.33-3.39), we can write the interaction Hamiltonian for nuclear spin-phonon coupling $\boldsymbol{X}$ as

$$\begin{aligned}\boldsymbol{X} =& \frac{A}{4}\sum_j [(I_{+j}I_{zj} + I_{zj}I_{+j})V_{-1j} \\ &+ (I_{-j}I_{zj} + I_{zj}I_{-j})V_{+1j} + I_{+j}^2 V_{-2j} + I_{-j}^2 V_{+2j}] \,, \end{aligned} \tag{5.4}$$

where

$$V_{\pm 1} = V_{xy} \pm iV_{yz} \tag{5.5a}$$

and

$$V_{\pm 2} = \frac{1}{2}(V_{xx} - V_{yy}) \pm iV_{xy} \,. \tag{5.5b}$$

The NAR1 transition is centered at the frequency $\omega_1 = \gamma_I H_z$; the NAR2 transition, at $\omega_2 = 2\gamma_I H_z$. From (5.1), the NAR1 second moment for spin

I may be computed [5.4]:

$$\begin{aligned}\langle \hbar^2 \Delta\omega^2 \rangle_{\text{av NAR1}} =& 2I(I+1)\sum_j a_{jk}^2 + 2I(I+1)\sum_j a_{jk}b_{jk} \\ &+ 3I(I+1)\sum_j b_{jk}^2 + \frac{1}{3}S(S+1)\sum_j \tilde{a}_{jk}^2 \\ &- \frac{4}{3}S(S+1)\sum_j \tilde{a}_{jk}\tilde{b}_{jk} \\ &+ \frac{4}{3}S(S+1)\sum_j \tilde{b}_{jk}^2 \\ &+ \frac{9}{7}[12I(I+1) - 17]N^{-1}\sum_j E_j^2 \ . \end{aligned} \quad (5.6)$$

Similarly, the computed NAR2 second moment is [5.4]

$$\begin{aligned}\langle \hbar^2 \Delta\omega^2 \rangle_{\text{av NAR2}} =& 2I(I+1)\sum_j a_{jk}^2 - 4I(I+1)\sum_j a_{jk}b_{jk} \\ &+ 6I(I+1)\sum_j b_{jk}^2 + \frac{4}{3}S(S+1)\sum_j \tilde{a}_{jk}^2 \\ &- \frac{16}{3}S(S+1)\sum_j \tilde{a}_{jk}\tilde{b}_{jk} \\ &+ \frac{16}{3}S(S+1)\sum_j \tilde{b}_{jk}^2 \\ &+ \frac{144}{7}[I(I+1) - 2]N^{-1}\sum_j E_j^2 \ . \end{aligned} \quad (5.7)$$

In (5.6) and (5.7)

$$b_{jk} = \gamma_i^2 \hbar^2 r_{jk}^{-3}\left(\frac{3}{2}\cos^2\theta_{jk} - \frac{1}{2}\right) \quad (5.8a)$$

and

$$\tilde{b}_{jk} = \gamma_I \gamma_S \hbar^2 r_{jk}^{-3}\left(\frac{3}{2}\cos^2\theta_{jk} - \frac{1}{2}\right) \ . \quad (5.8b)$$

r_{jk} is the vector from the j to the k nucleus, and θ_{jk} is the angle between the r_{jk} vector and the z-direction. For comparison with (5.6) and (5.7), the second moment of the spin–photon resonance [NMR or NAR(A-R)] centered at ω_1 may be computed [5.4] using (5.1):

$$\begin{aligned}\langle \hbar^2 \Delta\omega^2 \rangle_{\text{av NMR}} =& 3I(I+1)\sum_j b_{jk}^2 + \frac{1}{3}S(S+1)\sum_j \tilde{a}_{jk}^2 \\ &+ \frac{4}{3}S(S+1)\sum_j \tilde{a}_{jk}\tilde{b}_{jk} \\ &+ \frac{4}{3}S(S+1)\sum_j \tilde{b}_{jk}^2 \\ &+ \frac{9}{5}[4I(I+1)-3]N^{-1}\sum_j E_j^2 \, .\end{aligned} \qquad (5.9)$$

It is instructive to compare the properties of the second moments given in (5.6), (5.7), and (5.9) for resonance lines observed at the same frequency. We neglect in the following comparisons the cross–terms containing $a_{jk}b_{jk}$ and $\tilde{a}_{jk}\tilde{b}_{jk}$.

(1) Dipole–dipole second moments for NMR and NAR1 are identical.
(2) For dipole–dipole interactions, the ratio of NAR1 to NAR2 second moments is two for like spins, one for unlike spins.
(3) For NMR and NAR(A-R), exchange second moments are zero for like spins, nonzero for unlike spins.
(4) The ratio of NAR1 to NAR2 second moments for exchange interactions is four for like spins, one for unlike spins.
(5) The quadrupole second moments for NAR and NMR obey the following inequality: NAR1 > NAR2 > NMR.

5.2.2 Second Moment Analysis in Rhenium Metal

In Section 3.6, the NAR pure quadrupole resonance experiment in rhenium metal [5.5] is discussed. The coupling between acoustic waves and the nuclear spins characterized by the quadrupole–split energy levels is via the dynamic nuclear electric quadrupole interaction; NAR1 and NAR2 resonances are observed for both the ^{185}Re and ^{187}Re nuclear spin systems. For both nuclear spin systems, the line shapes and line widths are the same under the same experimental conditions. Surprisingly, a first derivative of a Gaussian line shape can be fit to the experimental first derivative (see Fig. 5.1) of the NAR1 and NAR2 resonances of both spin systems. (A Gaussian line shape is the result of the pseudo–exchange interaction between a measured nuclear spin and the possible large number of nuclear spin orientations of first and second neighbors.) Because of the low (10/1) signal-to-noise ratio after 48 hours of signal averaging, the method of analysis is to fit a first derivative of a Gaussian shape function to each measured resonance and to determine the resonance second moment from this Gaussian fit. (The propagation direction and the angle θ, which is between the propagation direction and the direction of the small external magnetic field H, are specified by Fig. 3.11.)

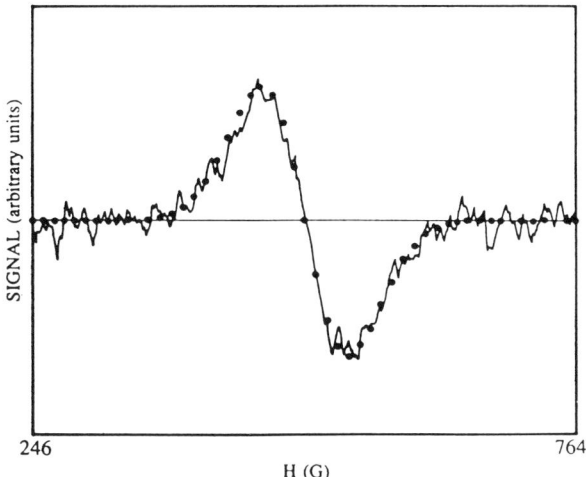

FIGURE 5.1. ^{187}Re NAR2 $\nu_{\beta'}$ transition first derivative absorption signal at $\theta = 90°$, 48 h signal averaging, 77.8 K, 500 G sweep in 5 min, 1.25 sec time constant, 34-G amplitude modulation at 37 Hz, and 39.117400 MHz presented as a solid line. The dots are a theoretical Gaussian first-derivative, plotted using the experimental peak-to-peak width, peak-to-peak amplitude, and resonance center as input parameters. (From Ref. 5.5)

At 77.8 K and $\theta = 90°$, the ^{187}Re NAR2 first derivative of the absorption peak-to-peak width is 73±4 G, with either a 17-G or 34-G amplitude modulation field; the 34 G amplitude was used in order to obtain the largest possible signal-to-noise ratio. At $\theta = 0°$, the ^{187}Re NAR1 first derivative of the absorption line shape has a peak-to-peak width of 98±4 G; at $\theta = 43.3°$, the NAR1 first derivative has a width of 64±4 G. These line widths are independent of magnetic field from 200 to 1200 G. Second moments can be computed from the measured line widths and are given in Table 5.1.

Table 5.1

Second Moment Determinations for ^{187}Re NAR

Θ (degrees)	S_2 (G^2)
0.0	2400
43.3	1150
90.0	1350

In a nonmagnetic single crystal metal, the line broadening mechanisms, which vary with the direction of the external magnetic field, are dipole–dipole, pseudo–dipolar and quadrupole broadening. An isotropic, pseudo–exchange broadening is also present in large Z metals.

That pseudo–exchange broadening must be present may be shown by comparing the ratio of those second moments determined from experiment at two different angles with those determined from theory for dipole–dipole, pseudo–dipolar and quadrupole interactions. The dipole–dipole second moment for a hexagonal $6\bar{m}2$ symmetry crystal is given in (5.10a) and (5.10b):

$$\langle \Delta \nu^2 \rangle = \frac{3}{8} \frac{\gamma^4}{(2\pi)^4} h^2 I(I+1)(m + n\cos^2\theta + o\cos^4\theta) , \qquad (5.10a)$$

where

$$m = \frac{1}{4} \sum_k r_{jk}^{-6}(11 - 30\zeta^2 + 27\zeta^4)$$

$$n = \frac{1}{2} \sum_k r_{jk}^{-6}(-15 + 126\zeta^2 - 135\zeta^4)$$

$$o = \frac{1}{4} \sum_k r_{jk}^{-6}(27 - 270\zeta^2 + 315\zeta^4) . \qquad (5.10b)$$

r_{jk} is the position vector from the j to the k nuclear positions, and ζ is the direction cosine of the angle between the r_{jk} and [0001] directions. The sums of m, n and o have been evaluated out to the 28th shell from a reference nuclear position. In units of a^{-6}, the results are $m = 24.36$, $n = -12.13$ and $o = 13.71$. The utilized lattice constants for hexagonal Re are $a = 2.708$Å and $c = 4.373$Å. The largest dipole–dipole second moment occurs when $\theta = 0°$. With the assumption of a Gaussian line shape, the expected dipole–dipole peak-to-peak line width is 5.0 G. Since the observed peak-to-peak widths are ten times larger, the dominant broadening is not dipole–dipole broadening.

5.2 THE METHOD OF MOMENTS

The pseudo–dipolar interaction constant B_{jk} is generally defined [5.6] to include the dependence on r_{jk}. B_{jk} is expected to become smaller at large distances as r_{jk}^{-3}, but its exact dependence is not known. The ratio of the distances from a reference nuclear position to atoms in the first two neighbor shells is 0.992587. The lattice sums m, n and o for the first two shell atoms are $m = 21.16$, $n = -15.92$ and $o = 18.94$. By using these values in (5.10a–b), one can compute quantities proportional to the second moments for pseudo–dipolar broadening for the first two neighbor shells, and one can compare ratios of these quantities to the ratios found experimentally at $\theta = 0°/90°$ and at $\theta = 0°/43.3°$. A comparison of such ratios shows disagreement of the experimental second–moment ratios with those computed assuming pseudo–dipolar broadening. The presence of the pseudo–dipolar interaction, together with a smaller dipole–dipole broadening, cannot explain the observed experimental line width anisotropy.

A possible source of electric field gradient that could produce the observed anisotropy in the observed line widths is a spread in the values of the nuclear electric quadrupole coupling constant due to crystal imperfections. In this case, the electric field gradient is along the [0001] direction, and the line broadening is proportional to $(3\cos^2\theta - 1)$. At $\theta = 54.75°$, this expression is zero, and the observed line width should reduce to the dipole–dipole width of 5.0 G if only the dipole–dipole and quadrupole broadenings are present. At $\theta = 54.7°$, the experimental absorption derivative peak-to-peak value is in the 75 to 80 G range.

One concludes that, in addition to dipole–dipole broadening, there must be pseudo–exchange with possible pseudo–dipolar and/or quadrupolar broadening. Which of these broadening interactions is dominant can be determined in the following way. We neglect the dipole–dipole interaction and equate the second moments determined from experiments at $\theta = 0°$, $\theta = 43.3°$ and $\theta = 90°$ to sums of second moments: (i) pseudo–exchange and quadrupole; (ii) pseudo–exchange and pseudo–dipole; (iii) pseudo–exchange, pseudo–dipole and quadrupole. For the case (i) of pseudo–exchange and quadrupole, one writes

$$S_2 = C^2 + D^2(3\cos^2\theta - 1)^2 , \qquad (5.11)$$

where S_2 is the second moment at the angle θ, C^2 is a constant that represents the pseudo–exchange second moment, and D^2 is a constant that represents the static quadrupole second moment. The values of C and D that give the best agreement with the second moment data derived from experiments at the three angles are $C = 32 \pm 2$ G and $D = 19 \pm 2$ G. A similar comparison for cases (ii) and (iii) give inconsistent results. It follows that the experimental line widths can be explained by a large isotropic pseudo–exchange (nuclear magnetic dipole–dipole exchange or nuclear electric quadrupole–quadrupole exchange) broadening; a smaller, anisotropic static nuclear electric quadrupole broadening with the electric field gradient in the [0001] direction; and the much smaller anisotropic dipole–dipole broadening. The fact that the experimental line shapes are close to Gaussian line shapes is consistent with the dominance of the pseudo–exchange interaction.

5.3 Generalized Bloch Equations

We now take up in some detail the interactions, represented earlier by $\mathcal{H}'(t)$ in (2.18), that are responsible for relaxation effects and for the line shape of the observed nuclear acoustic resonance spectrum. The approach generally taken to incorporate relaxation and line width effects into the formulation of magnetic resonance is to assume the validity of the Boltzman transport equation applied to spins, the Bloch Equations [5.1,5.2].

We recall that a spin $I = \frac{1}{2}$ can in general be characterized by three vector operators: I_z, $I_+ = I_x + iI_y$ and $I_- = I_x - iI_y$. For a system with many such spins, one uses the corresponding values M_z, M_+ and M_- for the total magnetization of the system at time t. For an external magnetic field applied in the z–direction, one obtains the independent equations of motion, or Bloch Equations:

$$\frac{d}{dt}\langle M_z \rangle = -\Gamma_z(\langle M_z \rangle - \langle M_z \rangle_0) \tag{5.12a}$$

$$\left(\frac{d}{dt} + i\omega_0\right)\langle M_\pm \rangle = -\Gamma_\pm \langle M_\pm \rangle . \tag{5.12b}$$

The term containing ω_0 ($= -\gamma H_0$) describes the precession of the transverse magnetization; Γ_z and $\Gamma_\pm$, designated equivalently as $1/T_1$ and $1/T_2$, are the spin–lattice and spin–spin relaxation rates, respectively; $\langle M_z \rangle_0$ is the instantaneous local equilibrium value of the longitudinal magnetization M_z; the corresponding equilibrium value for M_+, and for M_-, is zero. The Bloch Equations are valid in the high temperature limit $\hbar\omega_0 \ll k_B T$ (is applicable to nuclear spin systems down to millidegrees K).

For a system of particles with $I > \frac{1}{2}$, it no longer suffices to write (5.12) with only two relaxation rates. To describe the quadrupolar relaxation processes, therefore, it is necessary to introduce additional relaxation rates or relaxation times. A closed and compact form of the equations of motion for arbitrary I is obtained by applying the irreducible spherical tensor operators of the nuclear multipole moments (see Section 3.1), yielding for the $(2I + 1)^2$ independent generalized Bloch Equations [5.7]:

$$\left(\frac{d}{dt} + im\omega_0\right) A_{lm} = -\Gamma_{lm}(\langle A_{lm} \rangle - \langle A_{lm} \rangle_0) , \tag{5.13}$$

where $\langle A_{lm} \rangle_0$ is the instantaneous local equilibrium value of A_{lm}. The derivation of (5.13) holds, again, in the high temperature limit and relies on the conditions for orthonormality of the A_{lm} (3.6). A physical interpretation of (5.13) proposes that each component $\langle A_{lm} \rangle$ corresponds to an independent normal mode of the spin system and relaxes at a rate Γ_{lm}, where $\Gamma_{lm} = \Gamma_{l-m}$. The $\langle A_{lm} \rangle$ may be visualized as normal modes of vibration of a spherical shell.

In the above formalism, the relaxation rates Γ_{lm} may represent interactions of the spin system with an external system, in contrast to the usual spin–spin interaction within a spin system. An example, to be discussed in more detail in Section 5.6, is a transition metal—such as tantalum—in which spin–spin relaxation may occur as a result of rapidly-diffusing hydrogen ions.

5.4 Quadrupole–Split High Field Case: Hamiltonian, Energy Levels, Transitions

The simplest application of the formulation outlined in Sections 3.1 and 5.3 is to the case of a spin $\frac{1}{2}$ system in which the only externally applied magnetic field is $\boldsymbol{H}_0$. This yields a Hamiltonian:

$$\mathcal{H} = -\hbar\gamma H_0 I_z = -\hbar\omega_0 I_z \ . \tag{5.14}$$

In the event the equations of motion for the A_{lm} are decoupled, we may speak of normal modes of the system. The resonant frequencies of the normal modes are readily found from the Heisenberg equation of motion [5.8]:

$$i\hbar\frac{\partial}{\partial t}A_{lm} = [A_{l,m}, \mathcal{H}] = \hbar\omega_0 m A_{lm} \ . \tag{5.15}$$

If we use a harmonic time dependence $e^{-i\omega_{lm}t}$ and substitute into (5.15),

$$\hbar\omega_{lm}A_{lm} = \hbar\omega_0 m A_{lm}$$

or

$$\omega_{lm} = m\omega_0 \ . \tag{5.16}$$

For this, simple case, we can make the analogy to the ringing of a perfect metal sphere in classical physics. The surface modes of the sphere are described by the spherical harmonic functions Y_l^m, with the frequencies ω_{lm} and decay rates Γ_{lm}. Because we started with a perfect sphere, the ringing modes are necessarily decoupled. Generalizing, the A_{lm} correspond to modes of frequency ω_{lm} with decay (or relaxation) rates Γ_{lm}.

The Hamiltonian, of course, must be a scalar quantity. All nuclear spins have a dipole moment that couples vectorially to the magnetic field—as in (5.14)—yielding a scalar Hamiltonian. As we know, spins greater than $\frac{1}{2}$ also have an electric quadrupole moment, which, to give a scalar Hamiltonian, must couple to the electric field gradient, which is a second-rank tensor. Thus, for the case of $I > \frac{1}{2}$, one writes

$$\mathcal{H} = -\boldsymbol{\mu} \cdot \boldsymbol{H} + \sum_{m=-2}^{+2} A_{2m} V_m , \qquad (5.17)$$

where V_m is the electric field gradient tensor, $\boldsymbol{\mu}$ is the magnetic dipole moment vector, and $\boldsymbol{H}$ is the external magnetic field vector. The situation represented by (5.17) corresponds to a deformed sphere. The surface modes $\mathcal{Y}$ are still describable in terms of normal modes, but the new modes are now linear combinations of the spherical harmonics:

$$\mathcal{Y}_{dsm} = \sum_m c_{lm} Y_l^m , \qquad (5.18)$$

where the subscript *dsm* refers to *deformed spherical modes* and the c_{lm} are the appropriate expansion coefficients. We recall that, when written in terms of the conventional spin operators, the nuclear quadrupole interaction has the form

$$\mathcal{H}_Q = \sum_{i,j} \left(\frac{eQ}{4I(2I-1)} \right) V_{ij} \left[\{I_i, I_j\} - \frac{2}{3} I(I+1)\delta_{ij} \right] , \qquad (5.19)$$

where

$$V_{ij} = \frac{\partial^2 V}{\partial x_i \partial x_j} .$$

V is the electric potential, and $\{I_i, I_j\}$ is the anticommutator. We know from our earlier treatment (Chapter 3) that only five components of V_{ij} and of the second bracket are independent. The Zeeman interaction can always be written in terms of only one operator, I_z, by rotating the coordinates, such that $\boldsymbol{H} = H\hat{\boldsymbol{z}}$. The quadrupole Hamiltonian, on the other hand, cannot in general be written in terms of one operator such as $\mathcal{H}_Q = \hbar\omega_q[I_z^2 - (1/3)I(I+1)]$, but must mix several operators. However, in the high-field case we are treating, the quadrupole frequencies ω_q are much lower than the Larmor frequencies, $\omega_q \ll \omega_0$. It is thus convenient to choose coordinates $\boldsymbol{H} = H\hat{\boldsymbol{z}}$ to diagonalize the Zeeman part of the Hamiltonian of (5.17). Since the general equation for $\mathcal{H}_Q$ is in terms of the crystallographic

5.4 QUADRUPOLE—SPLIT HIGH FIELD CASE

axes, and since the external magnetic field H_0 is in an arbitrary direction, a rotational transformation is often necessary:

$$\mathcal{H} = -\hbar\omega_0 I_z + \hbar \sum_{m=-2}^{+2} \omega_{qm} A_{2m} , \qquad (5.20)$$

where the ω_{qm} are related to the V_{ij}. We have made use of the property that rotations of the irreducible operator leave l unchanged:

$$A_{lm} \to \text{rotation} \to \sum_{lm'} A_{lm'} \, a^{(l)}_{m'} , \qquad (5.21)$$

where the general form of $a^{(l)}_{m'}$ is given in [5.3].

Because $\omega_{qm} \ll \omega_0$ in the high-field Hamiltonian (5.20), we treat $\mathcal{H}_Q$ as a perturbation. The first-order correction to the Zeeman energy levels is given by the diagonal terms $\langle m|\mathcal{H}_Q|m\rangle$. The A_{2m} are proportional to $(I_{\pm}^m)$; therefore only the $m = 0$ diagonal terms survive. Setting $\omega_{q0} = \omega_q$, we obtain from (5.20) expressions equivalent to those given in (1.21):

$$\mathcal{H} = -\hbar\omega_0 I_z + \hbar\omega_q \left(I_z^2 - \frac{1}{3} I^2 \right) \qquad (5.22a)$$

and

$$E_m = -\hbar\omega_0 m + \hbar\omega_q \left(m^2 - \frac{1}{3} I(I+1) \right) . \qquad (5.22b)$$

In Fig. 5.2, the energy levels are drawn for the cases of $I = 1$ and $I = \frac{3}{2}$ (also see Fig. 1.3). In the case of half-integer spin I, we see that the $\frac{1}{2} \leftrightarrow -\frac{1}{2}$ splitting is unchanged by the quadrupolar interaction (in the present approximation), whereas for integral I, all splittings are changed. The observed absorption lines corresponding to Fig. 5.2 are shown schematically in Fig. 5.3:

From our discussion of the Bloch Equations in Section 5.3, the absorption line corresponds to a delta function if relaxation is ignored; otherwise, it has a Lorentzian shape. If $\omega_q \neq 0$, distinct lines (> 1) are observed, provided that all atoms see the same electric field gradient.

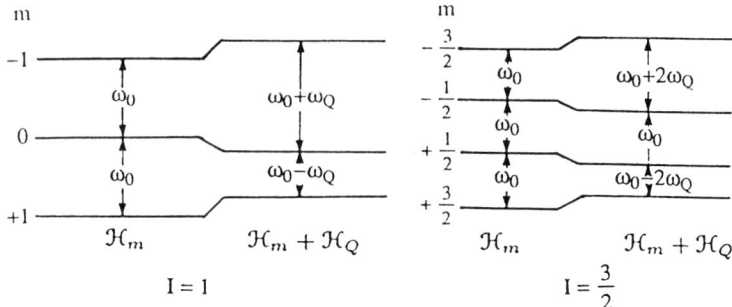

FIGURE 5.2. Energy levels for the cases of $I = 1$ and $I = 3/2$. (Also see Fig. 1.3.)

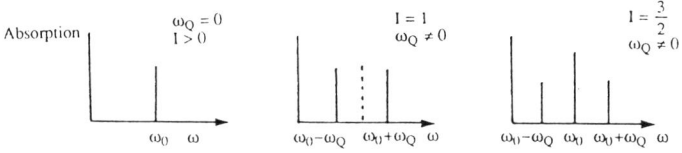

FIGURE 5.3. A schematic representation of the expected absorption lines corresponding to transitions between the energy levels of Fig. 5.2.

5.5 Line Shape of the Quadrupole–Split High Field Line: Intraspin Cross Relaxation

In this section we develop the theory capable of explaining the observed anomalies in NAR1 and NAR2 line shapes. In real solids, we must take into account (i) the fact that, due to strains, the electric field gradient differs at different lattice sites, necessitating an average over the expected distribution of electric field gradients; and (ii) the effects of relaxation, which result in Lorentzian line shapes, the widths of which may be comparable to the resonance line separations. For the case in which the relaxation is due to phonons ($\equiv$ lattice vibrations), the phonon field $\phi_m(t)$ is expressed in terms of strains $\varepsilon_{ij} = \partial u_i/\partial x_j$, which form a second–rank tensor. Because the Hamiltonian is a scalar, the spin–phonon (sp) coupling must be through

5.5 LINE SHAPE, INTRASPIN CROSS RELAXATION 177

the $l = 2$ A's:

$$\mathcal{H}_{sp} = \sum_{m=-2}^{m} A_{2m}\phi_m(t) \ . \tag{5.23}$$

As *Fedders* [5.3] has shown, the normal modes are always the A_{lm}. That is, there is no coupling between the A_{lm}'s due to the spin–phonon interaction; each A_{lm} decays at its own rate Γ_{lm}, where Γ_{lm} is the decay rate or inverse lifetime of the (lm) mode.

We first consider only spins that decay via an independent fluctuating field, and we later take spin-spin interactions into account. We assume, as above, that the static quadrupole splittings are much less than the Zeeman splitting, so that only the diagonal part of the static quadrupolar field needs to be used. For spins excited by an electromagnetic field, we use the Hamiltonian

$$\mathcal{H} = -\hbar\omega_0 I_z + \hbar\omega_q \left(I_z^2 - \frac{1}{3}I(I+1)\right) + \frac{1}{2}\hbar\omega_{11}(I_+ + I_-) \ , \tag{5.24}$$

where $\omega_{11} = \gamma H_1$, and H_1 is the amplitude of the RF magnetic field. All time-dependent quantities are assumed to vary as $e^{-i\omega t}$. In addition, we assume that the spins decay exponentially:

$$\frac{d}{dt}\langle A_{lm} \rangle = \Gamma_{lm}[\langle A_{lm} \rangle - \langle A_{lm} \rangle_0] \ . \tag{5.25}$$

Indeed, the analysis of Section 5.3 applies in the approximation that ω_q^2/ω_0 is much less than the ω_0 and the Γ_{lm}'s. For this case—applying (5.15), assuming an $e^{-i\omega t}$ time dependence and including the relaxation time—one obtains

$$\omega\langle A_{lm} \rangle = \frac{1}{\hbar}\langle [A_{lm}, \mathcal{H}] \rangle - i\Gamma_{lm}(\langle A_{lm} \rangle - \langle A_{lm} \rangle_0) \ , \tag{5.26}$$

a modified Bloch equation. In the last term, relaxation is towards a local, instantaneous equilibrium. Since we are assuming only linear response and since the driving term is proportional to A_{11} and $A_{1,-1}$, only the $\langle A_{lm} \rangle$ with $m = \pm 1$ are affected. $\langle A_{lm} \rangle$'s with different m's are not coupled.

For the terms of the commutator in (5.26), one obtains

$$[A_{l1}, -\hbar\omega_0 I_z] = +\hbar\omega_0 A_{l1}$$
$$[A_{l1}, \hbar\omega_q\{I_z^2 - (1/3)I(I+1)\}]$$
$$= \hbar\omega_q[D(l,1)A_{l-1,1} + D(l+1,1)A_{l+1,1}]$$

and

$$\left[A_{l1}, \frac{1}{2}\hbar\omega_{11}(I_+ + I_-)\right] = -\frac{1}{2}\hbar\omega_1\sqrt{l(l+1)}A_{l0}, \tag{5.27}$$

where the D's are constants that may be determined [5.3] by the Wigner–Eckart theorem [5.1]. Substituting into (5.26), one obtains

$$(\omega - \omega_0 + i\Gamma_{l1})\langle A_{l1}\rangle + \omega_q[D(l1)\langle A_{l-1,1}\rangle + D(l+1,1)\langle A_{l+1,1}\rangle]$$
$$= i\Gamma_{l,1}\langle A_{l1}\rangle_0 - \frac{1}{2}\omega_{11}[l(l+1)]^{\frac{1}{2}}\langle A_{l0}\rangle_0, \tag{5.28}$$

which is a Bloch-like equation.

As an example, we consider the electromagnetic case, where $l = 1$. For $\omega_q = 0$, (5.28) reduces to the usual Bloch Equation. For $\omega_q \neq 0$, m remains a good quantum number, and l is mixed with $l \pm 1$ by ω_q. In (5.28), the only nonzero $\langle A_{lm}\rangle_0$'s are $\langle A_{10}\rangle_0$, $\langle A_{11}\rangle_0$ and $\langle A_{20}\rangle_0$. The right-hand side of (5.28) can be computed as follows:

$$\begin{aligned}\langle A_{lm}\rangle_0 &= \frac{\text{Tr}(A_{lm}e^{-\beta\mathcal{H}})}{\text{Tr}1} \\ &\simeq \frac{\text{Tr}[A_{lm}(1 - \beta\mathcal{H})]}{2I+1} \\ &= \frac{-\beta\text{Tr}(A_{lm}\mathcal{H})}{2I+1},\end{aligned} \tag{5.29}$$

where the high-temperature approximation is used for the exponential in (5.29), where the 1 in the expansion is replaced by A_{00}, and where $\beta = 1/(k_B T)$. By using the orthonormality condition for the trace product of A_{lm} given in (3.7), we find the last equality in (5.29). In turn, $\langle A_{10}\rangle_0$ can be evaluated by recognizing that

$$I_z = \sqrt{\frac{2}{3}}A_{10} \tag{5.30}$$

and inserting (5.30) into (5.31):

$$\langle A_{10}\rangle_0 = \beta\text{Tr}[A_{10}(-\hbar\omega_0)I_z] = \sqrt{\frac{2}{3}}\beta\hbar\omega_0. \tag{5.31}$$

In a similar way, one can show that

$$\langle A_{11}\rangle_0 = \sqrt{\frac{1}{3}}\beta\hbar\omega_1 \tag{5.32}$$

5.5 LINE SHAPE, INTRASPIN CROSS RELAXATION

and

$$\langle A_{20} \rangle_0 = \frac{\sqrt{2}}{3}\beta\hbar\omega_q . \tag{5.33}$$

It is most convenient to normalize the solution to (5.28) as

$$\chi_{11}(\omega) = \frac{\sqrt{6}a_1\langle A_{11}\rangle}{\beta\hbar\omega_{11}} , \tag{5.34}$$

where χ_{11} is a generalized magnetic susceptibility, and $a_1 = [I(I+1)]^{-\frac{1}{2}}$. Fedders [5.3] defines the spectral shape function as the imaginary part of $\chi_{11}(\omega)/\omega$, which is proportional to the experimentally observed line shape. Equation (5.28) comprises a set of $2I$ coupled equations that may be conveniently written in matrix form as

$$[\alpha] = [M]^{-1} \cdot [f] , \tag{5.35}$$

where $[\alpha]$ and $[f]$ are $2I \times 1$ column matrices.

For the choice of $I = 1$, there are two elements: $l = 1$ and $l = 2$ for

$$[\alpha]_{l1} = A_{l1}(\omega) \tag{5.36}$$

and

$$[f]_{l1} = (3^{-\frac{1}{2}}\omega_{1,1}\beta\hbar)[(-\omega_0 + i\Gamma_{11})\delta_{l1} + \omega_q\delta_{l2}] . \tag{3.37}$$

The quantity $[M]$ is a 2×2 symmetric matrix whose only nonzero elements are

$$[M]_{ll} = \omega - \omega_0 + i\Gamma_{l1} \tag{5.38}$$

and

$$[M]_{12} = [M]_{21} = \omega_q . \tag{5.39}$$

From the solution of the matrix equation (5.35), we find, using (5.34), that

$$\chi_{11}(\omega) - 1 = -\frac{\omega\Omega_{21}}{D_{11}} , \tag{5.40}$$

where
$$\Omega_{lm} = \omega - m\omega_0 + i\Gamma_{lm} \tag{5.41}$$

and
$$D_{11} = \Omega_{11}\Omega_{21} - \omega_q^2 . \tag{5.42}$$

There are often static quadrupole fields due to a random distribution of defects in solids. A small concentration of defects usually leads to a Lorentzian distribution of field gradients [5.9]. The theoretical susceptibilities needed to compare with experiment are therefore given by (5.40) averaged over a Lorentzian distribution of ω_q. We average (5.40) over the function

$$\frac{\omega_1}{\pi(\omega_q - \omega_2)^2 + \omega_1^2} , \tag{5.43}$$

where $\omega_1 = \omega_{11}$, and $\omega_{21} = \omega_2$. With this average, (5.40) becomes

$$\chi_{11}(\omega) - 1 = -\left(\frac{\omega\Omega_{21}}{2b_1}\right)\left(\frac{1}{b_1 + i\omega_1 - \omega_2} + \frac{1}{b_1 + i\omega_1 + \omega_2}\right), \tag{5.44}$$

where
$$b_1 = (\Omega_{11}\Omega_{21})^{\frac{1}{2}} . \tag{5.45}$$

The imaginary part of (5.44) divided by ω is proportional to the experimental absorption signal. In taking the imaginary part of (5.44), cross–terms involving the products $\omega_1\omega_2$ are created; this is the source for $I = 1$ electromagnetic coupling intraspin cross relaxation. The computed imaginary part of $\chi_{11}(\omega)/\omega$ is shown in Fig. 5.4. The high frequency side of the resonance is shown with the resonance center on the left of each panel in the figure. We observe that in each panel of Fig. 5.4, intraspin cross relaxation is most noticeable near the resonance center, but that its effects are produced in the tails of the resonance as well. One of the most striking effects is the dip at the resonance center in the right–hand panels.

The $I = 1$ spectral shape function discussed above is, in fact, a special case. For other spins, the spectral shape function contains intraspin cross relaxation terms, even without assuming a Lorentzian distribution of electric field gradients.

5.5 LINE SHAPE, INTRASPIN CROSS RELAXATION

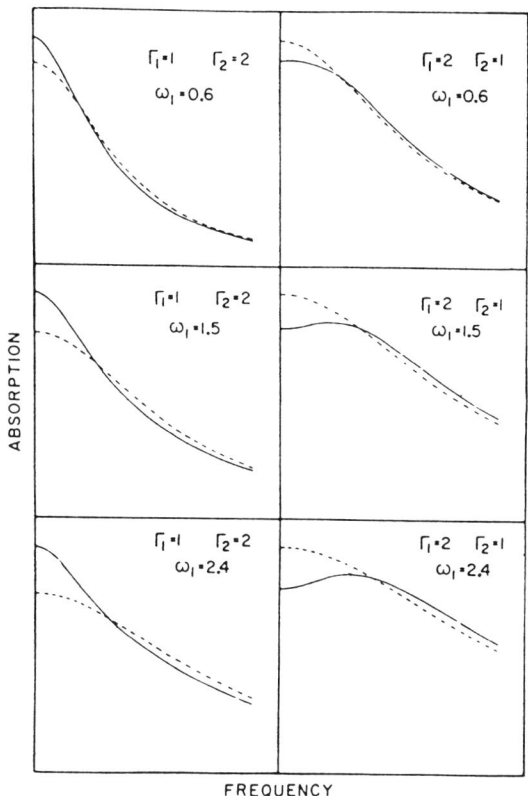

FIGURE 5.4. Imaginary part of $\chi_{11}(\omega)/\omega$ in arbitrary units versus $\omega - \omega_0$ for different values of Γ_1, Γ_2 and ω_1. All frequencies are in the same units, and the length of the horizontal x-axis is 4.5 of these units. The solid curve is computed including intraspin cross relaxation, while the dashed curve is computed without including intraspin cross relaxation. (From Ref. 5.3)

The equations for acoustically excited spin systems are quite similar to those for the electromagnetic case. For acoustic $\Delta m = 1$ transitions, we replace (5.24) with

$$\mathcal{H} = -\hbar\omega_0 I_z + \hbar\omega_q \left(I_z^2 - \frac{1}{3}I^2\right)$$

$$+ \frac{1}{2}\hbar\omega_{21}(\{I_+, I_z\} + \{I_-, I_z\}),\tag{5.46}$$

while for the $\Delta m = 2$ transitions, we replace (5.24) with

$$\mathcal{H} = -\hbar\omega_0 I_z + \hbar\omega_q \left(I_z^2 - \frac{1}{3}I^2\right) + \frac{1}{2}\hbar\omega_{22}(I_+^2 + I_-^2).\tag{5.47}$$

The values ω_{21} in (5.46) and ω_{22} in (5.47) are frequencies related to the dynamic electric quadrupole interaction, which are identified in *Fedders and Lu* [5.7]. The solutions of (5.26) with the Hamiltonians of (5.46) and (5.47) yield a set of $2I - 1$ equations similar to those developed in (5.26–5.45). With these solutions, the effects of intraspin cross relaxation for dynamic quadrupole coupling can be investigated.

The results developed in this section, which are applied in Section 5.6.2, depend on two critical assumptions:

(1) that ω_q/ω_0 is small enough that only the diagonal part of the static quadrupole Hamiltonian is needed;
(2) that the spin decay rates Γ_{lm} are not affected by the static quadrupole Hamiltonian.

If we assume that the decay of the spins is dominated by their interaction with some independent fluctuating field, then the Γ_{lm} can only depend on the spin Hamiltonian via the spin energy levels [5.7]. In this case, the Γ_{lm} are virtually unchanged by the static quadrupole Hamiltonian, as long as $\omega_0 \gg \omega_q$. The intraspin cross relaxation is Γ_{lm}, and the second-order quadrupole splitting is proportional to ω_q^2/ω_0. Therefore the effects of intraspin cross relaxation dominate second-order quadrupole splitting when $\Gamma_{lm} \gg \omega_q^2/\omega_0$.

In the presence of spin–spin interactions, the situation is made more complex, because the fluctuating field that each spin experiences is due to other spins, not to an independent mechanism. We must solve this case self–consistently for the spectral shape function. *Myles and Fedders* [5.10] have developed a method for generating integral equations for spin spectral functions. This method can be generalized to include both cases of spin-spin interactions and static fields. We discuss this generalized method of Myles and Fedders in Section 5.6.

5.6 Spin Relaxation by Dynamic Quadrupole Interaction

5.6.1 General Theory of Relaxation

Fluctuating electric field gradients cause nuclear spins ($I > \frac{1}{2}$) to relax. These fluctuating electric field gradients can be due to phonons, conduction electrons or other charged mobile carriers. An effective spin–phonon Hamiltonian based on quadrupole electric field gradients has been used to calculate electron spin relaxation rates for electromagnetic probes [5.7]. The response of the same system, in general, to an acoustic probe will be different. A general theory of spin relaxation via quadrupole electric fields due to either acoustic or electromagnetic probes, in which the results are expressed in terms of correlation functions of the quadrupole electric field gradient, has been developed [5.11]. This general theory allows a determination of the relaxation rates appropriate for both electromagnetic and acoustic spin resonance experiments on nuclear and electron spins.

In describing the dynamics of the spins, we use the irreducible tensor operators A_{lm} for a spin of magnitude I at a given site. Electromagnetic fields couple through the $l = 1$ A_{lm}, acoustic waves through the $l = 2$ A_{lm}. For simplicity, we use the short hand notation $A_\alpha = A_{lm}$, $\alpha = (lm)$. We assume the high temperature limit, in which all the spin energies are small compared to $k_B T$.

We define a set of diagonal two–point, time–dependent correlation functions

$$G_\alpha(t) = \langle A_\alpha(t) A_\alpha^+(0) \rangle \Theta(t)$$
$$G_\alpha(\omega) = \int_{-\infty}^{\infty} e^{-i\omega t} G_\alpha(t)\, dt , \qquad (5.48)$$

where the $\langle x \rangle$ denotes the average of x in the canonical ensemble, and $\Theta(t)$ is the step function. The $G_\alpha(\omega)$ can be related to other frequency–dependent correlation functions such as susceptibilities or structure functions by means of a fluctuation dissipation theorem [5.12]. As an example, the components of the usual dynamic electromagnetic susceptibility are proportional to G_{1m}, while the acoustic susceptibility due to dynamic quadrupole (or spin–phonon) coupling are proportional to G_{2m}. The complex self–energy $\Sigma_\alpha(\omega)$ is defined implicitly through

$$G_\alpha(\omega) = \frac{i}{\omega - \omega_\alpha - \Sigma_\alpha(\omega)} , \qquad (5.49)$$

where

$$\Sigma_\alpha(\omega) = \Pi_\alpha(\omega) - i\Gamma_\alpha(\omega) . \qquad (5.50)$$

In the above equations, $\Gamma_\alpha(\omega)$ is a frequency-dependent line width or decay rate; $\Pi_\alpha(\omega)$ is a frequency-dependent frequency shift, which in practice is almost always negligibly small. The frequency is $\omega_\alpha = m_\alpha \omega_0$, where $\alpha = (l_\alpha\, m_\alpha)$ and $\omega_0 = \gamma H_0$. If $\Gamma_\alpha(\omega)$ is a slowly varying function of ω, as is always the case for quadrupole coupling, then $\Gamma_{10}(\omega_0)$ is the decay rate for the z-components of the magnetization, usually denoted by $1/T_1$. Likewise, $\Gamma_{11}(\omega_0)$ is the decay rate for the transverse components of magnetization, usually denoted by $1/T_2$. In the acoustic ($l = 2$) case, there is no such well-recognized notation, and we label the relaxation rates by $\Gamma_{2m}(m\omega_0)$. The quantity $\Gamma_{22}(2\omega_0)$ refers to the decay rate of $\langle A_{22} \rangle$ or the line width of the $\Delta m = \pm 2$ resonance, while $\Gamma_{21}(\omega_0)$ refers to the decay rate of $\langle A_{21} \rangle$ or the line width of the $\Delta m = \pm 1$ resonance.

Following earlier arguments [5.13], one can write a very general expression for the decay rate:

$$\Gamma_\alpha(\omega) = \frac{1}{2\hbar^2} \int_{-\infty}^{\infty} dt\, e^{-i\omega t} \frac{\langle [A_\alpha(t), \mathcal{H}'][A_\alpha^+(0), \mathcal{H}'] \rangle}{\langle |A_\alpha|^2 \rangle} , \qquad (5.51)$$

where $\mathcal{H}'$ is the interaction Hamiltonian. To lowest order in $\mathcal{H}'$, we assume that the thermal average of the product of the spin and quadrupole field operators is equal to the product of the thermal average of the spin operators times the thermal average of the quadrupole operators. The correctness of this assumption is discussed by *Myles and Fedders* [5.10].

We consider a set of noninteracting nuclear or electronic spins in an external magnetic field $\boldsymbol{H}_0$, which, in terms of spherical polar coordinates, is at an angle determined by θ and ϕ relative to the coordinate system of the lattice. The spins are coupled to dynamically fluctuating quadrupole electric field gradients by the quadrupole interaction given in (5.19). This equation can be rewritten in terms of the irreducible tensor operators, with $\boldsymbol{H}_0$ defining the z-axis and with

$$V_{ij} = \frac{\partial^2 V}{\partial x_i \partial x_j}$$

5.6 SPIN RELAXATION BY DYNAMIC QUADRUPOLE FIELDS

still in the coordinate system of the lattice:

$$\mathcal{H}_Q = \sum_{i,j;m} \frac{A}{2} V_{ij} d(i,j;m) A_{2m} , \qquad (5.52)$$

where $d(i,j;m)$ is the transformation tensor that transforms A_{2m} to the lattice coordinate system [5.11]. After evaluating the commutators in (5.51) and using the independence of the thermal averages of the spin and quadrupole field, we find

$$\Gamma_\alpha(\omega) = \Re \int \frac{d\omega'}{2\pi} \sum_{iji'j'm\gamma} \left(\frac{A}{2\hbar} C(\alpha 2 m\gamma)\right)^2$$
$$\times d(i,j;m) d^*(i',j';m) F_{iji'j'}(\omega - \omega') G_\gamma(\omega') , \qquad (5.53)$$

where the constant C is given by *Fedders* [5.11], and the value γ is the index for the summation expression for the transformation of A_{2m}. $F_{iji'j'}(\omega)$ is the Fourier transform of the electric field gradient correlation function

$$F_{iji'j'} = \langle V_{ij}(t) V_{i'j'}(0) \rangle \Theta(t) . \qquad (5.54)$$

For situations of high symmetry, (5.54) can be considerably simplified. For cubic symmetry, and by using the fact that $\nabla^2 V = 0$, it is easy to show that

$$F_{iiii}(\omega) = \frac{2}{3} F_1(\omega) \qquad (5.55)$$

$$F_{iijj}(\omega) = -\frac{1}{3} F_1(\omega) \qquad i \neq j \qquad (5.56)$$

and

$$F_{ijij}(\omega) = F_{ijji}(\omega) = \frac{1}{2} F_2(\omega) , \qquad i \neq j . \qquad (5.57)$$

All other $F_{ijkl} = 0$. In general, the F terms are slowly varying in the frequency range where $G_\gamma(\omega)$ is large (within the range $|\omega - \omega_0| < \Gamma_\alpha$). In this case, the integration can be carried out, and for a cubic crystal, one obtains

$$\Gamma_\alpha(\omega) = \Re \sum_{i,j;m\gamma} \left(\frac{A}{2\hbar} C(\alpha 2 m\gamma) |d(i,j;m)|^2 \right.$$
$$\left. \times [\delta_{ij} F_1(\omega - \omega_\gamma) + (1 - \delta_{ij}) F_2(\omega - \omega_\gamma)]\right) . \qquad (5.58)$$

The quadrupole electric field correlation function $F(\omega)$ has real and imaginary parts:

$$F(\omega) = F''(\omega) - iF'(\omega) , \qquad (5.59)$$

where F' is an odd function of ω, and F'' is an even function of ω. If we assume the characteristic frequencies of $F(\omega)$ are much greater than ω_0, then F'' can be replaced by $F''(0)$ unless $F''(\omega)$ vanishes. If we assume a quadrupole field due to harmonic phonons, $F''(0)$ does not vanish, since the density of states is proportional to ω^3, and (5.58) can be written as

$$\Gamma_\alpha(\omega) = \sum_{i,j;m\gamma} \left(\frac{A}{2\hbar} C(\alpha 2 m \gamma)^2 |d(i,j;m)|^2 \right.$$
$$\left. \times [\delta_{ij} F_1''(\omega = 0) + (1 - \delta_{ij}) F_2''(\omega = 0)] \right) . \qquad (5.60)$$

5.6.2 Application of the Relaxation Theory to the Tantalum Hydrogen System

For the case of the tantalum single crystal with hydrogen impurities, $F_i''(0)$ have been calculated [5.14]. At room temperature in tantalum metal in the α phase, in which the atomic concentration of hydrogen impurities is less than 1%, the interstitial hydrogen atoms lose their electrons to the host conduction band and move as hydrogen ions from one interstitial site in the tantalum lattice to another. The interstitial sites that are closest to a given tantalum atom occupy sites with tetrahedral symmetry at a distance of r_n given by

$$r_n = \frac{\sqrt{5}}{4} a_0 , \qquad (5.61)$$

where a_0 is the lattice constant for the tantalum single crystal. There are 24 such first–neighbor interstitial sites. The second–neighbor sites, also 24 in number, are located a distance of $(\sqrt{13}/4)a_0$. We expect that $F_i''(0)$ should depend on the distance r and go as r^{-6}. This means that the contribution from second–neighbor interstitial sites to $F_i''(0)$ at a particular tantalum sites is 1/20th the contribution from the first–neighbor sites.

We next compute $F_{iji'j'}$ from the field gradient due to a hydrogen ion present at a first–neighbor tetrahedral site. The electric field is that due to a single electron, and we denote the proper quadrupole coupling constant by $\Omega_n = (eQ/\hbar)(eq)_n$. Thus when a hydrogen ion occupies interstitial site f, the component of the electric field gradient is

$$V_{ij} = -\frac{1}{2}(eq)_n (3\gamma_{if}\gamma_{jf} - \delta_{ij}) , \qquad (5.62)$$

5.6 SPIN RELAXATION BY DYNAMIC QUADRUPOLE FIELDS

where γ_{if} is the direction cosine from the fth interstitial site to the ith lattice direction. The hydrogen ions jump from one interstitial site to another, changing the direction and magnitude of the electric field gradient (EFG) at the tantalum site. We describe the time dependence of this ion jumping by $\zeta(t)$ (of order 1) and write the time dependent field gradient as

$$V_{ij}(t) = -\frac{1}{2}(eq)_n \sum_f (3\gamma_{if}\gamma_{jf} - 1)\zeta_f(t) . \quad (5.63)$$

The correlation function $G(fgt)$ can be written

$$G(fgt) = \frac{6}{c}\langle \zeta_f(t)\zeta_g(0)\rangle , \quad (5.64)$$

where $G(fgt)$ is normalized to 1 if one finds a single hydrogen ion in site f at time t, and in site g at time $t = 0$. The 6 is related to the first–neighbor site number, and c is the hydrogen ion concentration. It is now possible to write down the Fourier transform of the real part of the EFG fluctuation spectral function as

$$F''_{iji'j'}(\omega = 0) = \frac{c}{6}\mathrm{Tr}\left(\frac{(eq)_n^2}{4}f_{iji'j'}(\omega = 0)\right) , \quad (5.65)$$

where

$$f_{iji'j'}(\omega = 0) = \frac{1}{\mathrm{Tr}}\int_0^\infty dt \left[\sum_{fg} G(fgt)(3\gamma_{if}\gamma_{jf} - \delta_{ij})\right.$$
$$\left.(3\gamma_{ig}\gamma_{jg} - \delta_{i'j'})\right] . \quad (5.66)$$

The magnitude of $f_{iji'j'}$ depends on the geometry of the interstitial sites and the jump rate. If we assume spherical symmetry on the average over all the tantalum sites, the correlation function becomes

$$G(fgt) = e^{-t/\tau_r}\delta_{fg} \quad (5.67)$$

and

$$f_1(\omega = 0) = f_2(\omega = 0) = \frac{144}{5} . \quad (5.68)$$

Finally, the relaxation rates can be computed as

$$\Gamma_{1m} = \frac{6(2I+3)}{5I^2(2I-1)^2}c\tau_r\Omega_n \quad (5.69)$$

and
$$\Gamma_{2m} = \frac{18(4I^2 + 4I - 7)}{5I^2(2I-1)^2} c\tau_r \Omega_n \ . \tag{5.70}$$

Since the jump rate increases with increasing temperature, we expect that relaxation rates also increase, resulting in a narrowing of the tantalum magnetic resonance line. This is observed experimentally [5.15].

5.6.3 Explanation of the Anomalous Tantalum NAR1 and NAR2 Line Shapes in the Tantalum–Hydrogen System

We can now explain the experimentally observed anomalous NAR1 and NAR2 line shapes in tantalum when it is in the α phase and there is a fluctuating electric field gradient due to hydrogen ions jumping between interstitial tetrahedral sites in the tantalum lattice. In most cases for the tantalum–hydrogen system, the observed first derivative of the absorption line is displayed and used for computation, because the presence of line broadening in the tails of the resonance makes accurate integration of the line shape first derivative impossible.

If we assume that the dynamic electric field gradient correlation functions are spherically symmetric, then the multipole relaxation rates Γ_{lm} depend only on l and can be denoted as Γ_l. *Fedders* [5.11] shows that for $I = \frac{7}{2}$ (^{181}Ta), $\Gamma_2 = 2.8\Gamma_1$, $\Gamma_3 = 5\Gamma_1$, $\Gamma_4 = 7\Gamma_1$, $\Gamma_5 = 8\Gamma_1$, $\Gamma_6 = 7\Gamma_1$ and $\Gamma_7 = 2.8\Gamma_1$. In the absence of any other mechanisms, NMR and NAR(A-R) coupling ($l = 1$) would give a Lorentzian line shape with a line width, in frequency units, of Γ_1; for the case of NAR1 and NAR2, coupling is via the $l = 2$ quadrupole moment with the relaxation rate (in frequency) $\Gamma_2 = 2.8\Gamma_1$. Thus both NAR1 and NAR2 line widths should be the same when measured in frequency units.

It is well known that the tantalum–hydrogen system in the α phase can have static–charged defects in addition to the mobile hydrogen ions. Because of the large quadrupole moment of tantalum (3.9 barn) and the electric field gradients produced by these defects, one no longer has the spherical symmetry assumed earlier. We assume a Lorentzian distribution of static electric field gradients, with a Lorentz width ω_q and an average shift of the mth energy level by $m^2\omega_q$. It is the presence of the static–charged defects that produces the observed differences between the NAR1 and the NAR2 line shapes. *Fedders* [5.11] computes the line shapes for the spin modes ($l = 2, m = 2$) and ($l = 2, m = 1$) appropriate for the NAR2 and the NAR1 line shapes. These shapes depend strongly on the ratio ω_q/Γ_1, which makes it impossible to predict a general line shape. Theoretical first–derivative line shapes are compared with observed first–derivatives by comparing position of maximum and minimum as a function

5.6 SPIN RELAXATION BY DYNAMIC QUADRUPOLE FIELDS

of ω_q/Γ_1. Since the line shapes are antisymmetric about the resonance center, one measures the position of the extrema from the resonance center in the direction of increasing frequency. Although one expects only one derivative minimum and no derivative maximum, in practice this is contravened because of intraspin cross relaxation.

In the case of the NAR1 first–derivative line shape, plotted in Fig. 5.5, there is both a maximum and a minimum. The maximum in the first derivative corresponds to a dip in the absorption spectrum. We denote the distance in Fig. 5.5 from the resonance center to the maximum divided by Γ_1 as D_{1d}; likewise, the distance from the resonance center divided by Γ_1 is denoted D_{1b} (b for broad). The derivative of the NAR2 spectra, plotted from the center of the resonance toward high frequency, may have one or two minima. The distance from the resonance center to the first minimum divided by Γ_1 is denoted D_{2n} (n for narrow); and the distance from the resonance center to the second minimum divided by Γ_1 is denoted D_{2b}.

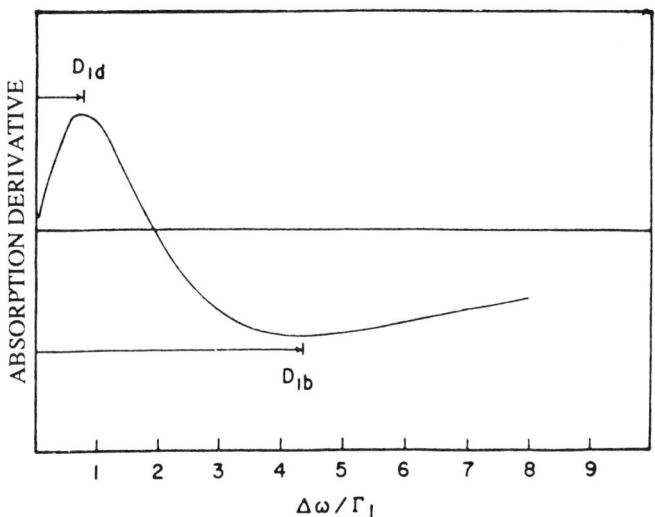

FIGURE 5.5. Typical NAR1 absorption first–derivative spectrum. This example corresponds to $\omega_q/\Gamma_1 = 1$. (From Ref. 5.9)

The results of calculations utilizing (5.24–5.47) are given in Figs. 5.6 and 5.7, which display D_{1d}, D_{1b}, D_{2n} and D_{2b} as functions of ω_q/Γ_1. For small electric field gradients, NAR1 has only a minimum; it acquires a maximum only if $\omega_q > 0.2\Gamma_1$. Analogously, for NAR2 the derivative line acquires a second minimum only if $\omega_q > 13\Gamma_1$. We note that D_{1d} and D_{2n} are approximately independent of ω_q, while D_{1b} and D_{2b} are approximately proportional to ω_q for large ω_q. D_{1b} and D_{2n} are often used to compare NAR lines and their ratio is also plotted in Figs. 5.6 and 5.7.

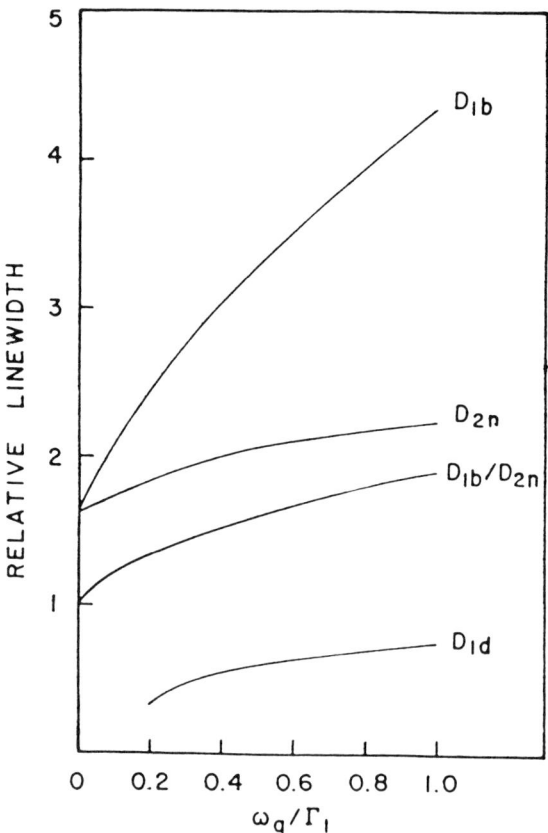

FIGURE 5.6. Positions of NAR1 and NAR2 absorption first-derivative extrema for $\omega_q < \Gamma_1$. The D's are defined in the text. (From Ref. 5.9)

5.6 SPIN RELAXATION BY DYNAMIC QUADRUPOLE FIELDS

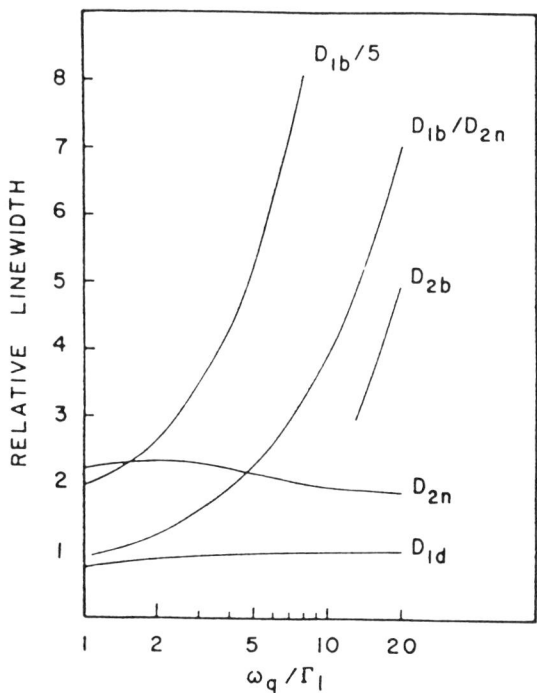

FIGURE 5.7. Positions of NAR1 and NAR2 absorption first-derivative extrema for $\omega_q > \Gamma_1$. The D's are defined in the text. (From Ref. 5.9)

The theoretical calculations can be compared with the experimental NAR investigations [5.16] in the temperature range from 200 K to 250 K, where the intrinsic decay is expected to be dominated by the mobile hydrogen ions. The results of the comparison are given in Table 5.2. The measured line widths of the NAR2 resonance, the assumption of a single value of ω_q for all temperatures and the results of Figs. 5.6 and 5.7 are the required inputs for the theoretical values relative to the NAR1 line shape and line width in Table 5.2. The agreement between theory and experiment is within 15%, which is as accurately as the pertinent experimental data can be determined.

Table 5.2

Experimental NAR1 and NAR2 Line Widths and
NAR 1 Theoretical Line Shape Properties in Tantalum

Temperature (K)	200	217	233	250
NAR2, (experimental, in G)	200.	140.	100.	73.
NAR1, broad line width (experimental, in G)	700.	500.	454.	414.
NAR1, broad line width (theoretical, in G)	664.	605.	444.	359.
NAR1, dip line width (experimental, in G)	120.	86.	66.	60.
NAR1, dip line width (theoretical, in G)	124.	90.	70	57.
ω_q/Γ_1	0.70	0.80	1.45	2.3

5.7 Dipolar and Quadrupolar Exchange Contributions to NAR Line Shapes

In high magnetic field NAR experiments with dynamic quadrupole coupling, one of the strongest resonances, measured by signal-to-noise ratio, occurs for ^{181}Ta in single crystal tantalum—not unexpected, since the ^{181}Ta quadrupole moment of 3.9 barn is one of the largest of all nuclei. NAR1 and NAR2 experiments in hydrogen–free tantalum single crystal that has undergone a careful high temperature (2500 C) high vacuum ($< 2 \times 10^{-10}$ torr) anneal yield line shapes with small Lorentzian tails (approximately Gaussian shapes) and line widths much larger than computed dipole–dipole widths. The line widths and line shapes are also independent of the angle of the applied magnetic field and are temperature independent in high field experiments [5.17].

Comparing niobium single crystals with tantalum single crystals—both of which are characterized by the same crystal structure—the ratio of Ta/Nb lattice constants is 1.00055, the ratio of Ta/Nb magnetic dipole moments is 0.3818, and the ratio of Ta/Nb quadrupole moments is 24.4. The high field NAR1, NAR2, and NAR(A-R) line widths have been measured in ^{93}Nb [5.18]; values of peak–to–peak first–derivatives are given in Table 5.3 for the directions of the magnetic field along three crystallographic axes. The method of moments is used to analyze the Nb line width second moments; one thereby finds that the major contribution is isotropic pseudo–exchange between the magnetic dipole moments, with a smaller pseudo–dipolar contribution that explains the anisotropy. Analysis shows that the electric quadrupole moment does not contribute significantly to the measured second moments for ^{93}Nb in niobium metal.

Table 5.3
NMR, NAR(A-R), NAR1 and NAR2 ^{93}Nb
Line Widths and Second Moments in Niobium Metal*

H_0 direction	[001]	[1$\bar{1}$1]	[1$\bar{1}$0]
NMR			
$\Delta H_{ptp}(G)$	4.3		9.1
$\langle \Delta H^2 \rangle (G^2)$	11. ±3		46. ±8
dipolar (G^2)	10.2	19.7	17.3
NAR(A-R)			
$\Delta H_{ptp}(G)$	3.9 ± 0.3	11.0 ± 0.5	8.6 ± 0.5
$\langle \Delta H^2 \rangle (G^2)$	12. ±3	50. ±8	40. ±8
NAR1			
$\Delta H_{ptp}(G)$	16.2 ± 0.5	23.0 ± 0.6	20. ± 0.8
$\langle \Delta H^2 \rangle (G^2)$	75. ±7	96. ±8	92. ±6
NAR2			
$\Delta H_{ptp}(G)$	8.9 ± 0.5	10.6 ± 0.4	10.9 ± 0.4
$\langle \Delta H^2 \rangle (G^2)$	20. ± 2.5	34. ±4	34. ±4

*(From Ref. 5.18)

The large ^{181}Ta NAR2 line shape, which is temperature independent, isotropic and approximately Gaussian, cannot be explained by isotropic pseudo–exchange between magnetic dipole moments or by pseudo–dipolar interactions, because of the almost identical lattice constants with niobium and the ratio of <1 of the Ta/Nb magnetic dipole moments. It is the value of 24.4 for the Ta/Nb quadrupole moment ratio that constitutes the major difference between tantalum and niobium. It is not surprising, therefore, that the explanation [5.14] of the hydrogen-free tantalum NAR1, NAR2 and NAR(A-R) line widths and line shapes invokes a novel line–broadening mechanism: pseudo–nuclear electric quadrupole exchange.

To describe theoretically the ^{181}Ta line shapes and line widths, the two-point correlation function of (5.48) is used in a generalization of the Heisenberg equation of motion based on (5.25):

$$\left(\omega - \sum_{lm}(\omega)\right) G_{lm}(\omega) + \omega_q[D(lm)G_{lm}(\omega) + D(l+1\,m)G_{lm}(\omega)] = i \, ,$$

(5.71)

where $G_{lm}(\omega)$ is the complex correlation function at zero wave vector ($q = 0$). We denote by $g_{lm}(\omega)$ the real part of $G_{lm}(\omega)$, and by $\tilde{g}'_{lm}(\omega)$ the first frequency derivative of the imaginary part of $G_{lm}(\omega)$. The $D(lm)$ have been discussed in conjunction with (5.24). ω_q is a frequency describing the inhomogeneous line broadening due to electric field gradients. After solving (5.71), each line shape function is averaged over all ω_q, with a Lorentzian distribution function $\rho(\omega_q)$ characterized by a width ω_Q:

$$\rho(\omega_q) = \frac{1}{\pi} \frac{\omega_Q}{\omega_q^2 + \omega_Q^2} \,. \qquad (5.72)$$

All frequencies ω in (5.71) are measured with respect to the resonant frequency $m\omega_0$, where $\omega_0 = \gamma H_0$. The self energy $\Sigma_{lm}(\omega)$ is defined following (5.50). For the case of ^{181}Ta NAR1, NAR2 and the NAR(AR) in hydrogen–free tantalum metal, the line widths of the correlation functions $g_{lm}(q\omega)$ are much larger in frequency than ω_q over almost all values of q in the first Brillouin zone. Therefore, the correlation functions themselves should predict the line shapes.

In Figs. 5.8 and 5.9 are shown, respectively, experimental NAR1 and NAR2 ^{181}Ta resonances observed at angles at which NAR(A-R) is zero. The experimental first derivatives are digitally integrated. The low–frequency side of the resonance is reflected about the resonance center so that both halves of the resonance are displayed from resonance center frequency to higher frequencies. Conversely, in Fig. 5.10 is shown the first derivative of NAR(A-R) χ' from resonance center frequency to higher frequencies. The NAR(A-R) resonance is observed at an angle at which both the NAR1 and NAR2 phase–sensitive detected resonances are antisymmetric, while the NAR(A-R) resonance is symmetric. Thus, by reflecting the measured combination NAR1 and NAR(A-R) first derivatives about the resonance center and by adding the two halves from center frequency to higher frequencies, the NAR(A-R) first derivative may be extracted.

In analyzing the data in Fig. 5.8, 5.9 and 5.10, we first note that the NAR1 and NAR2 line shapes and line widths are independent both of the crystal's orientation with respect to the magnetic field and of temperature. This implies that the dominant interactions in the spin system are also isotropic and temperature independent. Two classes of interactions are consistent with these observations: isotropic spin–spin interaction between tantalum nuclear spins and inhomogeneous broadening due to an isotropic distribution of electric field gradients generated by crystal defects. Although electric field gradients propagated by elastic forces are essentially anisotropic, electric field gradients from charged impurities and screened by isotropic conduction electrons generate isotropic electric field gradients.

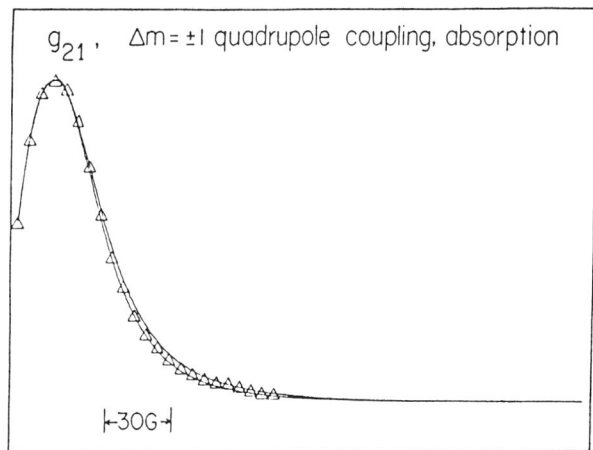

FIGURE 5.8. Two halves of the experimental NAR1 absorption line shape for ^{181}Ta in hydrogen–free tantalum metal are represented by the smooth curves plotted versus magnetic field with the resonance center to the left. The theoretical g_{21} line shape values are plotted at the triangle centers. (From Ref. 5.17)

We note further that both the NAR1 and NAR2 line shapes fall off exponentially for large frequencies, except for a small Lorentzian–like tail. A reasonable extrapolation of the tail back to the origin yields a line shape that includes only a small fraction of the spectral weight of the observed line shapes. Thus, from the evidence so far, we conclude that the dominant interaction in the tantalum sample is isotropic (dipolar or quadrupolar) exchange and that there is a small, but non–negligible, amount of isotropic inhomogeneous line broadening with a Lorentzian distribution of electric field gradients. We note that the dip at the resonance center in Fig. 5.8, not present in Fig. 5.9, can be explained by intraspin cross relaxation when there is a distribution of small electric field gradients on top of a larger homogeneous line width contribution.

With respect to Fig. 5.10, we note that the isotropic (dipolar or quadrupolar) exchange Hamiltonian commutes with $I_\pm$ or A_{11} (summed over all of the lattice sites) and that this interaction, by itself, contributes nothing to the NAR(A-R) resonance. Thus, an anisotropic (dipolar or quadrupolar) spin–spin interaction is needed to explain the width of NAR(A-R). Comparing Figs. 5.8 and 5.9 with Fig. 5.10, we see that the anisotropic spin–spin interactions must be small compared with the isotropic spin–spin interactions.

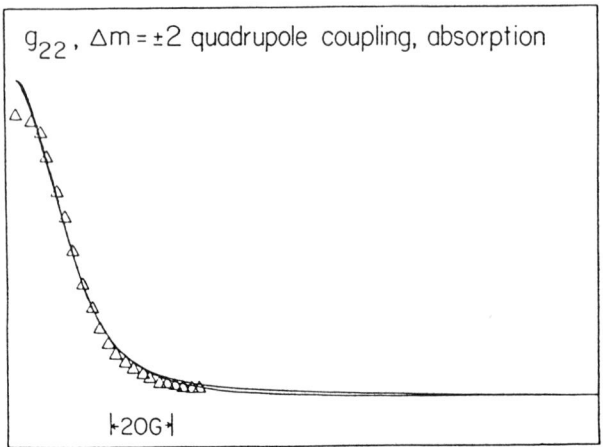

FIGURE 5.9. Two halves of the experimental NAR2 absorption line shape for ^{181}Ta in hydrogen-free tantalum metal are represented by the smooth curves plotted versus magnetic field with the resonance center to the left. The theoretical g_{22} line shape values are plotted at the triangle centers. (From Ref. 5.17)

Applying the solutions of (5.71) and averaging over ω_q, one obtains fits given by the triangles in Figs. 5.8, 5.9 and 5.10. The parameters used in the fit are given in Table 5.4. The quantity V is a measure of the strength of the isotropic quadrupole exchange. ω_Q is a measure of the strength of the inhomogeneous broadening due to an isotropic distribution of electric field gradients. ∇V is a measure of the anisotropic spin–spin interactions. We observe that the isotropic quadrupole exchange dominates over the other sources of line broadening and is in agreement with the experimental line width and line shape temperature independence and isotropy.

Table 5.4
Parameters Used in the Fit of Figures 5.8, 5.9 and 5.10*

parameter	($\times 10^4$ sec^{-1})	(G)
$V/\hbar$	4.2	
$V/\hbar\gamma$		13.2
ω_Q	0.95	
ω_Q/γ		2.98
$\nabla V/\hbar$	0.92	
$\nabla V/\hbar\gamma$		2.90

*(From Ref. 5.17)

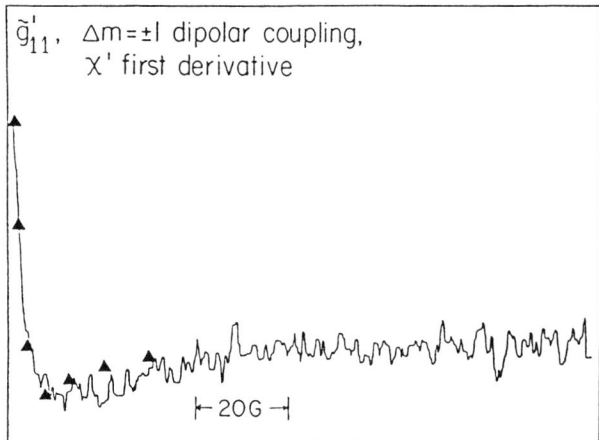

FIGURE 5.10. Half of the experimental NAR(A-R) χ' first-derivative for ^{181}Ta in hydrogen-free taltalum metal corresponds to the continuous line plotted versus magnetic field with the resonance center to the left. The theoretical shape $\tilde{g}'_{11}$ is plotted at the centers of the solid triangles. (From Ref. 5.17)

The argument that isotropic quadrupolar exchange is the dominant spin–spin interaction for ^{181}Ta in tantalum metal is interesting for several reasons:

(1) We know of no other nuclear spin system in which the spin–spin interaction is not totally dominated by the dipole moments of the spin;
(2) We know of no other nuclear spin system in which the spin–spin interactions are dominated by isotropic terms.

5.8 Relaxation by the Intrinsic Direct Process in Van Vleck Paramagnets

As mentioned in Section 2.2, enhanced nuclear magnetism occurs in paramagnetic ions with a singlet state and a nonzero nuclear spin. A much-studied system is that of ^{165}Ho $I = \frac{7}{2}$, in HoVO$_4$, a Van Vleck paramagnet whose crystal structure is tetragonal. The enhancement of the nuclear resonance frequency arises from the cross terms between the electronic Zeeman interaction and the magnetic hyperfine interaction. A detailed discussion of enhanced nuclear acoustic resonance in Van Vleck paramagnets is given in Section 9.3.

We recall from Chapter 1 that the absorption of phonons at the resonant nuclear frequency in NAR occurs via the same direct process that is at least in part responsible for the spin–lattice relaxation at very low temperatures. In their discussion of enhanced NAR in HoVO$_4$, *Bleaney et al.* [5.19] use the appropriate expression *intrinsic direct process* to distinguish their measured T_1, (obtained from resonant acoustic attenuation) from the *direct process* rate (which may be due predominantly to the presence of paramagnetic impurities) and from the much faster relaxation rates at higher temperatures due to the Orbach process. This ability to measure the intrinsic direct process contrasts with the saturation method of measuring relaxation rates in magnetic and acoustic resonance experiments, in which observation of the rate of recovery following saturation yields only the overall relaxation rate, which is dominated by the fastest processes. (In some cases, of course, the presence of more than one relaxation process may be detected through differing dependencies on time.) Direct relaxation corresponds to the exchange of resonant quanta between the thermal phonons and the magnetic system; the same process is involved in acoustic magnetic resonance between the externally generated acoustic phonons and the magnetic system. Thus, even in the presence of much faster relaxation processes, the intrinsic direct spin–lattice relaxation rate may be obtained from acoustic magnetic resonance (AMR) measurements.

A general derivation of the enhanced nuclear spin–strain Hamiltonian for crystals of tetragonal symmetry, such as HoVO$_4$ and PrVO$_4$, is outlined in Section 9.3 [5.20]. A state in which two spin levels are occupied, and in which no unresolved splittings are present from quadrupole interactions, is assumed. In terms of this Hamiltonian, the acoustic absorption coefficient may be written

$$\alpha(\omega) = \frac{n\pi\omega}{2\hbar\rho v^3} \langle i|\mathcal{H}|j\rangle^2 g(\omega) , \qquad (5.73)$$

where n is the difference in population per unit volume between two energy levels i and j. ρ is the crystal density, $g(\omega)$ is the line–shape parameter, and $\langle i|\mathcal{H}|j\rangle$ is the matrix element of the spin–strain Hamiltonian per unit strain. If only two levels are occupied,

$$n = n_0 \tanh\left(\frac{\hbar\omega}{2k_B T}\right) , \qquad (5.74)$$

n_0 being the number of spins per unit volume. For each phonon mode the intrinsic spin–lattice rate is given by

$$\frac{1}{T_1} = \frac{\omega^3}{2\pi\hbar\rho v^5} \coth\left(\frac{\hbar\omega}{2k_B T}\right) \langle i|\mathcal{H}|j\rangle^2 . \qquad (5.75)$$

Comparing (5.73) and (5.75), we obtain

$$\alpha T_1 = \frac{n_0 \pi^2 v^2}{\omega^2} \tanh^2\left(\frac{\hbar\omega}{2k_B T}\right) g(\omega) . \qquad (5.76)$$

Under conditions such that $(\hbar\omega/2k_B T) \ll 1$, the acoustic absorption coefficient becomes

$$\alpha = \frac{n_0 \pi \omega^2}{4\rho v^3 k_B T} \langle i|\mathcal{H}|j\rangle^2 g(\omega) \qquad (5.77)$$

and (5.76) similarly reduces to

$$\alpha T_1 = \frac{n_0 v^2}{16} \left(\frac{\hbar}{k_B T}\right)^2 g(\omega) . \qquad (5.78)$$

In HoVO$_4$, for longitudinal waves at 800 MHz propagated along the [100] axis of a single-crystal specimen, with the magnetic field in the (001) plane, a value of $T_1^{-1} = 10^{-3} T$ sec^{-1} is obtained for the intrinsic direct spin-lattice relaxation process [5.19]. This rate is to be compared with the measured total rate of 11 sec^{-1} at 0.2 K, attributable to the effect of paramagnetic impurities, and with the Orbach rate of $(2 \times 10^9 e^{-30/T})$ sec^{-1}. The enhanced NAR measurements are made between 1.6 K and 4.2 K, a temperature region in which the Orbach process for T_1 is six to seven orders of magnitude greater than the measured intrinsic direct T_1. The enhanced NAR technique thus provides a method of determining the intrinsic direct T_1 in enhanced nuclear paramagnets in the presence of much faster processes.

CHAPTER 5 REFERENCES

5.1 C. P. Slichter. *Principles of Magnetic Resonance* Third Enlarged and Updated Edition (Springer–Verlag, Berlin 1990).

5.2 A. Abragam. *The Principles of Nuclear Magnetism* (Clarendon Press, Oxford, 1961).

5.3 P. A. Fedders. "Spin–Resonance Line Shape Changes Induced by Intraspin Cross Relaxation." *Phys. Rev. B* **11**, 995–1000 (1975).

5.4 R. K. Sundfors. "Exchange and Quadrupole Broadening of Nuclear Acoustic Resonance Line Shapes in the III–V Semiconductors." *Phys. Rev.* **185**, 458–472 (1969).

5.5 R. K. Sundfors. "Determination of the Magnitude and Sign of the ^{185,187}Re Nuclear Electric Quadrupole Coupling Constants Using Nuclear Acoustic Resonance." *Phys. Rev. B* **42**, 1922–1928 (1990).

5.6 N. Bloembergen, T. J. Rowland. "Nuclear Spin Exchange in Solids: ^{203}Tl and ^{205}Tl Magnetic Resonance in Thallium and Thallium Oxide." *Phys. Rev.* **97**, 1679–1698 (1955).

5.7 P. A. Fedders, E. Y. C. Lu. "Theory of Ultrasonic Spin Echoes." *Phys. Rev. B* **8**, 5156–5162 (1973).

5.8 L. I. Schiff, *Quantum Mechanics* (McGraw Hill, New York 1955).

5.9 P. A. Fedders. "Comment on Anomalous Nuclear Acoustic Resonance Line Shapes in Ta with Mobile Impurities." *Phys. Rev. B* **14**, 4221–4223 (1976).

5.10 C. W. Myles, P. A. Fedders. "Dynamical Two–Point Correlation Functions in a High–Temperature Heisenberg Paramagnet." *Phys. Rev. B* **9**, 4872–4881 (1974).

5.11 P. A. Fedders. "Theory of Dynamic Quadrupole Spin–Relaxation–Application to Ta Nuclei with Mobile Hydrogen Impurities." *Phys. Rev. B* **10**, 4510–4514 (1974).

5.12 L. P. Kadanoff, P. C. Martin. "Hydrodynamic Equations and Correlation Functions." *Ann. Phys.* **24**, 419–469 (1963).

5.13 P. A. Fedders. "Microscopic View of Exchange Narrowing." *Phys. Rev. B* **3**, 2352–2356 (1971).

5.14 B. Ströbel. "Study of the Effects of the Nuclear Quadrupole Interaction in Metals Using Nuclear Acoustic Resonance." Ph. D. Dissertation, Universität Konstanz, 1979 (unpublished).

5.15 T. H. Wang. "Ultrasonic Resonance and Non–Resonance Studies of Single Crystal Tantalum with Known Hydrogen Impurities." Ph. D. Dissertation, Washington University, 1974 (unpublished).

5.16 B. Ströbel, K. Läuger, H. E. Bömmel. "Nuclear Acoustic Resonance in Tantalum with Low Concentrations of Hydrogen Impurities." *Appl. Phys.* **9**, 39–46 (1976).

5.17 P. A. Fedders, R. K. Sundfors. "Quadrupole Exchange Contributions to the ^{181}Ta Nuclear Acoustic Resonance Line Shapes in Pure Ta Single Crystal." *Phys. Rev. B* **19**, 1345–1350 (1979).

5.18 J. Pellison, J. Buttet. "Nuclear Acoustic Resonance Determination of the Gradient–Electric Tensor and Indirect Nuclear–Spin Interactions in Niobium." *Phys. Rev. B* **11**, 48–59 (1975).

5.19 B. Bleaney, G. A. D. Briggs, J. F. Gregg, G. H. Swallow, J. M. R. Weaver. "Enhanced Nuclear Acoustic Resonance in HoVO$_4$." *Proc. R. Soc. London A* **388**, 479–486 (1983).

5.20 B. Bleaney, J. F. Gregg. "Enhanced Nuclear Acoustic Resonance: Some Theoretical Considerations." *Proc. R. Soc. London A* **413**, 313–327 (1987).

CHAPTER 6
BONDING IN INSULATORS, SEMICONDUCTORS AND METALS: STRAIN–ELECTRIC FIELD GRADIENT TENSORS

6.1 Introduction

6.2 Ionic and Covalent Effects in Insulators:
The Sternheimer Antishielding Factor

6.3 Semiconductors

6.4 Antishielding in Metals

    6.4.1 Cubic Metals

    6.4.2 Noncubic Metals

References

6. BONDING IN INSULATORS, SEMICONDUCTORS AND METALS: STRAIN–ELECTRIC FIELD GRADIENT TENSORS

6.1 Introduction

Although the information given by nuclear acoustic resonance is often similar to that given by nuclear magnetic resonance, there are a number of areas in which NAR contributes uniquely. An example, which we shall discuss in this chapter, is the observation of the nuclear electric quadrupole interaction in cubic metals. Of special interest in these measurements is the electric field gradient (EFG), which is—as we have mentioned in a preliminary way in Section 3.4—a valuable probe for understanding electronic structure and related properties of atoms, molecules and solids. In the present chapter we discuss the contributions of NAR to the measurement and interpretation of EFG's in ionic, semiconductor and metallic solids. We begin by recalling the relationship between the strain–electric field gradient tensor S and the electric field gradient.

As we have seen in Chapter 3, a nucleus with a nonzero quadrupole moment Q can be used in solid state physics as a probe for the anisotropic part $\nabla_i \nabla_j \phi$ of the gradient of the electric field at the site of the nucleus:

$$V_{ij} = \nabla_i \nabla_j \phi - \frac{1}{3} \delta_{ij} \nabla^2 \phi , \qquad (6.1)$$

where ϕ is the scalar electric potential. It has become common in the literature to call this traceless second–rank tensor V_{ij} simply the *electric field gradient*, or EFG tensor. The isotropic part of V_{ij}, $\frac{1}{3}\delta_{ij}\nabla^2\phi$, is proportional to the local charge density. It does not contribute to the nuclear electric quadrupole interaction, and it is generally omitted.

The EFG's have been studied both at regular nuclear sites in a great number of noncubic solids and at nuclear sites near defects in cubic and noncubic solids. The principal methods used in this field are conventional hyperfine spectroscopy, such as NMR or nuclear quadrupole resonance (NQR), and methods of nuclear solid state physics, such as perturbed angular correlation of γ radiation. These methods fail, however, for nuclei surrounded by a perfect cubic lattice, since in cubic point symmetry the (traceless) tensor V_{ij} vanishes.

In order to observe nuclear electric quadrupole interaction effects in cubic crystals, the cubic point symmetry can be broken by slightly deforming the lattice unit cell. For small strains, as we have seen in Chapter 3, the strain–induced EFG, V_{ij}^s, may be represented in linear approximation by

$$V_{ij}^s = \sum_{k,l} S_{ijkl}\varepsilon_{kl} , \qquad (6.2)$$

where ε is the symmetrized strain tensor, and $\boldsymbol{S}$ is a fourth–rank tensor with nonvanishing components even in cubic symmetry. We refer to (6.2) as the strain EFG or simply the $\boldsymbol{S}$–tensor.

For cubic crystals with a rectangular coordinate system axes along the cubic axes of the unstrained crystal, the components of the traceless EFG tensor $\widehat{V} = (V_{ij} - \delta_{ij}\operatorname{Tr}\boldsymbol{V}/3)$ can be written as

$$\widehat{V}_{ii}^s = (S_{11} - S_{12})\hat{\varepsilon}_{ii} ; \qquad \widehat{V}_{ij}^s = 2S_{44}\hat{\varepsilon}_{ij} , \qquad i \neq j , \qquad (6.3)$$

where S_{11}, S_{12} and S_{44} are the three distinct components of the $\boldsymbol{S}$–tensor (in Voigt notation), and $\hat{\varepsilon}_{ij} = (\varepsilon_{ij} - \delta_{ij}\operatorname{Tr}\varepsilon/3)$ are the components, in first order, of the volume conserving part of the strain tensor. Volume deformations are represented by $\operatorname{Tr}\boldsymbol{V}^s = (S_{11} + 2S_{12})\operatorname{Tr}\varepsilon$, which, however, do not contribute to the electric quadrupole interaction. The number of independent $\boldsymbol{S}$–tensor components required to describe the electric quadrupole interaction, therefore, is two: $(S_{11} - S_{12})$ and S_{44}. The former refers to volume–conserving linear dilatations and the latter to shear distortions of the cubic lattice cells.

Because the $\boldsymbol{S}$–tensor characterizes the linear response of the EFG to a locally applied strain, it is, like the elastic constant tensor, a physical property of the unstrained crystal under consideration. This is true regardless of the magnitude of the strain actually applied, as long as the linear response range is not exceeded.

S-tensor components may be obtained by observing the quadrupole broadening of an NMR line width due to an applied static uni-axial stress or due to the presence of internal strains from a known concentration of dislocations or impurity atoms in a crystal. This technique does not work well in metals, due to the poor signal-to-noise of NMR in bulk metals. Nuclear acoustic resonance has the following advantages:

(1) that it can be applied effectively in metals, and
(2) that for certain particular directions of acoustic wave propagation and polarization in a cubic crystal, the elastic strain amplitudes do not enter into the determination of the S-tensor components.

The S-tensor components can be obtained from the NAR signal intensity for different crystallographic orientations of a single crystal specimen with respect to the external magnetic field. As we shall see in what follows, all of the quantities entering into the experimental determination of the product of quadrupole moment and the S-tensor component can be determined with experimental precision greater than ±5%.

6.2 Ionic and Covalent Effects in Insulators: The Sternheimer Antishielding Factor

The calculation of field gradients in ionic crystals is often made with the assumption that the ionic crystal consists of nonoverlapping spherically symmetric ions. Indeed, in early investigations the field gradients at the central-ion nucleus were evaluated as being due to point monopoles and dipoles on the lattice sites. These point charges were further assumed to be totally external to the ion containing the nucleus in question, and, as we shall see below, a multiplying factor γ_∞ was used to incorporate the antishielding of this ionic contribution by the core electrons of the central ion. Considering first the field gradient produced by charges external to the given ion, the value of any component of the field gradient tensor V_{ij} due to a point charge $n_k e$ at the position $\mathbf{r}_k$ is

$$V_{ij_k} = -\frac{n_k e}{r_k^5}(r_k^2 - 3x_{i_k}x_{j_k}) \ . \tag{6.4}$$

The total $V_{ij_{\text{ionic}}}$ is then the sum over all the point charges (ions) in the lattice, excluding the ion at the origin:

$$V_{ij_{\text{ionic}}} = -e \sum_k \frac{n_k}{r_k^5}(r_k^2 - 3x_{i_k}x_{j_k}) \ . \tag{6.5}$$

In many lattices the series (6.5) is slowly convergent because of the alternating signs of the charges. The lattice sums have been calculated for a number of crystal structures, some of which are given in *Cohen and Reif* [6.1].

6. FIELD GRADIENT TENSORS

The simple ionic model represented by (6.5) must be modified to take into account the distortion of the charge distribution associated with the ion in question. Since a spherically symmetric charge distribution has $V_{xx} + V_{yy} + V_{zz} = 0$, it appears that the filled shell (core) electrons of the nucleus in question need not be considered. *Sternheimer* [6.2] and others have shown, however, that the external EFG polarizes the ion. This can be seen by considering the interaction energy of an external charge e at a distance R_{ext} from the nucleus in question with an electron at position r, θ, ϕ with respect to the nucleus and the radius vector $\boldsymbol{R}$. Expanding the energy in powers of r, we have

$$-\frac{e^2}{R_{ext}} = -\frac{e^2}{R} - \frac{e^2 r \cos\theta}{R^2} - \frac{e^2 r^2 (3\cos^2\theta - 1)}{2R^3} - \cdots . \tag{6.6}$$

The first term cannot distort the electron distribution, the second term is responsible for the ordinary dipole polarization, and the third term produces a quadrupole polarization. This distortion of the electron shells associated with the quadrupolar nucleus adds a contribution to the total field gradient, which is conventionally written as

$$V_{ij} = V_{ij}^o [1 - \gamma(r)] , \tag{6.7}$$

where V_{ij}^o is the field gradient computed from a known distribution of ionic charges, and r is the distance from the charge e to the nucleus. For r much greater than the ion radius, γ becomes independent of r and is denoted by γ_∞, generally referred to as the Sternheimer antishielding factor. There are actually two terms in the electron energy that are proportional to Q/R^2 and, therefore, to which γ_∞ may be attributed. One is the interaction of the electronic quadrupole moment induced in the electron shells of the ion (by the aspherical nucleus) with the external electric field gradient. The other has been mentioned above: the interaction of the nuclear quadrupole moment with the electric field gradient arising from the distortion of the electron shells by the external charge. Choosing which term is calculated (to obtain the theoretical γ_∞) is only a matter of convenience, because the terms are equivalent and numerically equal.

A great deal of effort over the years has gone into the calculation of $(1-\gamma_\infty)$ [6.2]. In the past, a major part of this effort in investigating γ_∞ was directed towards the response of the electrons on free ions (as contrasted to ions in a solid) to the influence of the external field gradient. This approach utilized both differential equation and variational procedures. There are two important effects, however, that should be considered for the quantitative analysis of field gradients for ions in the solid state: first, the influence of the crystal field on the radial character of the electronic orbitals and the consequent alteration of their response to the external electric field; and, second, the influence of the interaction of the electrons among themselves on γ_∞ (a factor that must also be considered for free ions). A variety of sophisticated approaches to the treatment of these effects in the calculation of γ_∞ has been made and are summarized by *Schmidt, Sen, Das and Weiss* [6.3]. These solid state effects for calculating γ_∞ were included by these investigators by using the Watson sphere model for ionic solids. A comparison of the calculated values of γ_∞ for free ions and for ions in the solid state for sequences of ions is given in Table 6.1. These calculations, as well as others in the literature, suggest that the presence of the neighboring ions in the crystal lattice causes an increase in the free-ion γ_∞ for positive ions and a decrease of γ_∞ for negative ions. For comparison, values of γ_∞ calculated by alternate procedures are given in Column 4.

Table 6.1

Antishielding Factors for Free Ions and Ions in the Solid State

Ion	Free Ion[a]	Crystal[a]	Crystal[b]
Li^+	0.25	0.26	
Na^+	-5.3	-5.5	-4.6
K^+	-20.0	-21.8	-17.3
Rb^+	-47.7	-52.8	-47.2
Cs^+	-95.2	-111.	-103.
F^-	-28.6	-12.5	-22.5
Cl^-	-68.8	-42.	-56.6
Br^-	-166.	-85.5	-123.
I^-	-299.	-162.	
Mg^{2+}	-3.5	-4.1	-3.4
Ca^{2+}	-14.1	-18.8	-14.
Sr^{2+}	-38.9	-47.8	-41.4
Ba^{2+}	-76.2	-111.	-94.2
Cu^+	-21.5	-25.2	-17.3
Ag^+	-43.1	-44.5	-34.9

[a][6.3] [b][6.4]

The earliest measurements of S-tensor components by NAR and ASNMR techniques were made on alkali halide crystals. In general, good agreement was obtained between theoretical calculations and experimental results for γ_∞ of positive ions. For negative ions, however, there was poor agreement between theory and experiment. It is now generally recognized that the extreme ionic model of separated non–overlapping ions is not satisfactory, because small departures from spherical symmetry of the ions produced by covalency effects (such as, for example, the transfer of electrons from the negative ions into p– or d–like orbitals of the positive ions) can produce very significant changes in the field gradient. Attempts were made to patch up the differences by grafting covalent effects onto the ionic model, without great success. *Kanashiro's* particularly thorough experimental measurement, using ASNMR techniques, of the S-tensor components in a series of alkali bromide crystals is characteristic of this early work [6.5]. Among the more complex insulator crystals studied by NAR or ASNMR techniques have been $NaClO_3$, $KTaO_3$ and $NaNO_2$. The last is of particular interest, since the experimental investigation of the S-tensor components [6.6] was complemented by detailed theoretical treatment of the origin of field gradients in the same crystal [6.7]. In particular, *Betsuyaku* derived a general expression for the EFG at the positive ion nucleus (Na^+) in an ionic crystal ($NaNO_2$) with polyatomic negative ions (NO_2^-) in his calculations. In light of the theoretical developments that were to follow, the characteristic conclusions reached by these researchers are worthy of note:

(1) The field gradient in an ionic crystal depends too sensitively on the electron charge distribution of the polyatomic ions adjacent to the nucleus under study to be predicted by the point multipole model, and, therefore,
(2) the evaluation of the field gradient needs a careful treatment of the charge distribution of the closest ions and must include overlap and covalent bonding effects.

In the elaboration of a more complete theoretical program, the emphasis is placed on achieving an unambiguous first–principle formulation of the Sternheimer antishielding effects in an ionic solid—considered not as an assembly of point ions surrounding a central ion whose nucleus possesses a quadrupole moment, but as ions whose nuclei and electronic charge distributions overlap and interact with one another substantially. The aim of this program is to develop a procedure for the quantitative study of the influence of these interactions upon the EFG in the ionic solid; such an effort would be equivalent to a study of the antishielding effects associated with each component of the interacting, overlapping charge distribution in the solid. Thus the net field gradient is written in the form

$$q = (1 - \gamma_\infty)q^o_{\mathrm{DRL}} + q_{\mathrm{cluster}}, \tag{6.8}$$

where q_{cluster} includes the total contribution due to all electrons and nuclei of the cluster that is composed of the central ion and the nearest–neighbor ligands, as modified from their free–ion configurations by overlap, charge transfer and other effects in the solid. q^o_{DRL} is the unshielded contribution due to charges on distant ions in the rest of the lattice outside the cluster. The major problem in a quantitative evaluation of the EFG q—and, therefore, of the S-tensor components—is of course the treatment of q_{cluster}. A detailed formalism of such a first–principles approach to the calculation of field gradients in ionic solids is given by *Beri, Lee, Das and Sternheimer* [6.8] and applied in particular to ^{57}Fe in Fe_2O_3.

6.3 Semiconductors

An extremely thorough and detailed study of gradient–elastic tensor components in single crystal germanium and in a series of III–V and I–VII semiconductors has been made [6.9]. We present here the procedures used in these studies as characteristic of the application of NAR to the study of electric field gradients in nonmetals.

All the crystals studied, except for elemental germanium, are of the zinc–blende structure, characterized by local T_d tetrahedral symmetry at the nuclear positions. From crystal symmetry (see Appendix C), we know that there are three different nonzero components of S: S_{11}, S_{12} and S_{44}. Only the two quantities $(S_{11} - S_{12})$ and S_{44}, however, occur in the matrix elements for cubic symmetry. In accordance with the discussion in Section 3.4, on assuming that the electric charge at the nuclear position does not change with strain, we can further set $S_{11} = -2S_{12}$.

The coupling mechanism in all the specimens studied is dynamic quadrupolar. The relationship between measured change in acoustic attenuation and the S-tensor components S_{11} and S_{44} for cubic symmetry and acoustic wave propagation along certain high symmetry axes is given in Section 3.4. In the experiments under review, only $\Delta m = \pm 2$ spin transitions are utilized—a choice based on the narrower NAR line widths characterizing α_{Q2} over α_{Q1}. The S-tensor component S_{44} is measured by propagating transverse acoustic waves with the propagation vector $\boldsymbol{k}$ along the [110] axis, polarization vector $\boldsymbol{\xi}$ along the [001] axis, $\boldsymbol{H}_0$ in the (001) plane, and with ϕ denoting the angle between $\boldsymbol{H}_0$ and $\boldsymbol{k}$ (see Fig. 3.8). The corresponding resonant attenuation is given by

$$\alpha_{2x'z'} = 4CB_2(QS_{44}\sin\phi)^2. \tag{6.9}$$

The S-tensor component S_{11} is also measured by utilizing transverse acoustic wave propagation with k along the [110] axis, ξ along the [1$\bar{1}$0] axis, H_0 in the (1$\bar{1}$0) plane, and χ the angle between H_0 and k. In this case, the resonant attenuation is given by

$$\alpha_{2x'y'} = \frac{9}{4}CB_2(QS_{11}\sin\chi)^2 . \tag{6.10}$$

Relative signs between S_{11} and S_{44} are determined by propagating longitudinal acoustic waves, with k along the [110] axis, H_0 in the (001) plane, and the resonant attenuation given by

$$\alpha_{2x'x'} = CB_2Q^2\left(\frac{3}{4}S_{11} - S_{44}\cos 2\psi\right)^2 . \tag{6.11}$$

In the above expressions,

$$B_2 = \sum_m F_m(I)_{\pm 2}^2$$

$$F_m(I)_{\pm 2} = [(I \mp m)(I \mp m - 1)(I \pm m + 1)(I \pm m + 2)]^{\frac{1}{2}}$$

and

$$C = \frac{\pi^2}{16(2I)^2(2I-1)^2(2I\pm 1)} \frac{N\nu^2 e^2 g(\nu)}{\rho v^2 k_B T} .$$

For all but the measurements on Ge, NAR absorption measurements are made at 10 MHz and 300 K by utilizing a standard marginal oscillator ultrasonic spectrometer (see Chapter 7). For germanium, the measurements are made at 20 MHz and 300 K. The absolute change in acoustic attenuation is determined by comparing the NAR absorption signal with a known calibrator signal. We obtain the area under the absorption line by numerically integrating the experimentally observed NAR first-derivative signal. Magnetic field modulation is used, with modulation amplitudes chosen to be no larger than $\frac{1}{8}$ ($\frac{1}{3}$ for Ge) the peak-to-peak experimental line width. The dynamic strain amplitudes are chosen to produce less than 1% (4% for Ge) saturation of the NAR signal. Further details of the experimental techniques used are given in Chapter 7.

Measured values of S_{11} and S_{44} for Ge, for a series of III–V compounds and for I–VII compounds are given in Columns 3 and 4 of Table 6.2. Values of the quadrupole moment used in evaluating S_{11} and S_{44} are given in Column 2.

6.3 SEMICONDUCTORS

Table 6.2
Measured values of S_{11}, S_{22}, and Related Quantities*

Nuclear Position	Q	S_{11}	S_{44}	$1-\gamma_s$	Ion State	$\frac{Z^*}{e}$	$\frac{16e}{3d^3}$	$S_{11\,\text{ion}}$	$\langle r^{-3}\rangle$
^{73}Ge	0.173[a]	3.4	12.5	7.8[b]	Ge^{4+}	+4		5.4	4.7[g]
^{27}AlSb	0.165	1.8	3.0	3.3[c]	Al^{3+}	+5	0.137	2.3	1.1[h]
^{69}GaP	0.199	6.7	5.8	10.5[d]	Ga^{3+}	+5	0.195	10.2	2.9[h]
^{69}GaAs	0.199	6.4	6.5	10.5[d]	Ga^{3+}	+5	0.175	9.2	2.9[h]
^{69}GaSb	0.199	4.3	5.7	10.5[d]	Ga^{3+}	+5	0.139	7.3	2.9[h]
^{115}InP	0.860	14.0	6.7	25.9[e]	In^{3+}	+5	0.156	20.2	4.3[h]
^{115}InAs	0.860	11.8	7.0	25.9[e]	In^{3+}	+5	0.142	18.4	4.3[h]
^{115}InSb	0.860	9.2	7.4	25.9[e]	In^{3+}	+5	0.116	15.0	4.3[h]
Ga^{75}As	0.27	9.3	18.8		As^{5+}	+3	0.175		6.9[h]
In^{75}As	0.27	9.2	17.7		As^{5+}	+3	0.142		6.9[h]
Al^{121}Sb	0.53	9.6	22.2		Sb^{5+}	+3	0.137		9.0[h]
Ga^{121}Sb	0.53	7.5	21.7		Sb^{5+}	+3	0.139		9.0[h]
In^{121}Sb	0.53	8.6	21.0		Sb^{5+}	+3	0.116		9.0[h]
^{63}CuCl	0.21	4.2	3.0	18.[d]	Cu$^+$	-1	0.199	3.6	6.3[i]
^{63}CuBr	0.21	3.2	5.3	18.[d]	Cu$^+$	-1	0.171	3.7	6.3[i]
^{63}CuI	0.21	3.5	...	18.[d]	Cu$^+$	-1	0.142	2.6	6.3[i]
Cu^{35}Cl	0.09	5.6	5.6	55.[c]	Cl$^-$	+1	0.199	10.9	6.8[g]
Cu^{79}Br	0.29	10.4	16.1	101.[f]	Br$^-$	+1	0.171	17.3	12.0[g]
Cu^{127}I	0.80	16.8	20.7	176.[f]	I$^-$	+1	0.142	25.0	14.9[g]

[a][6.10] [b][6.12] [c][6.11] [d][6.12] [e][6.13] [f][6.14] [g][6.15] [h][6.16] [i][6.17]

*$S_{11\,\text{ion}}$ is computed using the quadrupole moment Q, antishielding factor $1-\gamma_s$, effective charge of the ion Z^*, and nearest neighbor distance d. S_{11}, S_{44}, $S_{11\,\text{ion}}$ and $16e/(3d^3)$ are in units of 10^{15} statCoulomb cm^{-3} and $\langle r^{-3}\rangle$ is in atomic units. The experimental uncertainty of QS_{11} and of QS_{44} is ±10% for ^{73}Ge and ^{27}Al; ±6% for ^{69}Ga, ^{63}Cu and ^{35}Cl; and ±4% for ^{115}In, ^{75}As, ^{121}Sb, ^{79}Br and ^{127}I. (From Ref. 6.9)

In the theoretical approach to evaluating the S-tensor components in these materials, it is assumed that the field gradient component V_{zz} can be written as the sum of two contributions: a point charge or ionic contribution, and an electronic or covalent contribution. The field gradient tensor components are expressed in orthogonal coordinates that are the principle axes of the tensor; therefore only diagonal components differ from zero. Specifically, we write V_{zz} at a nuclear position in terms of contributions from the charge distributions of the four nearest–neighbor atoms (considered as point charges) and from the bonding orbitals between the atom (with nucleus at the origin) and the four nearest neighbors:

$$V_{zz} = \sum_{i=1}^{4} \frac{(3\gamma_i^2 - 1)Z^*(1 - \gamma_s)}{d^3}$$

$$+ \sum_{i=1}^{4} \frac{(-1)(3\gamma_i^2 - 1)(2e)(1 - R_s)}{2}$$

$$\times \left\langle \phi_i \left| \left[\frac{(3\cos^2\theta_i - 1)}{r_i^3} \right] \right| \phi_i \right\rangle, \qquad (6.12)$$

where γ_i is the direction cosine from the z principle axis to the ith bond direction, e is the electron charge magnitude, and Z^* is the effective ion charge of the nearest–neighbor atom including the sign of the charge. γ_s and R_s are the Sternheimer antishielding and shielding terms for the solid, d is the first–neighbor distance, ϕ_i is the bond orbital wave function, and r_i and θ_i are the coordinates to the ith electron, expressed with respect to an orthogonal coordinate system with its z-axis along the ith bond direction.

In (6.12) the factor $\frac{1}{2}(3\gamma_i^2 - 1)$ is due to the transformation of the electron part of V_{zz} from a coordinate system with z-axis along the ith bond direction to the z-direction of the principal axes. The factor of 2 in the electronic part (6.12) comes from assuming two electrons per bond. The minus sign is inherent in the field–gradient definition.

The Sternheimer term $(1 - \gamma_s)$ is introduced to correct V_{zz} at a nuclear position of an ion due to the interaction of the ion with its nuclear quadrupole moment and with the external point–charge electric field gradient. We include the term $(1 - R_s)$ as a shielding correction to account for the shielding effect of the core electrons in the atom whose nucleus is at the coordinate origin.

The first term in (6.12) is the ionic term, written only for the first–neighbor shell; the second term is the electronic contribution. For atomic orbitals, ϕ_i in the term $\langle \phi_i | (3\cos^2\theta_i - 1)/r_i^3 | \phi_i \rangle$ can be replaced by the orbitals on the atom at whose nucleus V_{zz} is being computed. Other contributions to $\langle \phi_i | (3\cos^2\theta_i - 1)/r_i^3 | \phi_i \rangle$ are two orders of magnitude smaller.

Values of $\langle \phi_i | (3\cos^2\theta_i - 1)/r_i^3 | \phi_i \rangle$ for various atomic orbitals are given in Table 6.3.

Table 6.3

Contribution to V_{zz} for Various Atomic Orbitals*

Orbital	$V_{zz}/(e\langle 1/r^3 \rangle)$
s	0
p_z	4/5
p_x	−2/5
p_y	−2/5
$d_{3z^2-r^2}$	4/7
$p_{x^2-y^2}$	−4/7
p_{xy}	−4/7
p_{xz}	2/7
p_{yz}	2/7

*(From Ref. 6.18)

For zero strain, we expect V_{zz} to be zero. The coupling between acoustic waves and the nuclear spin system in an NAR experiment involves a linear dependence of field gradient on strain. By expanding (6.12) in a Taylor series in a specific component of strain, we can identify S_{11} and S_{44} as the expansion coefficients that are linear in strain. In order to relate the field gradient expression of (6.12) to the experimental strain, we choose V_{zz} to be along a [100] crystal direction, and we transform V_{ij} to a non-principal axis coordinate system, with x' along [110], y' along [1$\bar{1}$0], and z' along [001]. We find S_{11} by expanding the $x'y'$ component of the transformed V_{ij} in a Taylor series in the shear strain $\varepsilon_{x'y'}$. The $\varepsilon_{x'y'}$ strain produces displacements that can be written in terms of their radial and angular contributions, relative to the bond directions and the first-neighbor positions. From Fig. 6.1 we see that only angular displacements of the four first-neighbor bonds and four first-neighbor positions contribute to S_{11}. Thus the covalent contribution to S_{11} involves, in first order, only angular rotations of the bonding orbitals about the nuclear position at the origin.

In measuring S_{11}, the $\varepsilon_{x'y'}$ strain wave with an acoustic wave velocity $v = [(c_{11}-c_{12})/2\rho]^{\frac{1}{2}}$ is propagated. *Harrison and Phillips* [6.19] have found good agreement between predicted bond-orbital-model values of the elastic constant $(c_{11}-c_{12})$ and measured $(c_{11}-c_{12})$ values. In this model the shear strain $\varepsilon_{x'y'}$ does not alter the tetrahedral angles at either anion or cation in zinc–blende structure compounds, but the bond to first–neighbor atoms is broken by the shear strain. The use of the Harrison–Phillips model with the strains that measure S_{11} leads to the conclusion that the electronic contribution to S_{11} is zero. Since the electronic part of S_{11} depends on angular changes in the orbitals of the atom at whose nucleus S_{11} is measured, the Harrison–Phillips model predicts there are no angular changes in these orbitals.

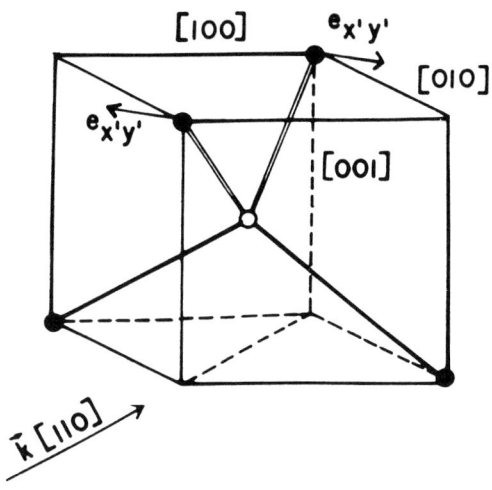

FIGURE 6.1. First–neighbor shear–strain directions relative to a center atom at whose nucleus the electric field gradients are measured.

For semiconductors, a computation has been made of the ionic contribution from point charges placed in successive shells of atoms at constant distances r from a nuclear position to a distance of 25 shells. By utilizing a model of a step dielectric function in which the first–neighbor shell is unscreened, but the second and succeeding shells are scaled by the low–frequency dielectric constant, the total ionic contribution is found to be close to that of the first–neighbor shell alone. The total ionic contribution to S_{11} is found to be

$$S_{11_{\text{ionic}}} = \frac{16}{3} \frac{Z^*(1-\gamma_s)}{d^3} . \qquad (6.13)$$

Computed values of $S_{11_{\text{ionic}}}$ are given in column 9 of Table 6.2; values of antishielding factors $(1-\gamma_s)$, first neighbor charges Z^*/e, and $\frac{16}{3}e/d^3$ are given in Columns 5, 7 and 8 of Table 6.2.

In Fig. 6.2 is plotted the experimentally measured values of S_{11} versus the calculated $S_{11_{ionic}}$ for diamond and zinc–blende compounds. The proportionality in Fig. 6.2 between $S_{11_{ionic}}$ and S_{11} for Ge for the III–V compounds and for the I–VII compounds supports the assumption based on the Harrison–Phillips model—that a simple point-charge, first-neighbor ion model explains S_{11} in diamond and zinc–blende compounds. The first-neighbor charges chosen in computing $S_{11_{ionic}}$ in order to obtain proportionality with S_{11} are +4 for Ge, +3 and +5 for the III–V compounds and +1 and −1 for the I–VII compounds. These charges are those for homopolar bonding for Ge and the III–V compounds without screening by the valence electrons in the chemical bonds. The charges chosen for the I–VII compounds, on the other hand, are for heteropolar bonding.

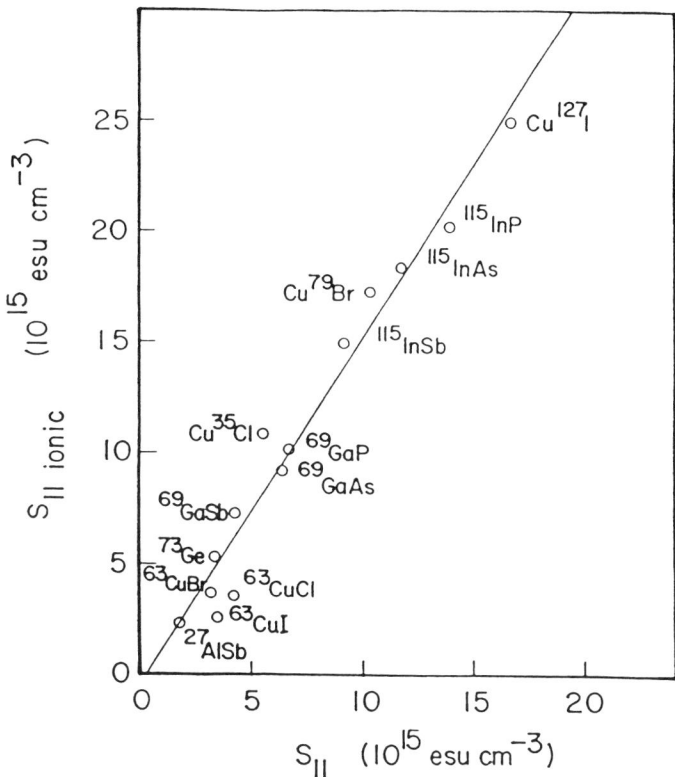

FIGURE 6.2. Plot of $S_{11\,ionic}$ vs S_{11} for diamond and zinc–blende compounds. The line is a linear regression using all S_{11} values. (From Ref. 6.9)

214   6. FIELD GRADIENT TENSORS

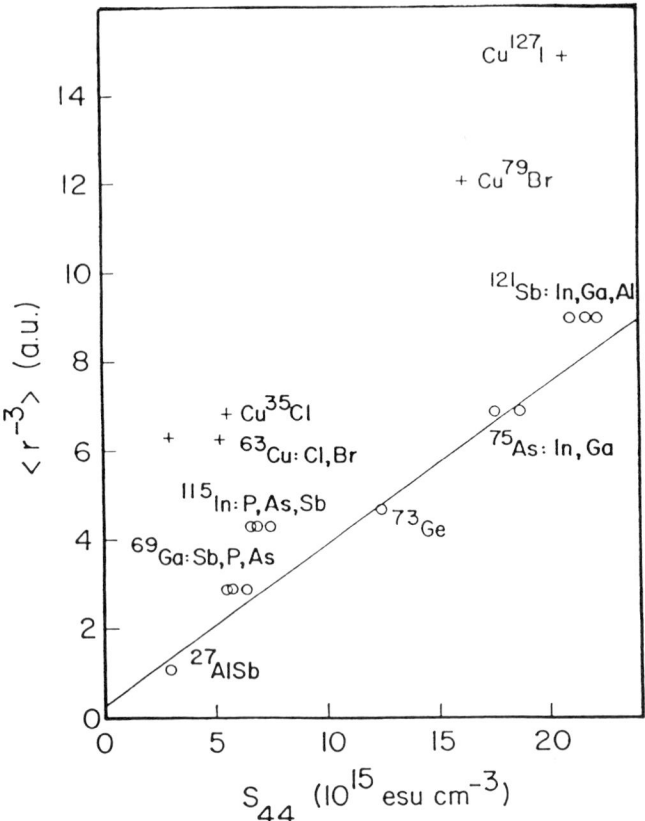

FIGURE 6.3. Plot of $\langle r^{-3} \rangle$ vs S_{44} for diamond and zinc–blende compounds. The circles are for Ge and III–V compounds; the crosses are for I–VII compounds. The line is a linear regression for values of ^{27}Al, ^{73}Ge, ^{69}Ga and ^{75}As, all of which have small Q and $1 - \gamma_s$ and, therefore, probable smaller ionic contributions to S_{44}. (From Ref. 6.9)

The similar S_{44} values at the same nuclear position in different III–V compounds suggest that the major contributions to S_{44} in these compounds are covalent. From (6.12) we may expect that the covalent contribution to S_{44} from different nuclear positions is proportional to $\langle r^{-3} \rangle$ for the p electron in the hybrid–bond wave functions. Computed values of $\langle r^{-3} \rangle$ are given in Column 10 of Table 6.2. In Fig. 6.3 we plot $\langle r^{-3} \rangle$ versus S_{44}. The approximate proportionality between $\langle r^{-3} \rangle$ and the Ge and III–V compound S_{44} in Fig. 6.3 indicates that the major contribution to S_{44} in these compounds is covalent, but that smaller ionic contributions may also be present. The I–VII S_{44} values, on the other hand, are far smaller than would be expected for those values of $\langle r^{-3} \rangle$ to which the Ge and III–V compound S_{44} are approximately proportional. One interpretation of Fig. 6.3 is that the ionic contributions to $S_{44_{exp}}$ slightly reduce its magnitude for those III–V atoms with larger Q and $(1-\gamma_s)$, and that they greatly reduce its magnitude for the I–VII compound atoms. This interpretation of S_{44} is consistent with the interpretation of Ge and the III–V compounds as having homopolar bonding, and of the I–VII compounds as having heteropolar bonding.

6.4 Antishielding in Metals

The electric field gradient in a metal is composed of contributions from the point charges on the positive ions in the lattice and from the conduction electron distribution. The ionic contribution is given by

$$q_{\text{ion}} = \sum_{N=1}^{\infty} \zeta e \left(\frac{3 \cos^2 \theta_N - 1}{r_N^3} \right) (1 - \gamma_\infty) , \qquad (6.14)$$

where ζ refers to the charge on the ions. The coordinates r_N and θ_N refer to the Nth ion, with respect to the ion at the origin that contains the nucleus for which the field–gradient is being calculated. When the summation over N in (6.14) is performed in a straightforward manner, it is only slowly convergent. Techniques have been developed, however, to make the summation process less time–consuming. Thus, in the method of plane–wise summation, one takes Fourier transforms of the sum in (6.14) and performs summations in k-space. Another procedure consists of dividing the metal into neutral units by superposing a uniform electronic background on the lattice charges. This approach does not alter the sum but allows one to interpret the field gradient as due to multipole moment arrays, rather than to point charges. The multipole moment approach results in a faster falling-off with r_N and, thus, a greater convergence than the sum in (6.14). As in the case of insulators and semiconductors, the quantity γ_∞, appropriate for the EFG due to a totally external charge, is the Sternheimer antishielding factor for the ions in the lattice.

Because of the continuous nature of the electronic density distribution in metals, the summation in (6.14) is replaced by an integration for the electronic contribution to the field gradient. The electronic contribution is then given by

$$q_{el} = -e \int \rho(\mathbf{r})[1 - \gamma(r)] \left[\frac{3 \cos^2 \theta - 1}{r^3} \right] d\tau . \qquad (6.15)$$

The charge density $\rho(r)$ is given by

$$\rho(r) = 2 \sum_{n\mathbf{k}} |\Psi_{n\mathbf{k}}(r)|^2 \, , \qquad (6.16)$$

where the summation over n refers to bands, and that over $\mathbf{k}$ refers to the states in each band that are occupied—that is, those lying below the Fermi surface. The factor of 2 in (6.16) takes care of the two spin states for each electron. We then rewrite (6.15) in the form

$$q_{el} = -e \sum_{n\mathbf{k}} \langle q \rangle_{n\mathbf{k}} \, ,$$

where

$$\langle q \rangle_{n\mathbf{k}} = \left\langle \Psi_{n\mathbf{k}}(r)(1 - \gamma(r)) \left| \frac{3\cos^2\theta - 1}{r^3} \right| \Psi_{n\mathbf{k}}(r) \right\rangle . \qquad (6.17)$$

The electronic contribution to the field gradient is obtained by summing over contributions from all occupied states. The quantity $\gamma(r)$ is the anti–shielding function, which depends on the distance of the conduction electron from the nucleus. For r going to infinity, $\gamma(r)$ reduces to γ_∞. For intermediate distances, $\gamma(r)$ must include exchange effects associated with the indistinguishability of conduction and core electrons when their wave functions overlap.

A physically useful picture of the origin of electric field gradients in metals—one that produces an empirical expression much used in analyzing NAR data—can be given. The electric field gradient at a nucleus arises from the distribution of charge (both electronic and nuclear) about it in space and is given by

$$eq = \int_{\text{all space}} \rho(r) \frac{1}{r^3} P_2^o(\cos\theta) d\tau \, . \qquad (6.18)$$

For convenience, we consider the EFG as arising from three separate sources. To aid in distinguishing these, we draw a sphere about the nuclear site at which we wish to evaluate q (see Fig. 6.4). The sphere radius can be chosen so that spheres for near neighboring atomic sites touch. The first EFG contribution is q_{latt}, arising from the region outside of the sphere, namely

$$eq_{\text{latt}} = 2 \int \rho(r) \frac{1}{r^3} P_2^o(\cos\theta) d\tau \, , \qquad (6.19)$$

where $\rho(r)$ is the nuclear plus electronic charge density. The integral can be replaced by a lattice sum over point charges and, higher multipoles can be centered at the various lattice sites. Multipoles as high as the hexadecapole can be experimentally significant.

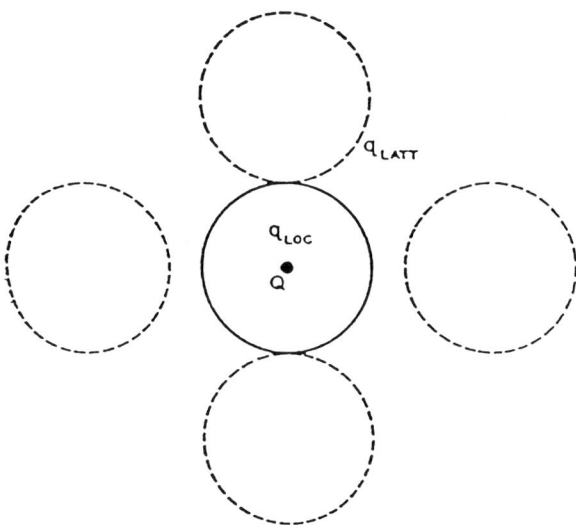

FIGURE 6.4. Atomic spheres for a hypothetical metal with q_{loc} arising from the conduction–electron charge within the sphere and q_{latt} from the nuclear and electronic charge outside. (From Ref. 6.20)

The second contribution comes from the conduction electrons within the atomic sphere:

$$eq_{\text{loc}} = 2 \int \rho_{\text{ce}}(r) \frac{1}{r^3} P_2^o(\cos\theta) d\tau \qquad (6.20\text{a})$$

$$= \sum_i \int \rho_i(r) \frac{1}{r^3} P_2^o(\cos\theta) d\tau \;, \qquad (6.20\text{b})$$

where ρ_i is the charge density of the ith conduction electron, and the sum is taken over the occupied conduction electron states.

Finally there are contributions from the closed–shell core electrons at the atomic site in question. In first approximation they are spherical and make no contribution to q, but they become distorted under the influence of their aspherical environment and interact, in turn, with the nuclear quadrupole moment. These distorted core electron interactions may be accounted for in terms of the Sternheimer antishielding factor $(1 - \gamma_\infty)$ and shielding factor

$(1-R)$, which is of the order of unity. In sum, the total field gradient from the three sources may be written as

$$q = q_{\text{latt}}(1-\gamma_\infty) + q_{\text{loc}}(1-R) . \tag{6.21}$$

As has been indicated previously, q_{latt} can be calculated in a straightforward manner, whereas attempts to calculate q_{loc} have generally not been successful. By plotting $eq' \equiv eq - eq_{\text{latt}}(1-\gamma_\infty)$ versus $eq_{\text{latt}}(1-\gamma_\infty)$, however, one is able to obtain a striking correlation between the lattice and the conduction electron contributions (see Fig. 6.5). The primary conclusion

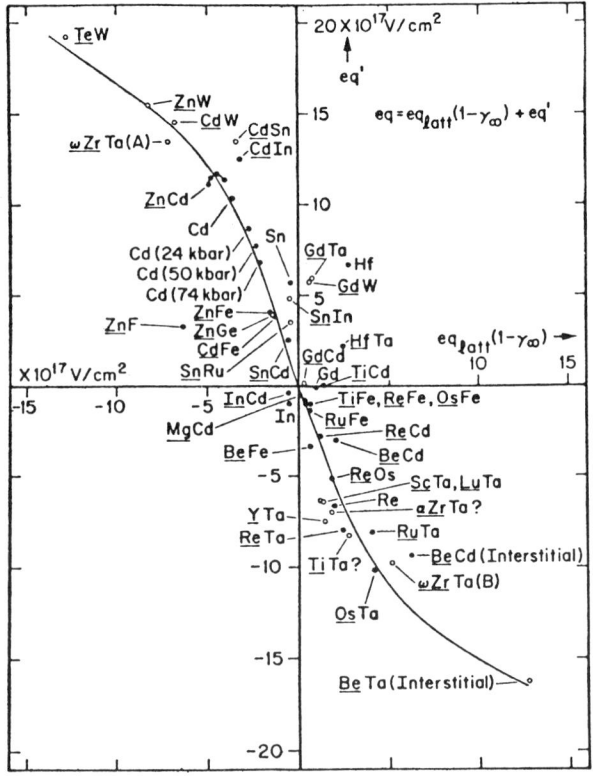

FIGURE 6.5. Correlation of ionic and extra–ionic electric field gradients in metals. Most values refer to room temperature. Underlined symbols refer to the host metal. Solid circles indicate data from e^2qQ values of known signs. For open circles, e^2qQ signs were unknown; locations of these points are predicted. (From Ref. 6.21)

to be drawn from the experimental systematic behavior demonstrated in Fig. 6.5 is that the major portion of the conduction electron field gradient should be universally correlated to, and should be of opposite sign to, $eq_{\text{latt}}(1 - \gamma_\infty)$. It is the dependence on the product $eq_{\text{latt}}(1 - \gamma_\infty)$ that is important in obtaining the indicated correlation. Thus the quantity $eq_{\text{latt}}(1 - \gamma_\infty)$ should be pictured as a measure of the nonsphericity of the atomic core of the interacting nucleus. This nonsphericity results both from the lattice gradient and from the deformability of the core. The implication, then, is that this nonsphericity systematically and proportionately influences the conduction electron distribution and thereby creates a large EFG of the opposite sign.

6.4.1 Cubic Metals

As noted above, since the electric field gradient vanishes at a nuclear site in cubic metals, most investigations, both theoretical and experimental, of EFG have dealt with noncubic metals and alloys or with impurity ion sites in cubic and noncubic metals. However, the presence of an acoustic wave destroys the cubic point symmetry and creates an acoustically induced dynamic electric field gradient (DEFG). NAR enables one to extend the EFG investigations to cubic metals and alloys. Indeed, most of the S-tensor measurements by nuclear acoustic resonance have been made in cubic metals, and in particular, transition metals and alloys.

To analyze S-tensor component measurements in NAR, it is convenient to write for cubic crystals, analogously to (6.21) for q,

$$S_{44} = S_{44_{\text{latt}}}(1 - \gamma_\infty) + S_{44_{ce}}(1 - R) \qquad (6.22a)$$

$$(S_{11} - S_{12}) = (S_{11} - S_{12})_{\text{latt}}(1 - \gamma_\infty) + (S_{11} - S_{12})_{ce}(1 - R), \qquad (6.22b)$$

where the subscript ce refers to conduction electrons. Further simplification is achieved for the lattice contributions by invoking the trace relation $S_{11} + 2S_{12} = 0$, which is satisfied when the charge density at the nucleus is zero. Furthermore, by calculating $S_{12_{\text{latt}}}$ from the relationship $V_{zz}^s = S_{12}\varepsilon_{xx}$, and $S_{44_{\text{latt}}}$ from $V_{zx}^s = S_{44}\varepsilon_{zx}$, one obtains $S_{12_{\text{latt}}} = S_{44_{\text{latt}}}$. The tensor components are thus related to $S_{11_{\text{latt}}}$ by the conditions

$$(S_{11} - S_{12})_{\text{latt}} = \frac{3}{2}S_{11_{\text{latt}}} \qquad (6.23a)$$

$$S_{44_{\text{latt}}} = -\frac{1}{2}S_{11_{\text{latt}}}. \qquad (6.23b)$$

The values of $S_{11_{\text{latt}}}$ may be calculated using the planewise–summation method first used by *De Wette and Schachner* [6.22] in order to improve the convergency of the lattice sum in the calculation of static EFG in hexagonal and tetragonal crystals. The authors utilized a uniform background lattice model, in which the conduction electrons give rise to a uniform negative charge density, and the ionic cores are considered as point charges. The unit cell of the crystal has to satisfy certain symmetry conditions in order for the method to be applicable. Let the unit cell of the lattice be given by the basic vectors a_1, a_2 and a_3; we require the (EFG)$_l$ be in the a_3 direction. This method can be applied in all cases where a_3 is perpendicular to both a_1 and a_2. Thus the (EFG)$_l$ can be evaluated along the c–axis of all crystals with monoclinic or higher symmetry. If the c–axis is a symmetry axis with lower than three–fold symmetry, a second (EFG)$_l$ is required to completely determine the field gradient tensor (with respect to its principal axes). This additional (EFG)$_l$ (for example, along a_1) can also be determined by the De Wette and Schachner method if a_1 is also perpendicular to a_2, which is the case for all crystals with orthorhomic or higher symmetry.

The uniform background lattice is a simple model for a metal. The conduction electrons are considered to be free, giving rise to a uniform charge distribution, and the ionic cores are considered as point charges. The charge density in such a lattice is

$$\rho(r) = \rho_e + Ze\sum_{\lambda}{}' \delta(r - r_\lambda) \,, \qquad (6.24)$$

where ρ_e is the background charge density, Ze is the net charge of the ionic cores, $\sum'$ is a sum over all lattice points excluding the nucleus at the origin, and r_λ designates a lattice point situated at a distance r_λ from the origin and forming an angle θ with the z–axis direction (symmetry axis). Charge neutrality of the elementary cell requires that $\rho_e = -Ze/V$, where V is the volume of the elementary cell (i.e., the volume allotted to one ion).

Following the uniform background lattice model, *Buttet* [6.23] writes

$$\begin{aligned}S_{11} =& 3Ze\sum_{\lambda} \frac{(5\cos^4\theta - 3\cos^2\theta)}{r_\lambda^3} \\ & - 3\frac{Ze}{V}\int \frac{(5\cos^4\theta - 3\cos^2\theta)}{r_\lambda^3} d^3r\end{aligned}$$

and obtains

$$S_{11_{\text{latt}}} = \frac{Z|e|}{a^3}C \,, \qquad (6.25)$$

where $Z|e|$ is the charge of the ion and a is the lattice constant. $C = 10.239$ for a fcc crystal, and $C = 5.758$ for a bcc crystal.

When appropriate, theoretically calculated values for $(1-\gamma_\infty)$ are used in combination with the computed values of $S_{11_{\text{latt}}}$ and the measured S-tensor component magnitude and relative signs, the conduction electron contribution $S_{ij_{ce}}$ to the S-tensor component can, in principle, be determined. Since the dynamic nuclear electric quadrupole transition probabilities (Section 3.4) are proportional to the square of linear combinations of $Q(S_{11}-S_{12})$ and QS_{44}, besides the relative sign of $S_{11}-S_{12}$ and S_{44}, only the absolute values $|Q(S_{11}-S_{12})|$ and $|QS_{44}|$ can be determined in NAR experiments. Assuming that the values of Q are known and that the above measurements and calculations have been made, one can extract the conduction electron contribution to the S-tensor components by utilizing (6.21). By applying the experimental systematics recorded in Fig. 6.5, however, we find that $(1-R)(S_{11}-S_{12})_{ce}$ must dominate $(1-\gamma_\infty)(S_{11}-S_{12})_{\text{latt}}$ and must be opposite in sign. (Caution: from the incompleteness and scatter of data in Fig. 6.5, it is obvious that there may be exceptions to this rule.) This observation [6.24] allows the magnitude and signs of $(S_{11}-S_{12})_{ce}$ and $S_{44_{ce}}$ to be determined.

From analyses based on the pair–potential model, it has been deduced that a substantial fraction of the charge density of the outer electrons of atoms is highly incompressible, but shape–deformable. It is therefore expected that phonon–induced changes in the shape of the electron spatial charge distribution play the dominant role in the volume–conserving part of the electron–phonon interaction. We recall that at a nuclear site and in a coordinate system having its axes along the cubic axes of the unstrained crystal, the components of the traceless DEFG tensor $\widehat{V}_{ij} \equiv (V_{ij} - \delta_{ij}\text{Tr}(\boldsymbol{V}/3))$ can be written as

$$\widehat{V}^s_{ii} = (S_{11}-S_{12})\hat{\varepsilon}_{ii}\,, \qquad \widehat{V}^s_{ij} = 2S_{44}\hat{\varepsilon}_{ij}\,,$$

where $\hat{\varepsilon}_{ij} \equiv (\varepsilon_{ij}-\delta_{ij}\,\text{Tr}(\boldsymbol{\varepsilon}/3))$ are the components of the volume–conserving part of the strain tensor. NAR experiments enable one to study the conduction electron response to the two fundamental shear deformations: *tetragonal shear* (*i.e.*, $\varepsilon_{zz} = -2\varepsilon_{xx} = -2\varepsilon_{yy} \neq 0$, $\varepsilon_{xy} = \varepsilon_{xz} = \varepsilon_{yz} = 0$) and, separately, *trigonal shear* (*i.e.*, $\varepsilon_{xx} = \varepsilon_{yy} = \varepsilon_{zz} = 0$, $\varepsilon_{xy} = \varepsilon_{xz} = \varepsilon_{yz} \neq 0$). Indeed, since the dynamic electric quadrupole interaction is invariant with respect to Tr$\boldsymbol{V}$, it follows that NAR investigations of the electronic contribution to the DEFG are sensitive only to the coupling of electrons to shear modes. Dynamic quadrupole NAR therefore enables one to study the coupling of shear modes to electrons without having to consider changes in the electron charge density caused by volume deformations of the lattice unit cells. The shape changes undergone by a bcc lattice cell under the influence of these shear distortions are shown in Fig. 6.6.

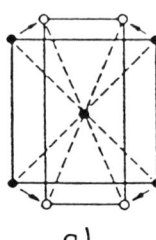 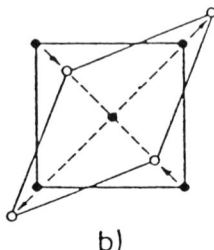

FIGURE 6.6. Tetragonal (a) and trigonal (b) distortions of a bcc lattice cell. (From Ref. 6.26)

Considering that for trigonal shear, as well as for tetragonal shear, a principal axes system (x, y, z) exists in which the DEFG tensor is diagonal, the conduction electron response to tetragonal and trigonal shear may be written in the form of ratios [6.25]:

$$r_{\text{tetr trig}} = \left[\frac{(\widehat{V}_{zz})^s_{\text{ce}}}{(1-\gamma_\infty)(\widehat{V}_{zz})^s_{\text{latt}}} \right] . \qquad (6.26)$$

In terms of the S_{ij}, (6.26) can be written as

$$r_{\text{tetr}} = \frac{(S_{11}-S_{12})_{\text{ce}}}{(1-\gamma_\infty)(S_{11}-S_{12})_{\text{latt}}} \qquad (6.27\text{a})$$

and

$$r_{\text{trig}} = \frac{(S_{44})_{\text{ce}}}{(1-\gamma_\infty)(S_{44})_{\text{latt}}} . \qquad (6.27\text{b})$$

The experimental results for the conduction electron response quantities r_{tetr} and r_{trig} for bcc transition metals are shown in Fig. 6.7. The most striking feature is that, within experimental error, r_{trig} is proportional to the density of states $N(E_f)$ at the Fermi energy, whereas r_{tetr} does not change with $N(E_f)$.

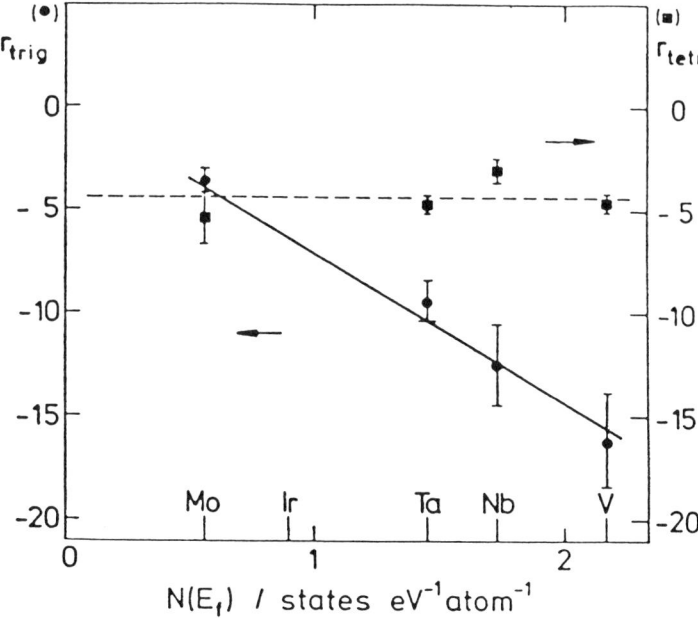

FIGURE 6.7. Electron–phonon coupling in bcc transition metals for tetragonal and trigonal shear modes. The open circles are predicted data for fcc transition metals, obtained on the supposition that fcc metals follow the bcc systematics. (From Ref. 6.25)

An explanation of these electron–phonon coupling effects in bcc transition metals has been given by *Müller et al.* [6.26]. They assume, first, that in response to an applied strain field the d-electron bonds follow the relative displacements of the next-nearest neighbors of the nucleus under question. Since it is clear from Fig. 6.6 that radial distortions of the next nearest neighbors, and therefore of the d-bonds, occur only for trigonal distortions of the cubic unit cell, it follows that r_{tetr} should vanish, and that r_{trig} should be linearly correlated with $N_d(E_f)$, provided that only d-electrons contribute to $(\widehat{V}_{zz})^s_{ce}$. In the presence of trigonal and tetragonal strains, however, the initially spherical shapes of the s-electron cloud are distorted into ellipsoids, thereby giving rise to s-electron transfer into higher orbitals (s–d transitions). It follows from this analysis that r_{tetr} should vary linearly with $N_s(E_f)$, whereas r_{trig} should be correlated linearly with both $N_s(E_f)$ and $N_d(E_f)$. Recalling that in transition metals, $N(E_f)$ in general is dominated by $N_d(E_f)$, and that $N_s(E_f)$ varies only slightly over the transition metal series, it follows, in agreement with the NAR results, that r_{tetr} is expected to remain constant, whereas r_{trig} is expected to correlate predominantly with $N_d(E_f)$.

From the experimental and theoretical work of [6.26] utilizing dynamic quadrupolar NAR on the electron–phonon interaction in bcc transition metals, one concludes that:

(1) For larger values of $N(E_f)$ and for trigonal distortions of the lattice cells, the coupling of electrons to shear modes is dominated by radial distortions of the d-bonds;

(2) For phonons associated with tetragonal distortions of the lattice cells, the coupling of electrons to shear modes is dominated by s-electron transfer into higher orbitals (s-d transitions).

6.4.2 Noncubic Metals

In noncubic metals, the static electric field gradient, SEFG, does not vanish at nuclear sites. Observers have carried out NAR experiments in those metals where either the available Zeeman energies are of the same magnitude as the quadrupole coupling constant energies ($\hbar\gamma H_0 \simeq e^2qQ$) or where the condition of pure quadrupole resonance exists, in which the Zeeman energies are much less than the quadrupole coupling energies ($\hbar\gamma H_0 \ll e^2qQ$).

When $\hbar\gamma H_0 \simeq e^2qQ$, the eigenstates can be obtained only by diagonalizing the full Hamiltonian

(1) for a given orientation of the magnetic field with respect to the electric field gradient symmetry axis and

(2) for a given ratio of Zeeman to quadrupole interactions.

The energy levels can then be computed numerically by diagonalizing the Hamiltonian. On the other hand, when $\hbar\gamma H_0 \ll e^2qQ$, the energy levels are given by the quadrupole Hamiltonian (1.13), with a first–order perturbation due to the Zeeman magnetic field.

For the case of dynamic nuclear electric quadrupole coupling in noncubic metals, the S-tensor defined by (3.46) has different nonzero components than in the cubic case. In general, for any crystal class, the S-tensor components can be determined by the methods of *Fumi and Ripamonti* [6.27, 6.28] by use of (3.46), which defines the S-tensor.

We have discussed (Section 3.6) the S-tensor and the expressions for the dynamic electric field gradient elements for different polarization directions and for both $\Delta m = \pm 1$ and $\Delta m = \pm 2$ transitions in single crystal rhenium. *Stachel and Bömmel* [6.29] measure and interpret the temperature dependence of the nuclear electric quadrupole coupling constant for ^{185,187}Re in single crystal rhenium metal, by using NAR under the condition that $\hbar\gamma H_0 \simeq e^2qQ$.

In noncubic metals that are not transition metals, it is determined empirically that the SEFG at the temperature T is given by

$$eq(T) = eq(0)(1 - BT^{\frac{3}{2}}) ,\qquad (6.28)$$

where B is a positive constant that depends on the metal. The d–electron contribution in transition metals may be expected to cause a different temperature dependence of the SEFG in transition metals.

A theoretical temperature dependence of $eq(T)$ in the transition metals is computed by *Piecuch and Janot* [6.30]—see (6.21). Their calculations, based on the tight–binding model, predict that the SEFG for rhenium should decrease with increasing temperature; however, these authors do not predict the functional form of the temperature dependence.

NAR resonance line position measurements on both ^{185}Re and ^{187}Re are carried out between 2 and 150 K. The temperature dependence of both the Knight shift and the SEFG are found from interpretation of this data. We write the measured SEFG as the sum (6.21) of lattice and electronic contributions. The lattice contributions for this hexagonal crystal are computed in (6.29) using the lattice–summing procedures of *DeWette*:

$$eq_{\text{latt}} = \frac{Z^+e}{4\pi\epsilon a^3}\left[0.0065 - 4.3584\left(\frac{c}{a} - 1.633\right)\right] .\qquad (6.29)$$

From the measured temperature dependence of the rhenium lattice constants [6.31], $eq_{\text{latt}}(T)$ is calculated from (6.29) to be

$$eq_{\text{latt}}(T) = eq_{\text{latt}}(0)(1 + 1.85 \times 10^{-7}T^{2.2}) .\qquad (6.30)$$

The data of Fig. 6.8 shows, first, that the experimentally determined SEFG increases with temperature—contradicting the theoretical prediction of Piecuch and Janot. Second, the experimentally determined points in Fig. 6.8 can be fitted with the exponential function

$$\left|\frac{e^2qQ}{h}\right|(T) = \left|\frac{e^2qQ}{h}\right|(0)(1 + 1.5 \times 10^{-7}T^{2.2}) .\qquad (6.31)$$

The dash–dot line in Fig. 6.8 is a plot of $eq_{\text{latt}}(T)$ of (6.30) times eQ/h. We note that both the fit to the experimental data (6.31) and the lattice contribution (6.30) have the same $T^{2.2}$ temperature dependence. The fact that the lattice contribution, (6.30), lies above the experimental points in Fig. 6.8 suggests that other contributions to the SEFG are present. The

Universal Correlation of *Raghavan et al.* [6.21] gives an empirical relationship between the total SEFG, eq, and the lattice part, eq_{latt}: namely,

$$eq = -k(1 - \gamma_\infty)eq_{\text{latt}} , \qquad (6.32)$$

where $k = 2$ to 3. From the above equation and from (6.21), the electronic contribution $eq_{el}(1 - R)$ is determined to be three to four times the lattice contribution $eq_{\text{latt}}(1 - \gamma_\infty)$. Therefore the electronic contribution is dominant in the SEFG; however, the temperature dependence up to 150 K is determined by the lattice contribution to the SEFG.

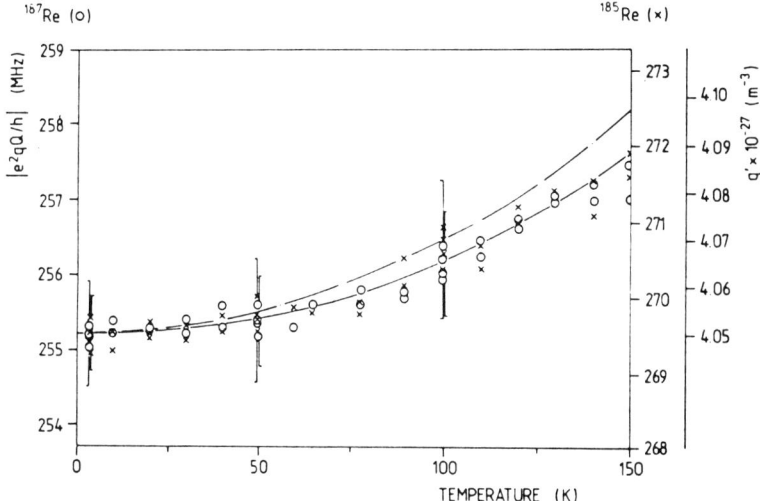

FIGURE 6.8. Temperature dependence of the nuclear quadrupole interaction in Re. The solid line represents an exponential function; the dash–dotted line represents the temperature behavior of the lattice gradient eq_{latt}, (6.31). (From Ref. 6.29)

CHAPTER 6 REFERENCES

6.1 M. H. Cohen, F. Reif. "Quadrupole Effects in Nuclear Magnetic Resonance Studies of Solids." *Solid State Physics*, Vol. 5, Ed. by F. Seitz and D. Turnbull (Academic Press, New York 1957).

6.2 R. M. Sternheimer. "Shielding and Antishielding of Nuclear Quadrupole Moments." *Z. Naturforsch* **41a**, 24–36 (1986).

6.3 P. C. Schmidt, K. D. Sen, T. P. Das, A. Weiss. "Effect of Self-Consistency and Crystalline Potential in the Solid State on Nuclear Quadrupole Sternheimer Antishielding Factors in Closed-Shell Ions." *Phys. Rev. B* **22**, 4167–4179 (1980).

6.4 R. E. Watson, R. M. Sternheimer, L. H. Bennett. "Trends in the Electric Quadrupole Fields at Dilute Impurity Sites in Transition Metal–Transition Metal Alloys." *Phys. Rev. B* **30**, 5209–5219 (1984).

6.5 T. Kanashiro. "Gradient–Elastic Tensor in Alkali Bromide Crystals." *J. Phys. Soc. Japan* **32**, 1270–1280 (1972).

6.6 T. Kanashiro, T. Ohno, M. Satoh. "Gradient Elastic Tensor in Ferroelectric Sodium Nitrite, $NaNO_2$." *J. Phys. Soc. Japan* **38**, 1293–1301 (1969).

6.7 H. Betsuyaku. "Charge Distribution and Nuclear Quadrupole Interactions in Ionic Crystals." *J. Chem. Phys.* **51**, 2546–2561 (1969).

6.8 A. C. Beri, T. Lee, T. P. Das, R. M. Sternheimer. "First–Principles Theory of Anti–Shielding Effects in the Nuclear Quadrupole Interaction in Ionic Crystals: Application to ^{57}Fe in Fe_2O_3." *Phys. Rev. B* **28**, 2335–2351 (1983).

6.9 R. K. Sundfors. "Nuclear Acoustic Resonance of ^{73}Ge in Single–Crystal Germanium; Interpretation of Experimental Gradient–Elastic Tensor Components in Germanium and Zinc–Blende Compounds." *Phys. Rev. B* **20**, 3562-3565 (1979).

6.10 A. F. Oluwole, S. G. Schmelling, H. A. Shugart. "Nuclear Spins, Nuclear Moments, and Hyperfine Structure of ^{69}Ge and ^{73}Ge and $^{3}P_2$ Hyperfine Structure of ^{71}Ge." *Phys. Rev. C* **2**, 228–237 (1970).

6.11 F. W. Landhoff, R. R. Hurst. "Multipole Polarizabilities and Shielding Factors from Hartree–Foch Wave Functions." *Phys. Rev.* **139**, A1415–A1425 (1965).

6.12 R. M. Sternheimer. "Shielding and Antishielding Effects in Various Ions and Atom Systems." *Phys. Rev.* **146**, 140–160 (1966).

6.13 R. M. Sternheimer. "Quadrupole Antishielding Effects of Ions." *Phys. Rev.* **159**, 266–272 (1967).

6.14 R. E. Watson, A. J. Freeman. "Quadrupole Antishielding Factors for Rare–Earth and Some Other Heavy Atoms." *Phys. Rev.* **135**, A1209–A1212 (1964).

6.15 C. Froese. "Hartree–Foch Parameters for the Atoms Helium to Radon." *J. Chem. Phys.* **45**, 1415–1420 (1966).

6.16 B. J. Wyluda. "Nuclear Spin–Lattice Relaxation Time in Germanium". *J. Phys. Chem. Solids* **23**, 63–65 (1962).

6.17 B. Bleaney, K. D. Bowers, M. H. L. Pryce. "Paramagnetic Resonance in Diluted Copper Salts. III. Theory and Evaluation of the Nuclear Electric Quadrupole Moments of ^{63}Cu and ^{65}Cu." *Proc. R. Soc. London A* **228**, 166–174 (1965).

6.18 G. A. Sawatzky, F. Van Der Woude. "Covalency Effects in the Hyperfine Interaction." *Colloque No. 6, Supplement J. Physique* **C6**, 47–60 (1974).

6.19 W. A. Harrison, J. C. Phillips. "Angular Forces in Tetrahedral Solids." *Phys. Rev. Lett.* **33**, 410–411 (1974).

6.20 R. E. Watson, A. C. Gossard, Y. Yafet. "Role of Conduction Electrons in Electric Field Gradients of Ordered Metals." *Phys. Rev. A* **140**, 375–388 (1965).

6.21 R. S. Raghavan, E. N. Kaufmann, P. Raghavan. "Universal Correlation of Electronic and Ionic Field Gradients in Non–Cubic Metals." *Phys. Rev. Lett.*, **34**, 1280–1283 (1975).

6.22 F. W. De Wette, G. E. Schachner. "Electric Field Gradients and Uniform–Background Lattices. II." *Phys. Rev.* **137A**, 92–94 (1965).

6.23 J. Buttet. "Determination of the Gradient–Elastic Tensor in Aluminum Using Nuclear Acoustic Resonance." *J. Phys. F Metal. Phys.* **3**, 918–925 (1973).

6.24 E. Fischer, V. Müller, D. Ploumbidis, G. Schanz. "Nuclear Acoustic Resonance Determination of the Electronic Contribution to the Electric–Field–Gradient Tensor in Molybdenum." *Phys. Rev. Lett.* **40**, 796–799 (1978).

6.25 V. Müller, E. J. Unterhorst, W. Neumann. "Nuclear Acoustic Resonance Measurements of the Electron–Phonon Interaction in bcc Transition Metals." *Phonon Scattering in Condensed Matter*, ed. W. Eisenmenger, K. Lassmann and S. Döttinger (Springer–Verlag, Berlin 1984), 322–324.

6.26 V. Müller, E. J. Unterhorst, W. Neumann, G. Schanz, C. Schubert. "Nuclear Acoustic Resonance Investigations of the Long–Wavelength Electron–Phonon Interaction in Vanadium." *Phys. Lett.* **99A**, 249–252 (1983).

6.27 F. G. Fumi, C. Ripamonti. "Tensor Properties and Rotational Symmetry of Crystals. I. A New Method for Group $3(3_z)$ and Its Application to General Tensors up to Rank 8*." *Acta Crystallogr. A* **36**, 535–551 (1980).

6.28 F. G. Fumi, C. Ripamonti. "Tensor Properties and Rotation Symmetry of Crystals. II. Groups with 1–, 2–, and 4–fold Principal Symmetry and Trigonal and Hexagonal Groups Different from Group 3*." *Acta Crystallogr. A* **36**, 551–558 (1980).

6.29 M. Stachel, H. E. Bömmel. "Temperature Dependence of the Nuclear Quadrupole Interaction and the Knight Shift in Rhenium Metal." *Appl. Phys. A* **30**, 27–32 (1983).

6.30 M. Piecuch, C. Janot. "Electrical Field Gradient in Pure Hexagonal Transition Metals." *Hyperfine Int.* **5**, 69–79 (1977).

6.31 V. A. Finkel', M. J. Palatnik, G. P. Kovtun. "X–ray Diffraction Study of the Thermal Expansion of Rubidium, Osmium and Rhenium." *Fiz. Met. Met.* **32**, 212–215 (1971). English trans: *Phys. Met. Metallogr.* **32**, 231–235 (1971).

CHAPTER 7
EXPERIMENTAL TECHNIQUES

7.1 Introduction

7.2 Continuous Wave Ultrasonics: Propagating Wave Model

7.3 CW Spectrometers
- 7.3.1 General Sensitivity Considerations
- 7.3.2 Marginal Oscillator Spectrometers
    - 7.3.2.1 Nature of Operation
    - 7.3.2.2 The MOUS: Marginal Oscillator Ultrasonic Spectrometer
    - 7.3.2.3 The TOUS: Transmission Oscillator Ultrasonic Spectrometer
- 7.3.3 Transmission and Transmission Bridge Spectrometers
- 7.3.4 Reflection Bridge Spectrometers
- 7.3.5 Sampled–CW Spectrometer

7.4 Signal Calibration

7.5 Transient NAR Techniques
- 7.5.1 Theory
- 7.5.2 Saturation Methods
- 7.5.3 Adiabatic Fast Passage
- 7.5.4 Rotating Frame Acoustic Relaxation
- 7.5.5 Pulse Echo

7.6 The Composite Resonator
- 7.6.1 Sample Preparation
- 7.6.2 Transducers and Bonds
- 7.6.3 CW Resonator Assemblies and Probes
    - 7.6.3.1 Sample Probes Used in Electromagnets
    - 7.6.3.2 Probes for Superconducting Magnets

7.7 Application: NAR in Rhenium

7.8 Application: Experimental Search for the Dynamic Nuclear Hexadecapole Interaction

References

7. EXPERIMENTAL TECHNIQUES

7.1 Introduction

The experimental techniques necessary for observing nuclear acoustic resonance combine those used in ultrasonic physics and those in nuclear magnetic resonance. In the familiar NMR technique, transitions among nuclear spin energy levels are induced by coupling electromagnetic energy to the nuclear spin system by means of an external RF coil that, when driven by a suitable RF oscillator, gives rise to an RF magnetic field. This field, denoted $\boldsymbol{H}_1$, interacts directly with the magnetic dipole moment of the nucleus. The NMR technique is not suitable for bulk conductors, since the externally generated RF magnetic field cannot penetrate beyond the RF skin depth of the conducting sample. In the technique of nuclear acoustic resonance, ultrasonic energy from an external source is introduced into the sample. Transitions among nuclear spin energy levels are induced by the effect of the ultrasound on an internal interaction (*e.g.*, the dynamic nuclear electric quadrupole interaction) or via an internally generated RF magnetic field $\boldsymbol{H}_1$ (dynamic Alpher–Rubin—or dipolar—interaction). Unlike electromagnetic energy, there is no special difficulty in introducing ultrasonic energy into bulk conductors.

In NAR the resonant coupling of ultrasound to the nuclear spin system is normally observed as a change of ultrasonic attenuation $\Delta\alpha$ or of the dispersion $\Delta v_a/v_a$ when an externally applied DC magnetic field is swept through the appropriate range at constant frequency. The very small changes characteristic of NAR absorption and dispersion of ultrasound are conventionally detected using continuous wave (CW) ultrasonic techniques. Techniques for making transient measurements in NAR are also reviewed below. Often, the solid sample is prepared with two opposite faces optically flat and parallel so that the ultrasonic waves propagated normal to these faces undergo multiple constructive interference, resulting in a frequency spectrum of standing wave (mechanical) resonances at frequencies (for an isolated sample) $\nu_n = nv/2l$, where v is the ultrasonic velocity, n is an integer, and l is the length of the sample. In practice, a piezoelectric transducer of the proper fundamental frequency and mode of vibration is bonded to one face of the sample. Although the analysis of the resulting composite resonator (transducer–bond–sample) is complex, the propagating wave model (Section 7.2) is capable of characterizing not only CW responses but also sampled–CW and pulse–echo responses.

For typical values of α, the acoustic (composite) resonator provides a signal enhancement, when the CW oscillator, which drives the piezoelectric transducer, is tuned to a frequency corresponding to the center of a mechanical resonance, of from one to several orders of magnitude, depending upon the (nonresonant) attenuation of the sample. Near a mechanical resonance, the acoustic resonator behaves electrically like a series RLC circuit. A change in velocity affects the reactive elements, shifting the resonant frequency by $\Delta v_a/v_a = \Delta \nu/\nu$; a change in attenuation changes the loss element according to $\Delta\alpha/\alpha = \Delta R/R$. Thus electrical techniques, as in NMR, can be used to measure ultrasonic absorption and dispersion—in particular (if sensitive enough)—NAR absorption and dispersion.

Depending upon the particular spectrometer used, one may monitor separately either the in–phase (A_1) or the out–of–phase (A_2) component of the acoustic resonator response, or one may simply measure the magnitude $|A| = (A_1^2 + A_2^2)^{\frac{1}{2}}$ without regard for the phase relationship of the response relative to the driving RF oscillator. For example, using phase–sensitive or bridge detection techniques, one may measure A_1 or A_2. The marginal oscillator ultrasonic spectrometer is sensitive only to A_1; the use of linear diode detection results in a measurement of $|A|$. Sensitivity consideration and NAR spectrometers are considered in detail in Section 7.3. Techniques for calibrating the sensitivity of NAR spectrometers are dealt with in Section 7.4. Because of its uniqueness, the SQUID NAR spectrometer is dealt with separately in Chapter 10. Transient NAR techniques are reviewed in Section 7.5. Preparation, construction and testing of acoustic composite resonators are discussed in Section 7.6. In Sections 7.7 and 7.8 we describe two specific applications in NAR of the techniques discussed in the present chapter.

Emphasis in the present text has been on the technique of NAR, in which only acoustic energy is introduced into the sample. NAR is thereby directly applicable to the study of bulk conductors as well as nonconductors. The closely related technique of acoustic saturation nuclear magnetic resonance (ASNMR), described briefly in Chapter 1, has been used extensively, especially in Russia, to investigate nonconductors. ASNMR utilizes both electromagnetic and acoustic energy in its studies. It is a powerful and sophisticated technique for studying spin–phonon interactions in insulators. The techniques of ASNMR, as well as a sampling of results obtained, are summarized in Chapter 8.

7.2 Continuous Wave Ultrasonics: Propagating Wave Model

As mentioned above, in continuous wave ultrasonics the sample is usually prepared with two opposite faces optically flat and parallel so that the ultrasonic waves propagated normal to these faces undergo multiple constructive interferences, resulting in a frequency spectrum of mechanical resonances. A typical swept–frequency mechanical resonance pattern, consisting of a broad transducer response envelope enclosing the approximately equally spaced acoustic standing wave resonances, is shown in Fig. 7.1. When the

CW oscillator frequency is adjusted to the center of a mechanical resonance, the electrical impedance may be represented by a parallel combination of a capacitive reactance of the quartz transducer (*e.g.*, approximately 100 pF at 12.5 MHz for a 0.95–cm–diameter X–cut transducer) and the real part of the composite resonator impedance (100 Ω for strong electrical coupling and low acoustic attenuation in the composite resonator). Mechanical resonance line widths can be as narrow as 100 Hz, *e.g.*, for low acoustic attenuation with a 10–MHz AT–cut 0.95–cm–diameter transducer.

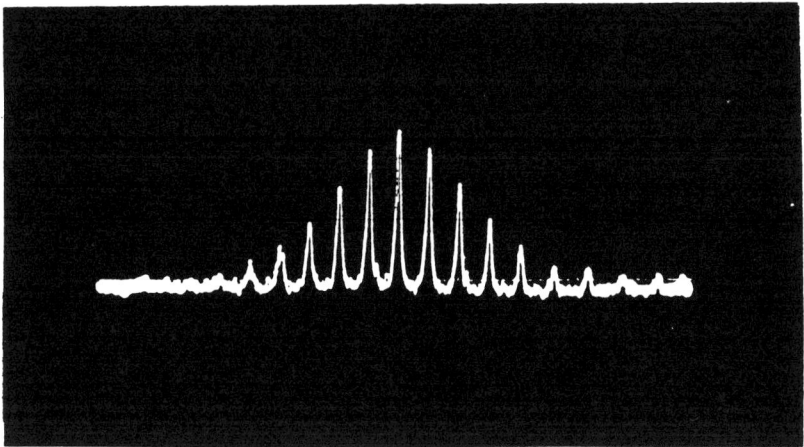

FIGURE 7.1. Typical swept–frequency mechanical resonance pattern.

Explicit expressions for the responses of acoustic resonators can be obtained by using a propagating wave model [7.1]. The model is of sufficient generality to include not only the CW but also the pulse–echo and sampled–CW responses. Because of its fundamental application to NAR (both CW and transient) and to ASNMR, we present this model in some detail. We limit the discussion to the case of an ultrasonic resonator in the form of an isolated specimen of length $l \equiv a/2$. We consider only instances in which the particle velocity can be expressed as a damped traveling wave $e^{-az}\cos(\omega t - kz)$, where $k = \omega/v$. A single transducer, affixed to the left face ($z = 0$) of the specimen, both provides the driving energy and monitors the resulting particle velocity $A(t)$ at $z = 0$. The method of analysis, which yields an expression for the amplitude and phase of the resulting particle velocity at $z = 0$, consists of summing the contributions to the particle velocity A at the $z = 0$ face resulting from waves that had been generated at $z = 0$ in the past and that have returned to $z = 0$ after multiple reflections from the end faces. In general, if either a purely longitudinal or purely transverse acoustic wave is incident upon a boundary, four waves—two longitudinal and two transverse—result. In the present section, however, only one-dimensional propagation and normal incidence will be treated. In these cases, no mode conversion occurs, and for an incident wave, longitudinal or transverse, only two waves of the same mode—one transmitted and one reflected—result. For the isolated specimen considered here, there are no transmitted waves. We assume that reflection at $z = 0$ and $z = a/2$ results only in a reversal of the direction of propagation.

A suitably coupled oscillator of frequency ω and constant amplitude is gated on at $t = 0$ and off at $t = t_d$. The driving particle velocity at $z = 0$ resulting from the action of the oscillator on the transducer is chosen arbitrarily to be of unit amplitude and zero phase relative to $\cos \omega t$. We introduce a shape function $\Delta(t)$ that describes the manner in which the oscillator is gated:

$$\Delta(t) = \begin{cases} 0, & t \leq 0 \\ 1, & 0 < t < t_d \\ 0, & t \geq t_d \, . \end{cases} \quad (7.1)$$

Using $\cos(\omega t - kz) = \Re[e^{i(\omega t - kz)}]$ and defining $\tau = a/v$ as the time required for the round trip of an acoustic wave in the specimen, one obtains for the (complex) particle velocity at $z = 0$

$$\begin{aligned}\tilde{A}(t) =& e^{i\omega t}[\Delta(t) + e^{-(\alpha a + ika)}\Delta(t - \tau) \\ &+ e^{-2(\alpha a + ika)}\Delta(t - 2\tau) + \cdots \\ &+ e^{-N(\alpha a + ika)}\Delta(t - N\tau) + \cdots] \, . \end{aligned} \quad (7.2)$$

The simple pulse–echo case, in which a pulse of electromagnetic energy generates an acoustic pulse that reflects back and forth within the specimen and that produces a signal each time the wave packet strikes the transducer, corresponds to the limit $t_d \ll \tau$. The spectrometer output consists, in this case, of a series of echoes equally spaced in time by τ. From (7.2), the Nth echo consists of a signal at the carrier frequency ω with well-defined phase (if $t_d \gg 2\pi/\omega$) modulated by an envelope whose amplitude is given by $e^{-N\alpha a}$.

The continuous wave case corresponds to the opposite extreme, $t_d \gg \tau$. Observation is begun only after a sufficient time has elapsed, so that a steady state condition has been reached. Under these conditions, the factors $\Delta(t - N\tau)$ in (7.2) are all simultaneously equal to unity. Under these circumstances, and for $\alpha > 0$, one may sum the geometric series to obtain

$$\tilde{A} = \frac{e^{i\omega t}}{1 - e^{-(\alpha a + ika)}} \, . \quad (7.3)$$

7.2 PROPAGATING WAVE MODEL

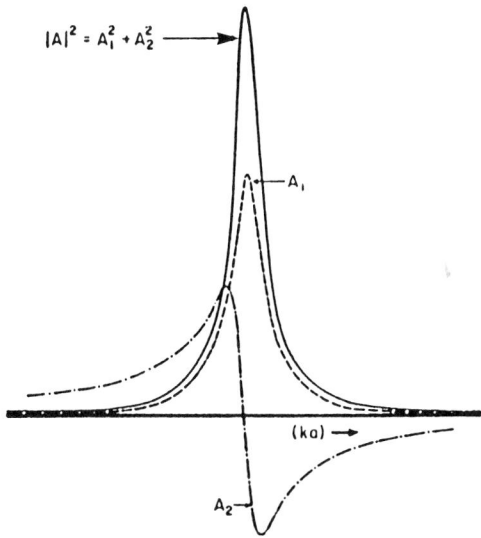

FIGURE 7.2. Plot of A_1, A_2 and $|A|^2 = A_1^2 + A_2^2$, in arbitrary units, as functions of ka in the vicinity of a mechanical resonance. The vertical scale for $|A|^2$ differs from that of A_1 and A_2. (From Ref. 7.1)

The resulting particle velocity $A = \Re[\tilde{A}]$ corresponding to the unit driving particle velocity $\cos \omega t$ is

$$A = A_1 \cos \omega t + A_2 \sin \omega t , \qquad (7.4a)$$

where

$$A_1 = \frac{(e^{\alpha a} - \cos ka)}{2(\cosh \alpha a - \cos ka)} \qquad (7.4b)$$

and

$$A_2 = \frac{\sin ka}{2(\cosh \alpha a - \cos ka)} . \qquad (7.4c)$$

The resulting particle velocity at $z = 0$ is seen to consist of a term (A_1) oscillating in phase with the driving oscillator and a term (A_2) oscillating in quadrature. A plot of (7.4a) as a function of frequency yields a set of equally spaced mechanical resonances whose frequencies correspond to the condition that the length of the crystal be equal to an integral number of half wavelengths. For the mth mechanical resonance, $ka = 2\pi m$ or

$\omega = \omega_m \equiv 2\pi m/\tau$. Figure 7.2 shows a plot of A_1, A_2 and $|A|^2 = A_1^2 + A_2^2$ in the region of one such mechanical resonance, where

$$|A| = \frac{e^{\frac{1}{2}\alpha a}}{\sqrt{2}(\cosh \alpha a - \cos ka)^{\frac{1}{2}}} \quad . \tag{7.4d}$$

Since A_1 is periodic in ka, it is convenient to translate the origin so that $ka = 0$ corresponds to the center of a particular mechanical resonance. In many cases of practical interest for CW ultrasonics, $\alpha a \ll 1$, so that, in the region of a particular mechanical resonance centered at $ka = 0$, (7.4b) to (7.4d) reduce to

$$A_1 \simeq \frac{\alpha a}{(\alpha a)^2 + (ka)^2} \tag{7.5a}$$

$$A_2 \simeq \frac{ka}{(\alpha a)^2 + (ka)^2} \tag{7.5b}$$

and

$$|A|^2 \simeq \frac{1}{(\alpha a)^2 + (ka)^2} \quad . \tag{7.5c}$$

From (7.5) one sees that the expressions for A_1, A_2 and $|A|^2$ are very good fits to Lorentz line shapes in the region of a mechanical resonance if $\alpha a \ll 1$. Off–resonance, the in–phase component of a Lorentz line shape (7.5a), goes to zero as $(ka)^{-2}$, while the exact expression for A_1 (7.4b) approaches a nonzero value ranging from 0.5 for specimens with very low-ultrasonic attenuation to 1.0 for specimens with high attenuation. In the limit of very high attenuation, for all frequencies, $A_1 = 1$, $A_2 = 0$, and no mechanical resonances occur. An alternate expression equivalent to (7.5a–c), for the frequency response in the vicinity of the particular mechanical resonance centered at $\omega = \omega_m$ is

$$A_1 \simeq \left(\frac{1}{\tau}\right) \frac{\omega_\alpha}{\omega_\alpha^2 + (\omega_m - \omega)^2} \tag{7.6a}$$

$$A_2 \simeq \left(\frac{1}{\tau}\right) \frac{\omega_m - \omega}{\omega_\alpha^2 + (\omega_m - \omega)^2} \tag{7.6b}$$

and

$$|A|^2 \simeq \left(\frac{1}{\tau}\right)^2 \frac{1}{\omega_\alpha^2 + (\omega_m - \omega)^2} \quad . \tag{7.6c}$$

Thus $\omega_\alpha \equiv \alpha v$ is seen to be one-half the natural, or *homogeneous*, linewidth $\Delta\omega$ of the mechanical resonance.

Experimental verification of (7.2), obtained using a transmission-type acoustic resonator assembly in order to minimize cross-talk between receiver and transmitter, is presented in Fig. 7.3. The three sets of oscilloscope tracings are obtained with a progressively longer—from (a) to (c)—horizontal time base. The upper trace in each set corresponds to the simple pulse–echo limit $t_d \ll \tau$; the lower trace, to the limit $t_d \gg \tau$. As can be seen from the lower trace of set (c), the t_d chosen is sufficiently long that steady state (*i.e.*, CW) conditions are reached. The CW condition continues until the oscillator is gated off at $t = t_d$, whereupon the oscillation decays back to zero according to (7.2). The frequency ω in Fig. 7.3 is chosen to correspond to the center of a CW acoustic standing wave resonance, *i.e.*, $ka = 2\pi m$, where $m = 1, 2, 3, \cdots$ in (7.2).

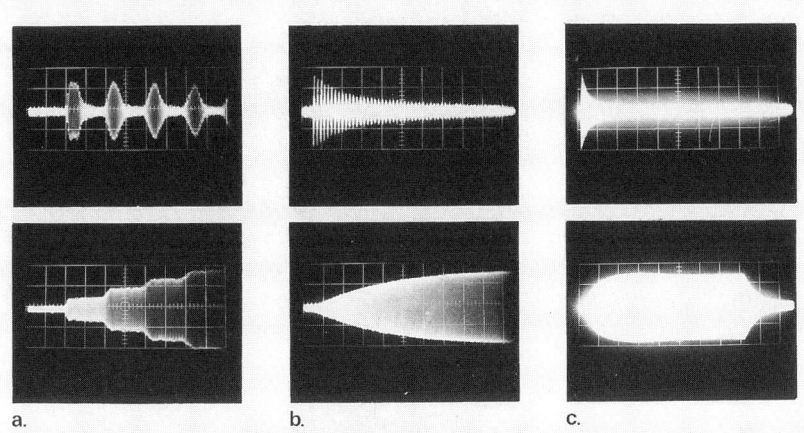

FIGURE 7.3. Three sets of oscilloscope tracings on progressively longer time bases from (a) to (c). Upper traces correspond to $t_d \ll \tau$; lower, to $t_d \gg \tau$. The frequency corresponds to that of the center of a standing wave acoustic resonance. (From Ref. 7.2)

One may observe a steady state response corresponding to (7.2) in the time domain as well as in the frequency domain. The frequency domain response consists of a set of acoustic standing wave resonances, each essentially described by a Lorentz line shape. The center of a particular mechanical resonance corresponds to a condition of complete phase coherence of the acoustic energy stored by multiple reflection within the specimen. The corresponding time domain response is defined as the decay from the steady–state condition. If the oscillator is tuned to the center of a mechanical resonance (*i.e.*, $\omega = \omega_m$), the decay beginning at $t = t_d$ consists of a series of discrete steps resulting from the turning off of the individual terms in (7.2). Although most CW spectrometers are designed to operate in the frequency domain, the sampled–CW spectrometer, described in detail in Section 7.3.5, is capable of operating in either the frequency or time domain observation modes.

Some features of the time and frequency domain modes of observation are illustrated in Fig. 7.4. Shown are the frequency domain response and the corresponding time domain decay of a harmonic resonance mode of an essentially ideal one–dimensional resonator. The 10.03–MHz response corresponds to the propagation of longitudinal waves in a 1.27–cm–length 1.59–cm–diameter cylinder of fused quartz whose end faces are prepared flat and parallel to optical specifications (flatness: $\frac{1}{10}$ wavelength of sodium light; parallelism: 12 min of arc). A 10–MHz X–cut quartz transducer is bonded to the specimen with silicone grease. In Fig. 7.4a are shown the frequency domain response and the corresponding (undetected) stepwise time domain response. The detected time domain response (exhibiting the stepwise decay) and an electronically generated exponential are shown in the photograph of Fig. 7.4b. The total sweep time for the photograph in (a) is about 1.1 msec; for the photograph in (b), about 0.1 msec. A magnified display of the difference between the ultrasonic and exponential decays is also presented in (b). The measured time between the steps is in good agreement with the observed acoustic pulse round trip time τ for the specimen.

7.3 CW Spectrometers

Several types of sensitive CW ultrasonic spectrometers have been used to observe and study NAR absorption and dispersion in solids. As mentioned above, depending upon the particular spectrometer used, one may monitor separately either the in–phase (A_1) or the out–of–phase (A_2) component of the acoustic resonator response, or one may measure the magnitude of $|A|$ without regard for the phase relationship of the response relative to the driving oscillator. Of the spectrometers reviewed below, RF reflection and transmission bridge spectrometers measure A_1 or A_2; marginal oscillator ultrasonic spectrometers are sensitive only to A_1; the sampled–CW spectrometer and simple (non–bridge) transmission spectrometers measure $|A|$. The use of a bolometer or other power–sensitive device (*e.g.*, a diode operated in the square law region) corresponds to a measurement of $|A|^2$. When, in the use of NAR spectrometers, a frequency corresponding to the center of a particular mechanical resonance is selected, $A_1 = |A|$.

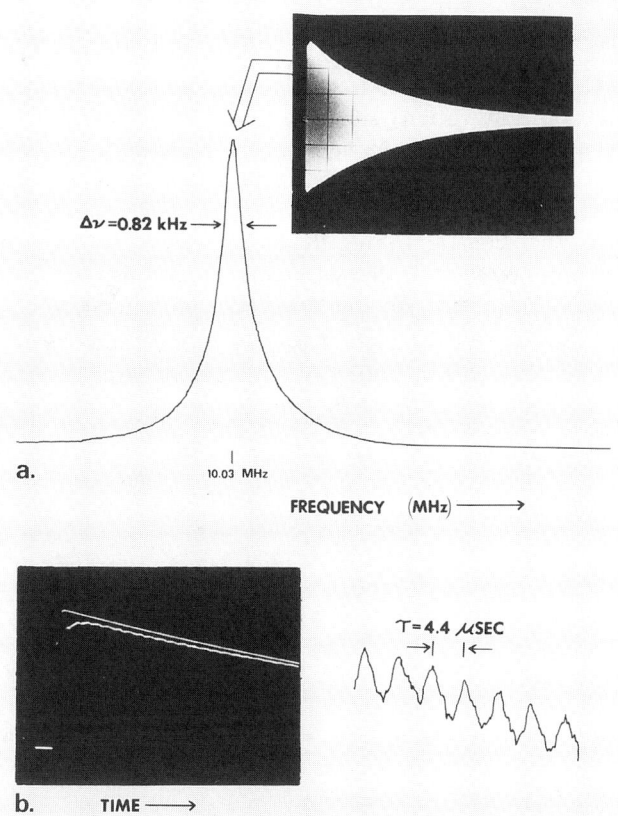

FIGURE 7.4. Frequency and time responses of the 10.03–MHz harmonic mode of an ideal one–dimensional fused quartz resonator: (a) frequency domain response and undetected time domain response; (b) detected time domain response shown with exponential comparator decay, after subtraction of exponential comparator decay. Total sweep time in the photograph of (a) is ~ 1.1 msec; in the photograph of (b), ~ 0.1 msec. (From Ref. 7.3)

7.3.1 General Sensitivity Considerations

An important function of the acoustic resonator is enhancing the sensitivity of the CW spectrometer to small changes in acoustic phase velocity and attenuation. In this respect, the composite resonator plays a role in NAR similar to that played by a high-Q RF coil in NMR or by a high-Q resonant cavity in electron spin resonance. A straightforward procedure for calculating the sensitivity–enhancement factors appropriate to CW NAR utilizes the propagating wave model discussed in Section 7.2. In practice, in an NAR experiment, one selects a particular mechanical resonance and measures the response at fixed frequency of small resonant changes in acoustic attenuation (absorption) and phase velocity (dispersion). Maintaining a fixed frequency of the acoustic signal is possible in most NAR experiments because the external magnetic field is the variable used to sweep out the NAR spectrum. In the SQUID-detected NAR technique, described in Chapter 10, the acoustic driving frequency is used to sweep out the resonant spectrum; however, further developments in this technique should also result in fixed-frequency, variable magnetic field operation.

For sufficiently small changes $\Delta\alpha$ (absorption) and Δk (dispersion) in acoustic properties corresponding to a change in some external parameter (*e.g.*, external magnetic field), the corresponding changes in the acoustic response may be written

$$\Delta A_1 \simeq \frac{\partial A_1}{\partial \alpha}\Delta\alpha + \frac{\partial A_1}{\partial k}\Delta k \tag{7.7a}$$

and

$$\Delta A_2 \simeq \frac{\partial A_2}{\partial \alpha}\Delta\alpha + \frac{\partial A_2}{\partial k}\Delta k \ . \tag{7.7b}$$

We define $S_A = \partial A_1/\partial\alpha$ and $S_D = \partial A_1/\partial k$ as absorptive and dispersive sensitivity-enhancement factors, respectively. Similar expressions hold for A_2; *e.g.*, a small change $\Delta\alpha$ is magnified by an amount given by the appropriate absorptive sensitivity enhancement factor $\partial A_2/\partial\alpha$. Evaluating S_A and S_D using (7.4b,c), one obtains [7.1, 7.4]

$$\frac{\partial A_1}{\partial \alpha} = -\frac{\partial A_2}{\partial k} = \frac{a}{2}\frac{1 - \cosh\alpha a \cos ka}{(\cosh\alpha a - \cos ka)^2} \tag{7.8a}$$

and

$$\frac{\partial A_1}{\partial k} = \frac{\partial A_2}{\partial \alpha} = -\frac{a}{2}\frac{\sinh\alpha a \sin ka}{(\cosh\alpha a - \cos ka)^2} \ . \tag{7.8b}$$

Similarly, for the amplitude response $|A|$,

$$\Delta|A| = \frac{\partial |A|}{\partial \alpha}\Delta\alpha + \frac{\partial |A|}{\partial k}\Delta k \ , \tag{7.9a}$$

where

7.3 CW SPECTROMETERS

$$\frac{\partial |A|}{\partial \alpha} = \frac{a(e^{-\frac{1}{2}\alpha a} - e^{\frac{1}{2}\alpha a}\cos ka)}{2\sqrt{2}(\cosh \alpha a - \cos ka)^{\frac{3}{2}}} \quad (7.9\text{b})$$

and

$$\frac{\partial |A|}{\partial k} = -\frac{ae^{\frac{1}{2}\alpha a}\sin ka}{2\sqrt{2}(\cosh \alpha a - \cos ka)^{\frac{3}{2}}} \,. \quad (7.9\text{c})$$

In the usual case in which $\alpha a \ll 1$ and measurements are made at frequencies close to a mechanical resonance, it is convenient to rewrite (7.6a–c) in terms of $k = \omega/v$ and $k_m = \omega_m/v$:

$$A_1 \simeq \frac{\alpha a}{(\alpha a)^2 + a^2(k - k_m)^2} \quad (7.10\text{a})$$

$$A_2 \simeq \frac{(k - k_m)a}{(\alpha a)^2 + a^2(k - k_m)^2} \quad (7.10\text{b})$$

and

$$|A|^2 \simeq \frac{1}{[(\alpha a)^2 + a^2(k - k_m)^2]} \,, \quad (7.10\text{c})$$

where $\omega_m = 2\pi mv/a$, corresponding to the mth mechanical resonance. The simplified sensitivity–enhancement factors, calculated using (7.10a–c), are:

$$\frac{\partial A_1}{\partial \alpha} = -\frac{\partial A_2}{\partial k} = \frac{a[a^2(k - k_m)^2 - (\alpha a)^2]}{[(\alpha a)^2 + a^2(k - k_m)^2]^2} \quad (7.11\text{a})$$

$$\frac{\partial A_1}{\partial k} = \frac{\partial A_2}{\partial \alpha} = -\frac{a[2\alpha a^2(k - k_m)]}{[(\alpha a)^2 + a^2(k - k_m)^2]^2} \quad (7.11\text{b})$$

$$\frac{\partial |A|}{\partial \alpha} = -\frac{a^2 \alpha}{[(\alpha a)^2 + a^2(k - k_m)^2]^{\frac{3}{2}}} \quad (7.11\text{c})$$

and

$$\frac{\partial |A|}{\partial k} = -\frac{a^2(k - k_m)}{[(\alpha a)^2 + a^2(k - k_m)^2]^{\frac{3}{2}}} \,. \quad (7.11\text{d})$$

Considering, specifically, the case in which the CW oscillator is tuned to a frequency corresponding to the center of a mechanical resonance, one finds $A_1 \simeq 1/\alpha a$, $\partial A_1/\partial \alpha = -1/\alpha^2 a$, and $\partial A_1/\partial k = 0$. The fractional change in A_1 is thus

$$\Delta A_1/A_1 \simeq [(\partial A_1/\partial \alpha)/A_1]\Delta \alpha \,.$$

For typical cases in solids studied by NAR, (αa) ranges from 0.1 to 0.001; therefore,

$$\Delta A_1/A_1 = -(1/\alpha a)\Delta \alpha a$$

ranges from -10 $\Delta\alpha a$ to -1000 $\Delta\alpha a$. Thus the acoustic composite resonator provides a signal enhancement of from one to several orders of magnitude, depending upon the (nonresonant) attenuation of the sample.

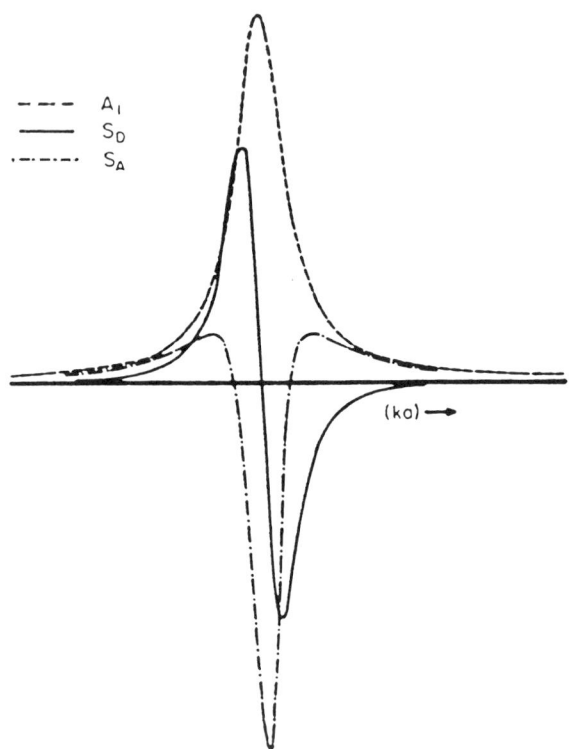

FIGURE 7.5. Absorption and dispersion sensitivities, with respect to A_1, superimposed upon a plot of A_1. (From Ref. 7.1)

The predicted absorption and dispersion sensitivity factors $S_A = \partial A_1/\partial \alpha$ and $S_D = \partial A_1/\partial k$ from (7.8) are shown in Fig. 7.5 for a range of frequencies near a particular mechanical resonance, together with a plot of A_1. The function S_D has zeros for $\cos ka = 1$. Thus if one is tuned to the center of a mechanical resonance, a small change in ultrasonic amplitude corresponds to pure absorption. The function S_A has zeros for $\cos ka = \text{sech}\,\alpha a$. For sufficiently small αa, this corresponds to a value of A_1 equal to approximately one half its maximum value. Tuning the carrier frequency to this point results in a pure dispersion signal. The dispersion sensitivity is down approximately 25% from its maximum value when tuned to this frequency. In some NAR experiments, this may be a small price to pay for obtaining a pure dispersion signal rather than a mixed dispersion–absorption signal.

The absorption and dispersion sensitivity-enhancement factors $\partial|A|/\partial\alpha$ and $\partial|A|/\partial k$ predicted by (7.9), together with a plot of $|A|$ over the same range, are shown in Fig. 7.6. Again the greatest sensitivity to absorption occurs at the frequency corresponding to the peak of the mechanical resonance; this also corresponds to a pure absorption signal. The factor $\partial|A|/\partial\alpha$ is negative for all frequencies, since an increase $\Delta\alpha$ in attenuation results in a decrease in $|A|$. There is no point on the $|A|$ response at which one obtains a pure dispersive signal.

7.3.2 Marginal Oscillator Spectrometers

7.3.2.1 Nature of Operation

A marginal oscillator is a feedback oscillator whose open loop gain G is a nonlinear function of the input voltage V [7.5]. As V increases, the gain decreases somewhat, as shown in Fig. 7.7a. For stable oscillation, the feedback condition $Gp = 1$ must be satisfied, where p is the feedback ratio; otherwise the oscillation level will increase $(Gp > 1)$ or decrease $(Gp < 1)$. Thus the feedback condition fixes the level of oscillation V. If p is such that the feedback condition cannot be satisfied for reasonable V, the marginal oscillator drops out of oscillation or begins to oscillate widely. The maintenance of stable oscillations in the presence of a feedback change p requires that

$$G\Delta p + p\Delta G = 0 \; ,$$

where

$$\Delta G = \frac{dG}{dV}\Delta V \; ,$$

so that

$$\Delta V = -\left(\frac{dG}{dV}\right)^{-1} G^2 \frac{\Delta p}{p} \; . \tag{7.12}$$

Thus the magnitude of the change of input voltage ΔV (or the magnitude of the change of the output voltage $G\Delta V$) depends on $(dG/dV)^{-1}$; the latter is largest where the amplifier is most linear, *i.e.*, at small V (Fig. 7.7b). For greatest sensitivity, the device is operated at the smallest sustainable oscillation level—whence the name *marginal oscillator*.

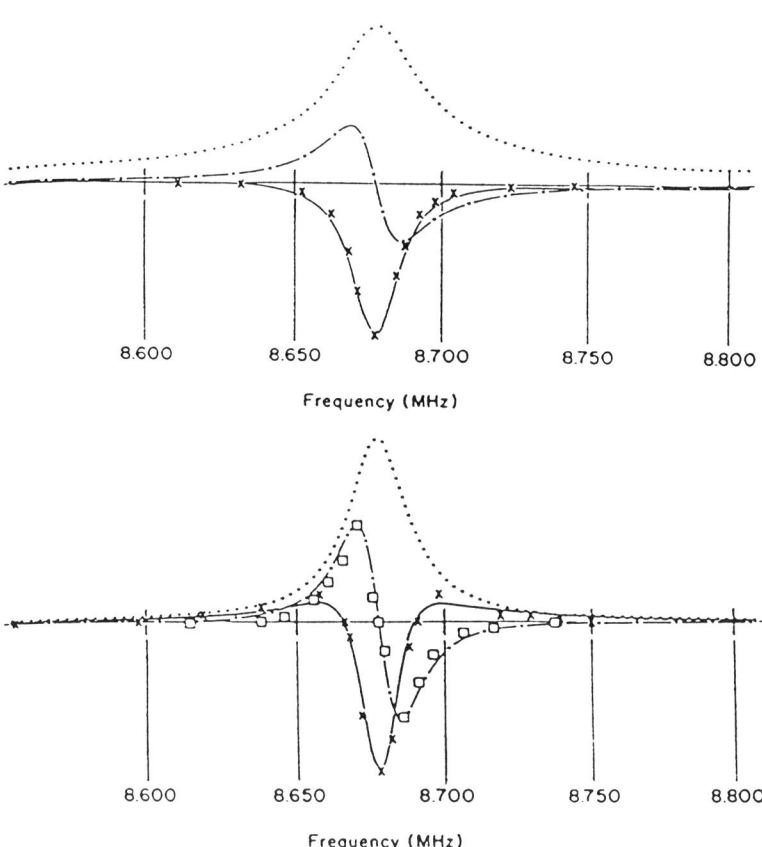

FIGURE 7.6. (a) Theoretical sensitivity–enhancement factors for attenuation ($\partial A/\partial \alpha$) (solid curve) and dispersion ($\partial A/\partial k$) (dash–dot) superimposed on a plot of $|A|$ (dotted curve). Experimental values are shown by X. (b) Theoretical line shapes for attenuation ($\partial A_1/\partial \alpha$) (solid curve) and dispersion ($\partial A_1/\partial k$) (dash–dot) superimposed on a plot of A_1 (dotted curve). Pure attenuation signals are obtained for the frequency corresponding to $\partial A_1/\partial k = 0$, while pure dispersion signals are obtained for $\partial A_1/\partial \alpha = 0$. Experimental values are indicated for attenuation (x) and dispersion ($\square$). (From Ref. 7.4)

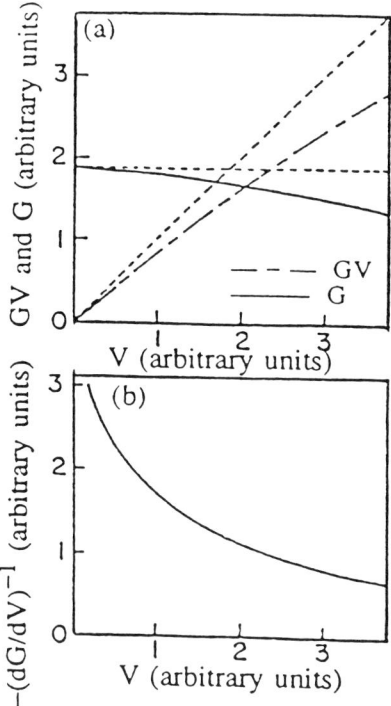

FIGURE 7.7. (a) Output voltage GV and gain V versus input voltage V for a hypothetical marginal oscillator. Dashed lines show GV and G for a linear amplifier. (b) Sensitivity factor $-(dG/dV)^{-1}$. (From Ref. 7.5)

7.3.2.2 The MOUS: Marginal Oscillator Ultrasonic Spectrometer

The marginal oscillator spectrometer, commonly used in NMR studies, has been adapted to the measurement of acoustic velocity and attenuation by the addition of a coupling network and the substitution of a composite resonator for the RF coil used in NMR. The coupling network converts the impedance minimum at the center of a mechanical resonance into an impedance maximum. Given sufficiently high Q of the mechanical resonance, the impedance maximum determines the RF oscillation level and locks the frequency. In applications to NAR, the small absorption of acoustic energy decreases the mechanical Q, resulting in a decrease in the level of RF oscillation.

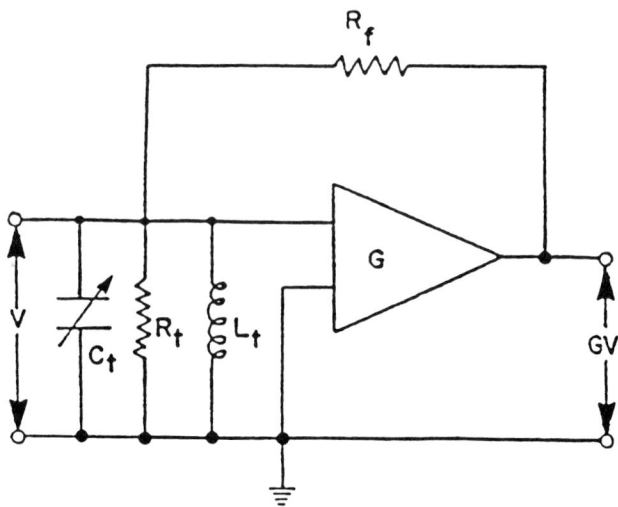

FIGURE 7.8. Idealized marginal oscillator circuit diagram. (From Ref. 7.6)

In the idealized marginal oscillator circuit diagram of Fig. 7.8, a parallel $R_t L_t C_t$ tank circuit is inserted into the feedback loop. For oscillations to be sustained, the phase shift around the loop must be a multiple of 2π; this constitutes a second feedback condition. The circuit oscillates at the tank's resonant frequency, at which the reactance is zero. Given a change in the resonant frequency due to a change in C_t or L_t, the frequency of oscillation shifts accordingly. As a result, the spectrometer can be tuned simply by changing C_t, and unintentional frequency drifts, such as those from thermal drift (due to the temperature dependence of the mechanical resonance) or microphonic changes of parasitic capacitances, are largely compensated for by the frequency tracking.

From the circuit diagram of Fig. 7.8, it is evident that the feedback ratio p increases with R_t. As a result, oscillation occurs at an impedance maximum rather than at a minimum. On the other hand, in ultrasonic experiments the impedance is a minimum at a mechanical resonance. Therefore a composite resonator cannot be substituted directly for the tank circuit of Fig. 7.8. This difficulty is overcome by introducing a double-tuned over-coupled RF transformer between the composite resonator and the input to the amplifier. When properly tuned, the transformer inverts the mechanical resonance, causing an impedance maximum to be presented to the marginal oscillator. Oscillation then occurs at the frequency of maximum ultrasonic amplitude in the sample.

Once the marginal oscillator is locked to a mechanical resonance frequency, changes in ultrasonic attenuation (via changes in the feedback ratio p) are reflected in the output voltage GV. This voltage is further amplified and detected. In many applications, magnetic field modulation and synchronous detection with a lock–in amplifier are used to increase the signal–to–noise ratio. The lock–in output can be recorded directly on a recorder or can be digitally averaged on a computer equipped with an analog–digital interface.

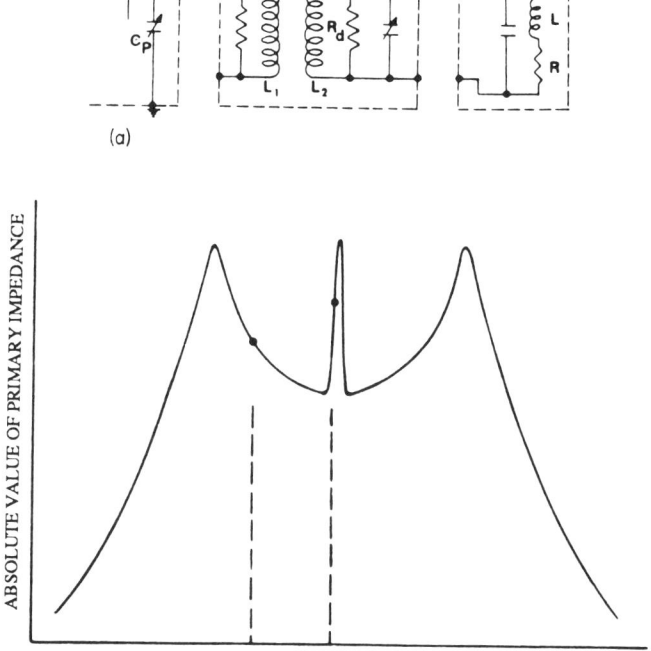

FIGURE 7.9. (a) Marginal oscillator ultrasonic spectrometer. (b) Impedance presented to the marginal oscillator by the coupling network. (From Ref. 7.7)

A block diagram of a MOUS is given in Fig. 7.9a, configured for an NAR experiment in which an electromagnet and magnet–modulating coils are utilized. The double–tuned coupling network, whose characteristics are shown in Fig. 7.9b, inverts the impedance minimum presented by the composite resonator to an impedance maximum at the tank circuit of the marginal oscillator. The FM circuit, consisting of local oscillator and a varactor, serves to tune the frequency of the marginal oscillator, which locks to a high-Q mechanical resonance peak. Phase–sensitive detection is used; the magnetic field, while being swept slowly through the value corresponding to nuclear acoustic resonance, is modulated at an audio frequency. The output of the lock–in amplifier is fed to a recorder, which outputs the derivative of the NAR absorption line. The calibrator, discussed in detail in Section 7.4, is used

(1) to check the MOUS for sensitivity, and
(2) to serve as the source of a standard signal for comparison with the observed NAR signal.

Sufficiently sensitive ($\sim 10^{-8} - 10^{-9}$ cm^{-1}) to detect nuclear acoustic resonance, the MOUS suffers from inherent nonlinearity, narrow–banded response, low power levels and limited frequency response. Over the years it has served, nevertheless, as a standard because of its extreme sensitivity to small changes in acoustic attenuation and because of the simplicity of the MOUS apparatus.

7.3.2.3 The TOUS: Transmission Oscillator Ultrasonic Spectrometer

A more versatile ultrasonic spectrometer utilizing the marginal oscillator mode is the transmission oscillator ultrasonic spectrometer (TOUS) [7.5], which compares with the MOUS in sensitivity but is free of many of the latter's limitations. Schematically shown in Fig. 7.10, the TOUS consists of an amplifier system that is caused to oscillate by providing it with a feedback path through the acoustic resonator. Pi matching networks provide impedance matching between the acoustic resonator and the electrical system. A tuning network selects the particular mechanical resonance and the point on that resonance at which the oscillator operates. The tuning network is a passive bandpass filter having a frequency response that is broad when compared with the width of a mechanical resonance, but is also sharp enough to discriminate between adjacent mechanical resonances. In practice, an inductor and a variable capacitor in parallel are used as a tank circuit. With loosely coupled input and output coils, the desired frequency response is achieved, the insertion loss being 10–20 dB at frequencies of the order of 10 MHz.

The loop contains two lossy elements, the tuning network and the acoustic resonator, both of which act as narrow bandwidth filters. This arrangement permits the distribution of gain around the loop, as shown in Fig. 7.10, with the result that lower noise is achieved. Only that fraction of the output noise of one amplifier that lies within the passband of the intervening narrow band element is seen by the input of the other amplifier. A signal splitter removes a portion of the signal at the highest RF level in the system for detection and signal processing. The variable attenuator, as will be discussed below, is adjusted to achieve marginal oscillation.

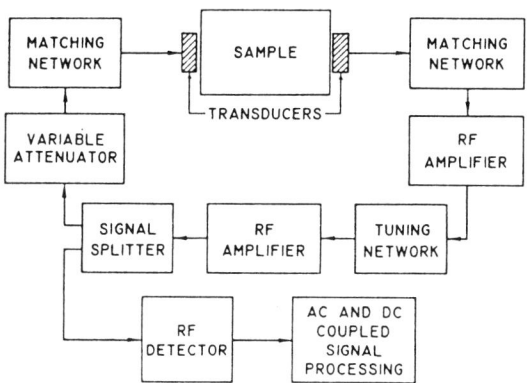

FIGURE 7.10. A Block diagram of a transmission oscillator ultrasonic spectrometer (TOUS). (From Ref. 7.5)

Stable oscillations of the TOUS occur at a level at which the phase shift around the loop is zero and the product of the gain G and the loss p is one. We assume that the condition on the phase shift is satisfied and that the gains and losses that are distributed around the loop can be ignored.

The enhanced sensitivity obtained by operating the TOUS under conditions of marginal oscillation can be understood on the basis of the following argument. If the output voltage V_{out} is plotted as a function of the input voltage V_{in} for a typical RF amplifier, a response curve similar to that shown in Fig. 7.7a is obtained. For sufficiently small input voltage, the response is nearly linear, and thus the gain $G = V_{out}/V_{in}$ (also shown in Fig. 7.7a) is nearly independent of V_{in}. Departure from linearity at higher RF levels is accompanied by a decrease in gain—*i.e.*, as the amplifier approaches saturation.

To be utilized for the TOUS, the small signal gain of the amplifier must exceed the attenuation; *i.e.*, the product of the linear region gain and the loss must be greater than 1.0. Under these conditions, oscillations, when initiated, grow in amplitude. The gain of the amplifier decreases with this increase in oscillation level. Stable oscillations occur when the amplitude has increased sufficiently, and thus the gain has decreased sufficiently that the product of the gain and loss equals 1.0. Stable oscillation cannot occur in a region where the amplifier is strictly linear.

The primary use of the TOUS is in the detection of very small changes in acoustic attenuation. A small increase in acoustic attenuation results in a corresponding small increase in the attenuation of the loop. This small increase in loop attenuation reduces the RF level at the input of the amplifier, and a slight increase in gain results. A new stable operating point is reached, since the extra loss due to the additional acoustic attenuation is compensated for by the addition gain. A measure of the sensitivity of the TOUS to small changes in acoustic attenuation α is the change in output level ΔV_{out} for a given $\Delta \alpha$. From the discussion above and from an inspection of Fig. 7.7a, one sees that the greatest change in V_{out} will occur in that region of the curve where the gain G is a rather slowly varying function of V_{in}. When dG/dV_{in} is small, the value of V_{in} (and thus the value of V_{out}) must decrease rather substantially in order to increase the gain sufficiently to compensate for the additional loss. Conversely, if dG/dV_{in} is large, only a relatively small change in output level results for the same change in acoustic attenuation $\Delta \alpha$.

From (7.12), the sensitivity of the TOUS to small changes in acoustic attenuation is approximately proportional to $-(dG/dV_{in})^{-1}$. Good sensitivity is achieved with amplifiers that are nearly linear and that saturate slowly. Sensitivity increases when the operating point is chosen at RF levels where the amplifier is more nearly linear. In practice, the operating point is chosen by adjusting the variable attenuator shown in Fig. 7.10.

A block diagram of a TOUS NAR spectrometer is shown in Fig. 7.11. The marginally-oscillating feedback loop, of which the composite resonator forms a part, is indicated on the diagram. The TOUS is claimed to be two orders of magnitude more sensitive than a standard CW transmission spectrometer.

7.3.3 Transmission and Transmission Bridge Spectrometers

Transmission and transmission bridge spectrometers, not operating as marginal oscillators, have been used to detect NAR. The composite resonator for a transmission spectrometer requires separate transmitting and receiving transducers bonded to opposite faces of the sample. Efficient shielding is also required to prevent RF leakage from the transmitter end to the receiver end. The standing wave responses for transmission, analogous to (7.4b–d) for the reflection mode, are:

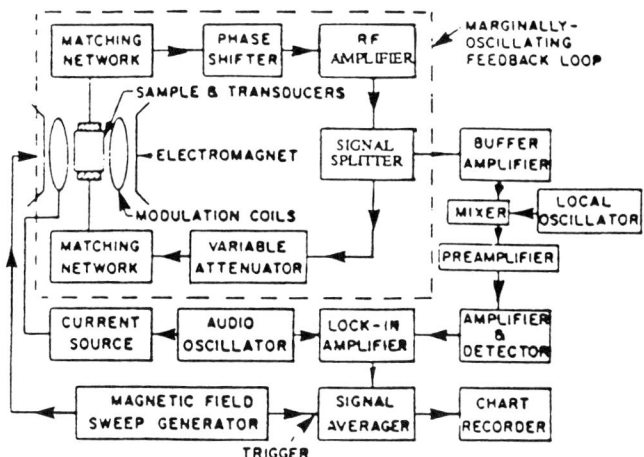

FIGURE 7.11. A block diagram of the TOUS spectrometer as adapted for NAR experiments. (From Ref. 7.5)

$$T_1 = \frac{\sinh(\alpha a/2)\cos(ka/2)}{\cosh\alpha a - \cos ka} \tag{7.13a}$$

$$T_2 = \frac{\cosh(\alpha a/2)\sin(ka/2)}{\cosh\alpha a - \cos ka} \tag{7.13b}$$

and

$$\begin{aligned}|T| &= \frac{1}{\sqrt{2}}\left[\frac{1}{\cosh\alpha a - \cos ka}\right]^{\frac{1}{2}} \\ &\equiv \frac{1}{2}\left[\frac{1}{\sinh^2(\alpha a/2) + \sin^2(ka/2)}\right]^{\frac{1}{2}},\end{aligned} \tag{7.13c}$$

where T_1 and T_2 are the in- and out-of-phase components of the transmission acoustic response, and $|T|$ is the magnitude of the transmission response.

A block diagram of an RF transmission spectrometer used to measure small changes in attenuation is shown in Fig. 7.12. The spectrometer is used in both frequency–modulated and magnetic field–modulated modes to study absorption and dispersion effects, *e.g.*, NAR of ^{19}F nuclear spin in antiferromagnetic RbMnF$_3$ (Chapter 9). The frequency stability required of the RF oscillator used in both modes of operation is dependent upon the background acoustic attenuation of the sample; a typical requirement is one part per million over a period of minutes. The receiver consists of a low noise, wide bandwidth preamplifier, followed by a wide band RF amplifier of variable gain. Magnetic field or frequency modulation produces an amplitude modulation of the carrier when the frequency and the DC magnetic field are set to satisfy the conditions for resonance. After RF detection, the resulting audio signal is fed into a phase–sensitive detector, the output of which drives a recorder.

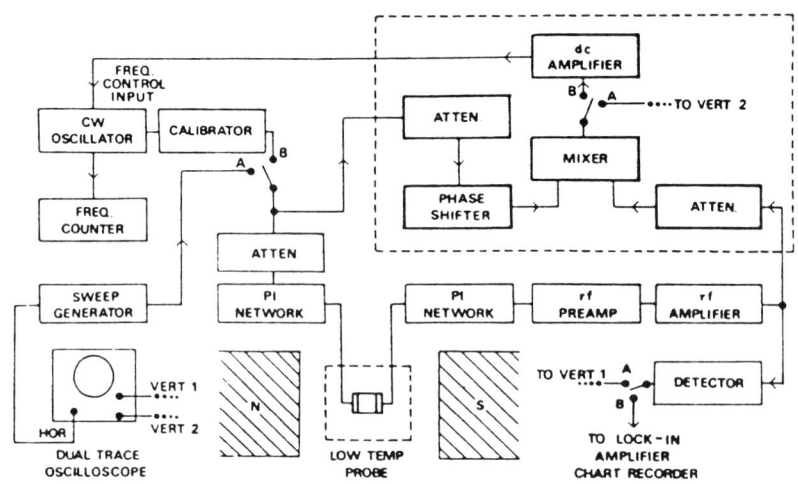

FIGURE 7.12. Sensitive CW transmission spectrometer.

Quantitative measurements are carried out with the aid of a calibrator consisting of a diode switch and an audio phase shifter. A known audio current at the modulating frequency is superimposed on the DC bias, thus amplitude–modulating the carrier. The audio phase shifter is used to adjust the phase of the calibrator signal so that it coincides with the phase of the transmitted acoustic signal. Knowledge of the power transmission versus bias characteristic of the diode switch enables one to calibrate absolutely a recorder deflection in terms of percent power modulation.

In order to minimize the noise and drift in the output that results from slight shifts in frequency (*e.g.*, those due to temperature changes or mechanical vibrations), provision is made in the spectrometer of Fig. 7.12 to frequency lock the spectrometer to a mechanical resonance (dashed line).

The dual balanced mixer produces a voltage proportional to the phase difference between a signal that has been transmitted through the acoustic probe and a reference signal taken directly from the CW oscillator. Once an operating point on the mechanical resonance is selected, any deviation in frequency results in a correction voltage, which is amplified and applied to the CW oscillator to return its frequency to the set point.

By the substitution of readily available commercial components, a spectrometer such as that shown in Fig. 7.12 can be used over a continuous range of frequencies, from a few megahertz through several GHz. The availability of wide bandwidth components, as well as the availability of broad band deposited thin film transducers, makes it feasible to cover a two–to–one frequency range without changing spectrometer components. The conversion from a transmission to a reflection operation mode is also straightforward.

7.3.4 Reflection Bridge Spectrometers

Next to the MOUS, the reflection bridge spectrometer has been most frequently used in nuclear acoustic resonance studies. *Müller* [7.8] and, more recently, *Sundfors* [7.9] have designed spectrometers that equal the marginal oscillator in sensitivity and are much more versatile in performance. The designs lend themselves readily to computer control of the spectrometer. A schematic of a relatively simple reflection bridge spectrometer, shown in Fig. 7.13, consists of a source of RF energy, a bridge built around a broadband hybrid junction and a standard superheterodyne receiver. A variety of RF sources may be used, depending upon the frequency range and power requirements. It is desirable that the RF source have swept frequency as well as CW capabilities for use

(1) in displaying mechanical resonance patterns,
(2) as an FM reflection spectrometer (analogous to the FM transmission spectrometer of Section 7.3.3), and
(3) in locking the spectrometer frequency.

Figure 7.13 shows a reflection bridge spectrometer where the bridge element is a circulator, which is used for frequencies greater than several GHz. The circulator of Fig. 7.13 only has three ports, but is the equivalent of four–port circulators used in electron paramagnetic resonance (EPR) experiments and of the hybrid junctions used in experiments for frequencies less than 2 GHz. The composite resonator is the direct analogue of the sample cavity in EPR. The analysis of the hybrid junction (a four–port circulator) is straightforward. When each of the arms is appropriately terminated, the RF signal enters the hybrid junction at port 1 and divides equally, half going to the sample arm via port 2 and half to the reference arm via port 3. The output signal coming from the acoustic specimen attached to port 2 is reversed in phase by 180° before being combined with the output signal from the reference arm port 3 to form a final output signal at port 4. The output of the bridge is amplified and detected using a superheterodyne receiver constructed of standard commercial components. Under favorable signal strength conditions, the CW frequency response of an acoustic resonator can be obtained by feeding the detected output into a differential voltmeter, which, in turn, drives a recorder. By appropriate adjustment of the attenuation and phase shift in the reference arm, either the in–phase (A_1) or the out–of–phase (A_2) component of the standing wave acoustic response can be obtained at port 4. A typical set of tracings, the in– and out–of–phase components of the acoustic response of a single crystal of InSb at a frequency of 10.20 MHz, is shown in Fig. 7.14. The frequency width at half–maximum is 600 Hz. The shunt capacitance that is associated with the transducer and transmission lines and that is not tuned out by the matching network contributes an additional term to the out–of–phase component A_2. The effect of imperfect tuning is evident in Fig. 7.14. Under favorable conditions, one also obtains $|A|$ by selecting a very large value for the attenuation in the reference arm, so that only information entering port 2 contributes to the output at port 4. Direct electromagnetic leakage from port 1 to port 4 renders this method impractical for all but very low–loss specimens. In the sampled–CW mode of operation described in Section 7.3.5, $|A|$ is observed without this leakage, since the transmitter is gated off when the receiver is on.

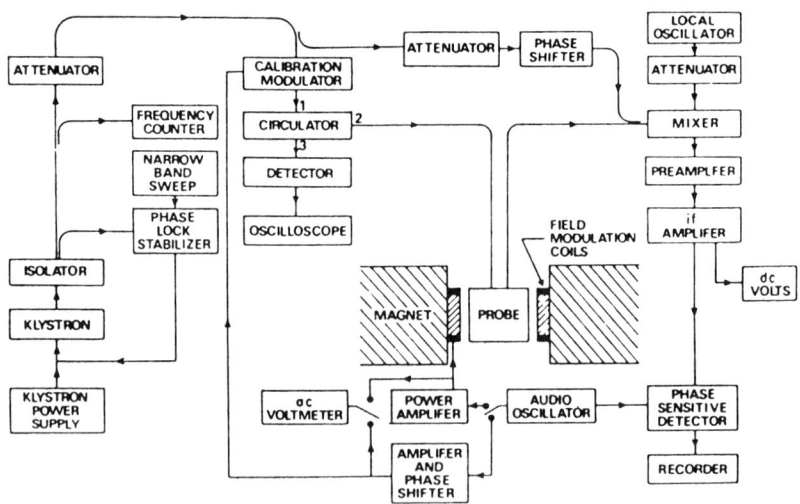

FIGURE 7.13. An X–band CW transmission bridge spectrometer. (From Ref. 7.10)

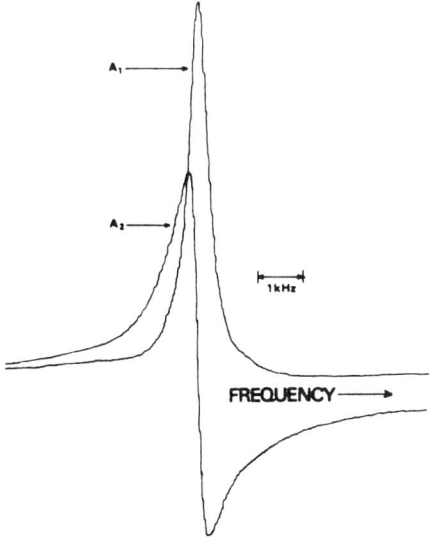

FIGURE 7.14. In–phase (A_1) and out–of–phase (A_2) components of ultrasonic response of single–crystal InSb at 10.20 MHz. (From Ref. 7.11)

The sophisticated computer–controlled reflection bridge spectrometer, shown schematically in Fig. 7.15, is constructed from off–the–shelf components and, with relatively slight adjustments, can be used in the frequency range between 5 and 400 MHz. Power input to the composite resonator can be varied from 0.01μW to 1.0 W with no decrease in sensitivity. The spectrometer is readily convertible for operation in a sampled–CW mode. The probe arm (port 2), consisting of a composite resonator, matching network and transmission line, in general presents a complex impedance to the hybrid junction. The reference arm (port 3) serves as a variable impedance adjusted to match the probe arm impedance, so that the voltages reflected from the two arms vectorially balance, resulting in a null output to the receiver arm (port 4). The hybrid junction thus acts to null the large steady–state voltage from the transmitter (port 1). A practical reference arm tuning element in the frequency range of from 5 to 40 MHz for approximately orthogonal tuning is a parallel combination of carbon or composition potentiometer (used as a variable RF resistance) and a variable air capacitor. The reference arm elements are adjusted to match the sample arm parallel impedance near a mechanical resonance center. The composite resonator is at the end of one of several, identical, 91.4–cm–long,

50 Ω, rigid–vacuum dielectric transmission lines that constitute part of the low temperature probe. At frequencies below 40 MHz, the transmission line transforms the real part of the composite resonator impedance to values approximating the actual real part of the composite resonator impedance. At frequencies above 40 MHz, the reference arm is terminated by 50 Ω, and a double–stub tuner is placed between the sample arm and the transmission line terminated by the composite resonator. Adjustment of the double–stub tuner for a minimum RF voltage at the hybrid junction output arm provides a two–variable tuning capability. The sensitivity is comparable to the parallel capacitor–resistor combination used below 40 MHz.

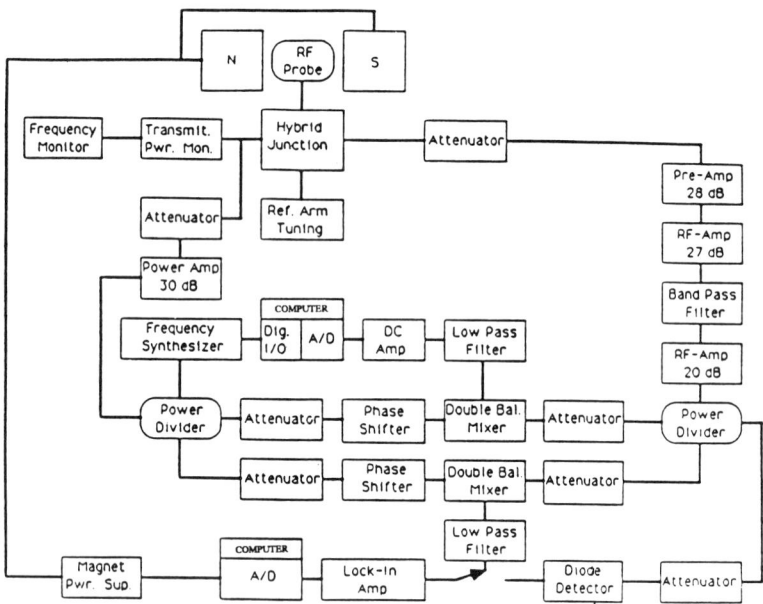

FIGURE 7.15. Computer–controlled CW reflection bridge spectrometer. (From Ref. 7.9)

The spectrometer in Fig. 7.15 is shown schematically configured for nuclear acoustic resonance at frequencies lower than 40 MHz. The output of the frequency synthesizer goes to a three–way power divider. Two of the three outputs of the power–divider serve as reference signals for an error loop and a signal loop; the third passes through a low–noise, stable 30–dB gain power amplifier, then through a 1–W, 130–dB, 1–dB step attenuator to the transmitter arm of the hybrid junction. A frequency counter monitors the frequency, while an RF voltmeter monitors the transmitter arm power.

In the receiver arm are a 1–W 130–dB, 1–dB step attenuator, followed by a very low noise, 28–dB gain RF preamplifier. Additional amplifiers raise the total receiver side gain by 70 to 90 dB before RF detection. At the output of the receiver amplifier chain is another three–way power divider; one of the three outputs feeds the error loop, a second goes to the signal loop, while a third goes to diode detection.

Two RF phase–sensitive detection loops connect the receiver section to the transmitter section. One of these loops provides for phase–sensitive detection of the signal of interest; the other provides an error signal, which can be used for frequency control to track in time any position on a mechanical resonance. From (7.4) the frequency response near a mechanical resonance at the output of the bridge for a close, but not exact, balance of the sample and reference arms is a mixture of an absorption signal (A_1) in phase (with the driving RF oscillator) and a dispersion signal (A_2) in quadrature.

The signal from any of the three output ports of the receiver power divider provides one of the inputs for a double–balanced mixer used as the RF phase detector. The second input to the phase detector comes from the transmitter section through a manually adjusted RF phase–shifter constructed from two quadrature (90°) hybrids and variable LC networks, as shown in Fig. 7.16. The output signal from the RF phase detector passes through an audio frequency low–pass filter to the lock–in detector. In actual use, the output from the low–pass filters is also sent to an oscilloscope for the purposes of adjusting the manual phase shifter for A_1 or A_2 and of monitoring the noise level.

A second output of the receiver power divider is used in the second RF phase–sensitive loop, which is identical to that described above. In this case, however, the RF phase–sensitive detector signal serves an an error signal. The error signal passes through a low–pass filter and DC amplifier, then to a analog–digital interface board in a microcomputer. A C–language program uses this error signal with 48 digital I/O channels in controlling the frequency of the frequency synthesizer. Adjustment of the phase shifter in the error signal loop allows the frequency synthesizer to track precisely any point on the mechanical resonance.

A third output from the receiver power divider passes to a channel used for observing the diode–detected signal. Since the receiver signal is a mixture of A_1 and A_2, this channel is generally used only for spectrometer testing. The receiver power level is also monitored in this channel.

7.3.5 Sampled–CW Spectrometer

Unlike acoustic pulse–echo techniques, CW acoustic techniques are inherently monochromatic and are suitable for examining arbitrarily thin

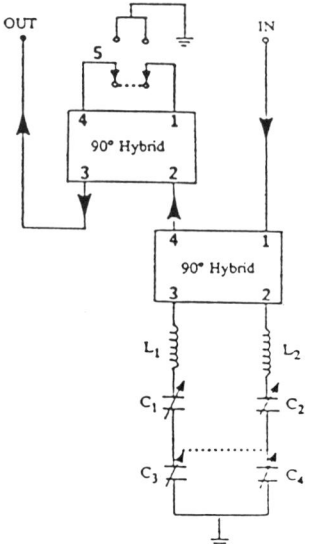

FIGURE 7.16. Phase shifter. The out termination is shifted from 0° to 180° relative to the in termination with the switch. In the two resonance circuits, L_1 and L_2 are air inductors constructed approximately equal for the desired frequency; C_1 and C_2 are 2–40-pF glass capacitors, and are adjusted to make the two resonance circuits track in phase together and to reduce the capacitance range of C_3 and C_4; C_3 and C_4 are two sections of a 20–300-pF air capacitor that track together and produce a variable phase change between the input and output terminations of from 0° to greater than 180°. The 90° hybrids are available in octave frequency ranges and in a flat–pack configuration suitable for a printed circuit board. (From Ref. 7.9)

specimens. Since the transmitter and receiver are on simultaneously, however, the CW technique is susceptible to cross–talk. A number of ingenious variations on the basic pulse–echo technique retain the important advantages of the basic pulse–echo scheme and, in addition, incorporate features such as monochromaticity and phase coherence to varying degrees. These techniques may be characterized as incorporating some aspects of the CW technique into essentially pulse–echo spectrometers. In contrast, the sampled–CW technique, described in the present section, incorporates advantages of the pulse–echo scheme—specifically, the complete isolation from cross–talk—into an essentially CW spectrometer. The sampled–CW technique is characterized by high sensitivity and is applicable to specimens of arbitrary size. It is especially well–suited to inherently single–ended (*i.e.*, reflection–type) applications.

With the addition, under computer control, of suitable RF gates placed before the input attenuators in the transmitter arm and after the attenuators in the receiver arm, and with the use of appropriate control programs, the spectrometer of Fig. 7.15 can be converted readily into a sampled–CW (SCW) spectrometer. If one wishes to make temperature–dependent measurements over a wide temperature range, for example, the necessity of bridge retuning can be eliminated (at the cost of slightly greater spectrometer complexity) by the use of a SCW spectrometer. Use of the SCW technique, further, minimizes power dissipation at low temperatures.

In the SCW technique the transmitter and receiver are gated so that they are never on simultaneously. The transmitter is gated on until a steady-state response is established in the composite resonator. The transmitter is then gated off. The receiver is gated on and samples the CW acoustic signal established in the resonator during the transmitter–on interval. The output of the transducer is proportional to the instantaneous acoustic particle velocity at the crystal face. One is thus enabled, as in the CW case, to distinguish among the responses A_1, A_2 and $|A|$ of (7.4). The SCW technique permits both frequency domain and time domain observation. Monitoring $|A|$ with diode detection in the time domain, one observes a step–wise decay of the steady state when the transmitter is turned off. Monitoring $|A|$ in the frequency domain, one observes the mechanical resonances. When the receiver is gated on during the first steps of the decay, both frequency and time domain responses closely approximate the steady–state response.

A schematic of a frequency–modulated SCW reflection spectrometer is given in Fig. 7.17. The RF carrier frequency ω of the transmitter is frequency modulated either internally or externally at an audio frequency ω_A and with FM peak deviation $\Delta\omega_A$ centered at ω. The values of ω_A and $\Delta\omega_A$ are chosen so as to make the modulation index $\delta \equiv \Delta\omega_A/\omega_A$ large compared to one. The frequency modulation produces an amplitude modulation of the signal at a frequency ω_A. The magnitude of the amplitude modulation is proportional to the slope of the mechanical resonance at the carrier frequency ω. As shown in Fig. 7.17, the modulated signal in the receiver arm is processed by a lock–in detector, which provides discrimination against noise. The reference signal for the lock–in detector is derived from the same audio frequency source used to FM modulate the transmitter.

In the SCW spectrometer of Fig. 7.17, the RF frequency generator, operated in its internal frequency modulation mode, provides the RF carrier modulation and the reference signal to the lock–in detector. The timing generator serves as a master clock for the spectrometer. Typical transmitter and receiver gates are on for 600 μsec with a repetition rate of 1.25 msec. During this cycle, steady-state standing waves are established such that the signal incident upon the receiver is the exponential time decay mode from steady state. The receiver signal is then amplified by a preamplifier and a main amplifier. The signal is diode-detected, filtered and passed on to the video sampler. The sampled-video output has a DC component proportional to the amplitude at a preselected step on the time domain decay. This signal contains the FM component and is fed into the preamplifier of the lock-in detector. The transmitter frequency is tuned for a null response on the lock-in detector, corresponding to the center of a mechanical resonance. This frequency is measured on a multimeter and recorded on a printer.

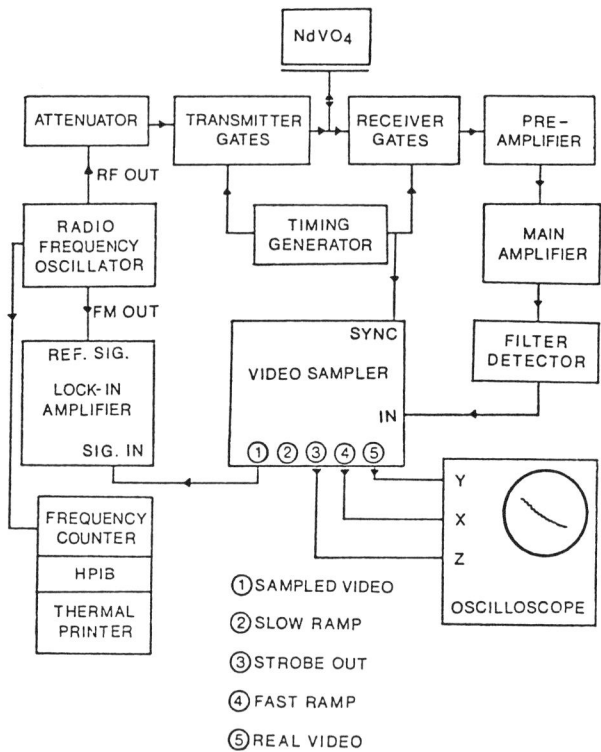

FIGURE 7.17. Frequency-modulated SCW reflection spectrometer schematic diagram. (From Ref. 7.12)

7.4 Signal Calibration

In order to compare quantitatively different NAR signals obtained by the use of the spectrometers reviewed in Section 7.3, an accurate means of calibration is required. Already mentioned in connection with the transmission spectrometer (Section 7.3.3) is a calibrator consisting of an audio phase shifter and a diode switch, whose power versus bias characteristic is tabulated. A similar device, incorporating a pin diode, has been utilized to calibrate a reflection bridge spectrometer.

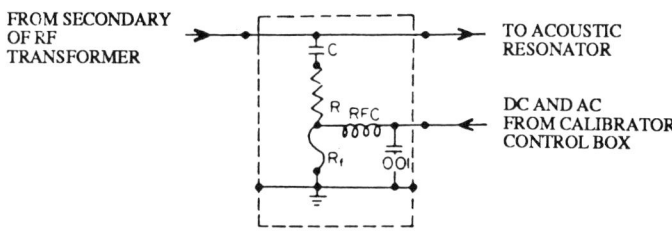

FIGURE 7.18. Smith-Sundfors calibrator for the MOUS. (From Ref. 7.13)

A calibrator designed for the MOUS (Section 7.3.2) but applicable to other NAR spectrometers is shown in Fig. 7.18. Specifically designed to be placed in the secondary of the MOUS RF transformer, the calibrator relies upon the use of a low current ($\sim$ 10mA) fuse, whose resistance R_f versus current i characteristic is closely linear over the range in which it is used (e.g., $dR_f/di \simeq 554\Omega/mA$ at $I_{DC} = 6.5$ mA). A known change in resistance $\Delta R_f = (dR_f/di)\Delta i$ is introduced by a small change in the current through the fuse. The admittance of the calibrator circuit of Fig. 7.18 is equivalent to that of a parallel RC network:

$$R_c = R_t \left(1 + \frac{1}{\omega^2 C^2 R_t^2}\right), \qquad C_c = \frac{C}{1 + \omega^2 C^2 R_t^2},$$

where $R_t = R + R_f$, and R is a a known resistance in the calibrator circuit. Thus a change ΔR_f results in a change of the effective calibrator resistance

$$\Delta R_c = \Delta R_f \left(1 - \frac{1}{\omega^2 C^2 R_t^2}\right). \qquad (7.14)$$

Since the values of the parameters involved in (7.14) are known, the resulting conductance change, ΔG_c, can be compared directly with the change in ΔG due to the resonant change in attenuation corresponding to a particular NAR resonance line. Another type of calibrator for NAR spectrometers consists of a single crystal CdS composite resonator, the acoustic properties of which are modulated with a light beam [7.4,7.14]. The CdS is photoconductive; thus small changes in intensity of light incident upon the specimen result in changes in the charge carrier system, in turn, affecting the acoustic properties of the resonator.

For metal specimens, a calibration procedure based on the nonresonant (DC) Alpher–Rubin interaction is applicable [7.15]. An advantage of this technique is its ability to monitor continuously the performance (sensitivity) of the NAR spectrometer. One recalls (Chapter 4) that the DC Alpher–Rubin interaction determines the magnetic field dependence of acoustic attenuation and acoustic velocity at room temperature, thereby resulting in an increase in both that is proportional to the square of the magnetic field. The expression for the change in attenuation is

$$\Delta \alpha = \frac{\sigma f(\theta)}{2\rho v c^2} \frac{\beta^2}{1+\beta^2} B^2 , \qquad (7.15)$$

where σ is the electrical conductivity, ρ the density, and θ the angle between the direction of acoustic wave propagation and the direction of the magnetic field. $f(\theta) = \sin^2\theta$ for longitudinal waves and $\cos^2\theta$ for transverse waves. Equation (7.15) has been verified as a function of angle, of frequency and of magnetic field for a wide variety of metals.

In the normative case of NAR spectrometers utilizing phase–sensitive detection with magnetic field modulation, the output signal is

$$S_{A-R} \propto \frac{1}{\alpha} \frac{d\Delta\alpha}{dB} \Delta B , \qquad (7.16)$$

where $\Delta\alpha = -(\Delta V/V)_{RF}\alpha$, α is the background attenuation, $(\Delta V/V)_{RF}$ is the relative change in RF output level corresponding to the change in attenuation $\Delta\alpha$ and ΔB is the amplitude of the modulation field. Thus the output signal resulting from the non-resonant Alpher–Rubin interaction is

$$S_{A-R} = \frac{C}{\alpha} \frac{\sigma f(\theta)}{\rho v c^2} \frac{\beta^2}{1+\beta^2} B\Delta B . \qquad (7.17)$$

All other factors being known, (7.17) allows an accurate determination of C, the calibration factor of the spectrometer. The signal S appears as a DC output from the lock–in detector; it is easily distinguished from other signals by its linear field dependence and its distinctive angular dependence. For an aluminum sample in a magnetic field of 20 kG and a modulation field of 10 G, the magnitude of S is approximately $10^{-5}C$, or five times that of the NAR absorption signal (for transverse waves) seen under the same conditions.

7.5 Transient NAR Techniques

Transient techniques are those most commonly used in nuclear magnetic resonance. Their use in acoustic saturation NMR is reviewed in Chapter 8. In NAR, however, they have not been used extensively, although they are potentially of importance. The theory of transient NAR effects, on the other hand, has been addressed in some detail.

7.5.1 Theory

Theory of transient effects in NAR is treated by *Kessel* in his early book on NAR, available in both Russian and German editions [7.16], and in a later brief review [7.17]. A particularly detailed theoretical treatment of acoustically induced transient effects has been given by *Fedders and Lu* [7.18], whose formalism we outline below.

For the case in which acoustic pulses couple to nuclear spins via their quadrupole moments, the usual Bloch equations, the foundation for discussion of transient effects in NMR, cannot be applied. Instead, one obtains a complicated set of equations, described by Kessel as corresponding to the rotation of a geometrical monster and by Fedders and Lu as making the physical interpretation of acoustic spin–echo experiments very difficult if not impossible. Fortunately, however, Fedders and Lu find that for spin-one systems at any temperature and for higher-spin systems in the high temperature limit ($k_B T > \hbar\Omega$, Ω being the Larmor frequency) the equations decouple in the usual (*i.e.*, NMR resonant) approximation, thereby enabling the use of semiclassical models to describe the effects of acoustic and/or electromagnetic pulses on the spin system. Under these conditions, the equations of motion for nuclear spins driven by an acoustic field can be cast into the form of the Bloch equations for the motion of an *effective* magnetization $\vec{\mathcal{M}}$ due to an *effective* magnetic field $\vec{\mathcal{H}}$:

$$\frac{\partial \vec{\mathcal{M}}}{\partial t} = \gamma \vec{\mathcal{M}} \times \vec{\mathcal{H}} . \tag{7.18}$$

As pointed out in Chapter 3, the coupling of acoustic fields to nuclear spin quadrupole moments is conveniently described in terms of the multipole operators A_{2m}. In the case in which the acoustic pulse couples to the dipole moments of the magnetization A_{1m}, as in the dynamic Alpher–Rubin interaction, the appropriate Hamiltonian is

$$\mathcal{H}_{em} = -2\hbar\gamma H_1 I_x \cos[\omega(t - x/v)] \tag{7.19}$$

for an acoustic pulse of velocity v traveling in the x-direction, where H_1 is proportional to the elastic strain. Equation (7.19) also applies to nuclear and electron spins in magnetically ordered materials (Chapter 9). The spin–phonon coupling of an acoustic pulse via the quadrupolar interaction is described by the Hamiltonian

$$\mathcal{H}_{ac} = -\hbar \sum_{m=-2}^{+2} \kappa_m e_m A_m^2 \cos[\omega(t - x/v)] \tag{7.20}$$

for an acoustic pulse of velocity v traveling in the x-direction. The factor κ_m and the elastic strain constant associated with the m transition e_m satisfy the relation

$$(\kappa_m e_m)^* = (-1)\kappa_{-m} e_{-m},$$

where κ_m, for a particular coupling, can be expressed in terms of familiar constants. To obtain explicit expressions for $\kappa_m e_m$, we use (3.37), (3.38), (3.48) and the S-tensor for a cubic crystal given by (D-23), and we write the Hamiltonian for the spin–phonon dynamic quadrupole interaction in a cubic lattice as

$$\mathcal{H}_Q = \frac{eQ}{2I(2I-1)} \left\{ (S_{11} - S_{12}) \sum_i \varepsilon_{i,i} \right.$$
$$\left. \times \left[I_i^2 - \frac{1}{3}I(I+1)\right] + S_{44} \sum_{i,i+1} \varepsilon_{i,i+1} \{I_i, I_{i+1}\} \right\}. \tag{7.21}$$

Comparing (7.21) with (7.20), one obtains

$$\kappa_{+2} e_2 = eQ \left[\frac{2}{15} \frac{(I+1)(2I+3)}{I(2I-1)}\right]^{\frac{1}{2}}$$
$$\times \left[\frac{1}{8}(S_{11} - S_{12})(\varepsilon_{xx} - \varepsilon_{yy}) - \frac{1}{4} i S_{44} \varepsilon_{xy}\right] \tag{7.22a}$$

$$\kappa_{+1} e_1 = eQ \left[\frac{2}{15} \frac{(I+1)(2I+3)}{I(2I-1)}\right]^{\frac{1}{2}}$$
$$\times \frac{1}{4} S_{44}(\varepsilon_{xz} - i\varepsilon_{yz}) \tag{7.22b}$$

and

$$\kappa_0 e_0 = eQ \left[\frac{1}{45} \frac{(I+1)(2I+3)}{I(2I-1)} \right]^{\frac{1}{2}}$$

$$\times \frac{1}{4}(S_{11} - S_{12}) \left(\varepsilon_{zz} - \frac{1}{2}\varepsilon_{xx} - \frac{1}{2}\varepsilon_{yy} \right). \quad (7.22c)$$

Equations (7.22a–c) are valid only if the external magnetic field points along a cube edge of the cubic lattice. Five sets of semiclassical Bloch vectors are defined, each set consisting of the expectation values of the components of A_{lm} multipole operators. The resulting Bloch–like equations of motion can be written

$$\frac{\partial \mathcal{M}_x}{\partial t} = \omega_0 \mathcal{M}_y - \omega_1 \mathcal{M}_z \sin(\omega t - \phi) \quad (7.23a)$$

$$\frac{\partial \mathcal{M}_y}{\partial t} = -\omega_0 \mathcal{M}_x + \omega_1 \mathcal{M}_z \cos(\omega t - \phi) \quad (7.23b)$$

$$\frac{\partial \mathcal{M}_z}{\partial t} = \omega_1 \mathcal{M}_x \sin(\omega t - \phi) - \omega_1 \mathcal{M}_y \cos(\omega t - \phi), \quad (7.23c)$$

where $\phi = 0$ for standing wave pulses and $\omega x/v$ for pulses traveling along the x-axis with velocity v. The five sets of Bloch vectors are summarized in Table 7.1, in which the angular brackets $\langle \rangle$ denote the thermal average of the quantity enclosed. For a detailed discussion of (7.23) and Table 7.1, the reader is referred to Reference [7.18]. In sum, the results of the Fedders–Lu treatment enable one to calculate the precession of the various components of dipole or quadrupole operators driven by the applied pulses (acoustic, electromagnetic or any combination thereof) just as one usually does (in NMR) with the magnetization vector in an RF magnetic field.

The above formalism describes the free precession of dipole and quadrupole tensors excited by acoustic (U) or magnetic (M) pulses. Fedders and Lu go further. Since more than one set of Bloch–like vectors are excited by either M or U pulses, and since a given component of the dipole or quadrupole tensor operator may couple to both M or U pulses, Fedders and Lu show that echo pulses arise by applying the following two-pulse sequences: MM, UU and UM.

Table 7.1

The Vector Components of the Effective Magnetization $\vec{\mathcal{M}}$ and the Values of the Effective ω_0 and ω_1 to be Used in the Bloch Equations Given by (7.23)*

Type	$\mathcal{M}_z$	$\mathcal{M}_x$	$\mathcal{M}_y$	ω_1	ω_0
M_1	$\langle A_{10} \rangle$	$\frac{\langle A_{1-1}-A_{11}\rangle}{\sqrt{2}}$	$\frac{i\langle A_{11}+A_{1-1}\rangle}{\sqrt{2}}$	γH_1	γH_0
M_2	$\langle A_{20} \rangle$	$\frac{\langle A_{2-1}-A_{21}\rangle}{\sqrt{2}}$	$\frac{i\langle A_{21}+A_{2-1}\rangle}{\sqrt{2}}$	$\sqrt{3}\gamma H_1$	γH_0
U_1	$\langle A_{10} \rangle$	$\frac{\langle A_{2-1}-A_{21}\rangle}{\sqrt{2}}$	$\frac{i\langle A_{21}+A_{2-1}\rangle}{\sqrt{2}}$	$\frac{|k_1 e_1|}{\sqrt{\frac{2}{3}J(J+1)}}$	γH_0
U_2	$\langle A_{10} \rangle$	$\frac{\langle A_{22}+A_{2-2}\rangle}{\sqrt{2}}$	$\frac{-i\langle A_{22}-A_{2-2}\rangle}{\sqrt{2}}$	$\frac{|k_2 e_2|}{\sqrt{\frac{1}{6}J(J+1)}}$	$2\gamma H_0$
U_3	$\langle A_{20} \rangle$	$\frac{\langle A_{1-1}-A_{11}\rangle}{\sqrt{2}}$	$\frac{i\langle A_{11}+A_{1-1}\rangle}{\sqrt{2}}$	$\frac{3}{2}|k_1 e_1|$	γH_0

*The cases are labeled by M or U according to whether the driving field is electromagnetic or acoustic. Cases M_1 and M_2 are valid for arbitrary spin. Cases U_1 and U_2 are valid for $J = 1$ or in the high temperature limit. Case U_3 is valid only for $J = 1$. (From Ref. 7.18)

Before passing on to an application of this formalism, it is useful to mention a technique for estimating the magnitude of the echo pulse. From (7.23) and ω_1 from Table 7.1, the values of $\mathcal{M}_x$ and $\mathcal{M}_y$ after a two-pulse sequence can be obtained. Since, further, the voltage amplitude of a magnetic pulse echo is known, by turning the Hamiltonian of (7.21) into an energy density, the amplitude of the acoustic echo pulse can be calculated. The contribution to the free energy density F from $\mathcal{H}_{ac}$ is

$$F = -\hbar \sum_m \kappa_m n_s \langle A_{2m}(x,t) \rangle e_m(x,t) , \qquad (7.24)$$

where n_s is the density of spins and $\langle A_{2m} \rangle$ is the thermal average of the appropriate multipole operator. Since only the $\Delta m = \pm 1$ or $\Delta m = \pm 2$ components are excited in a given experiment, the other components can be neglected. Assuming that the effective magnetization

$$\mathcal{M}_+ = (\mathcal{M}_x + i\mathcal{M}_y)/\sqrt{2}$$

describes a traveling pulse of height $\mathcal{M}_{\max}$ and of any width, it follows that the maximum-induced strain component ε_{max} is given by

$$\varepsilon_{\max} = \hbar n_s |\kappa_m \mathcal{M}_{\max}| \frac{ql}{2\rho v^2} , \qquad (7.25)$$

where q is the wave number of the acoustic wave, and l is the length of the sample.

As an application of this formalism, we consider the $\Delta m = \pm 2$ transitions of ^{115}In nuclear spins ($I = \frac{9}{2}$) in InAs for the case of two U_2 pulses of longitudinal waves propagated along the cube x-axis, with the external magnetic field along the z-axis. Substituting (7.22a) into the expression for $\omega_1(U_2)$ in Table 7.1 and using the experimental value

$$Q(S_{11} - S_{12}) = 23.9 \times 10^{-9} \text{statcoulombs cm}^{-1} ,$$

one obtains $\omega_1 = 3.3 \times 10^8 e_{xx}$ sec^{-1}. For a strain amplitude of 10^{-6}, a time of 4.7×10^{-3} sec is required for a 90° pulse. Using $\rho = 5.67$ g/cm^3, $v = 3.83 \times 10^5$ cm/sec, and $n_s = 1.72 \times 10^{22}$ cm^{-3}, the maximum strain amplitude of the echo pulse from (7.25) is

$$\varepsilon_{\max} = 7.32 \times 10^{-9} \mathcal{M}_+(ql) .$$

For a magnetic field of 10^4 G, the $\Delta m = \pm 2$ resonance is at a frequency of 1.87×10^7 Hz; the value of $\langle A_{01} \rangle$ is 3.06×10^{-4} at a temperature of 4.2 K. Thus for a crystal 1 cm long, one obtains a maximum strain $\varepsilon_{\max} = 4.8 \times 10^{-10}$.

7.5.2 Saturation Methods

Transient NAR techniques have been utilized to determine nuclear spin–lattice relaxation times in crystals characterized by magnetic dipolar coupling (Al in Al–Zn alloy) and by dynamic electric quadrupole coupling (Ta). The two methods used involve partially saturating the NAR signal and measuring the change in signal amplitude as a function of time. For the latter procedure, multiple–scan direct detection, rather than the conventional field modulation and lock–in detection method, is used to monitor the signal in times that are short compared to T_1 [7.19,7.20].

In the progressive saturation method, the acoustic power is set to cause partial saturation of the NAR line, and the magnetic field is repetitively scanned through the resonance during a time shorter than T_1. The consecutive resonances are recorded as a function of time in one curve that is stored in the memory of a signal averager. In this way, several resonances of decreasing amplitude can be observed and displayed in one experimental curve. After waiting several T_1's for the nuclear spin system to reach equilibrium, the procedure is repeated so that signal averaging can be used. From the resulting decrease in amplitude of the successive resonances as a function of time, T_1 can be determined. This method has been analyzed [7.21,7.22] and applied to NMR. The method, as applied to NAR, results in a theoretical expression, of formidable complexity for the heights of the sequence of resonances, from which T_1 is obtained after comparison with the experimental data.

To simplify the analysis of the data, a second method, saturation and recovery, is utilized. In this method, the acoustic power is set to produce strong saturation, and the magnetic field is set at the center of the resonance line for a period corresponding to many T_1's. Thus the spin system is prepared in a configuration far from its equilibrium value. The field is then quickly moved off resonance. After a variable time delay t, the field is swept back through the resonance, and the signal height is measured. Thus a measurement is made of the recovery of the resonance height toward its thermal equilibrium value. A plot of signal height versus t yields a determination of T_1 (Fig. 7.19).

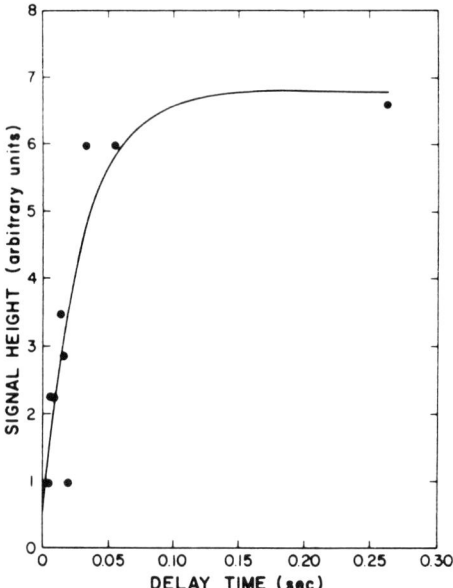

FIGURE 7.19. Data for T_1 measurement in tantalum using the saturation and recovery method. The points represent data taken as a function of delay time at $T = 2.0$ K. The solid line corresponds to an exponential with $T_1 = 0.0294$ sec. (From Ref. 7.20)

Utilizing the NAR(A-R) resonance, the two saturation methods have been applied to the measurement of T_1 in a single crystal of aluminum containing 0.87 at.% Zn. Measured values of $T_1 = 0.72$ sec at 2.2 K and $T_1 = 0.60$ sec at 2.6 K are obtained, corresponding to a $T_1 T = 1.58$ sec K and 1.56 sec K respectively. These compare with the value of pure Al at low temperatures of $T_1 T = 1.80$ sec K as measured by NMR. The saturation recovery method is also applied to the measurement of ^{181}Ta NAR2 T_1 at 2.0 K, with a resultant value of $T_1 = 0.029$ sec (see Fig. 7.19).

The superconducting quantum interference detector (SQUID) method of detecting NAR, described in detail in Chapter 10, can be used to measure T_1 in solids. A saturating acoustic pulse, short compared to T_1 at the appropriate resonance frequency, is applied to the sample. A superconducting pick-up coil, linked to the RF SQUID, with its axis parallel to the external magnetic field, detects the DC decay of the z-component of the total spin magnetization, from which T_1 can be determined in the laboratory reference frame. Results obtained for T_1 of ^{181}Ta in single crystal Ta in the temperature range 1.58–4.2 K are given in Chapter 10.

7.5.3 Adiabatic Fast Passage

In nuclear magnetic resonance, the adiabatic condition is met if

$$dH_0/dt \ll \gamma H_1^2 \; ,$$

where dH_0/dt is the time rate of change of the magnetic field in the z-direction, and H_1 is the amplitude of the rotating field in the xy-plane. Under this condition, the component of the nuclear magnetization along the effective field direction is a constant in time [7.23] and follows adiabatic changes in the direction of the effective field. In an NMR adiabatic fast passage experiment, if we start from a value of H_0 far above the resonance where the effective field is practically parallel to $\boldsymbol{H}_0$, and if we go through the resonance to the other side, far below the resonance, the magnetic moment $\boldsymbol{M}$, which is initially along $\boldsymbol{H}_0$, remains continuously parallel to the effective field and ends up antiparallel to $\boldsymbol{H}_0$. In solids, a weaker requirement for adiabatic fast passage is [7.23]

$$\frac{1}{T_1} \ll \frac{1}{H_1} \left| \frac{dH_0}{dt} \right| \ll |\gamma H_1| \simeq \frac{1}{T_2} \; . \tag{7.26}$$

Kessel and Shakirzyanov [7.24] predict that in the case of strong dynamic nuclear electric quadrupole coupling to a nuclear spin system, the expectation value of

$$Q_0 = \sum_j [3 I_{zj}^2 - I(I+1)] \tag{7.27}$$

will precess—as does the magnetization in an NMR experiment—in the rotating frame and will not decay under T_2 processes, but under T_1 processes. They call this process, which is observed in adiabatic fast passage experiments with saturating acoustic pulses, *Q-trapping* or *acoustic spin locking*. The experiment is carried out by using large-amplitude, long acoustic pulses at a frequency difference from the resonance frequency given by $\Delta\omega$. The magnetization is then measured as a function of $\Delta\omega$ by looking at the free induction decay following a 90° M pulse. Results of an experiment utilizing ^{23}Na in a NaCl single crystal are shown in Fig 7.20.

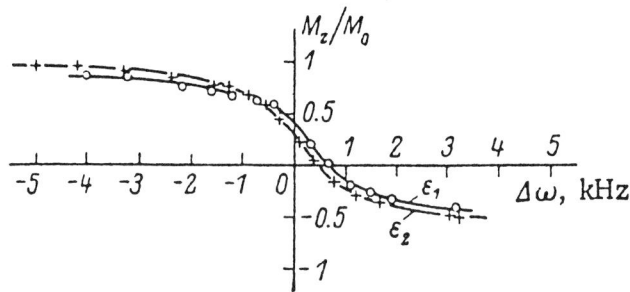

FIGURE 7.20. Dependence of M_z/M_0 on $\Delta\omega$. (From Ref. 7.25)

With the SQUID acoustomagnetic spectrometer (Chapter 10), one can directly observe the magnetization changes in adiabatic fast passage experiments. Ultrasonic energy coupled to a nuclear spin system by the dynamic electric quadrupole interaction changes the DC magnetization along the z-direction, which is the direction of the external magnetic field. In a SQUID experiment, one measures voltage changes proportional to these changes in magnetization. III–V semiconductors with small amounts of substitutional impurities have spin–lattice relaxation times dominated by the Raman process near room temperature and by the direct process at very low temperatures. At 4.2 K the spin–lattice relaxation time for ^{115}In in InP is approximately 10^6 sec. Since the adiabatic passage condition is very easy to meet for such a long relaxation time, a large adiabatic fast passage $\Delta m = \pm 2$ SQUID-detected magnetization change is observed as frequency $\Delta\omega$ is swept from far below resonance to far above, or vice-versa [7.26]. In general, the SQUID acoustomagnetic spectrometer allows the measurement, using adiabatic fast passage, of T_1's that are longer than the acoustic decay time (typically a msec).

Adiabatic fast passage is also discussed in Section 8.4.3.

7.5.4 Rotating Frame Acoustic Relaxation

It is possible in NMR to transform nuclear spins from Zeeman order to dipolar order through adiabatic demagnetization in the rotating frame (ADRF). The rotating reference frame has its z-axis parallel to $\boldsymbol{H}_0$; the xy-axes are perpendicular to $\boldsymbol{H}_0$ and rotate at the Larmor frequency of a nuclear spin system. One way to use ADRF is to create the pulse sequence of

(1) a 90° pulse of a large RF magnetic field $\boldsymbol{H}_1$ (perpendicular to $\boldsymbol{H}_0$) that rotates the nuclear spin magnetization perpendicular to $\boldsymbol{H}_0$ in the rotating frame, immediately followed by
(2) a long, smaller amplitude $\boldsymbol{H}_1$ *spin-locking* pulse, which has its RF phase shifted by 90° from the initial $\boldsymbol{H}_1$ saturating pulse.

At the beginning of the spin–locking pulse in the rotating frame, the nuclear spin system magnetization is perpendicular to $\boldsymbol{H}_1$. The magnetization experiences the $\boldsymbol{H}_1$ magnetic field, not the $\boldsymbol{H}_0$ magnetic field, and is said to be *spin-locked* to $\boldsymbol{H}_1$. As time passes after the beginning of the spin-locking pulse, the magnetization in the rotating frame decays toward $\boldsymbol{H}_1$ with an exponential time constant $T_{1\rho}$, where $T_2 \leq T_{1\rho} \leq T_1$. $T_{1\rho}$ can be measured as the free induction decay following the turn–off of the spin–locking pulse.

Kanert and Münter [7.27] have carried out an NMR rotating frame spin–lock experiment, as described above, but their experiment includes a low-frequency, time–varying stress that is applied to the single–crystal magnetic resonance specimen and that modulates the dynamic nuclear electric quadrupole interaction and results in $T_{1\rho}$ having a strong dependence on this dynamic stress. For ^{23}Na magnetic resonance in NaCl at 300 K, the acoustic stress frequency is in the kHz range; therefore the rotating frame experiment allows the observation of NAR in insulators at very low acoustic frequencies using the $T_{1\rho}$ free induction decay.

7.5.5 Pulse Echo

The use of sophisticated CW and transient techniques for observing NAR is normally required because of the typically small resonant attenuation coefficient ($\sim 10^{-5}$ m^{-1} or less). Several examples of extremely large resonant attenuation coefficients in NAR are known: *e.g.*, ^{55}Mn in antiferromagnetic RbMnF$_3$, measured at 650 MHz, and ^{165}Ho in the enchanced nuclear paramagnet HoVO$_4$ at 800 MHz. NAR in these magnetic materials is discussed in some detail in Chapter 9. To detect such large acoustic attenuation changes one may have resort to the standard ultrasonic pulse–echo technique, the mainstay for ultrasonic measurements since its initial use in the mid–1940's. Descriptions of pulse–echo spectrometers are given in the standard texts on ultrasonics. The spectrometers themselves are available from commercial suppliers.

A brief summary of the pulse–echo technique utilized in the enhanced NAR experiments follows [7.28]. Power from a signal generator at frequencies up to 1 GHz is amplified and, when necessary, fed to a frequency doubler. A single–pole, single–throw PIN switch is utilized to generate 200 nsec pulses at repetition rates of 1 kHz. The pulses are further amplified and transmitted to the sample in the helium cryostat by a coaxial line. A deposited ZnO transducer is used as a wide band acoustic generator and receiver. Incident and return pulses are separated by a single–pole, double–throw PIN switch with a speed of 5 nsec. The return pulses, amplified and detected by a diode, are displayed on an oscilloscope screen. A peak detector gate monitors the signal amplitude of the selected echo. The signal is measured by a multimeter or fed to the Y–input of an XY recorder. The X–input is derived from the magnetic field applied to the sample. Velocities are determined from the time delay (typically a μsec) of an echo and from known path lengths of the specimen. The errors may be as large as 5%, but very small changes ($\sim 10^{-9}$) in the time of flight may be detected by phase comparison with a CW signal introduced from the transmitter. For measurements of acoustic absorption, the best signal–to–noise ratio is obtained by monitoring the echo immediately before that with amplitude e^{-1} of the first echo. Step attenuators of 10 dB and 1 dB, calibrated at each frequency, are used.

7.6 The Composite Resonator

In NAR experiments the specimen being studied is usually part of a composite resonator, consisting of transducer, bond and sample. The most common geometrical configuration is that of a specimen with two faces prepared to be flat and parallel, with the transducer bonded to one of the parallel faces (for reflection spectrometers) or with transducers bonded to both parallel faces (for transmission spectrometers). The effects of the transducers and bonds on the measured properties of the standing wave pattern (see Fig. 7.1) have been analyzed theoretically and measured experimentally [7.11]. In the present section we briefly describe the preparation of specimens, the types of transducers and bonds commonly used in NAR, and two typical acoustic resonator assemblies and probes.

7.6.1 Sample Preparation

The requirements for sample preparation and inspection are similar to those for use in conventional high–frequency (nonresonant) CW ultrasonics [7.11]: optically flat and parallel pairs of surfaces on a bulk specimen specifically oriented with respect to crystal axes and achieved with minimum damage to the specimen. As an illustrative example, the sequence for preparing a single crystal of tungsten might be the following:

(1) Orient by means of a goniometer and X–ray crystallographic apparatus.
(2) Cut the desired faces using spark–cutting equipment.
(3) With diamond or alumina powder on a cast iron lap, rough grind the sample mounted in a special jig. The length is 0.160 cm.
(4) Etch for 45 min in a solution of 1 HF + 5 HNO_3 + 1 H_2O. The length is 0.155 cm. Large etch pits are visible under a microscope.
(5) Grind on a cast iron lap with 9 μm diamond abrasive. The length is 0.152 cm.
(6) Etch for 15 min in 2 HF + 4 HNO_3. Length is 0.143 cm. Etch pits are still visible. An additional 15–min–etch results in an improved etch pattern. The length is 0.136 cm.
(7) Grind with 9 μm diamond on cast iron, followed by polishing with 3 μm diamond on a lead lap. Use progressively finer grit size, if necessary, on lead or wax laps.
(8) Measure on a Newton's rings apparatus to insure that the faces are flat to within one fringe (1/2 wavelength of light). Measure on an auto–collimator or Millimess (parallel–measuring device) to insure that the faces are parallel to one wave–length of light. Final length at 21.0 C is 0.1283 cm.

The techniques of cutting, orientation, etching, grinding and polishing are well documented in the optical literature and in the literature of metallography and of CW ultrasonics. The procedure for a specimen, such as an alkali halide or semiconductor crystal, may differ markedly from that detailed above for tungsten. One should refer to the literature for proper precedents in handling a particular specimen.

7.6.2 Transducers and Bonds

The characteristics required of transducers in NAR do not differ from those in CW and pulsed high–frequency ultrasonics. Exciting RF voltages for CW are considerably lower than those for pulse–echo ultrasonics, placing a premium on uniform, low resistivity electrodes, with only modest provision against electromagnetic breakdown of the transducer. At frequencies in the RF and low UHF range, quartz piezoelectric transducers, operated either at a fundamental or odd harmonic frequency, are used. Alternatively, ceramic transducers are available at lower frequencies. X–cut quartz transducers are used to generate longitudinal waves; AC, BC, AT or BT transducers are used to generate transverse waves (Y–cut transducers should be avoided for this purpose because they do not generate pure shear waves). For metals, the technique of direct EM generation in a magnetic field, not necessitating a bonded transducer, is sometimes advantageous.

Bonding a quartz or ceramic transducer to the flat face of a specimen remains today more of an art than a science. The type of bond and the temperature range over which it can be used depend critically on

(1) surface characteristics and the nature of the specimen, especially its thermal expansivity (sometimes, as in the case of SnTe, on the anisotropy of the thermal expansivity); and

(2) whether transverse or longitudinal waves are to be propagated.

For use over a wide range (*e.g.*, 1.5 K–300 K) grease bonds such as Nonaq stopcock grease, silicone grease or silicone 200 fluids (of various viscosities, depending on temperature and specimen) have been found effective. Some, such as Nonaq, have to be wrung in; others, such as light silicone grease, will not function if wrung in. Among other bonds that have proven useful are salol (phenyl salicylate) at room temperature, Canada balsam, clear glyptal, various epoxy resins and (for solids such as the alkali halides with high thermal expansivity below room temperature) a bond of trichloropropene (or similar low freezing point organic liquid) formed at $\sim$ 100 K. At UHF and microwave frequencies, evaporated thin film CdS or ZnO transducers are often used. These are vapor–deposited directly onto the specimen and are characterized by extremely broad frequency bandwidth.

7.6.3 Acoustic Resonator Assemblies and Probes

For acoustic resonator assemblies and probes, often designed to be used at low temperatures, one requires in general:

(1) careful shielding of the receiver from the transmitter and of the sample probe from the transmitter;

(2) acoustic bonds capable of withstanding the desired temperature changes;

(3) firm electrical contact to the transducer electrodes;

(4) mechanical mounting of the specimen so that differential thermal expansion (contraction) does not deform the sample or cause a loss of electrical or thermal contact;

(5) proper impedance matching at the transmitter and receiver transducers so as to minimize power dissipation, power demands on the transmitter and sensitivity demands on the receiver.

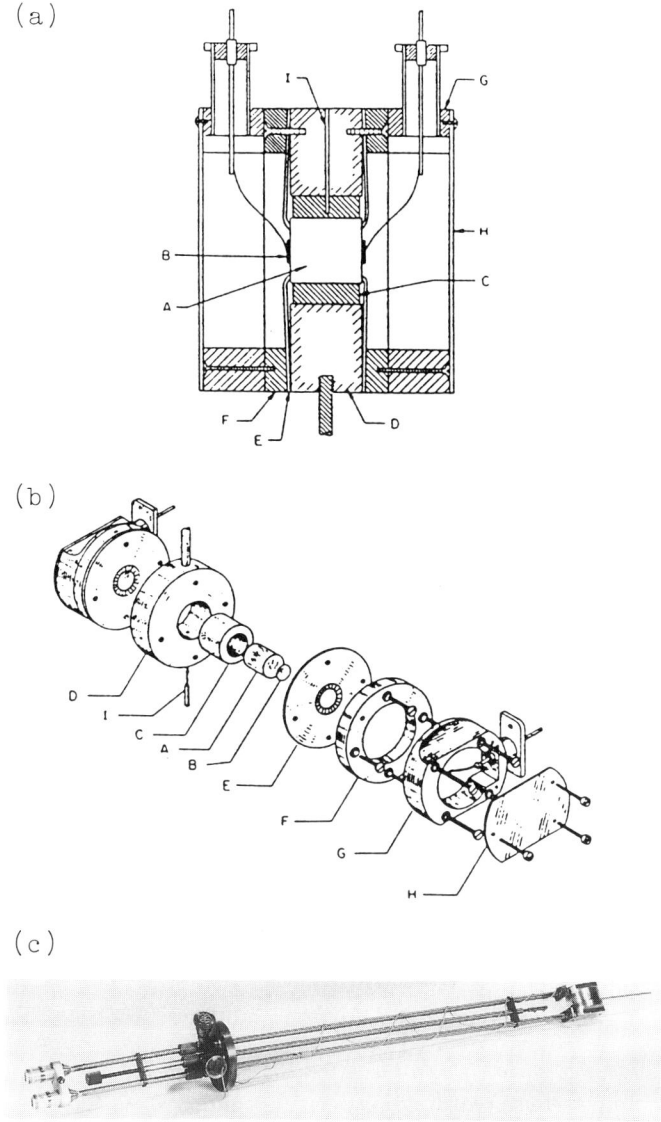

FIGURE 7.21. RF transmission–type resonator assembly (a) and (b) and low–temperature probe (c). (From Ref. 7.11)

7.6.3.1 Sample Probes Used in Electromagnets

In Fig. 7.21 is shown a typical NAR transmission probe suitable for use at frequencies under ~ 100 MHz and over the temperature range ~ 1.5 K to 300 K. In Figs. 7.21a and 7.21b the resonator assembly is shown. In Fig. 7.21c is shown the entire low–temperature probe. The resonator assembly consists of the sample A, to which are bonded piezoelectric transducers B; the sample is held in an epoxy resin holder C, frozen into a beryllium–copper ring D. Connection is made between the center conductors of the coaxial cables and the transducers by means of very flexible pure silver wire or straps, which are carefully soldered to the silver or gold plating on the exposed surfaces of the transducers. Gold–plated Be–Cu finger washers E are used to provide ground contact to the sample, on the end faces of which a plating of silver or gold has been evaporated. The washers also provide shielding against RF leakage between the transmitter and receiver ends of the assembly. The probe is completed by the coaxial coupling ring G and cover H. Beryllium–copper or copper has been found to be preferable to brass because of the presence of ferromagnetic impurities in brass. Provision is made for temperature measurement and control as low as liquid helium temperatures by means of platinum, germanium or carbon–glass resistors, and a gold 2.1 at.% cobalt–versus–copper thermocouple. The thermocouple is inserted into channel I in the Be–Cu and epoxy resin holders. The assembly can be heated by passing an electric current through 2–W resistors cemented to the resonator assembly shield; a copper or brass rod attached to the bottom of the probe acts as a heat leak to the liquid–helium or liquid–nitrogen bath. The entire assembly is detachable from the gold–plated stainless steel coaxial cables, which are used to insert the assembly into a low–temperature dewar. The entire low–temperature probe is shown (without radiation shields) in Fig. 7.21c.

At frequencies greater than $\sim$ 100 MHz, it is usually necessary to improve on the RF probe described above by

(1) further increasing electromagnetic shielding to minimize leakage;
(2) improving the method of impedance matching, usually by incorporating tuning devices into the probe so as to match at a point very near to the resonator assembly;
(3) using broad–band transducers.

In Fig. 7.22a is shown a sample assembly suitable for use between 0.2 and 2 GHz and at temperatures between 1.5 and 300 K. The sample, gold–plated on each end face and then plated with cadmium sulfide transducers, is held loosely with GE 7031 varnish in an epoxy holder. The gold film contacts a piece of indium foil around the edge of the sample face. Contact is assured between the indium foil and the grounded gold electrode by using silver paint to connect the two. A spring–loaded button is used for the hot electrode. The center conductor of the coaxial line is silver–plated metal tubing. The outer conductor is silver–plated brass. All of the metal used is of 0.0508-cm or less wall thickness. Since this probe was used in CW acoustic magnetic resonance experiments, it was necessary to use high-resistivity materials, such as brass, in order that the ac magnetic field used for modulation could penetrate the specimen without excessive loss and phase shift. Typically, the electromagnetic isolation between the transmitter end and the receiver end of the sample assembly is measured to be greater than 80 dB at 1.3 GHz.

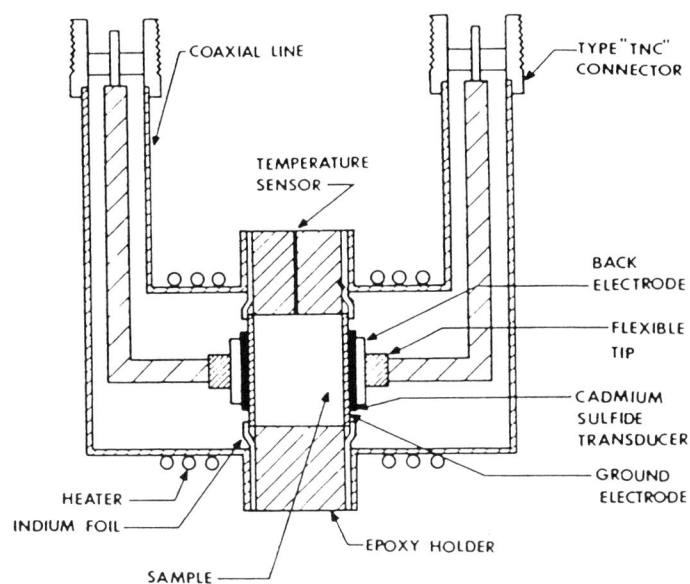

FIGURE 7.22A. Ultrasonic resonator assembly for use of frequencies between 0.2 and 2 GHz. (From Ref. 7.11)

The associated low-temperature probe of Fig. 7.22B incorporates stub stretchers that permit efficient impedance matching close to the composite resonator over a wide range of frequencies. The entire probe assembly is made vacuum tight by the use of O-ring seals at the top and by soldering on (using low-temperature solder) the stainless steel can after the resonator has been connected to the probe. The coaxial lines are made of thin-walled, silver-plated, non-magnetic stainless steel or monel tubing. The resonator assembly is attached by means of gold-plated TNC coaxial connectors. A small-diameter, thin-wall stainless-steel tube is provided to bring in electrical leads for temperature sensors and heaters and to provide a means of evacuating the probe.

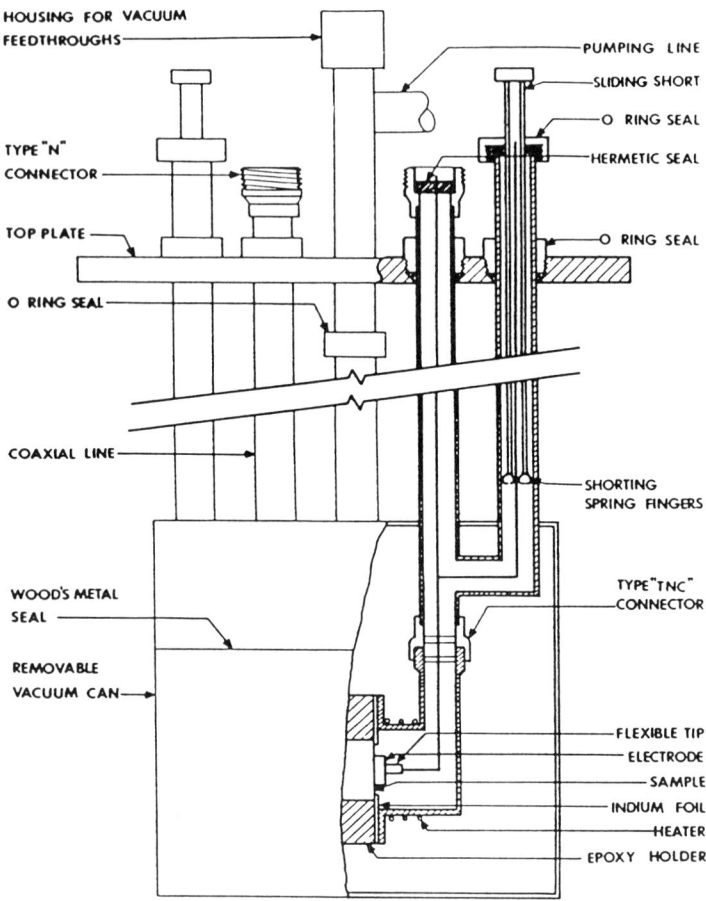

FIGURE 7.22B. Schematic of UHF and low–microwave frequency transmission-type probe for use between 0.2 and 2 GHz. (From Ref. 7.11)

7.6.3.2 Probes for Superconducting Magnets

Use of superconducting magnets in NAR experiments requires low temperature probes that have the following additional requirements over those used with electromagnets:

(1) reproducible, precise rotation of the sample in the sample probe, since the superconducting magnetic field direction cannot easily be varied;

(2) probe diameter restrictions inside the cylindrical superconducting magnet (typically 5.08-cm diameter) that limit the size of composite resonators and place a space restriction on what can be located at the top plate of the probe.

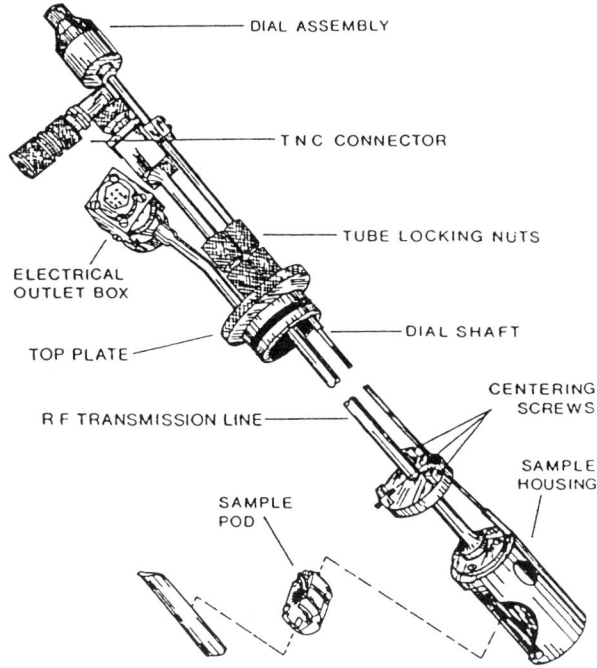

FIGURE 7.23. Ultrasonic probe for use in reflection experiments in the superconducting magnet. (From Ref. 7.12)

A typical ultrasonic probe for use in a superconducting magnet for acoustic and NAR reflection experiments is shown in Fig. 7.23. An air or vacuum dielectric transmission line is constructed with two concentric non-magnetic stainless steel tubes of wall thickness 0.0152 cm. The inner tube is 0.318-cm diameter and the outer tube 0.795-cm diameter, resulting in a transmission line impedance of 50 Ω. At one end, the outer (ground) transmission tube is terminated at the sample housing. The inner tube (RF contact) joins a contact mounted on the sample housing, which forms one side of a rubbing contact; the other side of the rubbing contact forms a spring finger (Fig. 7.24), which is mounted on the sample pod and makes contact with the transducer on the composite resonator in the sample pod. On the sample pod is a spur gear (Fig. 7.24), which engages with a worm gear mounted on the sample housing, as shown in Fig. 7.25. Running parallel to the transmission line, in Fig. 7.23, is a dial shaft with the worm gear at one end and a center dial assembly at the other. Below the top plate, the dial shaft passes through a 0.635 cm diameter, 0.0152 cm wall dial assembly base tube. Both the outer transmission line, and the dial assembly base tube pass through vacuum locking nuts, which allow adjustment of the probe length below the top plate.

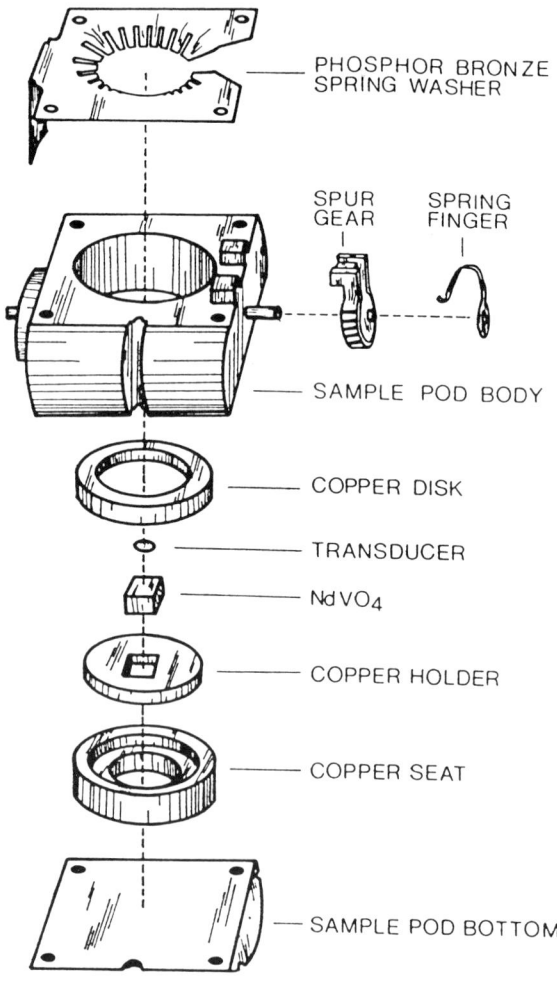

FIGURE 7.24. Illustration of the sample pod indicating how the sample is mounted. (From Ref. 7.12)

7.6 THE COMPOSITE RESONATOR

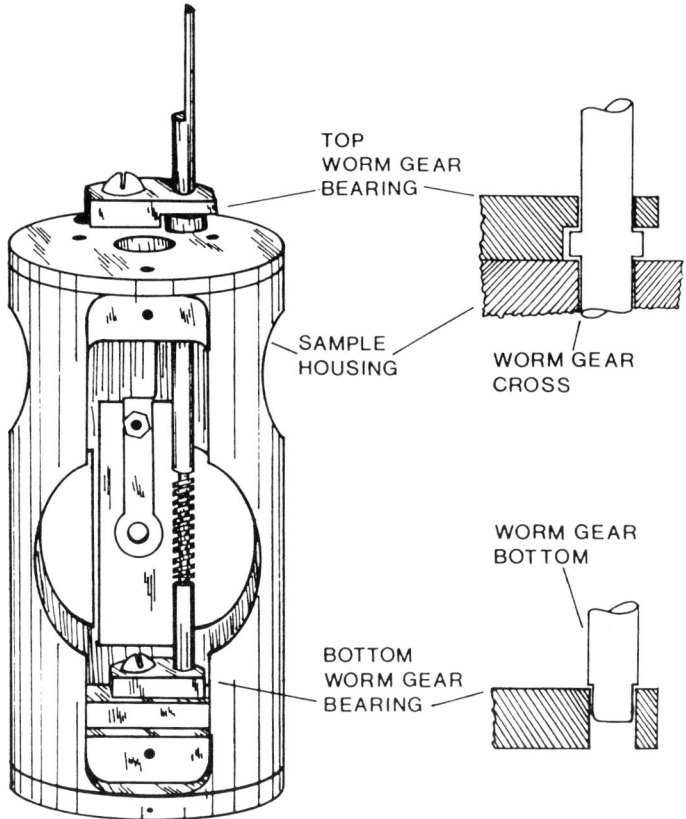

FIGURE 7.25. Sample housing showing how the worm gear is mounted in the worm gear bearing. (From Ref. 7.12)

Above the top plate, the transmission line is terminated in a hermetic TNC connector. The dial assembly base makes a screw vacuum connection to the dial assembly base tube, and the dial shaft passes through an O-ring compression fitting before it attaches to the rotating part of the dial. Orientation of the sample pod can be set within 1° with the dial assembly. Temperature sensors at the probe bottom are connected to 30-gauge manganin wire. These wires run up a 0.64-cm-diameter thin wall stainless steel tube through the top plate of the probe to a hermetically sealed connector. A brass plate is mounted approximately 10.16 cm above the sample housing to which centering screws are attached to center the sample probe in the dewar bore. The sample housing and sample pod are made from oxygen-free copper.

To provide RF shielding, a particular sample, *e.g.*, of $NdVO_4$, is glued with silver paint into a copper holder (shown in Fig. 7.24) which sits on the sample pod bottom. Other components of the sample pod are also assembled with silver paint.

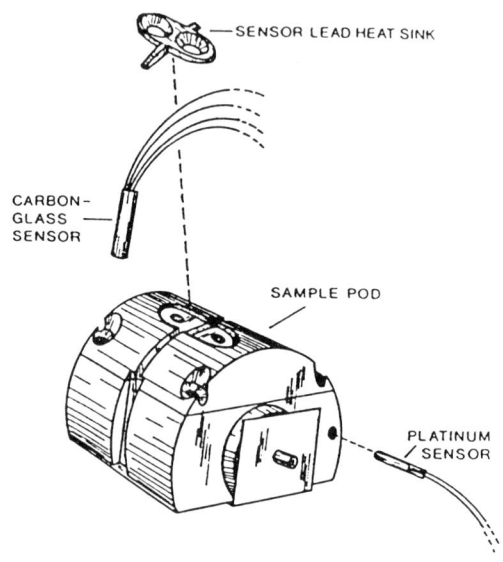

FIGURE 7.26. Carbon–glass and platinum temperature sensor placement in the sample pod. (From Ref. 7.12)

A clearer picture of the components of the sample housing is given in Fig. 7.25, which shows the worm gear attachments and the important spring contact. In Fig. 7.26, we show the temperature sensor placement on the sample pod. We point out that at the top of the sample housing, the inner transmission tube is soldered to the phosphor bronze spring contact, which is in rubbing contact with the phosphor bronze spring finger in Fig. 7.24. The spring finger is insulated from the spur gear by the Delrin inlay, shown in Fig. 7.27.

An obvious disadvantage of the superconducting magnet probe is the absence of impedance matching in the sample probe between the composite resonator and the 50 Ω transmission line. However, with the use of wide band 180° hybrid junctions (1 MHz to 1 GHz), as discussed in Section 7.3.4, one can carry out effective impedance matching at the top of the probe. For MOUS experiments below 35 MHz, which use a double–tuned over–coupled

RF transformer at the sample probe top, as described in Section 7.3.2.2, the lumped constant equivalents of the probe transmission line and the composite resonator become part of the secondary resonance circuit of the transformer.

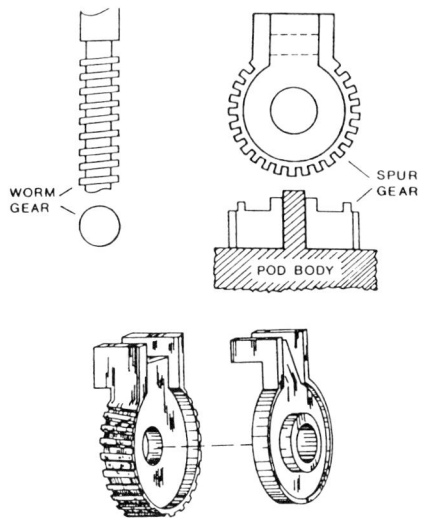

FIGURE 7.27. Final brass worm gear with a Delrin inlay for a brass spur gear. (From Ref. 7.12)

7.7 Application: NAR in Rhenium

The techniques utilized in obtaining the results described in Sections 3.6 and 5.2.1 serve to illustrate aspects of current practice in experimental NAR.

The rhenium crystal [7.29], approximately 1.3–cm–diameter by 1.3–cm long, is oriented and ground with a set of parallel faces whose perpendicular is within $\pm 1°$ of the [0001] (z–direction). Scratches along the sides of the crystal indicate the [$\bar{1}$100] (y–direction) to be within $\pm 1°$, with x defined to be the [11$\bar{2}$0] direction. A composite resonator is constructed using a 0.95–cm diameter quartz 12.5–MHz AT–cut transducer, Nonaq grease bond and the rhenium crystal. The transducer polarization direction is chosen along the x– or y–direction and the propagation direction along the z–direction. For both polarizations, the plane of rotation of the magnetic field is the plane defined by [11$\bar{2}$0] and [0001], as shown in Fig. 3.11. The composite resonator, mounted in a low–temperature probe with 50 Ω transmission lines, is connected to the sample arm of a commercial hybrid junction, which is the bridge element of the computer–controlled ultrasonic reflection bridge spectrometer described in Section 7.3.4.

From the measured coupling constants of rhenium metal, the expected zero magnetic field quadrupole transition from $\frac{3}{2}$ to $\frac{1}{2}$ spin energy levels is approximately 38.38 MHz for ^{185}Re. The transducer is driven at its third harmonic response, which for a composite resonator temperature < 180 K, where the Nonaq bond is frozen, consists of mechanical standing wave resonances spread in frequency from 35 to above 41 MHz. Typical resonances near the center of the third harmonic response at 77.8 K have frequency widths of 2 KHz. The real part of the composite resonator impedance at the mechanical resonance center is approximately 800 Ω. The low temperature sample probe is placed in an electromagnet with 30.5–cm–diameter cylindrical pole caps with a 7.62–cm air gap. The magnet rotates around a vertical axis and uses Hall probe magnetic field regulation and control. Magnetic field modulation coils are in coil forms surrounding the cylindrical pole caps, and the modulation field amplitude is calibrated at 300 K using the ^{1}H NMR in doped water for values of the magnetic field from 500 to 1200 G.

The condition for quadrupole resonance in a small magnetic field is determined by choosing a mechanical resonance center at the appropriate frequency for a particular angle θ, so that the magnetic field can be chosen to meet one of the four quadrupole resonance transitions shown in Fig. 3.9, but can still qualify as a small field $[(\gamma/2\pi)H \ll e^2qQ/h)]$. In order to display the resonance, the magnetic field is swept about the center field, while magnetic field modulation is used at 37 Hz with an amplitude of 17 or 34 G. Phase–sensitive detection is adjusted for 37 Hz with a 1.25 sec time constant. The first derivative absorption and dispersion resonance signals are digitized and averaged. Typical magnetic field sweep times are 300 sec. Sweep ranges are 250 to 500 G for the very broad $\frac{3}{2}$ to $\frac{1}{2}$ transitions. RF phase–sensitive detection is utilized so that both absorption and dispersion signals can be studied and so that averaging will be effective for 1 pass signal–to–noise ratios < 0.2. The choice of NAR absorption or dispersion signals is made by adjusting the frequency to the center of a mechanical resonance and then tuning the phase shifter in the signal arm of the spectrometer for, respectively, the in–phase or out–of–phase mechanical resonance response, as displayed on a monitoring oscilloscope using audio frequency modulation of the mechanical resonance.

Signal amplitudes for the NAR quadrupole resonance signals are dependent on frequency, magnet angle, numerical factors in the NAR transition probability, temperature, natural abundance of the nuclear spin system, acoustic power level, bond properties of the composite resonator, magnetic field modulation amplitude and skin depth of the modulation field. For the conditions described above, the best ^{187}Re NAR signal-to-noise ratio of the first derivative of the dispersion for a single pass at 77.8 K and $\theta = 90°$ is 0.3. At 4.2 K, the signal–to–noise ratio increases by a factor of 6 over the 77.8 K value. The 4.2–K resonances, however, are seen under conditions of saturation; the Re crystal skin depth prevents complete penetration of the 37–Hz modulation field, and the resonance sweep baselines are altered by the DC Alpher-Rubin effect. Most data, therefore, is taken at 77.8 K. The electrical power applied to the composite resonator is kept at the value 3.24 μW, a power level that does not show the effects of saturation.

The direction of [11$\bar{2}$0] is determined experimentally from study of the ^{187}Re NAR signal dependence on magnetic angle near $\theta = 90°$. At $\theta = 90°$, the $+\frac{3}{2}$ and $-\frac{3}{2}$ energy levels coincide, and the equal frequencies $\nu_{\beta'}$ and ν_α produce $\Delta m = \pm 2$ NAR signals centered at the same resonance magnetic field. If $\theta = 89°$ or $91°$, the $+\frac{3}{2}$ and $-\frac{3}{2}$ levels are split, and the frequencies $\nu_{\beta'}$ and ν_α are not equal, so that the observed signal consists mainly of two $\Delta m = \pm 2$ transitions separated by 50 G at a field of 1000 G. The effective NAR line width change as a function of θ is so large that the [11$\bar{2}$0] can be determined within $\pm \frac{1}{2}°$.

The precision of the measured quadrupole coupling constants of Re (Section 3.6) is limited by the unusually large width of the nuclear acoustic resonances, necessitating a heroic 48 hours of signal averaging to obtain a 10/1 signal–to–noise ratio. At 77.8 K and $\theta = 90°$, the ^{187}Re $\Delta m = \pm 2$ first derivative of the absorption peak–to–peak line width is 73 ± 4 G, with either a 17 G or 34 G amplitude modulation field; the 34 G amplitude is used to obtain the largest possible signal–to–noise ratio. At $\theta = 0$, the ^{187}Re $\Delta m = \pm 1$ first derivative of the absorption line shape peak–to–peak width is 98 ± 4 G. At $\theta = 43.3°$, the first derivative of the $\Delta m = \pm 1$ absorption line shape peak has a width of 68 ± 4 G. These line widths are independent of magnetic field over the range investigated, from 200 to 1200 G.

The line shapes and line widths for ^{187}Re and ^{185}Re are the same. As shown in Fig. 7.28, a first derivative of a Gaussian line shape is readily fit to the experimental first derivative of the ^{187}Re absorption line. Because the tails of the resonance absorption line shape exceed the sweep window of 500 G at $\theta = 0°$, the narrower dispersion signal is used in determining the quadrupole coupling constants (Table 7.2). The first derivative of the dispersion signal is a symmetric resonance structure, as shown in Fig. 7.29 for ^{185}Re at 77.8 K and $\theta = 90°$. With careful tuning of the RF phase shifter in the signal arm of the spectrometer it is possible to obtain a dispersion signal with symmetrical baselines that is centered in the magnetic field sweep window.

Table 7.2

Magnitudes of the Measured Nuclear Electric Quadrupole Coupling Constants for ^{185}Re and ^{187}Re*

Temperature (K)	Nucleus	e^2qQ/h (MHz)
77.8	^{187}Re	255.920 ± 0.013
77.8	^{185}Re	270.966 ± 0.013
4.2	^{187}Re	255.478 ± 0.013
4.2	^{185}Re	270.430 ± 0.013

*(From Ref. 7.29)

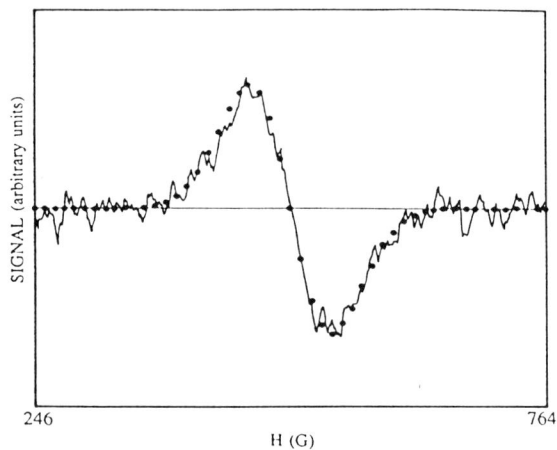

FIGURE 7.28. ^{187}Re NAR $\Delta m = \pm 2$ $\nu_{\beta'}$ transition first derivative absorption signal at $\theta = 90°$, 48 hours signal averaging, 77.8 K, 500–G sweep in 5 minutes, 1.25–sec time constant, 34–G amplitude modulation at 37 Hz, and 39.117400 MHz. The dots correspond to a theoretical Gaussian first derivative, plotted using, the experimental peak–to–peak width, peak–to–peak amplitude and center position. (From Ref. 7.29)

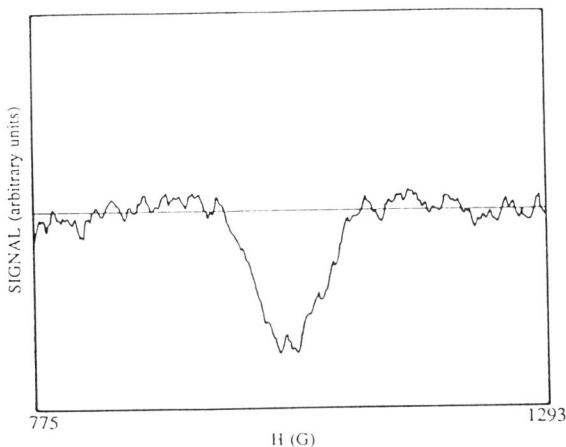

FIGURE 7.29. ^{185}Re NAR $\Delta m = \pm 2$ $\nu_{\beta'}$ transition first derivative dispersion signal at $\theta = 90°$, 48 hours signal averaging, 77.8 K, 500-G sweep in 5 minutes, 1.25-sec time constant, 34-G amplitude modulation at 37 Hz, and 39.117400 MHz. (From Ref. 7.29)

7.8 Application: Experimental Search for the Dynamic Nuclear Hexadecapole Interaction

A discussion of the nature of the dynamic nuclear electric hexadecapole interaction is given in Sections 3.8 and 3.9. In searching for a suitable candidate nucleus in which this interaction may be large, one again confronts ^{185,187}Re. A value of the Re hexadecapole moment has been estimated to be 0.015×10^{-48} cm^4. The value of the hexadecapole antishielding factor is estimated to be 50,000. This is to be compared with the Re nuclear quadrupole antishielding factor, $1 - \gamma_\infty = 45$. Assuming only ionic contributions from the 12 nearest neighbors in Re to both the hexadecapole coupling T-tensor and the quadrupole coupling S-tensor, one finds that the ratio of the dynamic hexadecapole coupling to the dynamic quadrupole coupling at the same frequency (30 MHz) and temperature (78 K) for a single pass through the assumed resonance line is 10^{-12}!

The seemingly hopeless task is made somewhat less hopeless, however, by the consideration of several additional factors. By reducing the temperature to 4.2 K and searching for the $\Delta m = \pm 4$ transitions for ^{187}Re at 90 MHz, one increases the probability of observing the dynamic HDI by a factor of 170. By signal averaging over longer periods of time, furthermore, the dynamic HDI can be increased by a factor of 50 to 100. Thus, in experiments that optimize temperature, frequency and signal enhancement, the ratio of HDI to NQI can be increased to 10^{-8} from the original estimate given above. Perhaps the most important enhancement factor relates to the discussion in Chapter 6 of lattice (ionic) and electronic contributions to the quadrupole coupling S-tensors and, specifically in the present instance, to the hexadecapole coupling T-tensors. The estimates above, resulting in such uninviting ratios, consider only ionic contributions. That this is unwarranted is indicated by many of the cases discused in Chapter 6; a case in point is ^{187}Au in gold metal, in which the measured S-tensor components are found to be 13 to 36 times the calculated lattice contributions. It is, indeed, the postulation of contributions substantially larger than the ionic in the dynamic HDI that is used to justify the rather extensive search by NAR techniques, described below, for the existence of the dynamic nuclear hexadecapole coupling. A class of materials in which enhancement of several orders of magnitude may exist is that of mixed–valence metals, in which a resonant enhancement of the electron density of states, associated with the localized nature of the f electron wave function, gives rise to a strong coupling between electrons and phonons [7.30]. An example of such a material is the rare earth mixed–valence compound $Sm_xY_{1-x}S$.

An initial attempt to observe the dynamic HDI utilizes the technique of SQUID–detected adiabatic fast passage (Section 7.5.3) in cubic ^{115}InP, in which studies utilizing dynamic NQI had previously been made. From the theoretical treatments in Section 3.7 and Appendices E and F, one can show that for cubic symmetry and propagation of longitudinal waves parallel to the [110],

$$V^{\pm 3} = \frac{1}{12}(G + J)\varepsilon_{xx} , \qquad (7.28a)$$

where

$$G = \frac{\sin\theta \cos^3\theta}{8}(T_1 - T_4) + \frac{3}{4}\sin^3\theta \cos\theta(T_3 - T_2) \qquad (7.28b)$$

and

$$J = 3\sin\theta \cos^3\theta\, T_5 - 9\sin^3\theta \cos\theta\, T_6 ; \qquad (7.28c)$$

$$V^{\pm 4} = \frac{1}{48}(K + L)\varepsilon_{xx} , \qquad (7.29a)$$

where

$$K = -\frac{1}{8}[(T_1 - T_4) - 6(T_3 - T_2)][3\sin^4\theta - 2\sin^2\theta + 3] \qquad (7.29b)$$

and

$$L = (\sin^4\theta - 1)(3\,T_5 - 9\,T_6) . \qquad (7.29c)$$

7.8 DYNAMIC HEXADECAPOLE INTERACTION

The composite resonator consists of an InP single crystal with parallel (110) faces, one of which is affixed to a 0.95-cm-diameter 5-MHz fundamental quartz X-cut transducer with a silicone grease bond. The crystal is oriented in the magnetic field, with $\boldsymbol{H}_0$ parallel to [001] and perpendicular to the direction of acoustic propagation. The magnitude of $\boldsymbol{H}_0$ is adjusted to the value corresponding to a $\Delta m = \pm 3$ transition at approximately 15 MHz. By using the SQUID acoustomagnetic spectrometer (described in Chapter 10) at 4.2 K, a SQUID-detected adiabatic fast passage signal is observed during a frequency sweep from 15.39 to 15.49 MHz (at the third harmonic of the transducer and in a frequency range between mechanical resonances). RF power applied to the transducer is 10 mW. The signal is observed at the limit of the sensitivity (equivalent to 10^{-7} fluxons) of the SQUID spectrometer with signal averaging. No signal is observed when the magnetic field is adjusted for a $\Delta m = \pm 4$ transition.

The experimental results appear to contradict the theoretical predictions of (7.28) and (7.29), which for $\boldsymbol{H}_0$ exactly perpendicular to the propagation, show $V^{\pm 3} = 0$, while $V^{\pm 4}$ is finite. The actual direction of $\boldsymbol{H}_0$, however, may have been as much as $\pm 2°$ off the perpendicular, in which case $V^{\pm 3}$ is not zero. For $\theta = 88°$, for example,

$$V^{\pm 3} \simeq 0.00218(T_3 - T_2) - 0.0262 T_6 \qquad (7.30a)$$

and

$$V^{\pm 4} \simeq -0.0104[(T_1 - T_4) - 6(T_3 - T_2)] + 0.0015(T_5 - 3T_6) . \qquad (7.30b)$$

Since $V^{\pm 3}$ and $V^{\pm 4}$ involve different combinations of the unknown tensor components T_{ij}, it is not impossible that at 88° one would observe a $\Delta m = \pm 3$ transition but not a $\Delta m = \pm 4$ transition.

We know, however, that the acoustic waves at frequencies between mechanical resonances should have spatial incoherence; in this case, there should be an average over the angle θ in (7.28) and (7.29). Again, the observation of a $V^{\pm 3}$ signal and not a $V^{\pm 4}$ signal is consistent with the different combination of T_{ij}'s in the expressions for the two transitions.

In an attempt to confirm the observation of a $\Delta m = \pm 3$ transition, resort is made to a conventional (as opposed to SQUID-detected) NAR experiment on ^{115}InAs at 78 K, using a sensitive MOUS. Advantages accruing to the MOUS technique are the following:

(1) The signal-to-noise ratio of a typical dynamic NQI $\Delta m = \pm 2$ response can exceed 20/1 in a single pass through the resonance.
(2) Angles between crystal directions and the external magnetic field can be determined within $\pm 0.5°$ precision and varied continuously greater than 180°.

(3) In addition, with the MOUS one is frequency-locked to the center of a mechanical resonance where the acoustic radiation has spatial coherence.

The MOUS is locked to the 10.41769-MHz mechanical resonance center of the composite resonator consisting of a 10-MHz, X-cut, 1.27-cm wrap-around quartz transducer, Canada balsam bond and InAs single crystal with a pair of parallel (110) faces. Propagation direction is [110], while the plane of the magnetic field rotation is $(1\bar{1}0)$. To check the apparatus, the magnetic field is first adjusted for the center of the $\Delta m = \pm 2$ NQI resonance and then swept with a DC magnetic field about this point by using field modulation. The magnetic field is then adjusted for the calculated center frequencies of the $\Delta m = \pm 3$ and $\Delta m = \pm 4$ HDI resonances, then again swept recurrently, signal averaging each sweep. For a single pass with audio frequency filtering time of 1.25 sec, the signal-to-noise ratio of the $\Delta m = \pm 2$ resonance is 10 to 1. Signal averaging sweeps up to 10,000 passes at a 1.25 sec time constant for both $\Delta m = \pm 3$ and $\Delta m = \pm 4$ find no indication of any dynamic nuclear HDI for ^{115}InAs.

Shear acoustic wave propagation in ^{121,123}Sb in GaSb is also used in the search for possible higher order NAR transitions. The composite resonator consists of a 1.27-cm-diameter, wrap-around AT cut quartz transducer, Nonaq grease bond and GaSb sample with parallel (110) faces. For the case of propagation direction along the [110] axis,

$$V^{\pm 3} = \frac{1}{12}[(T_1 - T_4) - 6(T_3 - T_2)]\left[\left(\frac{3\sin^2\theta - 1}{4}\right)\cos\theta\right]\varepsilon_{xy} \quad (7.31a)$$

and

$$V^{\pm 4} = \frac{1}{12}[-(T_1 - T_4) + (T_3 - T_2)]\frac{\sin\theta\cos^2\theta}{4}\varepsilon_{xy} . \quad (7.31b)$$

The direction of polarization is parallel $[\bar{1}10]$. The angle θ is measured from the propagation direction in the $(1\bar{1}0)$ plane to the magnetic field direction. For propagation in the [110] direction and shear polarization in the [001] direction,

$$V^{\pm 3} = \frac{\cos\theta}{12}(3\cos^2\theta - 5/2)(3T_6 - T_5)\varepsilon_{xz} \quad (7.32a)$$

and

$$V^{\pm 4} = \frac{\sin\theta}{12}(\sin^2\theta - 1/2)(3T_6 - T_5)\varepsilon_{xz} . \quad (7.32b)$$

The magnetic field is again rotated in a $(1\bar{1}0)$ plane. For a single pass, with a 1.25 sec sweep constant, the signal-to-noise ratio of the $\Delta m = \pm 2$ NQI resonance line is 40 to 1. Searches are made for the dynamic HDI for both $\Delta m = \pm 3$ and $\Delta m = \pm 4$ resonances at both shear polarizations and appropriate resonance magnetic fields by signal averaging as many as 1000 passes. The result is negative. The runs are repeated after changing the fundamental transducer frequency to 30 MHz, thus increasing the transition probability by a factor of nine. Again, no evidence is found for dynamic hexadecapole coupling for ^{121,123}Sb in GaSb.

In general, NAR dispersion signals saturate less rapidly with power (or other variables, *e.g.*, temperature) than do absorption signals. The MOUS, however, does not easily permit observation of dispersion signals. Resort is therefore made to the computer–controlled reflection bridge spectrometer of Section 7.3.4. Experiments are carried out on ^{121}Sb in GaSb at 30 MHz and 78 K, using both shear directions specified above, over a range of 80 dB in power. No $\Delta m = \pm 3$ or $\Delta m = \pm 4$ dynamic HDI transitions are found. Similarly, negative results are obtained for HDI in ^{181}Ta in single crystal tantalum metal at 78 K in a series of particularly detailed runs. The composite resonator is this case consists of a 0.95–cm wrap–around, 9.3–MHz, quartz AT–cut transducer bonded with silicone grease to a (110) face of the Ta crystal. Both shear directions (7.31) and (7.32) are investigated with both absorption and dispersion. The single pass ^{181}Ta $\Delta m = \pm 2$ dynamic NQI signal–to–noise ratio, with a 1.25 sec time constant, is 100 to 1. Signal averaging is used with the same time constant, for as many as 600 passes. Attempts to observe HDI of ^{115}In and ^{121}Sb in a single crystal of InSb similarly gives negative results.

Observation of HDI in ^{185,187}Re in Re metal is attempted at 78 K and at 4.2 K. As detailed in Section 3.6, for this hexagonal crystal the sixth–rank tensor T_{ijklmn} has 11 independent components,, with a total of 175 nonzero components. It can be shown that $V^{\pm 4}$ is given by

$$V^{\pm 4} = \frac{1}{6}(T_{i8} - 3T_{i9})\sin^3\phi\varepsilon_{13} , \qquad (7.33)$$

where the propagation direction of acoustic waves is along the [0001] axis, the polarization direction is [11$\bar{2}$0], ϕ is measured from the propagation direction in the plane defined by [0001] and [11$\bar{2}$0], and T_{i8} and T_{i9} are two of the independent components of the T_{ijklmn} tensor. The composite resonator consists of a 0.64–cm–diameter, 25–MHz, AT–cut quartz transducer operated at its third harmonic, a silicone grease bond and Re crystal with a pair of flat, parallel (0001) faces. From Fig. 3.9 and the accompanying discussion, we know that both $\Delta m = \pm 1$ and $\Delta m = \pm 4$ transitions are allowed between Zeeman split static quadrupole $\frac{5}{2}$ and $\frac{3}{2}$ levels. The zero–field splitting is 80.850 MHz for ^{185}Re and 76.56 MHz for ^{187}Re. The $\Delta m = \pm 1$ NQI absorption and dispersion signals are found for both nuclear spin systems with the use of a 1.25 sec time constant. Single pass signal–to–noise ratios are 1/1. With signal averaging of 600 passes at the appropriate magnetic fields for the $\Delta m = \pm 4$ transitions, no dynamic HDI is found for either spin system in Re.

CHAPTER 7 REFERENCES

7.1 J. G. Miller, D. I. Bolef. "Sensitivity Enhancement by the Use of Acoustic Resonators in CW Ultrasonic Spectroscopy." *J. Appl. Phys.* **39**, 4589–4593 (1968).

7.2 J. G. Miller, D. I. Bolef. "A Sampled–Continuous Wave Ultrasonic Technique and Spectrometer." *Rev. Sci. Instrum.* **40**, 915–920 (1969).

7.3 J. G. Miller, D. I. Bolef. "Sampled–CW Study of Inhomogeneous Ultrasonic Responses in Solids." *J. Appl. Phys.* **41**, 2282–2293 (1970).

7.4 J. S. Heyman, J. G. Miller. "Verification of Sensitivity Enhancement Factors for CW Ultrasonic Resonators." *J. Appl. Phys.* **44**, 3398–3400 (1973).

7.5 M. S. Conradi, J. G. Miller, J. S. Heyman. "A Transmission Oscillator Ultrasonic Spectrometer." *Rev. Sci. Instrum.* **45**, 358–360 (1974).

7.6 G. Mozurkewich. "Acoustic Magnetic Resonance Investigations Utilizing Direct, Backward Wave, and SQUID Detection." Ph.D. Dissertation, Washington University, 1981 (unpublished).

7.7 D. I. Bolef. "Interaction of Acoustic Waves with Nuclear Spins in Solids." *Physical Acoustics*, Vol **4A**, Ed. W. P. Mason, 113–182 (Academic, New York 1966).

7.8 V. Müller. "CW Acoustic Bridge Spectrometer for Direct Detection of Nuclear Acoustic Resonance." *J. Phys. E* **8**, 127–130 (1975).

7.9 R. K. Sundfors. "An Ultrasonic, Continuous Wave, Computer–Controlled Reflection Bridge Spectrometer." *Rev. Sci. Instrum.* **61**, 2239–2242 (1990).

7.10 D. A. Rudy. "Microwave CW Acoustic Studies of Paramagnetic Crystalline Solids." Ph.D. Dissertation, Washington University, 1969 (unpublished).

7.11 D. I. Bolef, J. G. Miller. "High Frequency Continuous Wave Ultrasonics." Chapter 3 *Physical Acoustics*, Vol. **8**, Ed. by W. P. Mason and R. N. Thurston, 95–201 (Academic Press, New York 1971).

7.12 J. Brown. "Finite Strain Effects and Magneto–Elastic Interactions in Neodymium Vanadate $NdVO_4$." Ph.D. Dissertation, Washington University, 1978 (unpublished).

7.13 W. D. Smith, R. K. Sundfors. "An Improved Calibration Technique for CW Ultrasonic and Nuclear Resonance Spectrometers."

Rev. Sci. Instrum. **41**, 288–290 (1970).

7.14 J. S. Heyman. "Ultrasonic Coupling to Optically Generated Charge Carriers in CdS: Physical Phenomena and Applications." Ph. D. Dissertation, Washington University, 1975 (unpublished).

7.15 J. R. Franz, M. E. Mullen. "High Sensitivity Nuclear Acoustic Resonance Spectrometer Incorporating a New Calibration Technique." *Rev. Sci. Instrum.* **48**, 531–532 (1977).

7.16 A. R. Kessel'. *Yadernuj Acusticheskij Resonance* (Verlag Nauka, Moskau 1969). German trans: Akustische Kernresonanz (Akademie–Verlag, Berlin 1973).

7.17 A. R. Kessel, A. F. Izmailov. "Some Remarks on the History of Acoustic Resonance in Russia." *Perspectives in Physics* Ed. by Y. Fu, R. K. Sundfors, and P. Suntharothok, 115–122 (World Scientific, Singapore 1992).

7.18 P. A. Fedders, E. Y. C. Lu. "Theory of Ultrasonic Echoes." *Phys. Rev. B* **8**, 5156–5162 (1973).

7.19 G. R. Ashton, D. K. Hsu, R. G. Leisure. "Comparison of Multiscan Direct and Lock–in Detection in Magnetic Resonance: Application to Nuclear Acoustic Resonance." *Rev. Sci. Instrum.* **51**, 454–458 (1980).

7.20 G. R. Ashton, D. K. Hsu, R. G. Leisure. "Direct Acoustic Measurement of Nuclear Spin Relaxation Times: Application to Aluminum and Tantalum." *Phys. Rev. B* **23**, 5681–5687 (1981).

7.21 D. C. Look, D. R. Locker. "Nuclear Spin–Lattice Relaxation by Tone Burst Modulation." *Phys. Rev. Lett.* **20**, 987–989 (1968).

7.22 D. S. Look, D. R. Locker. "Time–Saving in Measurement of NMR and EPR Relaxation Times." *Rev. Sci. Instrum.* **41**, 250–251 (1970).

7.23 A. Abragam. *The Principles of Nuclear Magnetizism.* (Clarendon Press, Oxford 1961).

7.24 A. R. Kessel', M. M. Shakirzyanov. "Capture of Quadrupole Moment of a Nucleus by a High Intensity Acoustic Wave." *Zh. Eksp. Teor. Fiz* **83**, 1100–1103 (1982). English trans: *Sov. Phys. JETP* **56**, 624–625 (1983).

7.25 V. A. Kirsanov, V. F. Tarasov, M. M. Shakirzyanov. "Dynamics of a Nuclear Spin System Under Acoustic Adiabatic Fast Passage Conditions." *Fiz. Tverd. Tela (Leningrad)* 27, 1554–1556 (1985); English trans: *Sov. Phys. Solid State* **27**, 938–939 (1985).

7.26 R. K. Sundfors. (unpublished).

7.27 O. Kanert, R. Münter. "Nuclear Acoustic Relaxation in the Rotating Frame." *J. Mag. Res.* **62**, 29–36 (1985).

7.28 B. Bleaney, G. A. D. Briggs, J. F. Gregg, C. H. A. Huan,

I. D. Morris, M. K. Wells. "Further Studies of the Enhanced Nuclear Magnet HVO$_4$. II. Experiments with Acoustic Waves." *Proc. Roy. Soc. (London) A* **416**, 75–81 (1988).

7.29 R. K. Sundfors. "Determination of the Magnitude and Sign of the ^{185,187}Re Nuclear Electric Quadrupole Coupling Constants Using Nuclear Acoustic Resonance." *Phys. Rev. B* **42**, 1922–1928 (1990).

7.30 A. Gerber, A. Herzenberg. "Resonant Enhancement of Ultrasound Attenuation in Mixed Valence Metals." *Phys. Rev. B* **37**, 740–744 (1988).

CHAPTER 8
ACOUSTIC SATURATION NMR AND DOUBLE RESONANCE

8.1 Introduction

8.2 Acoustic Saturation of NMR

    8.2.1 Theory

    8.2.2 Experimental Verification

        8.2.2.1 Pulse Saturation Technique

        8.2.2.2 A Necessary Digression: Measurement of Acoustic Energy Density

        8.2.2.3 Application: ASNMR of ^{23}Na and ^{35}Cl in NaCl

8.3 Continuous Wave ASNMR

8.4 Transient ASNMR

    8.4.1 Acoustic Nuclear Spin Echo

    8.4.2 Acoustic Saturation of NMR in a Rotating Reference Frame

    8.4.3 Acoustic Nuclear Adiabatic Fast Passage

8.5 Double Acoustic Magnetic Resonance

References

8. ACOUSTIC SATURATION NMR AND DOUBLE RESONANCE

8.1 Introduction

The acoustic saturation technique for investigating nuclear spin–phonon interactions, although seemingly less direct in its approach, is in many ways inherently more powerful and more generally applicable (with the important exception of conductors) than NAR techniques for the following reasons:

(1) It is often easier to observe an acoustic effect on an existing NMR resonance line than to search *ab initio* for an unknown NAR resonance line.

(2) High acoustic powers are available in acoustic saturation NMR (ASNMR) to render an otherwise small effect (*e.g.*, NAR) observable.

(3) The accessibility of high acoustic power in ASNMR also obviates the restriction (that exists in NMR) to specimens whose background acoustic attenuation is low.

(4) Double resonance techniques, to be described in Section 8.5, are often natural extensions of the ASNMR technique.

(5) Transient nuclear spin resonance techniques, so fruitful in conventional NMR applications, are more readily accessible to ASNMR than to NAR.

(6) Combining the above, it is much more likely that ASNMR may become a useful adjunct to NMR for imaging purposes.

Important caveats to items (2) and (3), it must be noted, are the well-known anharmonic and nonlinear effects attendant on the use of high acoustic powers in the specimen. One of the longest running (indeed, acrimonious) controversies among those doing ASNMR has been over the question of the isotropy, with angle θ (the angle between the direction of propagation of acoustic energy and the direction of the external magnetic field), of the ASNMR results, which are contrary to those obtained by NAR and to those predicted theoretically. This question is resolved by preparing specimens with a pair of flat and parallel faces perpendicular to a crystal direction in which one can propagate pure transverse or longitudinal acoustic waves; then propagating a saturating acoustic pulse (i) with a pulse length in time greater than the distance between parallel faces divided by the acoustic velocity and (ii) with frequency adjusted to the center of one of the mechanical resonances. In this case, the acoustic energy density of acoustic waves traveling perpendicular to the parallel faces is contained in a right circular cylinder with radius approximately equal to the radiating area of the piezoelectric transducer. Under typical conditions in NAR, the quality factor of this mode of oscillation can range from 10^3 to 10^4. There are also other non-resonance modes, which because of diffraction and reflection from the crystal–air surfaces, have directions other than a direction perpendicular to the faces and effective quality factors less than one.

The departure from the theoretically predicted anisotropy is principally due to operating at a frequency not on one of the mechanical resonances of the composite resonator, which results in spatially incoherent acoustic waves. (i) The mode with direction perpendicular to the composite resonator faces is no longer a resonance mode and has a quality factor of less than one; it is no larger than other modes which due to diffraction and reflections will have directions other than the direction perpendicular to the faces. (ii) With the frequency and magnetic field adjusted for a particular $\Delta m = \pm 2$ transition, there will be some directions of the spatially incoherent acoustic waves which have finite transition probabilities.

8.2 Acoustic Saturation of NMR

8.2.1 Theory

We assume a system of N identical spins distributed among equally spaced energy levels. The magnitude of the nuclear magnetization is determined by the competition among the externally applied, periodic–in–time fields (RF magnetic field and ultrasonic field) and the relaxation processes (*e.g.*, quadrupolar or paramagnetic [spin diffusion to paramagnetic impurities]), which tend to restore the populations of the spin levels to thermal equilibrium with the lattice. These interactions connect the spin system with the lattice. In addition, the spin–spin interaction, although not changing the net magnetization of the spin system, acts to maintain a Boltzmann distribution among the energy levels occupied by the spins. The spin–spin interaction can accomplish this through the simultaneous opposite flip of two spins, because of the equidistance of the levels. In many solids, the spin–spin relaxation time T_2 associated with the spin–flip process is much shorter than the spin–lattice relaxation time. As a result, the spins are maintained in thermal equilibrium with one another, even when they are not in equilibrium with the lattice. This is equivalent to saying that the spin system may be in equilibrium at a temperature (*spin temperature*) that is different from that of the lattice.

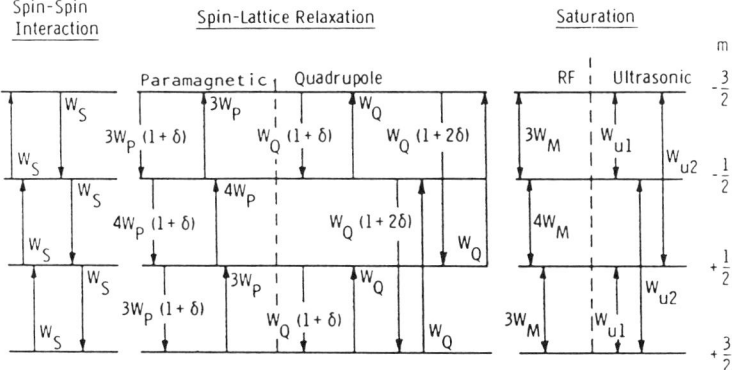

FIGURE 8.1. Transition probabilities among a set of four, equally spaced nuclear spin energy levels, $I = 3/2$.

In typical saturation experiments, the nuclear spin system in a solid is irradiated with ultrasonic energy at twice the Larmor frequency. The transition probability W_{u2} can be made large enough to saturate the $\Delta m = \pm 2$ resonance in a time short compared to T_1. The populations of the equally spaced levels of the nuclear spin in magnetic fields are inferred, using pulsed NMR techniques, by examination of the magnetization M_z both before and immediately after exposure to the ultrasonic waves. We now proceed to a derivation of the spin magnetization saturation ratio M_z/M_0 due to the presence of ultrasound in the solid. In Fig. 8.1 the equally spaced high-field levels of nuclei (*e.g.*, ^{23}Na or ^{35}Cl in NaCl) with spin $\frac{3}{2}$ in a magnetic field are given. The saturation transition probabilities for RF and W_M, as well as those for ultrasonic waves, W_{u2} and W_{u1}, are the same from higher to lower energy levels as from lower to high energy levels. The relaxation transition probabilities (*e.g.*, due to quadrupolar W_Q or paramagnetic W_P interactions with the thermal lattice vibrations) are slightly greater (by a factor $\delta \simeq h\nu/k_B T \ll 1$) from the higher to lower energy levels than from the lower to higher energy levels; *i.e.*, the relaxation mechanism tends to produce a Boltzmann distribution. Spin–spin transition probabilities W_S, equal for transitions up or down, change the net number of spins in any level only when the spin levels do not have a Boltzmann distribution of populations among themselves at a temperature T_S, which may differ from T. When the spin-spin transition rate is much faster than the transition rates due to the saturating and spin–lattice relaxation processes, a Boltzmann distribution of population among the spin levels is maintained according to the relationship

$$n_{\frac{3}{2}} - n_{-\frac{3}{2}} = 3(n_{\frac{1}{2}} - n_{-\frac{1}{2}}), \tag{8.1}$$

where $n_{\frac{3}{2}}$, $n_{\frac{1}{2}}$, $n_{-\frac{1}{2}}$ and $n_{-\frac{3}{2}}$ are the populations of the levels m of a nucleus with $I = \frac{3}{2}$. The magnetization of any level is given by $M_m = m\gamma\hbar n_m$; the total magnetization in the magnetic field direction, by $M_z = \sum_m m\gamma\hbar n_m$. Applying (8.1), we obtain

$$M_z = 5\gamma\hbar(n_{\frac{1}{2}} - n_{-\frac{1}{2}}). \tag{8.2}$$

The rate of change of the magnetization of each spin level due to the action of acoustic energy at the $\Delta m = \pm 2$ resonance frequency is given by

$$\dot{M}_{\frac{3}{2}} = -\frac{3}{2}\gamma\hbar W_{u2}(n_{\frac{3}{2}} - n_{-\frac{1}{2}})$$

$$\dot{M}_{\frac{1}{2}} = -\frac{1}{2}\gamma\hbar W_{u2}(n_{\frac{1}{2}} - n_{-\frac{3}{2}})$$

$$\dot{M}_{-\frac{1}{2}} = -\frac{1}{2}\gamma\hbar W_{u2}(n_{-\frac{1}{2}} - n_{\frac{3}{2}})$$

$$\dot{M}_{-\frac{3}{2}} = -\frac{3}{2}\gamma\hbar W_{u2}(n_{-\frac{3}{2}} - n_{\frac{1}{2}}). \tag{8.3}$$

The rate of change of total magnetization due to the acoustic input is therefore

$$\dot{M}_z = \sum_m \dot{M}_m = 2\gamma\hbar W_{u2}(n_{\frac{3}{2}} - n_{-\frac{3}{2}} + n_{\frac{1}{2}} - n_{-\frac{1}{2}}). \tag{8.4}$$

Using (8.1), one obtains

$$\dot{M}_z = 8\gamma\hbar W_{u2}(n_{\frac{1}{2}} - n_{-\frac{1}{2}}) ;$$

then, substituting from (8.2),

$$\dot{M}_z = \frac{8}{5}W_{u2}M_z . \quad (8.5)$$

Under steady-state conditions, this is equal to the rate of change of magnetization due to the spin–lattice relaxation process,

$$\dot{M}_z = \frac{(M_0 - M_z)}{T_1} , \quad (8.6)$$

where M_0 is the equilibrium magnetization in the absence of acoustic energy. We have assumed that a single relaxation time T_1 characterizes the system. For steady-state conditions in the presence of acoustic energy, or when the applied acoustic pulse is sufficiently long that equilibrium is established, we may equate (8.5) and (8.6) to obtain

$$M_z = \frac{M_0}{1 + (8/5)W_{u2}T_1} . \quad (8.7)$$

Since the NMR signal amplitude is proportional to M_z, we can write our result in the form

$$\frac{M_z}{M_0} = \frac{A}{A_0} = (1 + 8W_{u2}T_1/5)^{-1} , \quad (8.8)$$

where A_0 is the NMR signal amplitude without acoustic energy, and A is the NMR signal amplitude in the presence of acoustic energy. From (8.8) it is evident that the degree of saturation, via the acoustic excitation, of the nuclear magnetization depends on the relative magnitudes of T_1 and W_{u2}: when acoustic power is increased, the acoustically induced spin transitions serve to equalize the level populations faster than spin–lattice relaxation can return them to their equilibrium values. As a result, W_{u2} increases and M_z/M_0 decreases.

Equation (8.8) gives the effect of applying acoustic energy at frequencies corresponding to $\Delta m = \pm 2$ transitions on the total magnetization for spin $I = \frac{3}{2}$. The extension to the general case of acoustic saturation (still at $\Delta m = \pm 2$ frequencies) of a nuclear spin system of arbitrary I (still assuming that the nuclear spin levels are equally spaced) is given by

$$\frac{M_z}{M_0} = \frac{A}{A_0} = \left[1 + \frac{8I(I+1) - 6}{5I(2I-1)}W_{u2}T_1\right]^{-1} . \quad (8.9a)$$

For $I = \frac{3}{2}$, (8.9a) reduces to (8.8); for $I = \frac{5}{2}$,

$$\frac{M_z}{M_0} = \left[1 + \frac{32}{25} W_{u2} T_1\right]^{-1} ; \qquad (8.9b)$$

for $I = \frac{7}{2}$,

$$\frac{M_z}{M_0} = \left[1 + \frac{8}{7} W_{u2} T_1\right]^{-1} . \qquad (8.9c)$$

Expressions have been derived for saturation of individual (resolved) components of NMR lines when irradiated (in the presence of RF power) with acoustic energy corresponding to single–frequency (W_{u1}) and double frequency (W_{u2}) transitions. The acoustic saturation of the central ($-\frac{1}{2} \leftrightarrow \frac{1}{2}$) resonance line is given by

$$\frac{n_{\frac{1}{2}} - n_{-\frac{1}{2}}}{n_0} = \left(1 + p_0 + \frac{32}{(2I+1)^2} W''_{u2} T_1\right)^{-1}, \qquad (8.10)$$

where

$$W''_{u2} = W_u\left(-\frac{3}{2} \leftrightarrow \frac{1}{2}\right) = W_u\left(\frac{3}{2} \leftrightarrow -\frac{1}{2}\right).$$

n_0 is the population difference between adjacent levels under conditions of thermal equilibrium. p_0 is a measure of RF power, where

$$p_0 = (pm/2) + m - 1, \quad \text{and} \quad \left(p = \frac{1}{2}\gamma^2 H_1^2 T_1 g(\nu)\right).$$

Thus, although the transitions $m = \frac{1}{2} \leftrightarrow m = -\frac{1}{2}$ are forbidden acoustically (and therefore unaffected by $\Delta m = \pm 1$ transitions), the central line can be saturated by applying acoustic power at twice the Larmor frequency. Expressions are also available for the satellite lines corresponding to transitions (for $I > \frac{1}{2}$) $\frac{5}{2} \leftrightarrow \frac{3}{2}$, $\frac{7}{2} \leftrightarrow \frac{5}{2}$ and $\frac{9}{2} \leftrightarrow \frac{7}{2}$. Equation (8.10) holds for nuclear spins that relax via quadrupolar or paramagnetic ion interactions. Saturation ratios have also been obtained [8.1] for the effect of irradiation by acoustic energy corresponding to a particular W_{u2} [e.g., $W''_{u2} \equiv W_u(\frac{1}{2} \leftrightarrow \frac{3}{2})$] transition (all other W_{u2} transitions being zero) on the individual satellite lines of nuclei of arbitrary spin. In Tables 8.1 and 8.2 are given the results of these calculations for the cases of W_{u2} excitation—quadrupole and magnetic relaxation, respectively.

8.2 ACOUSTIC SATURATION OF NMR

Table 8.1
Quadrupole Relaxation*

Ultrasonic transition probabilities	Spin	$n_{\frac{1}{2}}-n_{-\frac{1}{2}}$	$n_{\frac{3}{2}}-n_{\frac{1}{2}}$	$n_{\frac{5}{2}}-n_{\frac{3}{2}}$	$n_{\frac{7}{2}}-n_{\frac{5}{2}}$	$n_{\frac{9}{2}}-n_{\frac{7}{2}}$
W''_{u2} only	3/2	S	S	—	—	—
	5/2	S	S	E	—	—
	7/2	S	S	E	S_s	—
	9/2	S	S	E	S_{vs}	E_{vs}
W'_{u2} only	5/2	UF	S	SU	—	—
	7/2	UF	S	SU	E	—
	9/2	UF	S	SU	E	S_{vs}
W'''_{u2} only	7/2	UF	E	S	SU	—
	9/2	UF	E	S	EU	SE
W''''_{u2} only	9/2	UF	S_{vs}	E	S	SU

*Effect of acoustic power on individual nuclear magnetic resonance lines. Variation of $(n_m - n_{m-1})$ versus U for NMR lines. U is a measure of acoustic power; $U = 1$ corresponds to the acoustic power for saturation. S indicates saturation, UF indicates unaffected, E indicates enhancement, SU indicates $S < 0$ for $U > 1$, EU indicates E for $U \leq 1$ and then S, SE indicates S for $U \leq 1$ and then E_s. The subscript s means small, and the subscript vs means very small. The ultrasonic transition probabilities W_{u2} are given by

$$W'_{u2} = W_{u_{\frac{5}{2}\to\frac{1}{2}}} = W_{u_{\frac{1}{2}\to\frac{5}{2}}} = W_{u_{-\frac{1}{2}\to-\frac{5}{2}}} = W_{u_{-\frac{5}{2}\to-\frac{1}{2}}},$$
$$W''_{u2} = W_{u_{-\frac{3}{2}\to\frac{1}{2}}} = W_{u_{\frac{1}{2}\to-\frac{3}{2}}} = W_{u_{\frac{3}{2}\to-\frac{1}{2}}} = W_{u_{-\frac{1}{2}\to\frac{3}{2}}},$$
$$W'''_{u2} = W_{u_{\frac{7}{2}\to-\frac{3}{2}}} = W_{u_{\frac{3}{2}\to-\frac{7}{2}}} = W_{u_{-\frac{3}{2}\to-\frac{7}{2}}} = W_{u_{-\frac{7}{2}\to-\frac{3}{2}}}$$

and

$$W''''_{u2} = W_{u_{\frac{9}{2}\to-\frac{5}{2}}} = W_{u_{\frac{5}{2}\to-\frac{9}{2}}} = W_{u_{-\frac{5}{2}\to-\frac{9}{2}}} = W_{u_{-\frac{9}{2}\to-\frac{5}{2}}}.$$

(From Ref. 8.1)

8. SATURATION NMR AND DOUBLE RESONANCE

Table 8.2
Magnetic Relaxation*

Ultrasonic transition probabilities	Spin	$n_{\frac{1}{2}}-n_{-\frac{1}{2}}$	$n_{\frac{3}{2}}-n_{\frac{1}{2}}$	$n_{\frac{5}{2}}-n_{\frac{3}{2}}$	$n_{\frac{7}{2}}-n_{\frac{5}{2}}$	$n_{\frac{9}{2}}-n_{\frac{7}{2}}$
W''_{u2} only	3/2	S	S	—	—	—
	5/2	S	S	UF	—	—
	7/2	S	S	UF	UF	—
	9/2	S	S_s	UF	UF	UF
W'_{u2} only	5/2	UF	S_s	SU	—	—
	7/2	UF	S	SU	UF	—
	9/2	UF	S	S	UF	UF
W'''_{u2} only	7/2	UF	UF	S_s	SU	—
	9/2	UF	UF	S_s	SU	UF
W''''_{u2} only	9/2	UF	UF	UF	S_s	SU

*Effect of acoustic power on individual nuclear magnetic resonance lines. Variation of $(n_m - n_{m-1})$ versus U for NMR lines. U is a measure of acoustic power; $U = 1$ corresponds to the acoustic power for saturation. S indicates saturation, UF indicates unaffected, SU indicates $S < 0$ for $U > 1$. The subscript s means small, and the subscript vs means very small. The ultrasonic transition probabilities W_{u2} are given by

$$W'_{u2} = W_{u_{\frac{5}{2} \to \frac{1}{2}}} = W_{u_{\frac{1}{2} \to \frac{5}{2}}} = W_{u_{-\frac{1}{2} \to -\frac{5}{2}}} = W_{u_{-\frac{5}{2} \to -\frac{1}{2}}},$$
$$W''_{u2} = W_{u_{-\frac{3}{2} \to \frac{1}{2}}} = W_{u_{\frac{1}{2} \to -\frac{3}{2}}} = W_{u_{\frac{3}{2} \to -\frac{1}{2}}} = W_{u_{-\frac{1}{2} \to \frac{3}{2}}},$$
$$W'''_{u2} = W_{u_{\frac{7}{2} \to \frac{3}{2}}} = W_{u_{\frac{3}{2} \to \frac{7}{2}}} = W_{u_{-\frac{3}{2} \to -\frac{7}{2}}} = W_{u_{-\frac{7}{2} \to -\frac{3}{2}}}$$

and

$$W''''_{u2} = W_{u_{\frac{9}{2} \to \frac{5}{2}}} = W_{u_{\frac{5}{2} \to \frac{9}{2}}} = W_{u_{-\frac{5}{2} \to -\frac{9}{2}}} = W_{u_{-\frac{9}{2} \to -\frac{5}{2}}}.$$

(From Ref. 8.1)

The derivation of (8.8) illustrates the role of spin–spin interactions in establishing and maintaining a Boltzmann distribution between the populations of a spin system with more than two energy levels. The steady–state distribution may not, in fact, correspond to an equilibrium distribution at the lattice temperature, but rather at a spin temperature T_s, where T_s is defined through the relation

$$\frac{n_m}{n_{m-1}} = e^{-(g\mu_n H_0)/(k_B T_s)} . \qquad (8.11)$$

8.2.2 Experimental Verification

8.2.2.1 Pulse Saturation Technique

The fundamentals of pulse acoustic saturation NMR are reviewed briefly in Chapter 1 (Section 1.7). A conventional crossed–coil NMR spectrometer, designed to investigate transient NMR effects, such as those described in Section 1.7, is shown in Fig. 8.2 (omitting the parts shown dashed).

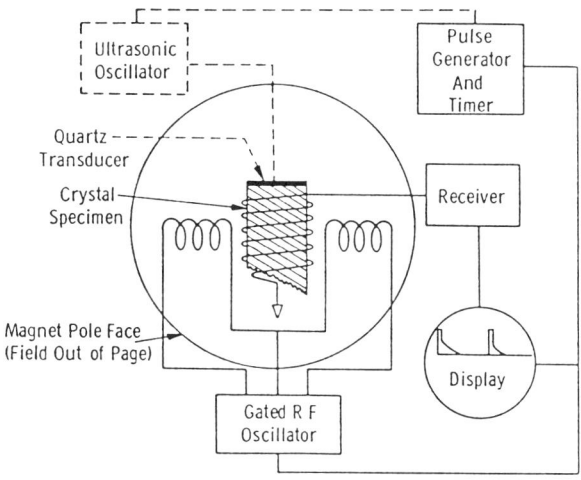

FIGURE 8.2. Schematic of crossed–coil NMR spectrometer for use in pulsed experiments. (From Ref. 8.2)

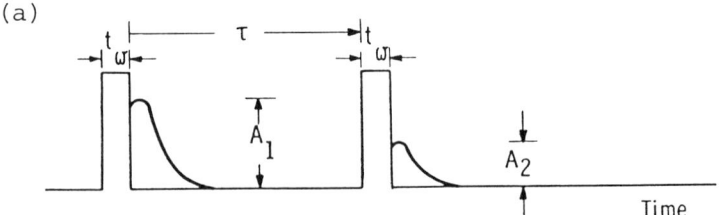

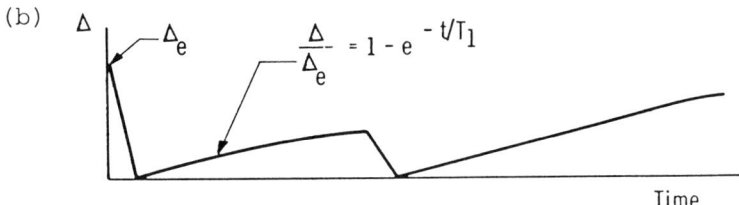

FIGURE 8.3. Transient technique for measuring nuclear magnetization by free nuclear induction decay. (From Ref. 8.3)

The separate transmitter and receiver coils surrounding the specimen are perpendicular to each other and to the axis of the DC magnetic field. In the case of pulsed NMR free induction decay, a short pulse of RF energy at the resonant frequency is applied to the transmitter coil. The pulse length and pulse amplitude satisfy the condition $\gamma H_1 t_w = \pi/2$ (90° pulse), where H_1 is the amplitude of the RF field, and t_w is the pulse length. The heights A_i of the nuclear induction transients, shown schematically in Fig. 8.3a, are proportional to the population differences (and therefore to the net magnetization M_z) at the beginning of the RF pulse. Following the first pulse is a transient of amplitude A_1. Granted that the spin system is in thermal equilibrium with the lattice at $t = 0$, A_1 is proportional to the population difference at thermal equilibrium. After a time τ, a second pulse of the same duration is applied, followed by the appearance of an induction signal of amplitude A_2. Figure 8.3b shows a plot of the instantaneous population difference between the two energy levels, between which transitions are being induced during the same interval. If $\Delta(0)$ is zero, as it is following the first 90° pulse, the instantaneous population difference Δ approaches the equilibrium value Δ_e at a rate given by the thermal relaxation time T_1:

$$\frac{\Delta(t)}{\Delta_e} = 1 - e^{-t/T_1} . \tag{8.12}$$

The heights A_i of the transients are proportional to the population dif-

8.2 ACOUSTIC SATURATION OF NMR

ferences at the beginning of the RF pulse; therefore,

$$\frac{A_2}{A_1} = \frac{\Delta(\tau)}{\Delta_e} = 1 - e^{-\tau/T_1} , \qquad (8.13)$$

where τ is the time interval between the trailing edge of the first pulse and the beginning of the second pulse. By varying τ, one can thus plot the recovery function $1 - e^{-\tau/T_1}$ from the initial amplitude of the free induction decay following the second pulse. If one plots $\ln(1 - A_2/A_1)$ versus τ, the thermal relaxation time T_1 is known from the slope of the resulting straight line.

In an acoustic saturation NMR experiment, the dashed portions shown in Fig. 8.2 are combined with the pulsed NMR spectrometer. An RF power oscillator, capable of itself being pulsed, is used to excite a transducer bonded to one end of the specimen. Often, as we shall see below, the ultrasonic frequency is adjusted to correspond to a $\Delta m = \pm 2$ transition at twice the NMR frequency ν_0 in order to minimize spurious excitation which is due to the RF driving the transducer. A typical sequence of pulses in a pulsed ASNMR experiment (on NaCl) is shown in Fig. 1.5, Section 1.7. The effect of the acoustic energy on the nuclear magnetization is measured by inserting an acoustic pulse of length τ_a between the two short RF pulses that sample the nuclear magnetization. Repetition of the pulse sequence is at the interval τ_R. For the case of ^{23}NaCl at 300 K, $\tau_a = 8$ sec and $\tau_R = 17$ sec. The free nuclear induction pulse heights A_i are observed to be a function of the amplitude and frequency of the RF voltage at the transducer. The transition probability W_{u2} entering into (8.9) is proportional to the energy density of ultrasonic waves in the specimen and, hence, to the square of the voltage applied to the transducer. One can therefore write (8.9) as

$$\frac{A}{A_0} = \frac{1}{1 + k_1 V^2} , \qquad (8.14)$$

where A_0 is the amplitude of the nuclear induction signal in the absence of ultrasonic excitation, A is the amplitude in the presence of ultrasonic excitation, and V is the peak voltage applied to the transducer. For the particular case of a longitudinal acoustic wave propagating along a cube axis, $\boldsymbol{H}_0$ perpendicular (Case I, Fig. 3.5, Section 3.4), assuming $S_{12} = -S_{11}/2$, one obtains for $I = \frac{3}{2}$

$$k_1 V^2 = \frac{8}{5} W_{u2} T_1 = \frac{81}{160} g(\nu) \left(\frac{A}{2\hbar}\right)^2 \eta_{\pm}^2 S_{11}^2 \varepsilon^2 T_1 . \qquad (8.15)$$

For completeness, in Fig. 8.4 we show a schematic of a sophisticated pulse ASNMR spectrometer. An RF oscillator is used to generate the NMR excitation pulses and to demodulate the received signals. A 180° hybrid power splitter (not shown) divides the signal, with half going to the RF pulse amplifier for the 90° excitation pulse, and half going to a mixer used to homodyne the incoming NMR signal. Gating for the RF pulses is provided by two cascaded mixers. The gate signal is a 5–V square pulse from the master pulse generator. The network, consisting of crossed diodes and a quarter wavelength of coaxial cable, acts as an automatic gating circuit designed to present a low impedance to high level signals while blocking low level signals at the same frequency; its purpose is to block leakage of RF power except when the excitation pulse occurs. The shunted crossed diodes, impedance matching circuit and attendant $\lambda/4$ cables similarly act to deliver excitation energy to the sample NMR coil. After the excitation pulse, the small NMR signal is directed into the RF portion of the receiver, consisting of low–noise RF amplifier, band–pass filters and homodyne mixer. The audio section and attendant signal averager are conventional.

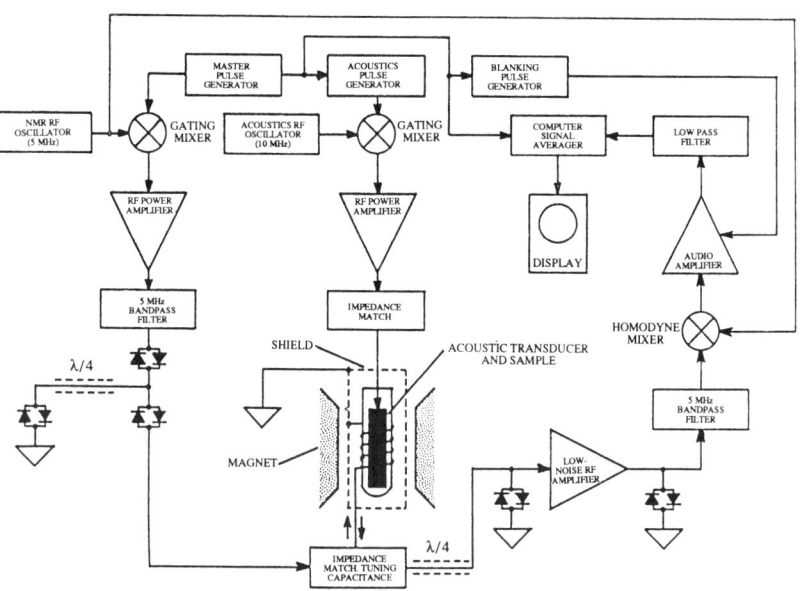

FIGURE 8.4. Block diagram of ASNMR spectrometer. (From Ref. 8.4)

A separate RF oscillator, with built–in attenuator, is used to drive the acoustic transducer. The output of an RF mixer provides the RF pulse used for acoustic excitation. Gating signals for the mixer are provided by a pulse generator slaved to the master pulse generator. The RF pulse from the mixer is amplified by an RF power amplifier and is impedance–matched to the acoustic transducer.

Returning to (8.14) and (8.15), the value of k_1 is obtained from the slope of the curve (A_0/A) versus V^2; V is the measured RF voltage at the transducer; $g(\nu)$ is known from the plot of (A/A_0) versus ultrasonic frequency at constant power around $2\nu_0$; T_1 is obtained independently of the ultrasound by pulsed NMR experiments. From expressions such as (8.15), therefore, one obtains the elements of the $\boldsymbol{S}$-tensor, provided one knows the amplitude of the acoustic vibration in the specimen. One evades the difficult task of evaluating the acoustic amplitudes ε_{ii} if one is willing to settle for ratios of q's or of S's—a strategem applicable to acoustically affected nuclei in the same specimen, $e.g.$, ^{23}Na and ^{127}I in NaI.

8.2.2.2 A Necessary Digression: Measurement of Acoustic Energy Density

The least well–known parameter necessary for a quantitative interpretation of the ASNMR data—$viz.$, (8.15)—is the acoustic energy density (or strain amplitude) in the specimen. Reasonably reliable methods of estimating strain amplitudes of acoustic waves in specimens are available. As is shown in standard texts on acoustics, for the case of air–backed, half wavelength quartz transducers radiating energy into a solid, the peak strain amplitude of the radiated acoustic wave is given by

$$\varepsilon_{\max} = \frac{\sqrt{2}}{v}\frac{2\beta V}{Z} , \qquad (8.16)$$

where V is the rms voltage applied to the transducers, Z is the acoustic impedance, and $\beta = e_{11}A/t$, where e_{11} is a piezoelectric constant of quartz (depending upon the specific cut). A and t are, respectively, the area and thickness of the quartz transducer. For the case of a standing wave pattern in the specimen (plane parallel reflecting ends) with the frequency at the center of a mechanical resonance,

$$Z = \rho v A \tanh(\alpha l) , \qquad (8.17)$$

where α is the attenuation coefficient of the specimen, and l is the specimen length. Losses in the quartz and specimen holder are neglected. In the case of a traveling wave (roughened end opposite the quartz transducer or matched absorber),

$$Z = \rho v A . \qquad (8.18)$$

An acoustic impedance is related to its equivalent electrical impedance by the expression

$$R = \frac{Z}{4\beta^2}. \quad (8.19)$$

Combining (8.16) and (8.19), one obtains

$$\varepsilon_{max} = \frac{V}{v}\sqrt{\frac{2}{RZ}}. \quad (8.20)$$

By measuring the electrical resistance R, therefore, one may obtain the strain amplitude of the acoustic wave. The effective resistance R may be measured by Q-meter or RF bridge techniques. Good agreement exists between this and other, independent methods of measuring strain amplitude.

In the above, brief review, we have assumed that the energy density, or strain amplitude, is uniform throughout the specimen. For the case of standing waves, the energy density goes from zero at a nodal point to a maximum one quarter wavelength later. There is, in addition, a slower radial variation. To take these variations into account, one must average (8.14) over the volume of the specimen after suitably expressing the strain as a function of radial distance r and distance z along the direction of the standing waves.

8.2.2.3 Application: ASNMR of ^{23}Na and ^{35}Cl in NaCl

A nuclear spin system well studied by ASNMR techniques is that of ^{23}Na and ^{35}Cl (both $I = \frac{3}{2}$) in crystalline NaCl. Transitions corresponding to W_{U2} result from application of ultrasound at twice the Larmor frequency. The populations of the four equally spaced levels of the nuclei in a magnetic field are inferred from measurements of A and A_0 from (8.14). In Fig. 8.5 we plot (8.14) together with the experimentally determined points for ^{23}Na and ^{35}Cl. The assumption made by *Abragam and Proctor* [8.5]—that spin-spin interactions must be taken into account when evaluating the results of the acoustic saturation of the spin levels—predicts a single time constant $(T_1^{-1}+8W/5)^{-1}$ characterizing the approach to a limiting value of A/A_0. In Fig. 8.6 is shown this approach to a limiting polarization for various applied voltages, with rapid saturation of the levels for high voltages occurring in times much shorter than T_1, measured to be 7.5 sec at room temperature in ^{23}NaCl.

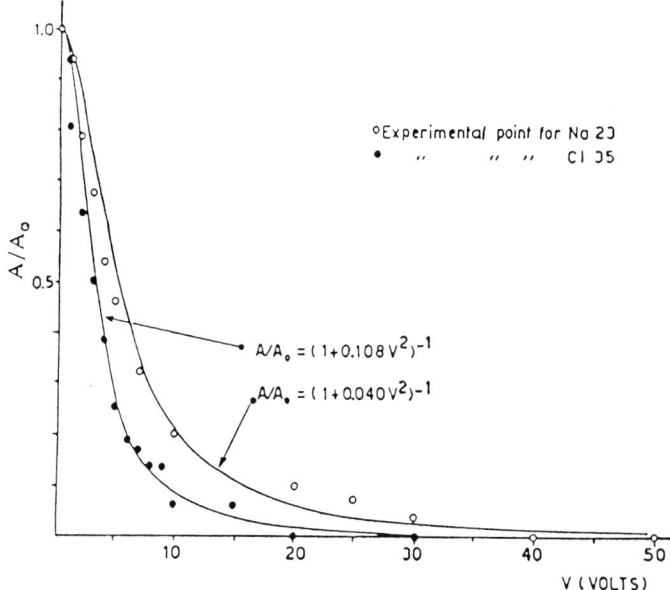

FIGURE 8.5. Attenuation of the magnetization of ^{23}Na and ^{35}Cl in NaCl as a function of voltage across the transducer; $\Delta m = \pm 2$ transitions. (From Ref. 8.5)

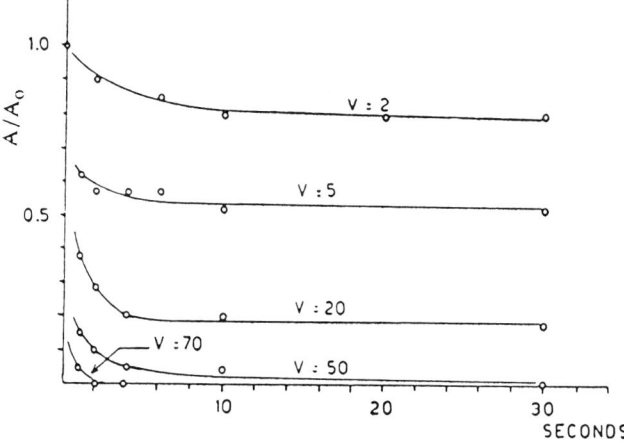

FIGURE 8.6. Magnetization of ^{23}Na as a function of time, measured from the beginning of the acoustic pulse, for several voltages applied to the transducer. (From Ref. 8.5)

8. SATURATION NMR AND DOUBLE RESONANCE

To obtain the line width of the acoustic saturation effect, one measures the ratio A/A_0 as a function of frequency, the acoustic power level being kept constant. Typical data for ^{23}Na in NaCl is shown in Fig. 8.7, where the full–width at half maximum is 4.0 kHz. This is to be compared with the NMR line width $\Delta\nu = 2.3$ kHz, corresponding to a $T_2 = 140\,\mu$sec. In practice, in an unstrained crystal of ^{23}NaCl, a frequency shift of 6 kHz or more from twice the resonant frequency has the same effect as with zero acoustic excitation.

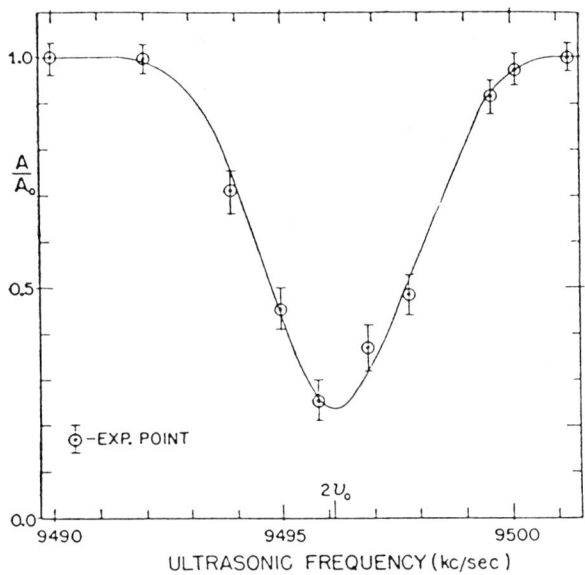

FIGURE 8.7. Line width of ^{23}Na in NaCl: A/A_0 as a function of acoustic frequency at $\Delta m = \pm 2$. (From Ref. 8.2)

8.3 Continuous Wave ASNMR

In continuous wave acoustic saturation techniques, the specimen can be a right cylinder with a piezoelectric transducer on one end face and an RF coil wound around the cylindrical face. The NMR resonance line is recorded using, for example, a marginal oscillator spectrometer (attached to the RF coil) without any acoustic excitation. The measurement is repeated, this time with a high–frequency voltage of constant amplitude and fixed frequency, applied to the transducer. The acoustic frequency corresponds, as in the pulse–saturation technique, to values near one of the resonance frequencies ν_0 and $2\nu_0$. The NMR line is recorded for constant amplitude, but varying frequency, of the acoustic wave. Each such series of measurements as a function of frequency may be made for several values of the voltage applied to the transducer. From these measurements one obtains

(1) the nuclear spin–phonon absorption as a function of A/A_0, where A_0 is the amplitude of the NMR signal without acoustic excitation;
(2) the nuclear spin–phonon linewidth as a function of acoustic frequency.

The specimen may either be prepared to have opposite ends flat and parallel, as in most NAR experiments, or to have the second end roughened and/or terminated in a matched acoustic absorber, so as to approximate the condition of pure traveling waves.

8.4 Transient ASNMR

The theory of transient NAR effects is reviewed in Section 7.5 of the previous chapter. The use of acoustic pulses, or of combinations of acoustic (U) and electromagnetic (M) pulses, to observe transient nuclear resonant effects is especially accessible to the ASNMR technique. From the earlier theoretical discussion, however, we know that a single acoustic pulse cannot, unlike the magnetic pulses in NMR, excite an observable free precession signal, but serves only to diminish the z-component of the macroscopic magnetization of the spin system. The results are more interesting, however, for 2 U pulses or for a combination of U and M pulses, when one satisfies the conditions

$$t_0 \ll T_2 \; , \; \tau < T_2$$
$$t_0 \ll \tau \; , \; t_0 \gg \frac{2\pi}{\omega_0} \qquad (8.21)$$

for pulses of duration t_0 and interval τ between pulses. These results are as follows:

(1) Two acoustic pulses with the same carrier frequency but traveling in opposite directions do not generate the transverse magnetization component M_{xy} that is necessary for observing transient effects.
(2) Two acoustic pulses U_1 and U_2, or a U_1 acoustic pulse combined with an M pulse, form a spin echo signal at $t = 2\tau$.

U_1 corresponds to $\Delta m = \pm 1$, U_2 to $\Delta m = \pm 2$.

8.4.1 Acoustic Nuclear Spin Echo

Acoustic spin echoes were first discovered in the paramagnetic electron spin systems in MgO. The large magnitude of the magneto–elastic coupling for the electron spin systems of Fe^{2+}, Fe^{3+}, Mn^{2+} and Ni^{2+} in MgO and the use of a frequency of 9 GHz, temperatures of 1.8 or 4.2 K, and very large acoustic power levels allow the observation, at the appropriate resonance magnetic fields, of small acoustic electron spin echoes after a pulse sequence of $U_1 U_1$ or $U_2 U_2$. In these experiments, the MgO sample is bonded to one end of a quartz resonator in the shape of a right cylinder. The other end of the quartz resonator is driven by a re–entrant microwave cavity (for U pulses), which also serves as the detector. A second microwave cavity resonator is located just outside the MgO quartz bond and can provide an M pulse. The microwave power for the U pulses is 2 kW, and, for the M pulse, 250 mW. The acoustic phonon echo for the sequence $U_2 U_2$ is shown in Fig. 8.8. Combination $U_1 M_1$ pulses also give acoustic spin echoes. For combination $U_1 U_1 M_1$ pulses, larger stimulated echoes have been seen.

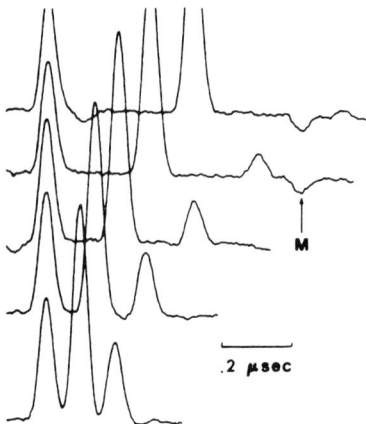

FIGURE 8.8. Phonon echoes from $\Delta m = \pm 2$ transition in Fe^{2+}:MgO. Pulse separation increases from bottom to top. The vertical scale is magneto–elastic stress, not energy. The two applied pulses are identical and, are ~10% higher than the second pulse of the bottom trace. The phonon echo signal at M in the upper two traces is due to an echo pulse that has made two round trips in MgO. This signal is weak and inverted as a consequence of phase–sensitive detection. (From Ref. 8.6)

Acoustic nuclear spin echo experiments have likewise been successful for those nuclear spin systems in which there is a very strong magneto–elastic (*i.e.*, dynamic quadrupole) coupling between elastic waves and nuclear spin systems. The ^{127}I nuclear spin system has a large quadrupole moment of -0.75 barn and a computed Sternheimer antishielding factor $(1-\gamma_\infty) = 175$

in I$^-$. A CsI single crystal in the form of a right cylinder is bonded to a 20–MHz quartz, X–cut transducer to form a composite resonator. An RF coil in the form of a solenoid surrounds the cylindrical surface of the CsI crystal, and the transducer is driven at an RF voltage of amplitude 50 V. This RF coil serves both to generate M_1 pulses and to detect electromagnetically the echoes. The free induction decay from the ^{127}I nuclear spin system in CsI has a signal–to–noise ratio of 1/3 at 300 K. From the combination $M_1 U_2$ pulses, an electromagnetic echo correlated to the presence of U_2 can be found [8.7]. Sequences of $M_1 U_2 U_2$ and $U_2 U_2 M_1$ also give electromagnetic echoes while sequences of $U_2 U_2$, $U_2 U_2 U_2$ and $U_2 M_1$ do not—in agreement with theory [8.7].

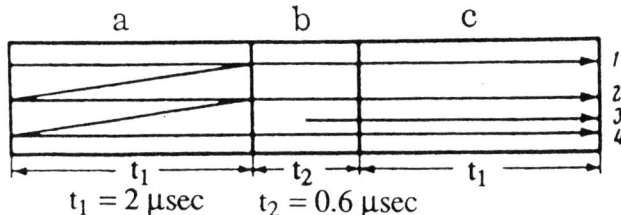

FIGURE 8.9. Passage of acoustic pulses through sample b; a is the transmitting piezoelectric transducer, and c is the receiving piezoelectric transducer. (From Ref. 8.8)

Acoustic nuclear spin echoes have been found in KMnF$_3$ and RbMnF$_3$, which have very large magneto–elastic coupling due to magnetic ordering. A sample in the shape of a right cylinder has both of its end surfaces bonded to lithium niobate transducers. The conditions for creating a $U_1 U_1$ pulse sequence followed by an ultrasonic nuclear spin echo for the ^{55}Mn nuclear spin system is illustrated in Fig. 8.9. They are the following:

(i) When the left transmitting transducer is driven with an RF pulse, an acoustic pulse is generated in the transducer and travels to the right to the transducer–sample interface. At the interface, part of the acoustic energy is transmitted and travels to the right as 1, arriving first at the receiving transducer–air interface on the right. The other part of the acoustic energy is reflected from the transducer–sample interface and travels to the left to the transmitting air–transducer interface. Here reflection occurs, causing acoustic energy to travel to the right to the transducer–sample interface, where part of the energy is transmitted as 2, to the receiving transducer, arriving $2t_1$ after 1.

314   8. SATURATION NMR AND DOUBLE RESONANCE

(ii) From the sample, an acoustic echo is created by previous acoustic traveling waves 1 and 2 traveling to the right through the sample. The acoustic echo traveling wave moves to the right to the receiver transducer as 3, arriving at a time of $4t_1$ after 1.

What an observer monitoring the receiving transducer sees, in a pulse sequence, is the first pulse and the first reflected pulse, followed by an acoustic nuclear spin echo on the background of the second reflected pulse, as shown in Fig. 8.10. By using electromagnetic detection with an RF coil around the sample–transducer arrangement described above, one also observes combinations of acoustic and electromagnetic nuclear spin echoes after combinations of U_1 and M_1 pulses.

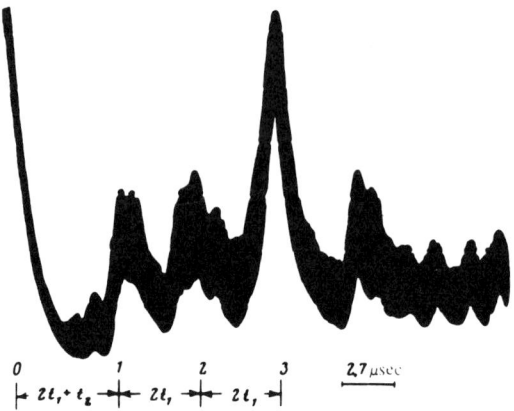

FIGURE 8.10. Acoustic echo signal 3 detected by receiving transducer. 1 and 2 are, respectively, the first and second acoustic pulses; 0 is the electromagnetic pulse exciting the transmitting transducer. (From Ref. 8.8)

8.4.2 Acoustic Saturation of NMR in a Rotating Reference Frame

It is often convenient in NMR to utilize a frame of reference rotating at a frequency ω equal to that of the linear oscillating field $\boldsymbol{H}_1$. In such a frame of reference, the spins are subject to a constant effective field $[(\omega_0 - \omega)^2 + \omega_1^2]^{\frac{1}{2}}/\gamma$, where $\omega_0 = \gamma H_0$ is the Larmor frequency, and $\omega_1 = \gamma H_1$. At exact resonance ($\omega = \omega_0$), in a reference frame which rotates with frequency ω, each spin is subject to the constant field $\boldsymbol{H}_1$. The nuclear magnetization parallel to this component of amplitude H_1 in the rotating frame persists (is *spin-locked* along $\boldsymbol{H}_1$) for a time T_1, the spin-lattice relaxation time. The magnetization can be induced to decay faster than T_1 by modulating either $\boldsymbol{H}_0$ or the amplitude of $\boldsymbol{H}_1$, thereby inducing transitions among the nuclear spin levels in the rotating frame. The modulation frequency capable of inducing $\Delta m = \pm 1$ transitions is $\omega_{M_1} = \gamma H_1$; for $\Delta m = \pm 2$ transitions, it is $\omega_{M_2} = 2\gamma H_1$. A major advantage of this technique is its use of acoustic driving energy at very low frequency. Acoustic excitation of nuclear spin energy levels in the rotating frame may be achieved through the dynamic multipolar interactions. Analogously to, for example, dynamic nuclear electric quadrupole coupling in the laboratory frame, in which frequencies $\omega = \gamma \hbar H_0$ are required, W_{Q1} and W_{Q2} transitions in the rotating frame are induced at frequencies in the kHz range, since the separation of the energy levels in the rotating frame is of order $\gamma \hbar H_1$.

The Hamiltonian $\mathcal{H}_1 = \mathcal{H}_1 + \mathcal{H}(t)$ in the rotating frame is

$$\mathcal{H} = -\gamma \hbar H_1 I_z + \frac{eQ}{4I(2I-1)} V_{zz}(t)[3I_z^2 - I(I+1)] . \tag{8.22a}$$

The electric field gradient V_{zz} can be expressed in terms of a driving force F applied in the [100] direction of a single crystal with cubic symmetry as

$$V_{zz} = -\frac{1}{2}(s_{11} - s_{12})S_{11} F \cos \omega_a t \tag{8.22b}$$

and as

$$V_{zz} = -\frac{1}{2}(s_{11} - s_{12})S_{11} \left[1 - \left(\frac{3}{2} - \frac{s_{44}}{s_{11} - s_{12}} \frac{S_{44}}{S_{11}} \right) \cos^2 \theta \right] F \cos \omega_a t , \tag{8.22c}$$

when the force F is applied in the [110] direction, where θ is the angle between $\boldsymbol{H}_0$ and the [1$\bar{1}$0] direction. In (8.22b) and (8.22c), s_{ij} are the appropriate elastic constants, and ω_a is the angular frequency of the acoustic driving force. Using (8.22), we may calculate W_{Q1} and W_{Q2} as was done earlier in the laboratory frame. The measured quantity in the present technique, however, is the spin-lattice relaxation time T_1^* in the rotating frame. Since the spin-spin interactions maintain a Boltzmann distribution among the spin levels at all times, we obtain [8.5]:

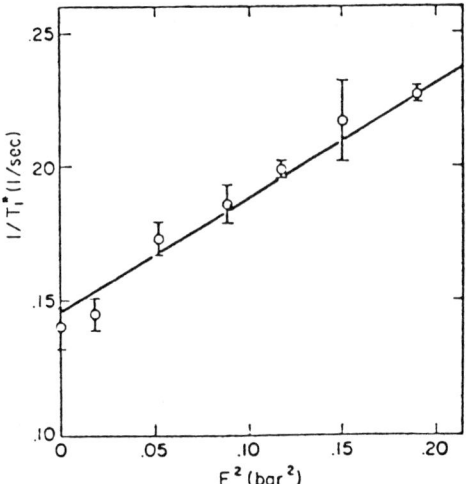

FIGURE 8.11. The nuclear spin relaxation rate in the rotating frame $1/T_1^*$ for ^{23}Na in NaCl versus the square of the acoustic amplitude. (From Ref. 8.9)

$$\frac{1}{T_1^*} = \frac{1}{T_1}\frac{H_1^2 + \alpha H_L^2}{H_1^2 + H_L^2} + \frac{1}{T_a}, \qquad (8.23)$$

where H_L is the local field due to the dipole–dipole interaction, and α is a numerical constant. For $I = \frac{3}{2}$, the acoustic driving term $1/T_a$ is given, for longitudinal strain along [100], as

$$\frac{1}{T_a} = \frac{3}{1280}\frac{e^2 Q^2}{\hbar^2}(s_{11} - s_{12})^2 S_{11}^2 F^2 g(\gamma H_1/2\pi) ; \qquad (8.24)$$

and for longitudinal strain along [110],

$$\frac{1}{T_a} = \frac{3}{1280}\frac{e^2 Q^2}{\hbar^2}(s_{11} - s_{12})^2$$
$$\left[1 - \left(\frac{3}{2} - \frac{s_{44}}{s_{11} - s_{12}}\frac{S_{44}}{S_{11}}\right)\cos^2\theta\right]^2 F^2 g(\gamma H_1/2\pi) , \qquad (8.25)$$

where $g(\gamma H_1/2\pi)$ is the line shape function, the peak of which occurs at $\omega_a/4\pi$. For ^{23}Na ($I = \frac{3}{2}$) in NaCl at room temperature, when the measured T_1 is 13 sec, the measured nuclear relaxation rate in the rotating frame $1/T_1^*$

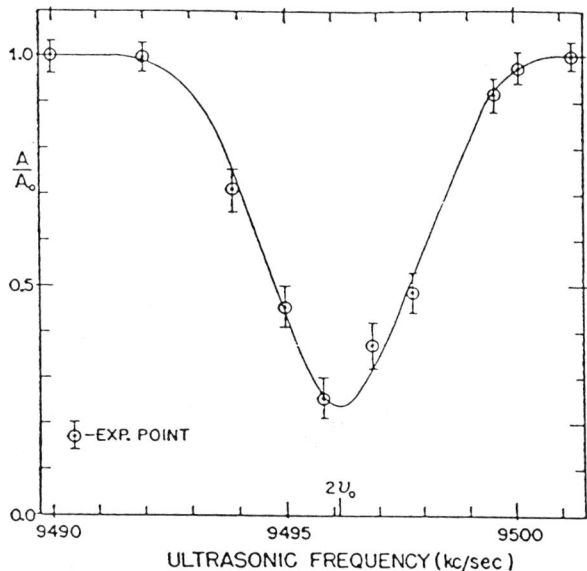

FIGURE 8.12. The acoustically induced nuclear spin relaxation rate $1/T_a$ versus $\gamma H_1/2\pi$. Acoustic amplitude of 0.44 bar at 4 kHz, applied along the [100] axes. (From Ref. 8.9)

is given in Fig. 8.11 for ν_a ($\equiv \omega_a/2\pi$) = 2 kHz. When H_1 is varied at fixed acoustic driving frequency, one obtains the line shape function shown in Fig. 8.12. The maximum value of $g(\gamma H_1/2\pi)$ is $g_{\max} = 1.26 \times 10^{-3}$ sec^{-1}. The orientation dependence of $1/T_a$ is given in Fig. 8.13, in which the solid line is drawn to fit the experimental points to (8.25), with $S_{44}/S_{11} = -0.30$. From the experimental data for NaCl, one obtains values in statcoul/cm^3 ($\times 10^{15}$) of $S_{11} = 2.8$ and $S_{44} = 0.81$.

8.4.3 Acoustic Nuclear Adiabatic Fast Passage

Under conditions of adiabatic fast passage (AFP), the magnitude of the field $\boldsymbol{H}_0$ (or frequency ω) is varied sufficiently slowly through resonance that the spin magnetization $\boldsymbol{M}$ follows the effective field $\boldsymbol{H}_{\text{eff}} = (\boldsymbol{H}_0 - \omega/\gamma) + \boldsymbol{H}_1$ adiabatically; *i.e.*, the orientation of $\boldsymbol{M}$, originally parallel to $\boldsymbol{H}_0$, is reversed as a result of passage through resonance, resulting in an inverted energy level population of the spins. The adiabatic condition may be written

$$(\gamma H_1)^{-1} \ll t ,$$

where t is the time to pass through the resonance half-width $\Delta\omega_0 (\equiv \gamma H_0)$.

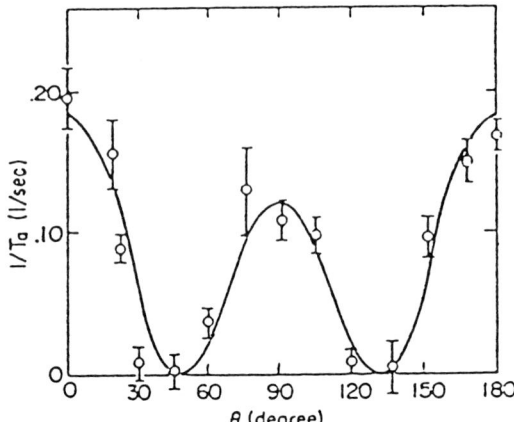

FIGURE 8.13. $1/T_a$ versus angle between $\boldsymbol{H}_0$ and $[1\bar{1}0]$; acoustic force applied in the [110] direction. The acoustic frequency is 4 kHz, $\gamma H_1/2\pi = 2$ kHz, and A is 0.48 bar. The solid line is the theoretical curve given by (8.26). (From Ref. 8.9)

Passage through the resonance by sweeping H_0 is entirely equivalent to sweeping ω; in both cases, inversion of the initial magnetization results, whichever the direction of the sweep.

To study acoustic adiabatic fast passage, the composite resonator is placed in the RF coil of a pulsed NMR spectrometer and the coil is located in an appropriate external magnetic field. The transducer is excited by a high power pulsed amplifier driven by an RF oscillator designed to provide variable time length sweeps at fixed repetition rate. We give results for ^{23}Na spins in single crystal NaCl with 20-MHz longitudinal waves propagated along a [100] axis. The magnitude and direction of the magnetization are determined by the free induction signal following the application of a 90° electromagnetic pulse. Acoustic fast passage is achieved by sweeping the ultrasonic frequency 16 kHz through the $\Delta m = \pm 2$ ^{23}Na spin resonance centered at 20.286 MHz in a magnetic field $H_0 = 0.9007$ T. The use of the $\Delta m = \pm 2$ acoustically induced transition minimizes the magnetic leakage effects from the sweep frequency excitation.

The variation of the magnetization with the time of passage ($\equiv$ sweep time) is shown in Fig. 8.14. Maximum inversion of the magnetization, amounting to 60% of the initial value, occurs for sweep time $\tau = 2$ sec and an applied voltage of 600 V. Recovery of the magnetization, characterizing inversion of the spin population, after the termination of the acoustic pulse, is shown in Fig. 8.15.

Adiabatic fast passage is also discused in Section 7.5.3.

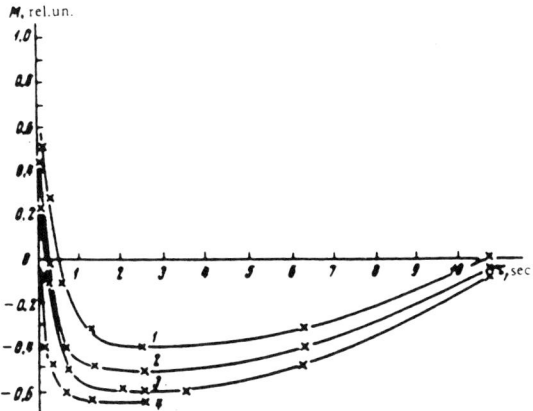

FIGURE 8.14. Dependence of ^{23}Na nuclear spin magnetization on the time of passage; acoustic drive voltages are: 1) 100 V, 2) 200 V, 3) 300 V, 4) 400 V. (From Ref. 8.10)

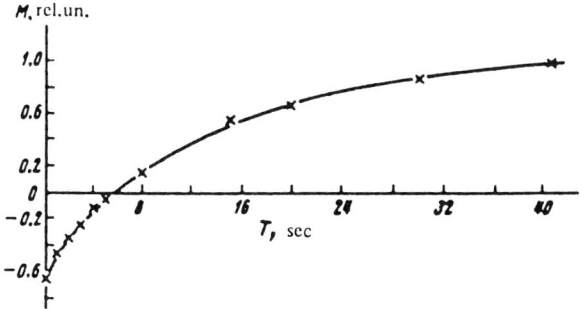

FIGURE 8.15. Recovery of ^{23}Na nuclear spin magnetization following termination of the acoustic pulse. (From Ref. 8.10)

8.5 Double Acoustic Magnetic Resonance

In NMR, the term double resonance connotes the simultaneous excitation of one resonant transition of a given spin system and the monitoring of another transition. In this sense, all of ASNMR constitutes acoustic–magnetic double resonance. In particular, in ASNMR we apply the term double resonance to one of two types of situations:

(1) nuclear–nuclear double resonance, in which a weakly coupled or low–abundance nuclear spin species is observed through the effect of a strong saturating acoustic driving field at the appropriate resonant frequency while monitoring with RF the resonance of a more abundant nuclear spin species;

(2) electron–nuclear double resonance, in which one acoustically induces nuclear–spin transitions while electron resonances are being observed (or vice–versa).

Both types of double resonance have been utilized in ASNMR.

One may schematically represent either form of double resonance, as is done Fig. 8.16. Two properly prepared spin systems a and b, with spin temperatures θ_a and θ_b, are mutually coupled by dipolar and other interactions characterized by time constant τ and are coupled to the lattice by interactions characterized by T_1. If the a species is at a different spin temperature than the b species, the whole system approaches a common spin temperature in a time of the order of τ. Given that $\tau \ll T_1$, the common spin temperature can be different from the lattice temperature. For the case of a spins (abundant) and b spins (rare), the common spin temperature may be not much different from the initial spin temperature of the a spins. To observe the effect of the (rare) b spins on the (abundant) a spins, one keeps the b spin hot—i.e., one keeps the b–spin resonance saturated. Energy then flows continuously from the b–spin system to the a spin. By monitoring changes in the a–spin temperature, one thereby detects the weak b–spin resonance.

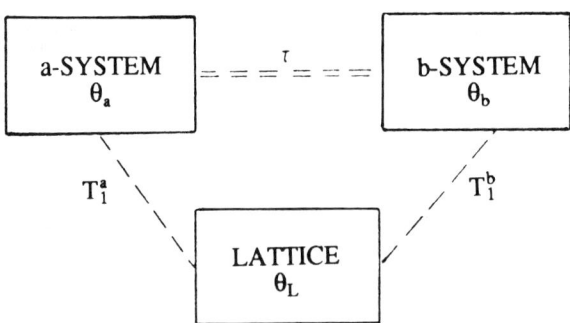

FIGURE 8.16. Schematic of the coupling of two spin systems with each other and with the lattice. (From Ref. 8.11)

Utilizing acoustic excitation, nuclear–nuclear double resonance has been observed in alkali halide crystals (e.g., $^7\text{Li}^{19}\text{F}$, $^{133}\text{Cs}^{127}\text{I}$), with signal enhancements of up to 50%. An extensively studied spin system utilized to investigate both nuclear–nuclear and nuclear–electron double resonance is that of Al_2O_3 doped with 0.05 at. % Cr^{3+} ions. The normal NMR signal of ^{53}Cr ($I = \frac{3}{2}$, abundance = 9.54%, ν_{res} = 2.406 MHz) is much weaker than that of ^{27}Al ($I = \frac{5}{2}$, abundance = 100%, ν_{res} = 11.094 MHz). Although the ^{27}Al and ^{53}Cr nuclear spin are weakly coupled, given the disparity of resonance frequencies, the dipole–dipole system of the Cr^{3+} ions provides a strong coupling. The $\Delta m = \pm 1$ and $\Delta m = \pm 2$ acoustically induced transitions of the ^{53}Cr spin system are deduced from changes in the intensity of the $\frac{1}{2} \leftrightarrow \frac{3}{2}$ NMR line of ^{27}Al. With the NMR frequency fixed at 3.570 MHz in a magnetic field of 3.210 kG, changes in intensity of the ^{27}Al NMR signal are $\simeq$ 80% for acoustic strains (at the ^{53}Cr $\Delta m = \pm 1$, $\Delta m = \pm 2$ frequencies) of 10^{-6}. In the case of electron–nuclear double resonance in the Al_2O_3:Cr^{3+} system, one monitors the electron spin resonance (ESR) line of $Cr^{3+}(S = \frac{1}{2} \leftrightarrow S = -\frac{1}{2})$ at a fixed microwave frequency as one sweeps the acoustic frequency through either a $\Delta m = \pm 1$ or a $\Delta m = \pm 2$ nuclear spin transition. A relative change of $\simeq$15% in ESR signal intensity is observed for $\Delta m = \pm 2$ transitions. Again, the coupling between the ^{27}Al nuclear spin and Cr^{3+} is assumed to occur via the Cr^{3+} - Cr^{3+} dipole–dipole pool. For a review of this work, we refer the reader to *Golenishchev-Kutuzov et al.* [8.12].

CHAPTER 8 REFERENCES

8.1 D. P. Tewari, G. S. Verma. "Acoustic Saturation of Nuclear Magnetic Resonance Lines." *Il Nuovo Cimento* **38**, 197–205 (1965).

8.2 W. G. Proctor, W. A. Robinson. "Ultrasonic Excitation of Nuclear Magnetic Energy Levels of ^{23}Na in NaCl." *Phys. Rev.* **104**, 1344–1352 (1956).

8.3 W. G. Proctor, W. H. Tantilla. "Influence of Ultrasonic Energy on the Relaxation of Chlorine Nuclei in Sodium Chlorate." *Phys. Rev.* **101**, 1757 (1956).

8.4 T. J. Hirsch. "Application of Acoustic Nuclear Magnetic Resonance to Medical Imaging." M. S. Dissertation, University of Arizona, 1986 (unpublished).

8.5 A. Abragam, W. G. Proctor. "Spin Temperature." *Phys. Rev.* **109**, 1441–1457 (1958).

8.6 D. R. Taylor, I. G. Bartlett. "Ultrasonic Spin Echoes in Fe:MgO." *Phys. Rev. Lett.* **30**, 96–98 (1973)

8.7 V. A. Golenishchev-Kutuzov, A. I. Siraziev, N. K. Solovarov, V. F. Tarasov. "Magnetoacoustic Excitation of Nuclear Spin Echo." *Zh. Eksp. Teor. Fiz* **71**, 1074–1082 (1976); English trans: *Sov. Phys. JETP* **44**, 562–566 (1976).

8.8 Kh. G. Bogdanova, V. A. Golenishchev-Kutuzov, A. A. Monakhov. "Acoustic Nuclear Spin Echo in the Antiferromagnets $KMnF_3$ and $RbMnF_3$." *Pis'ma Zh. Eksp. Teor. Fiz* **25**, 292–295 (1977); English trans: JETP Lett. **25**, 268–270 (1977).

8.9 M. Hanabusa, T. Yamaguchi. "Acoustic Excitation of Nuclear Magnetic Energy Levels in the Rotating Frame." *J. Phys. Soc. Jap.* **26**, 901–905 (1969).

8.10 V. F. Tarasov, V. A. Golenishchev-Kutuzov. "Inversion of Magnetization of a Nuclear Spin System by Ultrasound in Adiabatic Fast Passage." *ZhETF Pis. Red.* **21**, 275–277 (1975); English trans: *JETP Lett.* **21**, 126–127 (1975).

8.11 E. P. Jones, S. R. Hartmann. "Steady-State Nuclear Double Resonance: An Application to the Study of the Quadrupole Resonances of ^{39}K, ^{40}K, and ^{41}K in $KClO_3$." *Phys. Rev. B* **6**, 757–771 (1972).

8.12 V. A. Golenishchev-Kutuzov, R. V. Saburova, N. A. Shamikov. "Double Magnetoacoustic Resonances in Crystals." *Usp. Fiz. Nauk* **119**, 201–222 (1976); English trans: *Sov. Phys. Usp.* **19**, 449–461 (1976).

CHAPTER 9.
MAGNETIC MATERIALS

- 9.1 NAR in Antiferromagnetic Insulators
  - 9.1.1 Theory of Magneto–Elastic Coupling to Nuclear Spins
    - 9.1.1.1 Magnetic Properties of a Cubic Antiferromagnet: $RbMnF_3$
    - 9.1.1.2 Silverstein–Fedders Theory
    - 9.1.1.3 Semiclassical Theory
    - 9.1.1.4 Angular Dependence and Equilibrium Conditions
  - 9.1.2 Coupling to Magnetic Nuclei
    - 9.1.2.1 ^{55}Mn in $RbMnF_3$
    - 9.1.2.2 ^{55}Mn and ^{57}Fe in Other Antiferromagnets
  - 9.1.3 Coupling to Nuclei of Nonmagnetic Atoms: ^{19}F in $RbMnF_3$
    - 9.1.3.1 Application of Semiclassical Theory: Role of Crystal Imperfections
    - 9.1.3.2 ^{19}F NAR in $RbMnF_3$
- 9.2 NAR in Ferromagnets
- 9.3 Enhanced Nuclear Acoustic Resonance
  - 9.3.1 Enhanced Nuclear Magnetism in Van Vleck Paramagnets
  - 9.3.2 Theory of Resonant Acoustic Coupling in an Enhanced System
  - 9.3.3 Experimental: ^{165}Ho in $HoVO_4$
- 9.4 Effect of Demagnetization on NAR Line Shapes in Bulk Metals
  - 9.4.1 Calculation of the Demagnetization Effect
  - 9.4.2 ^{183}W NAR Line Shape
    - 9.4.2.1 The ^{183}W Experiment
    - 9.4.2.2 Comparison with Theory

References

9. MAGNETIC MATERIALS
9.1 NAR in Antiferromagnetic Insulators
9.1.1 Theory of Magneto–Elastic Coupling to Nuclear Spins

At least three types of nuclear spin–phonon interactions have been proposed for NAR in antiferromagnets. These interactions can be explained in terms of an ultrasonic perturbation of the energy of the coupled nuclear and electronic spins of an antiferromagnet. The energy per unit cell of this system is given by

$$E = E_{ex} + E_z + E_a + E_h , \qquad (9.1)$$

where E_{ex} is the electronic exchange energy, E_z is the nuclear Zeeman energy,

$$E_a = \frac{K}{S^4} \sum_{i>k} S_i^2 S_j^2 \qquad (9.2)$$

is the anisotropy energy, and

$$E_h = A\boldsymbol{I} \cdot \boldsymbol{S} + A'\boldsymbol{I}' \cdot \boldsymbol{S} \qquad (9.3)$$

is the hyperfine energy. In these equations, $\boldsymbol{S}$ and $\boldsymbol{I}$ refer to the electronic and nuclear spins of a magnetic atom, respectively, and $\boldsymbol{I}'$ refers to the nuclear spin of a neighboring nonmagnetic atom. Examples of magnetic and nonmagnetic atoms are, respectively, ^{55}Mn and ^{19}F in the cubic antiferromagnet RbMnF$_3$. The subscripts i and j ($= 1,2,3$) indicate components of the vectors. The factor K is the anisotropy constant, A is the hyperfine coupling constant between $\boldsymbol{I}$ and $\boldsymbol{S}$, and A' is that between $\boldsymbol{I}'$ and $\boldsymbol{S}$. The NAR interaction first proposed by *Buishvili and Giorgadze* [9.1] involves a modulation of the hyperfine interaction. The interaction energy may be obtained from an expansion of E_h in terms of the crystalline strain e_{ij}:

$$E_{int} = \sum_{i>j} \wedge_{ij} I'_i S_j e_{ij} . \qquad (9.4)$$

The coupling constant $\wedge_{ij}$ is expected to be significant only for an interaction between the electronic spin of a magnetic atom and the nuclear spin of a neighboring, nonmagnetic atom.

A magneto–elastic NAR mechanism involving a two–step process was proposed by *Silverstein* [9.2] for uniaxial antiferromagnets (*e.g.*, MnF$_2$) and extended by *Fedders* [9.3] to cubic antiferromagnets (*e.g.*, RbMnF$_3$). The magneto–elastic coupling arises from a modulation of the anisotropy energy. The interaction energy is

$$E_{int} = \frac{b_1}{S^2} \sum_{i=1}^{3} S_i^2 e_{ii} + \frac{b_2}{S^2} \sum_{i>j} S_i S_j e_{ij} , \qquad (9.5)$$

where b_1 and b_2 are the magneto–elastic coupling constants. The time–dependent magneto–elastic interaction produces a modulation of the effective anisotropy field. This modulation causes a rocking of the sublattice magnetization, which, in turn, couples energy to the nuclei via the hyperfine field. Silverstein has described this interaction in terms of quasi–particles as follows:

> The mechanism with which the coherent lattice energy is absorbed by the nuclear spin system is a two–step process in which a phonon first excites a virtual spin wave via the magnetoelastic coupling and then a nuclear spin is flipped through the decay of the virtual spin wave via the off–diagonal matrix elements of the electron–nuclear hyperfine interaction (from [9.2], page 997).

This magneto–elastic NAR mechanism applies to the nuclei of both magnetic and nonmagnetic atoms.

A third magneto–elastic interaction, proposed by *Shrivastava and Stevens* [9.4], may be thought of as a modulation of the exchange interaction. This modulation causes a rocking of the sublattice magnetization, which then couples to the nuclear spins in the manner described above. Thus modulation of the exchange interaction is another way to initiate the two–step NAR process. In the case of effective anisotropy field modulation, ultrasonic energy is coupled to the electrons via linear magnetostriction; for exchange interaction modulation, the coupling mechanism is volume magnetostriction.

In light of the above, it interesting to consider the relationship between NAR and nuclear spin–lattice relaxation processes in antiferromagnets. We know that for dynamic nuclear electric quadrupolar coupling, the NAR coupling mechanism is the inverse of that for nuclear spin–lattice relaxation. In antiferromagnets, there exists a relationship between NAR and spin–lattice relaxation. However, it not always as straightforward as for dynamic nuclear electric quadrupole coupling. In antiferromagnetic materials, nuclear spin–lattice relaxation derives from the magnetic dipolar interaction between a nuclear spin and the surrounding atomic moments. A possible mechanism for spin–lattice relaxation is that of direct modulation of the hyperfine interaction by thermal phonons. Energy is thereby transferred directly from the nuclear spins to the lattice, as shown schematically by arrow 1 in Fig. 9.1.

This relaxation mechanism is simply an inverse process of the Buishvili–Giorgadze NAR mechanism. A second type of relaxation mechanism does not involve a direct interaction between nuclear spins and lattice phonons; nuclear spin energy is transferred to the lattice via an intermediate electronic state shown schematically by Arrows 2 and 3 in Fig. 9.1. The atomic spin system can take up the nuclear spin energy without the intervention of lattice vibrations, by means of a rearrangement of the atomic spins rather than by a change in the total magnetic moment of the atomic spin system. This energy is eventually transferred to the lattice; however, the quanta involved in this relaxation of the atomic spin system are much larger than the nuclear spin quanta. The nuclear spin relaxation is independent of the atomic spin relaxation time. Since the process does not directly involve phonons with the nuclear spin system, we see that there is, in the case of antiferromagnets, no necessarily simple relationship between NAR and nuclear spin–lattice relaxation.

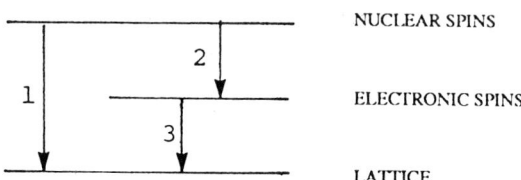

FIGURE 9.1. Diagram showing the path of energy transfer in nuclear spin relaxation processes. (From Ref. 9.5)

Since the nomenclature peculiar to a discussion of spin–phonons in magnetic materials is both more complex and differs substantially from that used in earlier chapters, in Table 9.1 we list appropriate symbols with basic definitions, with particular relevance to a cubic antiferromagnet, *e.g.*, RbMnF$_3$ [9.5].

9.1.1.1 Magnetic Properties of a Cubic Antiferromagnet: RbMnF$_3$

The magneto–elastic theories reviewed in the following two subsections have been applied in greatest detail to the configuration of a cubic antiferromagnet. Their predictions have been verified most extensively in NAR studies of ^{55}Mn and ^{19}F in RbMnF$_3$.

RbMnF$_3$ is a cubic antiferromagnet with a Néel temperature of $T_N = 83$ K. The magnetic properties of this antiferromagnet can be described largely in terms of internal fields: the anisotropy field H_A, the hyperfine field H_N, the nuclear polarization field H_{NE} and the exchange field H_E. The anisotropy field is strongly dependent upon crystalline impurities; H_A has been observed to vary from 4.59 to 0 Oe, as cobalt doping is increased from 0 to 620 ppm. Even in crystals grown without intentional impurities, H_A has been found to vary from 3 to 4.59 Oe. The hyperfine field $\boldsymbol{H_N}$ is parallel to the sublattice magnetization $\boldsymbol{M_0}$. The hyperfine frequency is

$\nu_N = \gamma_N H_N/2\pi = 686.2$ MHz at 4.2 K, where γ_N is the nuclear gyromagnetic ratio for ^{55}Mn. The ^{55}Mn nuclear polarization field seen by the electrons is $H_{NE} = (9.43/T)$ Oe and is also parallel to $\boldsymbol{M}_0$. The exchange field is $H_E = 0.816 \times 10^6$ Oe for $T = 4.2$ K.

The calculated magnitude of the sublattice magnetization M_0 for each sublattice is 305 Oe at the zero temperature limit, where the thermal average of the electronic spins $\langle S \rangle = \frac{5}{2}$. The orientation of the sublattice magnetization is essentially perpendicular to the applied field $\boldsymbol{H}_0$ when $H_0 > (2H_E H_A)^{\frac{1}{2}} (\sim 2$ kOe). When in this spin–flopped state, the magnetization $\boldsymbol{M}_0$ is canted toward $\boldsymbol{H}_0$ by a small angle $t = H_0/2H_E$. All measurements discussed in Section 9.1.2.1, below, are taken in the spin-flopped state.

RbMnF$_3$ has been the object of extensive studies of ^{55}Mn nuclear magnetic resonance and antiferromagnetic resonance (AFMR). The strong hyperfine field in RbMnF$_3$ puts the nuclear resonance frequency in the UHF region; the small anisotropy field puts the AFMR frequency in the X-band microwave region. The combination of small H_A and large hyperfine constant A causes these nuclear and electronic modes to be strongly coupled. This strong electron–nuclear coupling gives rise to a large frequency pulling of the nuclear modes by the electronic modes. As for any antiferromagnet in a spin–flopped configuration, RbMnF$_3$ has two nondegenerate AFM modes. The resonance frequencies of the field–dependent (ω_{e+}) and the field–independent (ω_{e-}) antiferromagnetic modes are given [9.6] as

$$\omega_{e+} = \gamma_e (H_0^2 + 2H_E H_{NE} + 3BH_E H_A)^{\frac{1}{2}} \qquad (9.6a)$$

and

$$\omega_{e-} = \gamma_e (2H_E H_{NE} + 3CH_E H_A)^{\frac{1}{2}} , \qquad (9.6b)$$

where γ_e is the electronic gyromagnetic ratio, and B and C are functions of the orientation of $\boldsymbol{H}_0$. The coupling of these AFM modes to the nuclei gives rise to two nondegenerate nuclear modes whose resonance frequencies are given [9.6] as

$$\omega_{n\pm} = \omega_N [1 - 2\gamma_e^2 H_E H_{NE}/(\omega_{e\pm})^2]^{\frac{1}{2}} , \qquad (9.7)$$

where $\omega_N (= \gamma_N H_N)$ is the unpulled ^{55}Mn resonance frequency. The ω_{n+} mode is field–dependent and the ω_{n-} mode is field–independent. *Only the ω_{n+} mode has been observed by NMR with the spins in a flopped configuration.*

Table 9.1
Symbol Definitions Used in This Chapter

M	=	the sublattice magnetization, which is sometimes called M_1 or M_2 when it is necessary to distinguish between the two magnetic sublattices
M_0	=	the static component of M
K	=	anisotropy constant
λ	=	exchange constant
H_A	=	$\frac{4}{3}\frac{K}{M_0}$ = effective anisotropy field for a cubic AFM, $H_A \parallel <111>$
H_E	=	λM_0 = effective exchange field
H_{NE}	=	polarization field of manganese nuclei seen by the electrons via the hyperfine interaction
H_N	=	hyperfine field parallel to M_0
H_0	=	the applied magnetic field
γ_e	=	the electronic gyromagnetic ratio
γ_n	=	the nuclear magnetic gyromagnetic ratio (for ^{55}Mn or ^{19}F, as the case may be)
ω_e	=	antiferromagnetic resonance frequency
ω_n	=	$2\pi\nu_n$ = pulled ^{55}Mn resonance frequency
ω_N	=	$\gamma_n H_N$ = unpulled ^{55}Mn resonance frequency
ω_{NE}	=	$\gamma_e H_{NE}$
ω_E	=	$\gamma_e H_E$
b_1, b_2	–	the magneto-elastic coupling constants
H_{ME}	=	effective magneto-elastic field
ψ	=	$(H_0, [001])$
θ	=	$(M_0, [001])$
ϕ	=	$(M_0, [100])$
t	=	tilt angle between $(M_1 - M_2)/2$ and M_1
Ω	=	volume of a unit cell
N	=	$1/\Omega$

9.1.1.2 Silverstein–Fedders Theory

In the magneto–elastic theory of *Silverstein* [9.2], extended by *Fedders* [9.3] and applied, in particular, to cubic antiferromagnets, coupling of ultrasound to the coupled system of electronic and nuclear spins proceeds via a modulation of the effective anisotropy field. This modulation causes a rocking of the sublattice magnetization, resulting finally in a coupling of energy to the nuclei via the hyperfine field. Fedders has applied this theory to calculate the acoustic dispersion and absorption in a cubic antiferromagnet. For a flopped spin configuration, and for $|\omega - \omega_n|$ and $|\omega - \omega_e|$ much greater than, respectively, the nuclear spin line width and antiferromagnetic resonance line width, the calculated acoustic dispersion is given by

$$\Delta\omega_\pm(s^{-1}) = \frac{2b_1^2 \gamma_e}{\rho M_0 v^2} \omega_E \omega \\ \times \left[\frac{2\omega_E \omega_{NE} \omega_N^2}{\omega_{e\pm}^4 (\omega^2 - \omega_{n\pm})} + \frac{1}{\omega^2 - \omega_{e\pm}^2} \right] \Gamma_\pm , \qquad (9.8)$$

where $\Gamma_\pm$ is a function dependent only on the orientation of $\boldsymbol{H}_0$ for a given acoustic propagation vector $\boldsymbol{k}$, ρ is the density, v is the acoustic velocity, b_1 is a magneto-elastic coupling constant, $\omega_E = \gamma_e H_E$, and $\omega_{NE} = \gamma_e H_{NE}$. This calculation is for the case of longitudinal waves propagating along a cube axis. The first term in the brackets is due to the nuclear spin–phonon interaction via the electrons; the second term is due solely to an electronic spin–phonon interaction. The corresponding NAR absorption near nuclear spin resonance is given by

$$\alpha_\pm = \frac{2\pi \gamma_e \omega_E^2 \omega_{NE} \omega_N}{\rho v^2 M_0 \omega_{e\pm}^4} \left(\frac{\omega_N}{\omega_{n\pm}} \right) b_1^2 \Gamma_\pm \omega \, g(\omega) , \qquad (9.9)$$

where $g(\omega)$ is the Lorentzian line shape function. The Silverstein–Fedders magneto–elastic theory is more than a theory of nuclear acoustic resonance. The contribution from the antiferromagnetic modes in the UHF region can be comparable to the nuclear acoustic dispersion.

9.1.1.3 Semiclassical Theory

Results similar to those reported by Fedders can be obtained from a semiclassical calculation similar to that originally suggested by *Silverstein* [9.2]. Although the method is less general than that used by Fedders, it gives physical insight into the coupling mechanism and aids in the construction of a simple model. By using this model, the angular function $\Gamma_\pm$ can be readily calculated for arbitrary field configurations.

The nuclear spin–phonon interaction is treated in terms of an effective RF field H_1 (perpendicular to I_z) interacting with the nuclear dipole moment. The interaction Hamiltonian is then

$$\mathcal{H}_I = \gamma_n \hbar \boldsymbol{I} \cdot \boldsymbol{H}_1 = \frac{1}{4}\gamma_n \hbar H_1 (I_+ e^{-i\omega t} + I_- e^{i\omega t}) , \qquad (9.10)$$

where H_1 is the field produced by time–dependent crystalline strains e_{ij}. Following the standard perturbation treatment of Chapter 2, one derives for the acoustic absorption coefficient α_n (in units of rad/sec)

$$\alpha_n(\sec^{-1}) = \frac{\pi N \hbar \omega \gamma_n^2 H_1^2 \langle I \rangle}{4\rho v^2 e_{ij}^2} g(\omega) , \qquad (9.11)$$

where N is the number of spins, and $\langle I \rangle$ is the thermal average of the spin angular momentum.

An expression for H_1 may be obtained by noting that the presence of ultrasonic strains gives rise to an effective magneto–elastic field $\boldsymbol{H}_{ME}$ given by

$$\boldsymbol{H}_{ME} = \nabla_M E_{ME} , \qquad (9.12)$$

where

$$E_{ME} = \frac{b_1}{M_0^2}\sum_{i=0}^{3} M_i^2 e_{ii} + \frac{b_2}{M_0^2}\sum_{i>j} M_i M_j e_{ij} . \qquad (9.13)$$

The nuclei, however, see a field much larger than $\boldsymbol{H}_{ME}$. As in the case of ^{55}Mn NMR in RbMnF$_3$, there is large enhancement of the induced RF field by the electrons. The time–dependent field seen by the nuclei comes from a magneto–elastic modulation of a large hyperfine field. The details of the model are most easily derived for a spin–flopped uni-axial antiferromagnet. It is convenient at this point to introduce a new quantity: H_c, the constraining field. The sublattice magnetization is constrained to its equilibrium configuration by the components of the internal field $\boldsymbol{H}_A$ and the applied field $\boldsymbol{H}_0$ that are parallel to $\boldsymbol{M}_0$; this field is called the *constraining field*. If a magneto–elastic field $\boldsymbol{H}_{ME}$ perturbs the sublattice magnetization, as shown in Fig. 9.2, then $\boldsymbol{M}_0$ would rock through an angle β given by $\beta = H_{ME}/H_c$. The effective $\boldsymbol{H}_{ME}$ perpendicular to $\boldsymbol{M}_0$ excites the electronic modes. Consider, for example, the effect of a component of $\boldsymbol{H}_{ME}$ parallel to $\boldsymbol{H}_0$, as shown in Fig. 9.2. This configuration of $\boldsymbol{H}_{ME}$ causes the sublattice to gyrate about an axis perpendicular to the plane of the figure. The constraining field is then

$$H_c(+) = H_A + H_0 \sin t = H_A + \frac{H_0^2}{2H_E} = \frac{\omega_{e+}}{2\gamma_e \omega_E} , \qquad (9.14)$$

9.1 NAR IN ANTIFERROMAGNETIC INSULATORS

where $\omega_{e+} = \gamma_e(H_0^2+2H_EH_A)^{\frac{1}{2}}$, the frequency of the field–dependent mode for a uni–axial antiferromagnet. A component of $\boldsymbol{H}_{ME}$ perpendicular to $\boldsymbol{H}_0$ and $\boldsymbol{M}_0$, on the other hand, causes the sublattice magnetization to gyrate about an axis parallel to $\boldsymbol{H}_0$. The constraining field in that case is

$$H_c(-) = H_A = \frac{\omega_{e-}^2}{2\gamma_e\omega_E}, \qquad (9.15)$$

where $\omega_{e-} = \gamma_e(2H_EH_A)^{\frac{1}{2}}$, the frequency of the field–independent mode for a uni–axial antiferromagnet. Equations (9.14) and (9.15) are extended to apply to a cubic antiferromagnet by using the definitions of $\omega_{e\pm}$ given in (9.6). The constraining field may always be found from the equation for the antiferromagnetic resonance frequency, since, in general, $(\omega_e/\gamma_e)^2 = 2H_EH_c$.

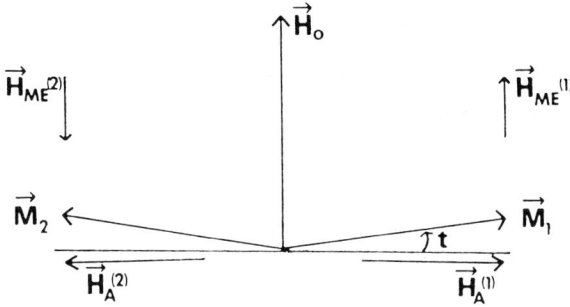

FIGURE 9.2. The electronic (+) mode ω_{e+} excited by a magnetoelastic field $\boldsymbol{H}_{\mathrm{ME}}$ in a spin–flopped antiferromagnet. $t\,(\simeq 0.2°)$ is the tilt angle of $\boldsymbol{M}$, and (1) and (2) indicate different magnetic sublattices. (From Ref. 9.6)

The perturbing field is given by

$$H_1(\pm) = H_N\tan\beta = \frac{2AS\gamma_e\omega_E}{\hbar\gamma_n\omega_{e\pm}^2}\boldsymbol{H}_{ME}\cdot\epsilon_\pm, \qquad (9.16)$$

where $\epsilon_+ = \boldsymbol{H}_0/H_0$, and $\epsilon_- = \boldsymbol{M}_0\times\boldsymbol{H}_0/M_0H_0$. The acoustic attenuation can now be written as

$$\alpha_n = \frac{\pi N\hbar\omega}{4\rho v^2 M_0^2}\left(\frac{2AS\gamma_e\omega_E}{\hbar\omega_{e\pm}^2}\right)^2 B_\pm^2\Gamma_\pm\langle I\rangle g(\omega), \qquad (9.17)$$

where we have set $(\boldsymbol{H}_{ME} \cdot \boldsymbol{\epsilon}_\pm)^2 = B_\pm^2 \Gamma_\pm e_{ij}^2/M_0^2$. By using conventional definitions of H_N, H_{NE} and M_0, the acoustic attenuation can be written in a form similar to that of *Fedders* [9.3]:

$$\alpha_n = \left[\frac{2\pi\gamma_e}{\rho v^2 M_0} \frac{\omega_E^2 \omega_{NE} \omega_N \omega}{\omega_{e\pm}^4} g(\omega)\right] B_\pm^2 \Gamma_\pm \ . \tag{9.18}$$

If $g(\omega)$ is replaced by a Lorentzian line for which $|\omega - \omega_n|$ is much greater than the nuclear linewidth, then (9.18) and Fedders's expression differ only by a factor of ω_N/ω_n. This difference arises because the correlation of nuclear and electronic spins is not included in the semiclassical model, as it is in Fedders's derivation.

The NAR coupling to the nuclear spin of the magnetic atom described by (9.18) is basically the same for any antiferromagnet. Only the angular dependence $B^2\Gamma$ and the exact form of ω_e depend on lattice symmetry. The NAR coupling described by (9.18) can be interpreted as resulting from the excitation of the nuclear mode ω_{n+} (ω_{n-}) via the corresponding electronic mode ω_{e+} (ω_{e-}). The magneto–elastic field that couples to the electronic magnetization differs in sign at different sublattices. The electronic mode is thus excited by the component of $\boldsymbol{H}_{ME}$ parallel to the major axis of precession of the mode. From these considerations, the NAR interaction should be strong for *both* nuclear modes. Nuclear magnetic resonance, on the other hand, is observable only for the ω_{n+} mode in the spin–flopped regime.

9.1.1.4 Angular Dependence and Equilibrium Conditions

The angular dependence $B^2\Gamma$ of the interaction of (9.18) may be evaluated explicitly by using (9.12), (9.13) and the expressions for ϵ_+ and ϵ_- given above. The angular factor is a function of the three angles shown in Fig. 9.3. The position of $\boldsymbol{H}_0$ in a given plane is specified by the angle ψ; the position of $\boldsymbol{M}_0$ is specified by θ and ϕ. It is convenient, however, to write the angular dependence in terms of a single, experimentally measurable angle ψ. To express θ and ϕ in terms of ψ, magnetization equilibrium conditions are needed. We derive the equilibrium conditions for the spin–flopped case for $\boldsymbol{H}_0$ anywhere in the (100) and ($1\bar{1}0$) planes.

When the electron spins are completely flopped, the electronic magnetizations lie in a plane perpendicular to $\boldsymbol{H}_0$; their position in this plane is determined solely by the anisotropy energy

$$E_A = \frac{K}{M_0^4}(M_x^2 M_y^2 + M_y^2 M_z^2 + M_z^2 M_x^2) \ , \tag{9.19}$$

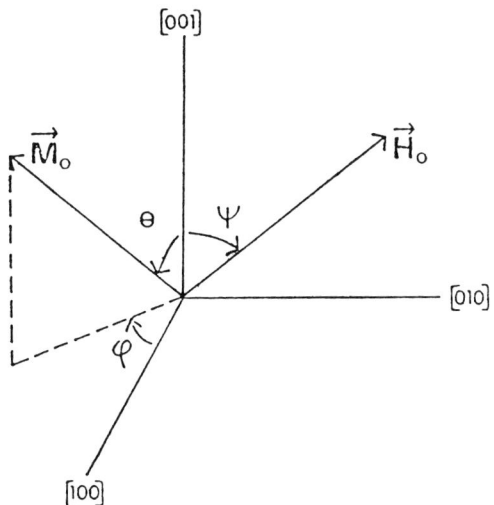

FIGURE 9.3. Spherical angles used to describe the orientation of the applied field $\boldsymbol{H}_0$ and the sublattice magnetization $\boldsymbol{M}_0$.

where K is the anisotropy constant for cubic RbMnF$_3$. The coordinates x, y and z are coincident with the [100], [010] and [001] crystal axes, respectively. The equilibrium conditions are most easily determined by specifying the spin orientation with respect to a transformed coordinate system (primed symbols), with the z'-axis parallel to $\boldsymbol{H}_0$, in which case $\theta' = 90°$ and ϕ' is the only variable. The equilibrium condition in the transformed system for $\boldsymbol{H}_0$ in the (100) plane is

$$\frac{\partial E_A(\psi, \phi')}{\partial \phi'} = (4 - \sin^2 2\psi)\sin^2 \phi' - 2 = 0 \ . \tag{9.20}$$

For $\boldsymbol{H}_0$ in the ($1\bar{1}0$) plane, the equilibrium condition is

$$\begin{aligned} 0 = \sin 2\phi' \{ &\sin^2 \phi' [1 + 4\cos^2 \psi - 3\cos^4 \psi] \\ &+ 2\cos 2\phi' [1 - (3/2)\cos^2 \psi] - 1 \} \ . \end{aligned} \tag{9.21}$$

For $0 \leq \psi \leq 54.7°$ with $\boldsymbol{H}_0$ in the ($1\bar{1}0$) plane, the stable equilibrium condition is given by $\sin 2\phi' = 0$; i.e., $\theta = 90° + \psi$. For

$$54.7° \leq \psi \leq 180° - 54.7° \ ,$$

the factor in curly brackets gives the stable solution.

With these equilibrium conditions and the associated transformation equations, the angular dependence $B_\pm^2 \Gamma_\pm$ can be written in the form shown in Table 9.2. The symbol e denotes the direction of polarization of the acoustic wave.

Table 9.2

Angular Dependence of the Magneto-Elastic NAR Interactions in a Cubic Antiferromagnet*

k_{long} $\parallel$	k_{trans} $\parallel$	$\hat{e}$ $\parallel$	H_0 in	ψ	$B_+^2 \Gamma_+$	$B_-^2 \Gamma_-$
[001]			(100)		$\frac{2b_1^2 S^2}{A}$	$\frac{8b_1^2 D s^4}{A^2}$
[001]			$(1\bar{1}0)$	$< 54.7°$	$b_1^2 S^2$	0
[001]			$(1\bar{1}0)$	$> 54.7°$	$\frac{b_1^2 E S^2}{B}$	$\frac{4b_1^2 E s^4(3c^4-7c^2+2)}{B^2}$
[011]			(100)		$\frac{2b_2^2 C^2}{A}$	$\frac{2(b_1-b_2 S^2)^2 D}{A^2}$
	[001]	[010]	(100)		$\frac{2b_2^2 C^2}{A}$	$\frac{2b_2^2 D S^2}{A^2}$
	[001]	[100]	(100)		$\frac{b_2^2 D c^2}{A}$	$\frac{b_2^2 c^4 s^6}{A^2}$
	[001]	[110]	$(1\bar{1}0)$	$< 54.7°$	$b_2^2 C^2$	0

*To simplify the table, we define $s = \sin\psi$, $S = \sin 2\psi$, $c = \cos\psi$, $C = \cos 2\psi$, $A = 4 - \sin^2 2\psi$, $B = 3\cos^4\psi - 10\cos^2\psi + 3$, $D = 2 - \sin^2 2\psi$ and $E = 1 - 3\cos^2\psi$. (From Ref. 9.6)

9.1.2 Coupling to Magnetic Nuclei

9.1.2.1 ^{55}Mn NAR in RbMnF$_3$

For ^{55}Mn NAR in RbMnF$_3$ at $T \simeq 4$ K and $H_0 > (2H_E E_A)^{\frac{1}{2}}$, both the field-dependent (+) and the field-independent (−) nuclear modes are observed to give rise to an extraordinarily intense acoustic absorption at resonance [9.6]. Indeed, the intensity of the absorption renders unnecessary the use of the usual sensitive NAR detection techniques. For $k \parallel [001]$ and H_0 along a cube axis, the NAR absorption vanishes. As ψ increases, the absorption increases markedly, from approximately 6 dB/cm for $\psi = 1°$ to 30 dB/cm for $\psi = 10°$. For $\psi = 45°$, the NAR absorption is a maximum and is so intense that the swept frequency pattern is obliterated, as illustrated in Fig. 9.4. The absorption at the center of the resonance is too great to be measured easily. Measurements in the wings of the (+) mode show that the maximum attenuation is much greater than 40 dB/cm.

9.1 NAR IN ANTIFERROMAGNETIC INSULATORS

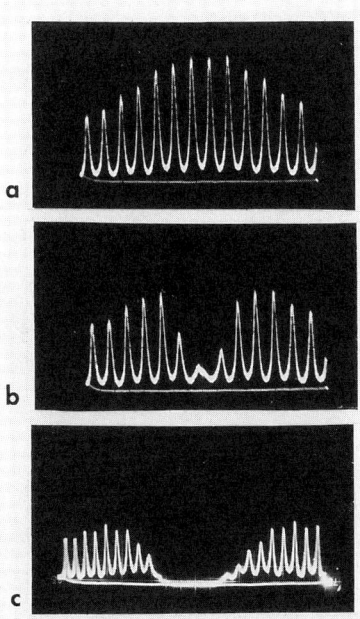

FIGURE 9.4. Swept–frequency oscilloscope traces of ultrasonic mechanical resonance patterns in RbMnF$_3$ for 650–MHz longitudinal waves propagated along [100] axis. Spacing between mechanical resonance peaks is $\approx$ 330 kHz. $k\|[100]$. $\phi \equiv \angle(\boldsymbol{H}_0, \boldsymbol{k})$ was (a) 0°, (b) 1° and (c) 10°. $H_0 = 6.0$ kOe. $T = 4.3$ K. (From Ref. 9.7)

The field–dependence of the NAR frequency ν_{n+} is shown (solid dots) in Fig. 9.5 for $T = 4.20$ K, $\psi = 0.3°$ and $\boldsymbol{H}_0$ in the (100) plane. For $\nu = 662.27$ MHz, $\psi = 0.7°$ and $T = 4.8$ K, the field at the center of the (+) mode absorption line is 6.857 kOe, and the field width at the half power points is 56 Oe.

NAR dispersion is measured as a function of frequency and field angle for $\boldsymbol{H}_0$ in the $(1\bar{1}0)$ and (100) planes. With the field in the $(1\bar{1}0)$ plane, only the (+) mode is observed when $\psi < 54.7°$. When $\psi \geq 54.7°$, the (−) mode appears as a strong absorption and dispersion ranging in frequency from 470 MHz to approximately 600 MHz. Hysteresis effects are evident. With the field in the (100) plane, the NAR dispersion as a function of frequency is measured near ν_{n+} for $H_0 = 6.00$ kOe and $T = 4.18$ K. Since the NAR coupling vanishes for $\psi = 0°$, the dispersion at $\nu(10.8°)$ is given by $\Delta\nu = \nu(10.8°) - \nu(0°)$. Dispersion versus frequency for the ν_{n+} mode is plotted in Fig. 9.6. Dispersion versus field angle ψ is shown in Fig. 9.7. Utilizing values of $H_A = 2.92$ Oe, $H_E = 0.810 \times 10^6$ Oe, $H_{NE} = (9.43/T)$ Oe and $\nu_N = 686.2$ MHz, a theoretical fit (solid line) is made to the ν_{n+} versus H_0 data in Fig. 9.5.

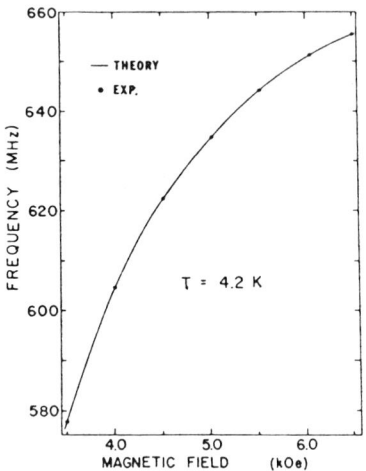

FIGURE 9.5. Nuclear acoustic resonance frequency ν_{n+} versus the applied field H_0. $\boldsymbol{H_0}$ is the (100) plane. $\psi = 0.3°$. (From Ref. 9.5)

The field-dependence of ν_{n+} comes from the pulling term in (9.7):

$$\nu_{n+} = \nu_N \left[1 - \frac{2 H_E H_{NE}}{H_0^2 + 2 H_E H_{NE} + 3 B H_E H_A} \right]^{\frac{1}{2}}.$$

The calculated frequency width of the (+) mode NAR absorption line is $\delta\nu_{n+} = 0.40$ MHz, in agreement with the (+) mode NMR linewidths. The NAR absorption linewidth of the (−) mode is 16±3 MHz, 40 times greater than that for the (+) mode.

The linewidth of the (+) mode in RbMnF$_3$ for $H_0 < 10$ kOe is partially explained on the basis of inhomogeneities in the antiferromagnetic spectrum. Such inhomogeneities, acting through the frequency–pulling mechanism, cause a broadening of the nuclear line width. Similarly, the large difference between the line widths of the (+) and the (−) NAR modes can be explained in terms of an inhomogeneous anisotropy field. Line width calculations assuming a ±0.30 Oe variation of H_A give $\delta\nu_+ = 0.36$ MHz and $\delta\nu_- = 19$ MHz, which are in good agreement with experiment. The inhomogeneities are attributed to the anisotropy field, since H_A has been shown to be strongly dependent upon crystalline impurities.

The NAR absorption line shape is approximately Gaussian. NAR dispersion measurements are used to obtain the line shape in the wings, several MHz away from the center of the resonance line, where inhomogeneous broadening should be absent. The dispersion shape in this region is found to be Lorentzian. An example is given in Fig. 9.6, in which the $\Delta\nu_{n+}$ versus ν data are well fitted by a Lorentzian curve over a range of 8 to 35 half line widths. Good agreement is found also between theory and the $\Delta\nu$-versus-ψ data if a Lorentzian line shape is assumed.

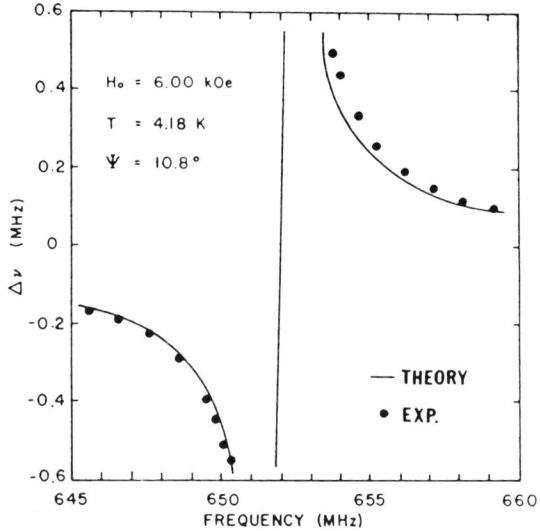

FIGURE 9.6. Calculated and observed dispersion ($\Delta\nu_{m+}$) versus frequency (ν) for $H_0 = 6.00$ kOe and $\boldsymbol{H}_0$ in the (100) plane. (From Ref. 9.5)

The observed intensities of the NAR absorption and dispersion lines are accounted for by magneto–elastic theory. For $H_0 = 6$ kOe, the peak absorption for the $(+)$ mode is calculated to be 10^3 dB/cm; for the $(-)$ mode, 10^2 dB/cm. Since measurements of dispersion are more readily and accurately made than those of absorption, quantitative comparison between theory and experiment is made for both NAR intensity and angular dependence of the dispersion. We illustrate this comparison with a single example: $\Delta\nu$ versus ψ with $\boldsymbol{H}_0$ in the (100) plane and $\nu(0°) = 671.05$ MHz.

In Fig. 9.7 are shown the values of the dispersion as a function of ψ for the (+) mode (dashed line) and (−) mode (dot–dashed line) calculated for (9.8), with the angular dependence taken from Row 1 of Table 9.2. The solid line, representing the sum of $\Delta\nu_+$ and $\Delta\nu_-$, shows excellent agreement with the experimental data. The only adjustable parameter in these calculations is b_1, which is set equal to 2.14×10^6 erg/cm^3 for the best fit. The data shown in Fig. 9.7 are taken under conditions such that $\nu_- \simeq 540$ MHz and $\nu_+ \simeq 655$ MHz, so that $\nu > \nu_{n+}$, $\nu > \nu_{n-}$, and the nuclear dispersion for both modes is positive.

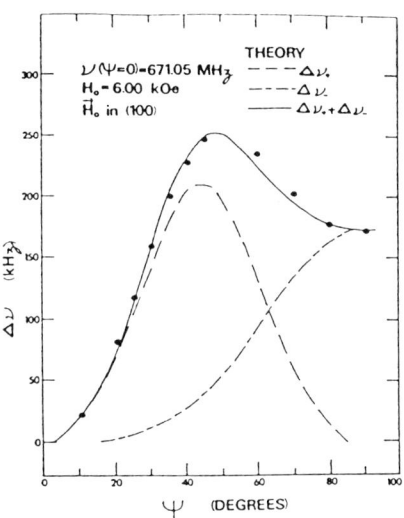

FIGURE 9.7. Calculated and observed dispersion ($\Delta\nu$) versus field angle (ψ) for $H_0 = 6.00$ kOe, $\boldsymbol{H}_0$ in the (100) plane, and $\nu(0°) = 671.05$ MHz. (From Ref. 9.5)

9.1.2.2 ^{55}Mn and ^{57}Fe NAR in Other Antiferromagnets

The form of results of the semiclassical model of Section 9.1.1.3 strongly suggests that the magneto–elastic theory may be generalized to apply to nuclei of magnetic and nonmagnetic atoms in any antiferromagnet. The parameter that varies most from specimen to specimen is H_1, which may be written as

$$H_1 = \frac{H_N}{H_C} \boldsymbol{H}_{ME} \cdot \boldsymbol{e} \ . \qquad (9.22)$$

The field $\boldsymbol{H}_1$ is the time–dependent component of the nuclear hyperfine field produced by a rocking of the sublattice magnetization $\boldsymbol{M}_0$. The ultrasonically induced magneto–elastic field $\boldsymbol{H}_{ME}$ exerts a torque on $\boldsymbol{M}_0$ causing it

9.1 NAR IN ANTIFERROMAGNETIC INSULATORS

to rock. The less M_0 is constrained by H_C, the greater the effect of H_{ME}. For many antiferromagnets, including RbMnF$_3$, the constraining field H_C is approximately equal to the anisotropy field H_A. The equation for NAR absorption of the nuclei of magnetic atoms can be written in a form that more clearly relates to these antiferromagnetic properties. Equation (9.17) can now be rewritten as

$$\alpha_n(\sec^{-1}) = \frac{2\pi^3 h I(I+1)}{\rho M_0 v^2 \gamma_e \langle S \rangle 3kT} \frac{B^2 \nu_N^4}{H_C^2} \Gamma g(\nu) , \qquad (9.23)$$

where it is assumed that the nuclear frequency pulling is negligible and that H_N is much larger than any applied field. From (9.23), the extraordinarily large NAR attenuation of a magnetic nucleus in an antiferromagnet may be attributed to the high nuclear resonance frequency of ^{55}Mn ($\nu_N = 650$–680 MHz) and to the small anisotropy field of RbMnF$_3$ (the effective anisotropy field at 4.2 K is $\simeq 6$ Oe). Indeed, any antiferromagnet that contains ^{55}Mn and has a small anisotropy field ($H_A \leq 300$ Oe) should exhibit a strong nuclear resonance (10–10^4 dB/cm). Among the materials that have these properties are MnTe ($H_A \simeq 30$ Oe at $T = 80$ K), KMnF$_3$ ($H_A \simeq 3$ Oe at $T = 4.2$ K), and CsMnF$_3$ ($H_A \simeq 4$ Oe at $T = 4.2$ K).

A strong ^{55}Mn NAR absorption has been observed in MnTe [9.8,9.9]. This large NAR signal has several properties that indicate that the magneto–elastic interaction is responsible for it. The frequency dependence ($\alpha_n \sim \nu_N^4$) and the field dependence of the NAR attenuation are in agreement with predictions of the magneto–elastic theory. The observed NAR attenuation ($\alpha_n = 290$ dB/cm for $\nu_N = 398$ MHz, $T = 187$ K, $H_0 = 0$) agrees with that calculated by using estimated values of the nuclear line width $1/T_2$, H_A and B: $1/T_2 \simeq 1$ MHz, $H_A(187$ K$) \simeq 7$ Oe, and $B \simeq 3.5 \times 10^6$ erg/cm^3.

Acoustic saturation NMR techniques (Chapter 8) have been utilized to observe acoustic coupling to ^{55}Mn in KMnF$_3$ and RbMnF$_3$ [9.10], as well as to ^{57}Fe nuclei in hematite (α-Fe$_2$O$_3$) [9.11]. NAR techniques have been used to study acoustic coupling to ^{55}Mn nuclei in manganese fluoro–phosphate glass, a spin glass [9.12].

9.1.3 Coupling to Nuclei of Nonmagnetic Atoms: ^{19}F in RbMnF$_3$

9.1.3.1 Application of Semiclassical Theory: Role of Crystalline Imperfections

The resonant frequency of the fluorine nuclei in antiferromagnetic RbMnF$_3$ is determined by the external field H_0 and the fields due to the manganese ions. The magnetization of these ions couples to the fluorines directly via a dipolar interaction and indirectly via an isotropic Fermi contact interaction. Approximately 98% of the hyperfine field H_N seen by a fluorine nucleus is due to its two nearest Mn neighbors and is given approximately by

$$H_N = \frac{A'}{\gamma_n \hbar} \sum_i \langle S \rangle_i . \tag{9.24}$$

The sum is over nearest neighbors, A' is the hyperfine constant for the coupling between a fluorine nucleus and the electrons of the nearest manganese ion, and γ_n is the ^{19}F gyromagnetic ratio. In the perfect cubic lattice of RbMnF$_3$, each fluorine has two Mn neighbors with spins oppositely oriented, making $\sum \langle S \rangle_i = 0$ for zero applied field. If $H_0 \neq 0$, however, the magnetizations no longer cancel exactly. Instead, they give rise to a small net magnetic moment that is parallel to $\boldsymbol{H}_0$. If the magnetizations are in a spin–flopped state (to which case we restrict our discussion), they are tilted by a small angle $t = H_0/2H_E$. The effective field seen by the fluorine nuclei is then

$$H_N = \frac{A'\langle S \rangle}{\hbar \gamma_n} 2t = \frac{A'\langle S \rangle H_0}{\hbar \gamma_n H_E} . \tag{9.25}$$

Because of lattice symmetry, $\boldsymbol{H}_N$ is parallel to $\boldsymbol{H}_0$. The fluorine nuclei are not at magnetically equivalent sites, unless the applied field $H_0 = 0$ or $\boldsymbol{H}_0 \parallel$ [111]. If $\boldsymbol{H}_0 \parallel$ [110], there are two nonequivalent sites with a population ratio of 2:1, resulting in a ^{19}F NMR spectrum of two lines. The larger line is referred to as the 2-F line, the smaller as the 1-F line.

The magneto–elastic nuclear spin–phonon interaction for ^{19}F nuclei can be rigorously derived in a manner similar to that for ^{55}Mn. The nuclear spin–phonon interaction again proceeds via the electronic (antiferromagnetic) modes. Since there are two electronic modes (+) and (−), there are two NAR modes. For a perfect crystal, the acoustic absorption coefficient due to the two nuclear modes when the system is in a spin–flopped configuration is given by (analogous to (9.9) for ^{55}Mn)

$$\alpha_\pm = \frac{2\pi Z \gamma_e}{\rho v^2 M_0} \frac{\langle I \rangle \omega_E^2 \omega}{\langle S \rangle \omega_{e\pm}^4} \omega_N^2 B^2 \Gamma_\pm g(\omega) . \tag{9.26}$$

In (9.26), $\langle I \rangle$ and $\langle S \rangle$ are the thermal averages of the ^{19}F and the electronic spin, respectively; Z is the number of fluorine nuclei per unit cell resonating at angular frequency ω. Calculations show that α_+ is some five orders of magnitude too small to explain the observed NAR and is furthermore characterized by a temperature and angular dependence quite different from that observed.

The expression for α_-, on the other hand, contains terms that give a temperature and angular dependence quite similar to that observed. The temperature dependence of the α_- mode is proportional to $(1/T)(BM_0/K)^2$, which agrees with that observed. However, in a perfect crystal, in which each fluorine atom lies equidistant from two Mn ions with oppositely oriented spins, the formalism shows that the effects from each Mn ion exactly cancel, giving $\Gamma_- \equiv 0$ and, therefore, $\alpha_- \equiv 0$. Considering a real crystal, the possibility exists that the NAR coupling takes place near sites of crystalline imperfections (vacancies, substitutions, dislocations, *etc.*) at which the crystalline symmetry is broken. The coefficient multiplying Γ_- in the equation for α_- is about 12 orders of magnitude greater than the corresponding coefficient in α_+. This fact lends credence to the hypothesis that ^{19}F NAR coupling occurs via the relatively few ^{19}F nuclei located near lattice imperfections. As in the case of ^{55}Mn above, the NAR coupling to the nuclear mode ω_{n-} (ω_{n+}) proceeds via its corresponding electronic mode ω_{e-} (ω_{e+}). The ultrasonically produced magneto–elastic field causes a rocking of the electronic magnetization that modulates the hyperfine field H_N. This process stimulates nuclear spin transitions if there is a component of the time–dependent hyperfine field perpendicular to I_z. The equilibrium configuration of the internal fields relevant to ^{19}F NAR is shown in Fig. 9.8.

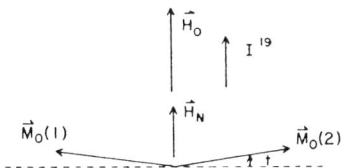

FIGURE 9.8. Equilibrium configuration of the internal fields of RbMnF$_3$. (From Ref. 9.13)

In the spin–flopped case, the electronic (+) mode is excited by rocking the sublattice magnetization in the plane of Fig. 9.8. The time–dependent field perpendicular to I_z is $H_N \tan\beta$, where β is the rocking angle. The electronic (−) mode, on the other hand, is excited by rocking the sublattice magnetization about $\boldsymbol{H}_0$. This causes no rocking of the field H_N that is seen by the vast majority of the fluorine nuclei, since $\boldsymbol{H}_N \parallel \boldsymbol{H}_0$ for these nuclei. The fields are collinear because of lattice symmetry. The fluorine nuclei are located equidistantly between two Mn ions with oppositely oriented magnetic moments. For fluorine nuclei near sites of crystalline imperfections, these conditions no longer hold, and, in general, an electronic dipolar or hyperfine field perpendicular to H_0 exists. We refer to this field as $H_{N\perp}$. Since the nonsymmetrically located ^{19}F nuclei contribute to the NAR spectrum, H_N^2 in (9.26) for the (−) mode should be replaced by

$$(H'_N)^2 = \frac{1}{N} \sum_i^N |H_{N\perp}(i)|^2 , \qquad (9.27)$$

where N is the number of unit cells in the crystal, and the sum is over one particular fluorine site per unit cell. With this modification, α_- is no longer identically zero.

9.1.3.2 ^{19}F NAR in RbMnF$_3$

We limit our treatment of ^{19}F NAR to the NAR line intensity, line shape and the concomitant demands on the magneto–elastic theory to obtain agreement with the experimental results. From the above discussion, the magnitude and peculiar line shape of the observed ^{19}F NAR spectrum can be accounted for by the (−) mode of the magneto–elastic interaction. For this NAR mechanism to be operative, however, certain fluorine nuclei must see a magnetic moment perpendicular to H_0. This condition obtains only where the lattice symmetry of RbMnF$_3$ is broken by lattice imperfections.

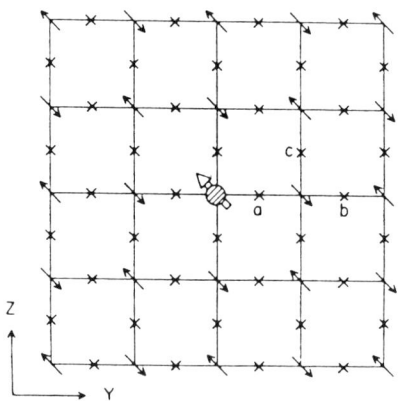

FIGURE 9.9. Diagrammatic sketch of the yz–plane of the RbMnF$_3$ lattice with a manganese vacancy. x denotes a F^{19} site; ↑, a Mn^{55} site; and ⊘, a vacancy. (From Ref. 9.13)

A simple type of lattice imperfection that can be shown to give rise to many of the observed effects is a manganese vacancy. For every missing Mn ion, there is an uncompensated magnetic moment in the crystal, that produces a dipolar field centered at the vacancy site. We assume here that this is the only effect of the vacancy. A model of this lattice vacancy is shown schematically in Fig. 9.9. In this figure, the magnetizations lie in the yz-plane and $\boldsymbol{H}_0$ is parallel to the x-axis. The six fluorines that are nearest to the vacancy (a site) see a large hyperfine field of $\simeq$ 100 kOe due to the unbalanced pair of sublattice magnetizations. The resonance frequencies of these fluorines are pushed into the UHF region ($\simeq$ 400 MHz for 100 kOe) and do not contribute to the observed 10–30-MHz NAR spectrum. The hyperfine field seen by the next–nearest neighbor is insignificant compared to the dipolar field resulting from the vacancy. This field is given by $\boldsymbol{B}(r) = [3(\boldsymbol{r} \cdot \boldsymbol{\mu})\boldsymbol{r}/r^5 - \boldsymbol{\mu}/r^3]$, where $\boldsymbol{r}$ is centered on the vacancy site, $\boldsymbol{\mu}$ is the dipolar moment of the Mn ion, and $\boldsymbol{\mu}$ is parallel to $\boldsymbol{M}_0$. The magnitude of the dipolar field seen by the next–nearest fluorine, 6.33Å away from the vacancy, is $\mu/(6.33\text{Å})^3$ or $514(\langle S \rangle/S_0)$ Oe. The component of $\boldsymbol{B}$ parallel to $\boldsymbol{H}_0$, $B_\parallel$, causes a shift in the nuclear resonance frequency. The component perpendicular to $\boldsymbol{H}_0$, $B_\perp$, corresponds to H'_N in (9.27). The fields seen by the next–nearest neighbor (site c shown in Fig. 9.9) are $B_\perp = 396$ Oe and $B_\parallel = 276$ Oe for $T = 77$ K.

An application of this model is illustrated in Fig 9.10. In Fig. 9.10(a) is shown the unperturbed ^{19}F nuclear resonance spectrum (*i.e.,* expected NMR spectrum) for $H_0 = 7$ kOe, $\boldsymbol{H} \parallel [110]$ and $T = 77$ K. The dashed and solid lines in Fig. 9.10(a) correspond, respectively, to the 1–F and 2–F lines. In light of the discussion regarding the orientation of $\boldsymbol{M}_0$, the predicted spectrum should be calculated by averaging over all possible orientations of $\boldsymbol{M}_0$. Considering the simplicity of the present model, we choose to approximate this averaging procedure by taking the sum of α(hard axis) and α(easy axis), corresponding to the weighting factor Γ_s. When the predicted NAR spectrum for $\boldsymbol{M}_0$ along a hard axis is added to that of Fig. 9.10(b), the $\alpha(\Gamma_s)$ spectrum of Fig. 9.10(c) is obtained. The ^{19}F NAR spectrum observed under conditions corresponding to those assumed in Fig. 9.10 is shown in Fig. 9.11. In Fig. 9.11(a) is shown the recorder tracing of the spectrum that is the derivative of the NAR absorption. In Fig. 9.11(b) is shown the curve obtained after integrating the curve of Fig. 9.11(a). The spectra of Figs. 9.11(b) and 9.10(c) show many similarities, demonstrating that it is reasonable to assume that lattice imperfections result in additional broad structure in the vicinity of the perturbed ^{19}F resonance lines. There is little similarity, on the other hand, between the spectra of Figs. 9.11(b) and 9.10(b). This may be taken as a further indication that $\boldsymbol{M}_0$ is not always in an easy configuration.

We consider, finally, the question of whether a reasonable density of lattice imperfections can account for the magnitude of the observed ^{19}F

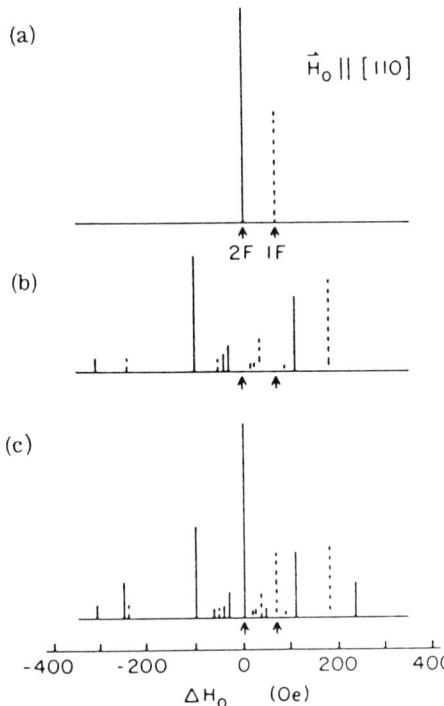

FIGURE 9.10. Calculated ^{19}F nuclear resonance spectrum for $T = 77.4$ K and $\boldsymbol{H}_0 \| [100]$. $\Delta H_0 = 0$ corresponds to $H_0 = 7.0$ kOe: (a) the "unperturbed" or NMR spectrum; (b) the NAR spectrum corresponding to $\alpha(\Gamma_-)$; (c) the NAR spectrum corresponding to $\alpha(\Gamma_s)$. The vertical scale in (c) is twice that in (a) and (b). The arrows indicate expected positions of the unperturbed 1-F and 2-F lines. (From Ref. 9.13)

NAR. From (9.26) and (9.27), one can show that α_- is given by

$$\alpha_- = \frac{1.28 \times 10^{-5}(H'_N)^2}{(H'_A)^2}, \tag{9.28}$$

where $H'_A [\simeq \frac{3}{2} B(\psi) H_A]$ is the projection of $\boldsymbol{H}_A$ on the sublattice magnetization. [The factor $(\omega_{e-}/\gamma_e)^2$ in (9.26) has been written as $(\omega_{e-}/\gamma_e)^2 = 3B(\psi)H_E H_A = 2H_E H'_A$.] We calculate α_- for the following conditions: longitudinal waves propagated along the [001] axis, $H_0 = 7$ kOe, $\boldsymbol{H}_0 \| [100]$, $T = 77$ K, $Z = 2$ and $\Gamma_- = 1$. The NAR attenuation at the center of the large unshifted line for the conditions stated above is $\alpha_- = 180$ Hz.

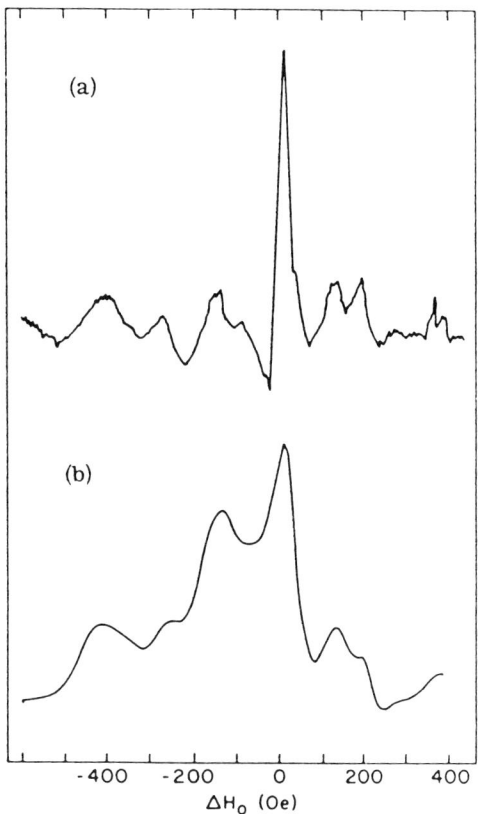

FIGURE 9.11. Flourine NAR spectrum for $H_0 = 7$ kOe, $\boldsymbol{H}_0 \| [110]$ $\boldsymbol{k}_l \| [110]$ and $T = 77.4$ K: (a) recorder tracing of derivative spectrum; (b) integrated (absorption) spectrum. (From Ref. 9.13)

Recalling that $H_A \simeq 10^{-3}$ Oe at 77 K, and assuming that all unit cells participate equally in the interaction, (9.28) gives $(H'_N)^2 \simeq 14$ Oe^2. Since only n unit cells with a vacancy are participating in the NAR interaction, $(H'_N)^2 = (n/N) \sum |B_\perp(i)|^2$, in which the sum includes all fluorines that are neighbors of an Mn vacancy and that contribute to the 2-F line. Calculations show that $\sum |B_\perp(i)|^2 \simeq 2 \times 10^6$ Oe^2 for each Mn vacancy. The fraction of unit cells that must have a vacancy to give the observed intensity is therefore given by $n/N = (H'_N)^2 / \sum |B_\perp(i)|^2 \simeq 7 \times 10^{-5}$. Considering that the impurity density in a typical sample is of the order of 10^{-4}, this is a reasonable order of magnitude.

The specific model of lattice imperfection assumed is not critical; other types of imperfections are capable of giving rise to a dipolar field of sufficient magnitude to account for the observed NAR. In particular, a single lattice imperfection undoubtedly affects the magnetization at neighboring sites. Thus a far smaller concentration of defects than assumed above may account for the observed NAR spectrum.

9.2 NAR in Ferromagnets

A theory of acoustic magnetic absorption and dispersion in bulk ferromagnets has been given by *Fedders* [9.14]. Because the general equations include both electronic and nuclear contributions, and because they are appropriate for a disk of ferromagnetic material with any symmetry, the results are extremely complicated. The treatment is based on Fedders' estimate that the indirect coupling of the nuclear spins to the lattice via the electron spin and the hyperfine interaction is several orders of magnitude stronger than any direct interaction, *e.g.*, dynamic dipole or dynamic quadrupole coupling. For Ni, Fe and Co, the magneto–elastic spin–lattice coupling is considerably larger than the electromagnetic (or Alpher–Rubin) coupling. In any event, the latter can be totally suppressed by using transverse waves propagating perpendicular to the plane of magnetization.

Fedders treats specifically the following three special cases:

(1) cubic material whose magnetic easy axes are $\langle 100 \rangle$ (like Fe) or $\langle 110 \rangle$;
(2) cubic material whose magnetic easy axes are $\langle 111 \rangle$ (such as Ni) or $\langle 110 \rangle$;
(3) hexagonal close–packed material whose magnetic easy axis is [0001] (such as Co).

The nuclear spin resonance frequencies for the case of the electronic magnetization M parallel to the external field H_0 are given by $\omega_n = \gamma_n(H_0 + H_N)$, where H_N is the hyperfine field due to the electronic spins acting on the nuclear spin. H_N is, of course, much greater than any reasonably applied H_0, but H_0 (for narrow–enough lines) may be used to change the resonant frequency. Since $H_N = f(T)$, however, one may choose to vary temperature to display the resonance line. Attenuation changes corresponding to nuclear acoustic resonant absorption have been calculated: $\alpha \simeq 10^{-7}$ cm^{-1} for Fe, 10^{-5} cm^{-1} for Ni and 10^{-2} cm^{-1} for Co. All of these estimates are at 4.2 K and use known values of NMR line widths.

9.3 Enhanced Nuclear Acoustic Resonance

9.3.1 Enhanced Nuclear Magnetism in Van Vleck Paramagnets

The magnetic properties of paramagnetic ions in solids have been extensively studied [9.15]. In a Van Vleck paramagnet, the singlet ground state yields no EPR spectrum. If the ion has a nuclear spin, however, the system can be investigated by NMR. In such paramagnetic systems, an enhanced nuclear magnetism occurs for the interaction of a nuclear spin $I \neq 0$ in an external magnetic field $\boldsymbol{B}$. The enhancement, electronic in nature, is induced through the hyperfine interaction between the nuclear and electron spins. We follow the description given by *Abragam and Bleaney* [9.16].

The nuclear magnetic moment $\boldsymbol{\mu}_I = \gamma_I \hbar \boldsymbol{I}$ polarizes the electronic shells of the paramagnetic ion, which then acquire an electronic magnetization $\boldsymbol{M}_I(\boldsymbol{r})$, which interacts with the applied magnetic field $\boldsymbol{B}$, resulting in an interaction energy bilinear in $\boldsymbol{I}$ and $\boldsymbol{B}$:

$$-\hbar \boldsymbol{I} \cdot \boldsymbol{\gamma}' \cdot \boldsymbol{B} , \qquad (9.29)$$

where $\boldsymbol{\gamma}'$ is a tensor whose axes and principal values depend upon the electronic structure of the particular ion in question. Writing $\boldsymbol{\gamma}' = \gamma_I \boldsymbol{K}$, where $\boldsymbol{K}$ is a dimensionless tensor, we may represent the total nuclear Zeeman energy as

$$Z_I = -\gamma_I \hbar \boldsymbol{I} \cdot (\boldsymbol{1} + \boldsymbol{K}) \cdot \boldsymbol{B} = \hbar \boldsymbol{I} \cdot \boldsymbol{\gamma} \cdot \boldsymbol{B} , \qquad (9.30)$$

where $-\gamma_I \hbar \boldsymbol{I} \cdot \boldsymbol{B}$, as usual, represents the direct coupling of the nuclear spin with the applied magnetic field $\boldsymbol{B}$.

The enhancement factor $(1 + K)$ results in an increase in the NMR frequency of several orders of magnitude for ions of the lanthanide group. SQUID NMR studies at 4.2 K of the lanthanide ion ^{169}Tm ($I = \frac{1}{2}$, $\mu = 0.2388\mu_n$) in TmPO$_4$, a tetragonal, rare-earth, zircon-structure crystal, for example, indicate that an applied magnetic field produces an effective field at the site of the Tm nuclei enchanced by a factor $(1 + K) \simeq 78$. A number of enhanced nuclear magnetic systems, particularly insulating paramagnetic crystals, have been studied by NMR techniques. All of the magnetic systems (excluding metals and intermetallic compounds) are Van Vleck paramagnets, in which the lowest electronic levels consist of a singlet ground state, characterized by the absence of a permanent magnetic moment, and low-lying excited states (typically 10–50 cm^{-1}).

Enhanced nuclear acoustic resonance of ^{141}Pr ($I = \frac{5}{2}$) in single-crystal PrF$_3$ at 4.2 K is characterized by anisotropic principal enhancement factors of approximately 3 and 8. For ^{165}Ho ($I = \frac{7}{2}$), the nuclear magnetic moment is 4.173 nm, and $\gamma_I/2\pi = 9.0$ MHz T^{-1}. In the Van Vleck paramagnet HoVO$_4$, this is enhanced by the hyperfine field generated by the induced moment to approximately 15 MHz T^{-1} parallel and 1529 MHz T^{-1} perpendicular to the [001] axis at 1.6 K. Extensive NAR studies of ^{165}Ho in HoVO$_4$ have been made over a frequency range of 0.8 to 1.6 GHz;

both $\Delta m = \pm 1$ and $\Delta m = \pm 2$ transitions are observed. For the case of longitudinal acoustic waves at 800 MHz, propagated along the [100] axis of of single crystal HoVO$_4$, with B in the (001) plane, the $\Delta m = \pm 1$ resonant absorption is inversely proportional to the absolute temperature, with a power attenuation coefficient of $4.5 \times 10^2 T^{-1} m^{-1}$.

9.3.2 Theory of Resonant Acoustic Coupling in an Enhanced System

A description of the mechanism of resonant coupling of acoustic waves to ^{165}Ho nuclear spins in HoVO$_4$ follows [9.17,9.18]. We recall (see above) that in a Van Vleck paramagnet there is no permanent magnetic moment in the electronic singlet ground state; it is the externally applied magnetic field B that induces an electron magnetic moment M. The effect of the acoustic strain is to modulate this induced moment in both size and direction. As a result of the magnetic hyperfine interaction, a field $B_e \gg B$ is set up to be proportional to M. With a change in the direction of M, therefore, there is a corresponding change in the direction of B_e, producing an oscillatory component of B_e normal to the applied external field B. As in the case of the other dynamic spin–phonon interactions, described in Chapters 3 and 4, net energy is transferred to the nuclear spin system from the acoustic wave when the resonant condition is satisfied, resulting in resonant acoustic absorption and dispersion.

The most thorough study of enhanced NAR has been made of ^{165}Ho in HoVO$_4$, a tetragonal paramagnet with the zircon structure, space group D_{4h}^{19} (I4/amd) and parameters $a_0 = 0.71214$ nm, $c_0 = 0.62926$ nm at 25°C [9.17,9.18]. The Ho^{3+} ion occupies a site of tetragonal symmetry. The lowest electronic levels consist of a singlet ground state, with a doublet lying higher in energy by 21 cm^{-1}. The enhanced nuclear spin Hamiltonian is

$$\mathcal{H} = \gamma_\parallel \hbar B_z I_z + \gamma_\perp \hbar (B_x I_x + B_y I_y) + P[I_z^2 - (1/2)I(I+1)], \quad (9.31)$$

where $\gamma_\parallel = 20$ MHz T^{-1}, $\gamma_\perp = 1523$ MHzT^{-1}, and $P/h = 25$ MHz for ^{165}Ho at a temperature of 1.6 K.

Calculation of the acoustic strain Hamiltonian $\mathcal{H}_{AC}$ requires the use of third–order perturbation theory. We note that in the case of the lanthanide group of ions with the configuration of 4f^n (4f^{10} for the ground state of the Ho^{3+} ion), a good approximation assumes that J is a good quantum number. Under this assumption, the electronic Zeeman energy $-(M \cdot B)$ can be written as $g_J \mu_B (B \cdot J)$ and the hyperfine energy as $A_J (J \cdot I)$. For a longitudinal acoustic wave propagated along the a–axis (chosen to be the x–axis), the acoustic strain gives rise to an extra term in the crystal field

9.3 ENHANCED NUCLEAR ACOUSTIC RESONANCE

of the form $(1/2)(J_+^2 + J_-^2)$ that has a matrix element within the excited doublet. Applying perturbation theory results in a strain Hamiltonian

$$\mathcal{H}_{ac} = A^+ \left[\frac{1}{2}\left(\frac{g_J\mu_B}{A_J}\right)(B_x^2 - B_y^2) + (B_x I_x - B_y I_y) \right.$$
$$\left. + \frac{1}{2}\left(\frac{A_J}{g_J\mu_B}\right)(I_x^2 - I_y^2) \right] . \quad (9.32)$$

Similarly, the strain operator $(1/2)(J_x^2 - J_y^2)$ produces the strain Hamiltonian

$$\mathcal{H}_{ac} = A^- \left[\frac{1}{2}\left(\frac{g_J\mu_B}{A_J}\right)(B_x B_y + B_y B_x) + (B_x I_y + B_y I_x) \right.$$
$$\left. + \frac{1}{2}\left(\frac{A_J}{g_J\mu_B}\right)(I_x I_y + I_y I_x) \right] . \quad (9.33)$$

In (9.32) and (9.33), the first term represents the change in the second-order electronic Zeeman energy. The second term represents the enhanced nuclear acoustic strain energy, while the last term corresponds to a change in the pseudo–quadrupole interaction that arises from the use of the magnetic hyperfine operator to the second degree. In (9.31), the quantity P, the overall effective quadrupole parameter, contains in addition to the pseudo–quadrupole contribution two contributions from the true nuclear electric quadrupole interaction: one from the electric field gradient of the 4f electrons, another from that of the crystal lattice. The associated strain-induced time–varying EFG's are much less effective in causing nuclear spin transitions than the mechanisms considered in (9.32) and (9.33). The quantities A^+ and A^- in (9.32–9.35) are complicated constants proportional to the acoustic strain [9.15].

The applied field $\boldsymbol{B}$ is at an angle ϕ to [100] in the (001) plane. The acoustic strain Hamiltonians (9.32) and (9.33) are transformed to a coordinate system in which $\boldsymbol{B}$ lies along the Z-axis and the X-axis is parallel to the former z-axis (the tetragonal axis). Equation (9.32) becomes

$$\mathcal{H}'_{ac} = A^+ \left\{ B(I_Z \cos 2\phi - I_Y \sin 2\phi) \right.$$
$$\left. + \frac{1}{2}\left(\frac{A_J}{g_J\mu_B}\right)[(I_Z^2 - I_Y^2)\cos 2\phi - (I_Y I_Z + I_Z I_Y)\sin 2\phi] \right\} ; \quad (9.34)$$

and (9.33), with angle ϕ changed by 45°, becomes

$$\mathcal{H}_{ac} = A^- \Big\{ B(I_Z \sin 2\phi - I_Y \cos 2\phi)$$
$$+ \frac{1}{2}\left(\frac{A_J}{g_J \mu_B}\right)[(I_Z^2 - I_Y^2)\sin 2\phi - (I_Y I_Z + I_Z I_Y)\cos 2\phi]\Big\} \,. \quad (9.35)$$

In (9.34) and (9.35), terms in I_Z and I_Z^2 are diagonal, representing changes in the static Hamiltonian that result from the acoustic strain. The term in I_Z represents a change in enhancement of the nuclear resonance frequency; that in I_Y gives rise to $\Delta m = \pm 1$ NAR transitions, the intensity of absorption increasing with the fourth power of frequency. The remaining terms (other than those in I_Z and I_Y) in (9.34) and (9.35) arise from the use of the magnetic hyperfine operator to the second degree and are similar in nature to the pseudo–quadrupole contributions to the parameter P in (9.31). The acoustic strain results in a change in the pseudo–quadrupole interaction even when there is no applied magnetic field $\boldsymbol{B}$. The term in $(I_Y I_Z + I_Z I_Y)$ also contributes to $\Delta m = \pm 1$ transitions, but is smaller than the term in I_Y by the factor $(A_J/g_J \mu_B B)$.

The term in I_Y^2 corresponds to NAR $\Delta m = \pm 2$ transitions, with intensity increasing with the square of frequency. The $\Delta m = \pm 2$ transitions are weaker than the $\Delta m = \pm 1$ transition by a factor of approximately $(A_J/2g_J \mu_B B)^2$. The strength of the absorption depends, as usual, on the direction of $\boldsymbol{B}$: for longitudinal acoustic waves along a [100] axis, the $\Delta m = \pm 1$ absorption is a maximum for $\boldsymbol{B} \parallel [110]$; conversely, the $\Delta m = \pm 2$ absorption maximum is displaced by 45° from the $\Delta m = \pm 1$ transition maximum.

Under conditions such that $(\hbar\omega/k_B T) \ll 1$, the acoustic absorption coefficient is given by

$$\alpha = \frac{n_0 \pi \omega^2}{4\rho v^3 k_B T} \langle i|\mathcal{H}_{ac}|j\rangle^2 g(\omega) \,, \quad (9.36)$$

where n_0 is the population per unit volume of the two energy levels i, j involved in the transition. Since (9.34) and (9.35) give the matrix elements for acoustic transitions within the enhanced nuclear system, the acoustic power absorption has the form

$$T^{-1}[D_1^+ \nu^2 \cos^2 2\phi + D_2^+ \nu^4 \sin^2 2\phi] \quad (9.37)$$

for longitudinal acoustic waves propagated along [100]. Along [110], the form is

$$T^{-1}[D_1^- \nu^2 \sin^2 2\phi + D_2^- \nu^4 \cos^2 2\phi] \,. \quad (9.38)$$

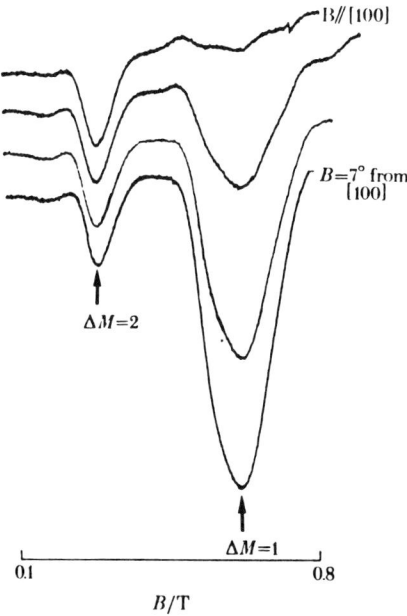

FIGURE 9.12. Enhanced nuclear acoustic resonance in HoVO$_4$ for longitudinal acoustic waves propagated along a [100] axis. For magnetic field B along the same axis, the ± 2 resonance line (at the lower field strength) has maximum intensity, falling as the direction of B moves away in the (001) plane. The intensity of the ± 1 line (in higher field) is zero for B along [100], but increases rapidly as B is rotated away from this axis, to reach a maximum along [110]. (From Ref. 9.18)

9.3.3 Experimental: ^{165}Ho in HoVO$_4$

For a magnetic field B in the (001) plane of HoVO$_4$ with a magnitude 0.5 to 1 T, the ^{165}Ho resonant frequency varies from approximately 0.8 to 1.6 GHz. We have assumed the value $\gamma_\perp / 2\pi = 1529$ MHzT^{-1} at 1.6 K. The line width for an undiluted crystal, due mainly to the unresolved quadrupole splitting, is 52 mT or 80 MHz; the line shape is approximately Gaussian. We note from the discussion above that since the resonance lines corresponding to $\Delta m = \pm 1$ and $\Delta m = \pm 2$ transitions have different angular dependences, their relative intensities vary with the direction of the applied magnetic field. For example, in Fig. 9.12, which is illustrative of the case of longitudinal waves propagated along [100], the intensity of the $\Delta m = \pm 1$ NAR line is almost zero when $B \parallel [100]$, but it increases rapidly as B is moved away from this axis. The $\Delta m = \pm 2$ NAR line, on the other hand, has maximum intensity with B along [100], diminishing slightly as B is rotated through a small angle.

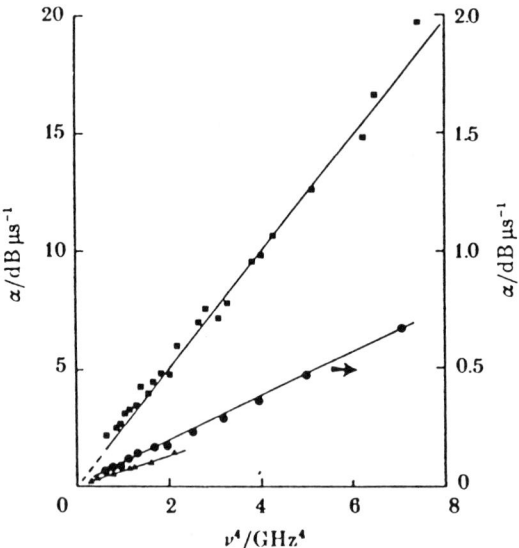

FIGURE 9.13. Measurements of the acoustic absorption for the ± 1 line as a function of frequency from 0.5 to 1 GHz. The plot against the fourth power of the frequency confirms the theoretical dependence on frequency. ■: HoVO$_4$, k-vector along [100], magnetic field $B \sim 10°$ from [100]. ●: HoVO$_4$, k-vector along [110], B along [100]. ▲: (0,1 Ho, 0.9 Y) VO$_4$, k-vector along [100], B along [110]. (From Ref. 9.18)

In Fig. 9.13 the experimental data confirm the dependence on the fourth power of NAR $\Delta m = \pm 1$ transitions for longitudinal waves propagated along both [100] and [110] axes. The variation of the intensity of absorption of $\Delta m = \pm 2$ transitions with ν^2 is similarly confirmed by experiments in which the frequency is varied from 0.8 to 1.6 GHz. Variation of absorption intensity with the inverse power of the temperature is also observed experimentally.

In Table 9.3, the experimental results for enhanced NAR of ^{165}Ho in HoVO$_4$ are presented in terms of the parameters D_1 and D_2 of (9.37) and (9.38). Numerical values represent absorption per meter, with frequency expressed in GHz. The values of D_1 in parentheses are estimated from the theoretically predicted ratio $D_1/D_2 = 0.0060$.

Table 9.3
Values of the Two Parameters D_1 and D_2 in (9.37) and (9.38)*

crystal	k-vector	D_1	D_2
HoVO$_4$	[100]	5.7	950
HoVO$_4$	[110]	(0.10)	17
(0.1 Ho, 0.9 Y)VO$_4$	[100]	(0.50)	83

*D_1 and D_2 represent the resonant absorption for longitudinal acoustic waves in concentrated and dilute HoVO$_4$. (From Ref. 9.18)

9.4 Effect of Demagnetization on NAR Line Shapes in Bulk Metals

The demagnetizing field of a bulk metal specimen which is used in many NAR studies, gives rise to a magnetic field at the nucleus that depends on the geometry and orientation of the sample. For substances with large bulk susceptibilities χ, the nonuniform demagnetization field may make a significant contribution to the local magnetic field variation over the spatial extent of the sample, thereby resulting in distortion of the observed magnetic resonance line shape. This demagnetization field effect is expected to be most marked in bulk metallic samples with large χ, such as Nb:H and W, both of which have a line shape that has been investigated in detail by NAR techniques. In Section 9.4.1, we outline a calculation of the demagnetization line shape for bulk specimens for the case of a paramagnetic cylinder magnetized along its symmetry axis. The most notable feature of the predicted resonance line is structure in the form of a satellite in the derivative line. The calculation takes into account the nonuniform radial distribution of acoustic energy in the sample and is generalized to apply to variable cylinder dimensions and orientation. Comparison of the demagnetizing field theory with experimental results for W is given in Section 9.4.2.

9.4.1 Calculation of the Demagnetization Effect

The microscopic magnetic field $\boldsymbol{H}_\mu$ observed by a nuclear spin at site $\boldsymbol{r}$ can be written

$$\boldsymbol{H}_\mu(\boldsymbol{r}) = \boldsymbol{B}_0 + \boldsymbol{H}_D + \frac{1}{3}(4\pi)\boldsymbol{M} + s\boldsymbol{M} \ . \qquad (9.39)$$

Assuming that the sample is homogeneous and has infinite spatial extent, the first term of (9.39), $\boldsymbol{B}_0(=\boldsymbol{H}_0)$, is the macroscopic magnetic field in the sample. The magnetic induction is $B = B_0 + 4\pi M$. The second term, H_D, is a correction to the field arising from the existence of boundaries and from nonuniformities in the sample (*e.g.*, if the sample is a powder). This term is called the *demagnetizing field* and will be discussed in detail below.

The first two terms together define the local, macroscopic magnetic field

$$H_a(r) = B_0 + H_D , \qquad (9.40)$$

which is an average over volumes large compared to the interatomic spacing. Nuclear spins sample the microscopic field, which fluctuates on an atomic scale because of the presence of more–or–less localizable magnetic dipoles. Traditionally, these dipoles are divided into two groups by an imaginary Lorentz sphere centered on the particular nuclear spin at r. The dipoles outside the sphere contribute the third term of (9.39); those inside contribute the fourth term. At sites of cubic symmetry, $s = 0$.

Inhomogeneous broadening of a magnetic resonance line results when any term of (9.39) varies spatially. Magnet inhomogeneity, for example, limits the uniformity of B_0. Inhomogeneity due to variation of sM and to broadening arising from discontinuities in M at grain boundaries in powder have been analyzed in the literature of NMR. Here we specifically consider variation of the term H_D across the macroscopic extent of the specimen.

Consider a sample with magnetization $M(r)$. The field distribution arising from $M(r)$ is equivalent to that from an Amperian current density,

$$J(r') = c\nabla \times M(r') , \qquad (9.41a)$$

whence

$$B(r) = \frac{1}{c}\int \frac{J(r') \times (r - r')}{|r - r'|^3} d^3 r' . \qquad (9.41b)$$

Alternatively, the field can be considered as arising from a bound pole density $\rho_b(r') = -\nabla \cdot M(r')$, in which case $H(r)$ is obtained from the volume integral

$$H(r) = \int \frac{\rho_b(r')(r - r')}{|r - r'|^3} d^3 r' . \qquad (9.41c)$$

B and H in (9.41) are departures from the applied field B_0. In a given situation, one or the other approach—(9.41b) or (9.41c)—may be preferred. For a paramagnet, the magnetization depends on the local macroscopic field $H_a(r)$ through

$$M(r) = \chi_v H_a(r) \simeq \chi_v B_0 = \text{constant vector}. \qquad (9.42)$$

Corrections are of order $\chi_v^2 B_0$, which is ordinarily less than 1 mG for most paramagnets in typical laboratory fields. When M is a constant within the sample, the volume integral reduces to a surface integral.

We specialize to the case of a right circular cylinder of radius a and length l, with cylinder axis $\equiv z$ axis. B_0 lies in the yz-plane at an angle θ to the cylinder axis (see Fig. 9.14). Assuming uniform magnetization, the current on the lateral cylindrical surface is

$$J_{\text{lat.}} = c\chi_v B_0(-\sin\theta\cos\phi\hat{z} - \cos\theta\sin\phi\hat{x} + \cos\theta\cos\phi\hat{y})\delta(\rho - a), \quad (9.43a)$$

and the currents on the faces are

$$J_{\text{face}} = \pm c\chi_v B_0 \sin\theta\hat{x}\,\delta\!\left(z \mp \frac{1}{2}l\right), \quad (9.43b)$$

where the upper sign applies to the face at $z = +\frac{1}{2}l$ and the lower sign to that at $z = -\frac{1}{2}l$. The radial and z integrals are calculated numerically, yielding H_μ at a fine array of points throughout the sample. A weighting function $\eta(H)$ is defined such that $\eta(H)\Delta H$ is proportional to the number of lattice sites at which the microscopic field H_μ falls within ΔH of H. This counting operation is the numerical equivalent of an integral over a δ function. In spherical coordinates,

$$\eta(H) = \int (r')^2 dr' \int \sin\theta' d\theta' \int d\phi'\,\delta[H - H_\mu(r',\theta',\phi')], \quad (9.44)$$

with appropriate limits.

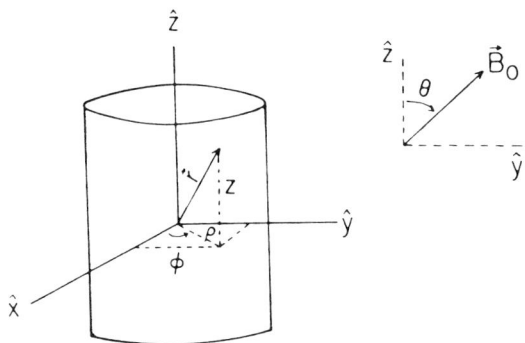

FIGURE 9.14. Coordinate system (x, y, z) and (ρ, ϕ, z) for a cylindrical sample. The orientation of B_0 in the yz-plane is shown. (From Ref. 9.19)

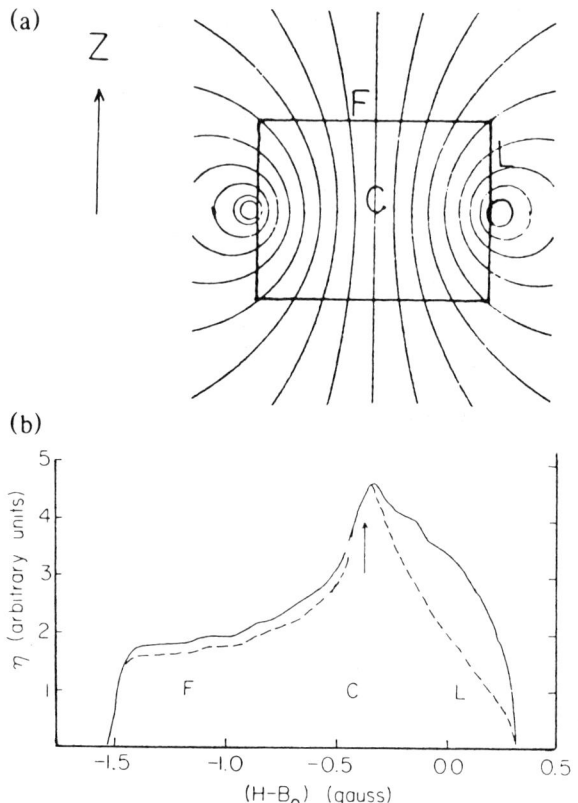

FIGURE 9.15. (a) Flux line diagram from a cylinder with axis parallel to B_0. Three regions are indicated: F(face), C(center) and L(lateral region). (b) Calculated weighting function η. Solid line: uniform acoustic energy density. Dashed line: acoustic energy density decreasing radially by a factor of 4. Contributions from regions F, C and L are indicated. (From Ref. 9.19)

For a right circular cylindrical sample oriented parallel to the magnetic field, with $\theta = 0$, $\eta(H_\mu)$ is shown in Fig. 9.15b. The vertical arrow in the figure is located at $H_\mu - H_0 = -0.38$ G, the value at the cylinder center C. To obtain η, the azimuthal integral may be calculated on a computer at an array of points spaced 1/100th of the sample radius apart. The complete cylindrical symmetry of this special orientation allows the calculation to be confined to the yz-plane. This situation is exactly equivalent to a solenoid of the same dimensions as the sample. The familiar flux-line sketch in

Fig. 9.15a shows the origins of the contributions to η. In the lateral region L, the field is stronger than at C, contributing to the high–field side of η. Near the faces F, the field is weaker contributing to the low–field shoulder.

The solid curve of Fig. 9.15b is obtained by weighting all lattice sites equally. The acoustic energy density for shear waves in the cylinder, however, is known to have a complicated radial distribution. For the ideal case of guided acoustic waves (transducer diameter and specimen diameter equal, with perfectly reflecting cylindrical surfaces) the distribution has been calculated and also experimentally measured by *McSkimin* [9.20]. The ideal guided wave case differs significantly, however, from the composite acoustic resonator utilized in NAR experiments: transducer diameter less than the specimen diameter, partially absorbing cylindrical surfaces, and with a propagation direction not always exactly coinciding with a pure mode direction. We therefore assume a less sophisticated model of a Gaussian radial energy distribution, which decreases by a factor of 4 from the center to the transducer edge. By weighting the contribution to η from each site accordingly, the dashed curve in Fig. 9.15b is obtained. The resonance line shape is given by the convolution

$$S(B_0) = \int_{-\infty}^{\infty} \Delta\alpha(H_\mu)\eta(H_\mu - B_0)dH_\mu , \qquad (9.45)$$

with $\Delta\alpha$ from (4.39).

9.4.2 ^{183}W NAR Line Shape

9.4.2.1 The ^{183}W Experiment

The transition metal tungsten is a particularly promising candidate for demagnetization studies. Its bulk susceptibility is $\chi_0 = 6.2 \times 10^{-6}$ cgs, and its small magnetic moment (0.1165 nuclear magnetons) allows use of high magnetic fields at moderate ultrasonic frequencies ($B_0 = 78.2$ kG corresponds to 14.0 MHz). For the effect of demagnetization on width to be observable, the nuclear resonance must have a very small intrinsic width. This is the case in tungsten, whose only magnetically active nucleus is the 14% abundant, spin $\frac{1}{2}$ ^{183}W nucleus.

The tungsten NAR experiment is performed with a marginal oscillator ultrasonic spectrometer at 78 K and 78.2 kG. With the use of a proton sample, the magnetic field can be shimmed to achieve a homogeneity better than 1 ppm over the sample volume. The single–crystal tungsten sample is a cylinder 0.615 cm long and 0.81 cm in diameter, with the crystallographic [110] axis 4° off the cylinder axis. A 0.64-cm-diameter quartz transducer bonded to the sample with silicone grease generates the 14–MHz shear waves.

The peak resonant attenuation with cylinder axis parallel to the applied field is $\simeq 1\times 10^{-8}$ cm^{-1}. The attenuation is measured by using the standard Smith–Sundfors calibrator and by comparing it with the observed nonresonant Alpher–Rubin signal. Synchronous detection and signal averaging are used to improve the signal-to-noise ratio. The trace shown in Fig. 9.16 is the sum of 83 sweeps with a three-second time constant, using a peak modulation width of 0.24 G at 37 Hz. The dominant feature in this first derivative trace has a peak-to-peak width of 0.45 G.

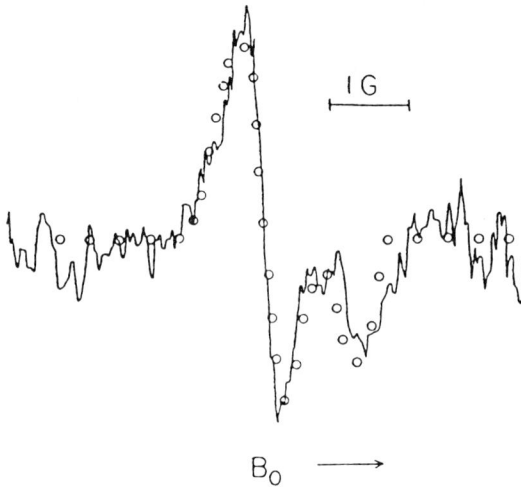

FIGURE 9.16. Solid line: experimental derivative curve of NAR attenuation by ^{183}W spins for 14.000 MHz shear waves (polarization parallel to [1$\bar{1}$0]), propagating along [110] in single-crystal tungsten. $B_0 = 78.21$ kG, $\theta = 0°$ and $T = 77.8$ K. Open circles: derivative line shape, $\Delta S/\Delta B_0$, calculated as described in text. (From Ref. 9.19)

The measured Knight shift is $(1.035\pm0.007)\%$. The field at which resonance occurs is measured by proton nuclear magnetic resonance (NMR). The tungsten shift is then calculated with respect to $\gamma/2\pi \equiv 177.160$ Hz/G. The cited uncertainty is an estimate of absolute error. The precision, as indicated by agreement between shifts calculated from separate traces, is substantially better.

9.4.2.2 Comparison with Theory

In the absence of inhomogeneous broadening, the shape of the resonant attenuation $\Delta\alpha(B_0)$ is expected to be a mixture of the real and imaginary parts of the dynamic susceptibility $\chi = \chi' + \chi''$; for dynamic Alpher–Rubin coupling in the high temperature limit, from (4.39), it should be

$$\Delta\alpha(B_0) \propto (1 - \beta^2)\chi''(B_0) - 2\beta\chi'(B_0) \ . \quad (9.46)$$

A value of $\beta = 0.04$ is expected under the cited experimental conditions. If we take χ'' to be Gaussian, for no value of β can we produce a line shape like that of Fig. 9.16. Because the only magnetically active tungsten nucleus is the 14% abundant ^{183}W, which has no quadrupole moment, the homogeneous width should be dipolar. Using the Van Vleck moments formalism and replacing the number of lattice sites N by $0.14N$ (= the number of ^{183}W nuclei), the predicted peak-to-peak width of a Gaussian line is $H_{pp} = 0.10$ G—less than the magnet inhomogeneity of 0.15 G. The hypothesis of a homogeneous line seems inadequate to explain either shape or width.

To apply the demagnetization theory of Section 9.4.2.1, as expressed in (9.45), we note that the NAR signal is observed with a synchronous detector. To account for finite audio modulation width, we assume square wave modulation of width H_m. Under these circumstances, the predicted detector output is

$$\frac{\Delta S}{\Delta B_0} = \frac{S(B_0 + H_m) - S(B_0 - H_m)}{2H_m} \ . \quad (9.47)$$

Figure 9.16 shows the calculated derivative line shape $\Delta S/\Delta B_0$ superimposed on the experimental resonance line. $\Delta S/\Delta B_0$ is obtained as described above, with $\beta = 0.04$, $H_{pp} = 0.15$ G (magnet inhomogeneity) and $H_m = 0.23$ G. Amplitude is adjusted for the best fit. The agreement is good on the down-field side and in the central section of the observed resonance curve. The up-field agreement is not as good, although the structure is qualitatively explained. Careful examination of several experimental traces reveals some shallow, broad structure in the wings of the observed resonance. This is reproducible but is not explained by the theoretical curve. (The dip at the right-hand end of the experimental trace is an artifact that does not reproduce.) If the distribution of acoustic energy is neglected, the shape is qualitatively the same, but the predicted width of the dominant structure is 50% wider than the measured width.

The effect of demagnetization is of particular interest in determining shifts (such as the Knight shift) in paramagnetic substances. The analysis

360   9. MAGNETIC MATERIALS

is simplest for infinitely long cylinders and infinitely thin discs. As a result of the calculation presented above, we are now in a position to examine the effect of the length-to-diameter ratio on the line shape.

Figure 9.17 shows the effect on line shape of cylinder length-to-radius ratio l/a for cylinder axis parallel to $\boldsymbol{B}_0$. The shapes are calculated from (9.45) and (9.39), using $\beta = 0$ and a Gaussian χ'' of width $H_{pp} = 0.1M$. Shapes are shown for two transducer radii, $a_t = 0.4a$ and $0.8a$. All nuclei with $\rho < a_t$ (*under* the transducer) are weighted equally, and those nuclei with $\rho > a_t$ are ignored (*i.e.*, the radial distribution of acoustic energy density is neglected). The vertical scales are adjusted so that all curves have the same amplitude. The horizontal scale is in units of the magnetization M. Shifts in the positions of the line peaks are suppressed.

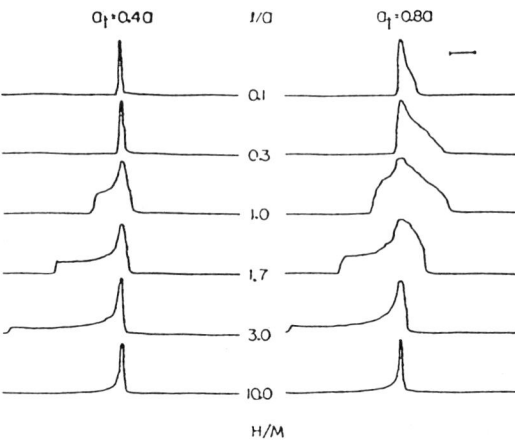

FIGURE 9.17. Calculated line profiles for several ratios of cylinder dimensions l/a. $\theta = 0°$. The scale of H is indicated at upper right in units of M. Profiles are shown for two values of transducer radius, $0.4\,a$ and $0.8\,a$. (From Ref. 9.19)

The absorption lines are narrowest for either very thin or very long samples. When l/a is small, the field varies most strongly in the radial direction, as indicated by the up–field shoulder in Fig. 9.17. In these cases, it is advantageous to use a small transducer. When l/a is large, the field varies mostly in the lengthwise direction, causing a down–field shoulder or tail that is not removed by using a smaller transducer. For $l/a = 1.0$ and $a_t = 0.8a$, considerable structure is seen on both sides of the central peak.

CHAPTER 9 REFERENCES

9.1 L. L. Buishvili, N. P. Giorgadze. "On the Nuclear Acoustic Resonance in Ferro- and Anti-ferromagnets." *Fiz. Tverd. Tela* **7**, 769-774 (1965); English trans: *Soviet Phys. Solid State* **7**, 614-617 (1965).

9.2 S. D. Silverstein. "Acoustic Nuclear Magnetic Resonance in Antiferromagnet Insulators." *Phys. Rev.* **132**, 997-1003 (1963).

9.3 P. A. Fedders. "Coupled Electronic Spins, Nuclear Spins, and Phonons in a Cubic Antiferromagnet." *Phys. Rev. B* **1**, 3756-3762 (1970).

9.4 K. N. Shrivastava, K. W. H Stevens. "Nuclear Acoustic Resonance in Ordered Magnetic Materials." *Phys. Soc. Lond., Jour. of Phys. C (Solid State Physics)* **3**, L64-L66 (1970).

9.5 J. B. Merry. "Nuclear Acoustic Resonance in Antiferromagnetic $RbMnF_3$.". Ph. D. Dissertation, Washington University, 1970 (unpublished).

9.6 J. B. Merry, D. I. Bolef. "Nuclear Acoustic Resonance of ^{55}Mn in Antiferromagnetic $RbMnF_3$." *Phys. Rev. B* **4**, 1572-1579 (1971).

9.7 J. B. Merry, D. I. Bolef. "Direct Detection of Nuclear Acoustic Resonance of a Magnetic Nucleus in an Antiferromagnet: ^{55}Mn in $RbMnF_3$." *Phys. Rev. Lett.* **23**, 126-128 (1969).

9.8 K. Walther. "^{55}Mn Nuclear Acoustic Resonance in MnTe." *Phys. Lett.* **32A**, 201-207 (1970).

9.9 K. Walther. "Magnetic Field Dependence of the ^{55}Mn Nuclear Acoustic Resonance in MnTe." *Phys. Lett.* **38A**, 149-150 (1972).

9.10 Kh. G. Bogdanova, V. A. Golenishchev-Kutuzov, F. S. Vagapova, A. A. Monakhov. "Double Acoustic Nuclear Antiferromagnet Resonance in $RbMnF_3$ and $KMnF_3$." *Zh. Eksp. Teor. Fiz.* **68**, 1834-1840 (1975); English trans: *Sov. Phys. JETP* **41**, 919-926 (1976).

9.11 R. A. Bagautdinov, Kh. G. Bogdanova, V. A. Golenishchev-Kutuzov, G. R. Enikeeva. "Acoustic NMR and Nonlinear Nuclear Spin Phenomena in Hematite." *Zh. Eksp Etor. Fiz.* **87**, 2008-2013 (1984); English trans: *Sov. Phys. JETP* **60**, 1158-1160 (1984).

9.12 P. Doussineau, A. Levelut, W. Schön, W. D. Wallace. "Nuclear Acoustic Resonance in an Insulating Manganese Spin-Glass." *Euro Phys. Lett.* **8**, 803-808 (1989).

9.13 J. B. Merry, P. A. Fedders, D. I. Bolef. "On the Coupling of Ultrasound to ^{19}F Nuclear Spins in Antiferromagnetic $RbMnF_3$." *Phys. Rev. B* **5**, 3506-3511 (1972).

9.14 P. A. Fedders. "Theory of Acoustic Resonance and Dispersion in Bulk Ferromagnets." *Phys. Rev. B* **9**, 3835-3844 (1974).

9.15 A. Abragam, B. Bleaney. *Electron Paramagnetic Resonance of Transition Ions.* (Claredon Press, Oxford 1970).

9.16 A. Abragam, B. Bleaney. "Enhanced Nuclear Magnetism: Some Novel Features and Prospective Experiments." *Proc. R. Soc. Lond. A* **387**, 221–256 (1983).

9.17 B. Bleaney, G. A. D. Briggs, J. F. Gregg, G. H. Swallow, J. M. R. Weaver. "Enhanced Nuclear Acoustic Resonance in $HoVO_4$." *Proc. R. Soc. Lond. A* **388**, 479–486 (1983).

9.18 B. Bleaney, G. A. D. Briggs, J. F. Gregg, C. H. A. Huan, I. D. Morris, M. R. Wells. "Further Studies of the Enhanced Nuclear Magnet $HoVO_4$ IV. The Resonant Magnetoacoustic Absorption." *Proc. R. Soc. Lond. A* **416**, 93–101 (1988).

9.19 G. Mozurkewich, H. I. Ringermacher, D. I. Bolef. "Effect of Demagnetization on Magnetic Resonance Line Shapes in Bulk Samples: Application to Tungsten." *Phys. Rev. B* **20**, 33–38 (1979).

9.20 H. J. McSkimin. "Propagation of Longitudinal Waves and Shear Waves in Cylindrical Rods at High Frequencies." *J. Acoust. Soc. Am.* **28**, 484–494 (1956).

CHAPTER 10
SQUID DETECTION OF NUCLEAR ACOUSTIC RESONANCE

10.1 Introduction
10.2 The SQUID Acoustomagnetic Effect
    10.2.1 Effects of External Flux
    10.2.2 Signal–to–Noise Ratio in RF SQUID Systems
    10.2.3 Use of DC SQUIDS
10.3 The SQUID Acoustomagnetic Spectrometer
10.4 Detection of Acoustic Composite Resonator Responses
10.5 Nuclear Spin–Phonon Interactions in Tantalum Metal
    10.5.1 CW SQUID–Detected NAR
    10.5.2 Transient SQUID–Detected NAR
        10.5.2.1 Spin–Lattice Relaxation Time
        10.5.2.2 Spin–Spin Relaxation Time
10.6 Nuclear Spin–Phonon Interactions in Antimony Metal
    10.6.1 CW SQUID–Detected ANQR
    10.6.2 Spin–Lattice Relaxation Times
    10.6.3 T_1 Determination by Adiabatic Fast Passage
10.7 SQUID–Detected Acoustic Magnetic Resonance: A Prognosis
References

10. SQUID DETECTION OF NUCLEAR ACOUSTIC RESONANCE

10.1 Introduction

An acoustic technique capable—as is conventional NAR—of detecting the resonant interaction of acoustic waves with nuclear spins is that based on the Superconducting Quantum Interference Device (SQUID) acoustomagnetic effect (AME), which relies upon the detection of changes in the component of magnetization parallel, to an applied, external magnetic field. This differs from conventional NAR, which relies directly upon changes in the acoustic absorption or acoustic dispersion of the acoustic waves introduced into the specimen. In SQUID AME, the acoustic waves generated by a transducer affect the magnetization of the sample via an internal spin–phonon interaction. The coupling of the elastic waves to the measured change in the susceptibility may be mediated by a number of effects: thermal, mechanical (*e.g.*, volume changes, distortion of the lattice), electronic and others. When transitions among the nuclear magnetic or nuclear electric quadrupole energy levels are induced by the acoustic energy, the allowed transitions are the same, whether the effects of the nuclear spin–phonon interaction are monitored directly by a change in acoustic absorption/dispersion (NAR) or by a change in the total z-component of magnetization induced (indirectly) by the acoustic waves. Analogous to the term SQUID–detected NMR, we may designate the use of SQUID AME to monitor nuclear spin–phonon interaction by the term SQUID–detected NAR.

Notable advantages of the SQUID AME technique are its high sensitivity and the ease with which transient, as well as steady–state phenomena, are accessed. Direct measurement of spin–lattice relaxation time T_1 is obtained by monitoring ΔM_z after the exciting energy is removed from the resonance line. ΔM_z decays to its equilibrium value with time constant T_1. Provided T_1 is longer than the acoustic decay time $T_{AC} = (\alpha v)^{-1}$, which is usually less than a msec, T_1 is measured directly from the decay. Detailed information about the shape of the decay is also obtained directly. The spin–spin relaxation time T_2 is determined from the measurement of ΔM_z as a function of the input acoustic power. To evaluate T_2, a knowledge of T_1 and the appropriate spin–phonon coupling parameters is required (S–tensor components for dynamic quadrupole coupling, electric conductivity for dynamic dipolar coupling, and the elastic strain magnitude for both). The maximum signal–to–noise ratio available with SQUID detection is independent of T_1 and T_2, making the technique especially useful for nuclear (or electron) spin systems with wide resonance line widths and long relaxation times. With respect to line widths, also, it is to be recalled (Section 4.3) that in the case of dynamic dipolar (A-R) coupling in metals, the signal is a mixture of χ' and χ'' in direct NAR detection; with SQUID detection, the signal is pure χ''. From this brief introduction, metallic spin systems not easily accessible to direct NAR should be measurable by SQUID–detected NAR: (a) nuclei with gyromagnetic ratios γ that are especially small (^{128}W, ^{109}Ag, ^{107}Ag, ^{197}Au, *etc.*); (b) metals, alloys, and metal hydrides in which the lines are broadened by quadrupolar or demagnetization effects (transition metal hydrides, Mg, W, *etc.*); (c) systems characterized by long spin–lattice relaxation times.

SQUID–detected NAR, in addition, avoids the rather stringent requirement present in most direct NAR experiments that the samples be prepared to exacting requirements of flatness and parallelism in order to form high-Q mechanical resonators. In some metals, inherently low acoustic Q's make direct detection difficult. Because the acoustic attenuation in metallic samples increases significantly below $T \simeq 40K$, further, the direct NAR technique often cannot take advantage of the T^{-1} increase in nuclear susceptibility. With SQUID-detected NAR, it is possible to use irregularly shaped samples that are not mechanically resonant. The acoustic Q is irrelevant, provided that sufficient acoustic energy is available to drive the spin system. The SQUID technique is also applicable to thin film samples evaporated onto the end of an acoustic buffer rod, on the other end of which a transducer is affixed.

10.2 The SQUID Acoustomagnetic Effect

It was proposed [10.1] in 1980 that the SQUID, which had for a decade or more been used to detect nuclear magnetic resonance, could be used to detect nuclear acoustic resonance. SQUID-detected NAR is based on the ability of the SQUID to detect the very small change in magnetization that occurs when resonant transitions cause a redistribution of spins among the energy levels. The SQUID relies upon the detection of changes ΔM_z in the component of magnetization parallel to the applied magnetic field. Thus it is sensitive to magnetic flux Φ itself, rather than (as in conventional NMR) to $d\Phi/dt$. When the SQUID is used, the rate at which ΔM_z changes is irrelevant; only the magnitude ΔM_z counts, except in T_1 measurements. The maximum magnetization available, when M_z is driven from its equilibrium value M_0 completely to saturation, is

$$\Delta M_z = M_0 = \chi_0 H_0 , \qquad (10.1)$$

where χ_0 is the static nuclear susceptibility, and M_0 is given by (1.11). The maximum change in magnetic flux through a loop of area A is thus

$$\Delta \Phi_{max} = (4\pi \chi_0 H_0)(A \eta t_x) , \qquad (10.2)$$

where η is the filling factor of order of magnitude $1/2$, and t_x is a flux transfer factor. An estimate can be made of the flux change Φ/Φ_0 expected at the RF SQUID for acoustic coupling to nuclear spins in several metals and two insulators. These are given in Table 10.1. The value of the fluxon Φ_0 is 2.068×10^{-7}G m^2. The ultimate rms noise in an RF SQUID is approximately $10^{-3}\Phi_0/\sqrt{\text{Hz}}$; thus the estimated values of Φ/Φ_0 in Table 10.1 should be observable. Presuming that M_z can be driven only 1% towards saturation, the signal-to-noise ratio for ^{183}W in a 1-Hz bandwidth, for example, should be nearly 10.

Table 10.1

Signal Φ_{max}/Φ_0 Available at the SQUID*

Spin System	f_0 in 10 kG (MHz)	χ_0 (cgs)	Φ_{max}/Φ_0
^{183}W metal	1.75	5.8×10^{-12}	0.094
^{197}Au metal	0.69	2.9×10^{-11}	0.46
^{207}Pb metal	8.9	1.1×10^{-10}	1.7
^{27}Al metal	11.09	1.8×10^{-8}	290.
^{27}Al$_2$O$_3$	11.09	7.0×10^{-9}	110.
^{29}SiO$_2$	8.46	1.9×10^{-11}	0.31

*From (10.2), assuming complete saturation, where $H_0 = 10$ kG, $A = 3.8$ cm^2, $\eta = 0.2$, and $t_x = 0.035$. (From Ref. 10.1)

Expression (10.1) represents the nuclear magnetization when all the spins contained in the sample are driven. The change in magnetization resulting from the acoustically induced transitions of some of the nuclear spins among some of the nuclear spin levels is the quantity detected in a SQUID AME experiment. The mechanism for coupling acoustic energy to the nuclear spin system remains the same as in an NAR experiment utilizing the same sample, e.g., dynamic quadrupole or dynamic dipole (Alpher–Rubin) interaction, as the case may be. As mentioned in the introduction, however, the magnetic effects (change in the magnitude of M_z) resulting (ultimately) from acoustic wave interaction with the nuclear spin system may, under certain circumstances, be mediated by other, even sequential, effects. An illustration of thermally mediated interactions is given in Section 10.4.

In metals, it is likely that the internal magnetic field created by the DC Alpher–Rubin mechanism contributes to the acoustomagnetic coupling. As discussed in Chapter 7, the DC Alpher–Rubin mechanism creates internal magnetic fields in the metal that are strongly dependent on the angle between the axial magnetic field and the (i) propagation direction, and (ii) polarization direction of the acoustic waves. We recall from Chapter 7 that the internal time–varying magnetic fields at low temperatures for longitudinal acoustic wave propagation are given by $\varepsilon H_0 \sin\theta$, where θ is the angle between the propagation direction and the external axial magnetic field. In both Cu and Ta, discussed below, the propagation direction is perpendicular to the magnetic field; thus the observation of acoustomagnetic effects in these bulk metals is consistent with ΔM_z changes due to the Alpher–Rubin internal magnetic fields. If so, measurements of ΔM_z in a known magnetic field can be used to give the mean strain ε for a particular power level in the sample. This, in turn, allows absolute determination of the transition probability in SQUID–detected NAR via the dynamic quadrupole coupling, as in the cases reviewed below of Ta (Section 10.5) and Sb (Section 10.6).

In the case of semiconductors, such as InP and GaAs—in both of which SQUID–detected, mechanical resonance acoustomagnetic effects have been observed—the acoustomagnetic interaction may be mediated by changes in donor electron concentration resulting from acoustically induced temperature changes. The InP and GaAs samples are characterized by large energy band gaps (approximately 1.5 eV), donor level concentrations of 10^{16} cm^{-3} and possible donor levels 0.01 to 0.02 eV below the conduction band. Since the number of donor atoms that have excess charge and the number of donor electrons in the conduction band are both strongly temperature–dependent near 4 K (where the data was taken), the observed AME responses may be due to changes with temperature of electron and donor atom susceptibility. As the acoustic frequency is varied to that corresponding to a mechanical resonance, the sample temperature increases; concomitantly, the donor electron concentration increases, and the number of donor atoms with excess charge decreases, resulting in a susceptibility increase that is reflected in the increase in magnetization detected at the SQUID.

10.2.1 Effects of External Flux

The use of the SQUID as a sensitive detector of magnetic flux is based on Josephson tunneling through a weak link between two superconductors. Of the two major types of SQUID magnetometers, the RF SQUID, formed by a superconducting ring containing one Josephson junction operated with an RF bias, was used by *Pickens* in the original SQUID detection of NAR [10.2]. The operating equation for the RF SQUID is

$$\Phi + LJ_c \sin\left(\frac{2\pi\Phi}{\Phi_0}\right) = \Phi_{\text{ext}}, \tag{10.3}$$

where Φ is the total flux through the hole in the superconducting ring, L is the self–inductance of the ring, J_c is the critical current in the ring, and Φ_{ext} is the applied external flux. For the case of a strong junction (the type of junction used in RF–biased SQUID magnetometers), $2\pi LJ_c/\Phi_0 > 1$. This case is illustrated in Fig. 10.1, where the total admitted flux Φ is plotted against the externally applied flux Φ_{ext} for the case $LJ_c = 4\Phi_0/2\pi$. The resulting curve for a strong junction is multivalued: as Φ_{ext} is increased from zero, Φ increases slowly until point B is reached. The jump from B to C corresponds to the increase of the fluxoid through the hole by one fluxon. Similar jumps occur as Φ_{ext} is further increased, *e.g.*, from point D to point

E. A reduction in Φ_{ext} from its value at point E to zero results in the flux following path EFGHA to zero. The different paths followed by increasing or decreasing flux are typical of hysterisis. This hysterisis only occurs in the strong junction case which is used, for example, in the RF-biased SQUID to load an RF-tank circuit.

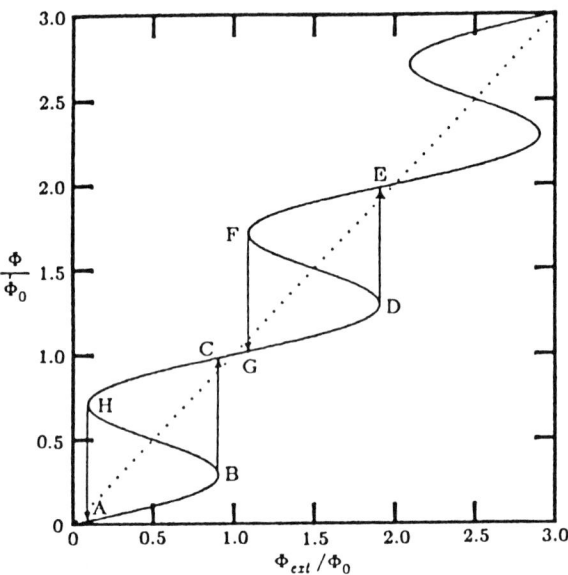

FIGURE 10.1. Total admitted magnetic flux versus externally applied flux for a strong Josephson junction. (Ref. 10.3)

A typical RF SQUID, thus, is formed by inductively coupling a superconducting ring containing a strong Josephson junction with an RF-tank circuit. Given an RF-voltage level high enough to drive the junction around a dissipative hysteresis loop, described above, the dissipation in the junction determines the Q of the RF-tank circuit, and, hence, the level of RF oscillation. As Φ_{ext} varies, therefore, the operating point of the system moves correspondingly, changing the RF-level. The RF level thus exhibits a periodicity corresponding to (10.3), resulting in a triangle–shaped dependence of the RF-level on applied external flux, with a period Φ_0, as shown in Fig. 10.2. Rounding of the maxima and minima occur because high frequency thermal fluctuations in the SQUID cause premature triggering of the quantum transitions.

10.2 THE SQUID ACOUSTOMAGNETIC EFFECT

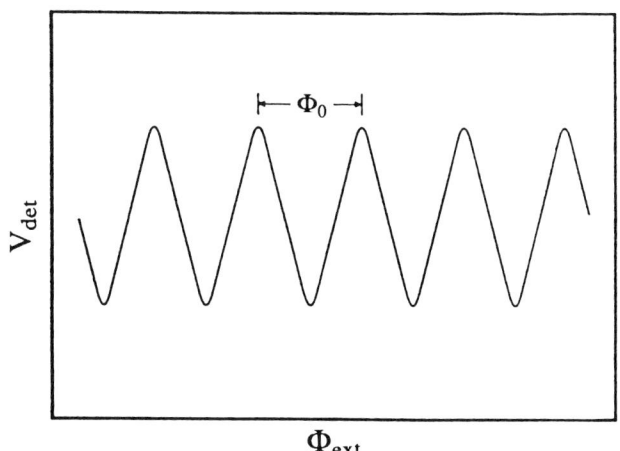

FIGURE 10.2. The "triangle" pattern observed in the RF voltage as a function of the applied external flux. (From Ref. 10.3)

10.2.2 Signal–to–Noise Ratio in RF SQUID Systems

The superconducting ring with a strong Josephson junction of

$$2\pi L J_c/\Phi_0 > 1 \ ,$$

described above, together with an RF-bias circuit, is used to measure $\Phi \ll \Phi_0$. This is accomplished by an arrangement termed a *flux-locked loop*, in which an audio-frequency magnetic field feedback loop is utilized to lock to an inflection point of the triangle pattern shown in Fig 10.2. (Experimental details of the flux-locked loop are given in the following section.) A commercial RF-biased SQUID system is capable of measuring $\Delta \Phi \simeq 10^{-3}\Phi_0$.

The SQUID, however, cannot discriminate the source of $\Delta\Phi$; thus microphonic variation in Φ due to the motion of parts of the apparatus relative to an inhomogeneous magnetic field is sensed, as well as the flux change of interest. The ultimate limit of flux sensitivity of a SQUID is determined by its inherent thermal noise. From Nyquist's theorem for a resistor R at temperature T, the mean square noise voltage is

$$\langle v_n^2 \rangle = 4k_B T R \Delta f \ , \tag{10.4}$$

where Δf is the frequency band width of the measuring instrument. The mean square noise current is

$$\langle i_n^2 \rangle = \frac{4k_B T \Delta f}{R} \ . \tag{10.5}$$

Since self-inductance is defined by $L = \Phi/i$, the mean square flux noise is

$$\langle \Phi_n^2 \rangle = \frac{4L^2 k_B T \Delta f}{R} . \tag{10.6}$$

For an RF–SQUID, $2\pi f = R/L$, where f is the RF frequency. Thus (10.6) becomes

$$\langle \Phi_n^2 \rangle = \frac{2L k_B T \Delta f}{\pi f_{RF}} . \tag{10.7}$$

For typical values of these parameters, one obtains

$$\langle \Phi_n^2 \rangle \simeq 2 \times 10^{-5} \Phi_0 \sqrt{\text{Hz}} .$$

This predicted noise level, however, is normally overwhelmed by contributions of the room temperature electronics; observed values are thus commonly higher:

$$\langle \Phi_n^2 \rangle \simeq 1 \times 10^{-3} \Phi_0 \sqrt{\text{Hz}} .$$

10.2.3 Use of DC SQUIDS

Recent work [10.4] in SQUID–detected NMR has emphasized the advantages of using DC SQUIDS rather than RF SQUIDS in resonance–related experiments. A review of and comparison between DC and RF SQUIDs has been given by *Clarke and Koch* [10.5]. Present commercially available DC SQUID systems may give improvement up to a factor of 100 in the speed of detection of ΔM_z and improvement up to 20 dB in sensitivity. The DC SQUID–based magnetic resonance spectrometer, like the RF SQUID–based AME spectrometer described in Section 10.3 below, relies on measurement of the longitudinal magnetization rather than of the transverse magnetization; the linear response of the flux-locked DC SQUID is thus capable of detecting signals over a frequency range from DC to tens of megahertz. The observed spectrometer noise, as in the case of the RF SQUID, is dominated by noise contributions from the room temperature electronics. The measured noise is on the order of $10^{-3} \Phi_0 / \sqrt{\text{Hz}}$.

10.3 The SQUID Acoustomagnetic Spectrometer

The SQUID acoustomagnetic spectrometer, shown diagrammatically in Fig. 10.3, may be divided into a data collection system and a control system. The data collection system consists of the SQUID susceptometer, flux transformer, a highly stable superconducting magnet, magnetic shielding and the RF acoustic driving system. The control system consists of a microcomputer, analog-digital and digital-analog interfaces and a digitally-controlled frequency synthesizer.

10.3 THE SQUID ACOUSTOMAGNETIC SPECTROMETER

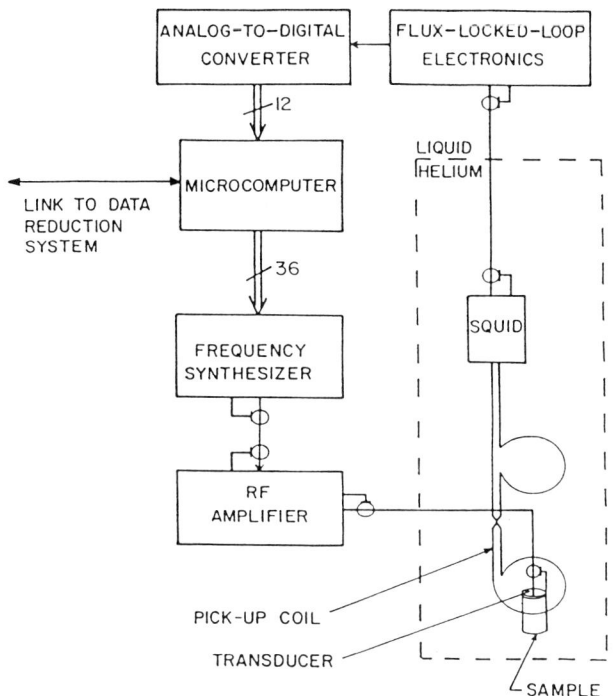

FIGURE 10.3. Schematic diagram of a SQUID acoustomagnetic spectrometer. The superconducting magnet, superconducting shields and complete flux–transformer circuit are not shown. (From Ref. 10.6)

The SQUID susceptometer consists of an RF–biased SQUID probe with flux–locked loop electronics. Magnetic flux present at the SQUID input is maintained at a single value by the SQUID control system, which incorporates a negative feedback system. When the magnetic flux couples to the SQUID from the sample changes, the SQUID control system compensates for this change by injecting more or less flux, thereby keeping the total flux at the SQUID locked to a single value. A voltage proportional to this compensation is output, reflecting the change in magnetization of the sample. The SQUID system has a full scale deflection of $\pm 5\Phi_0$ and an output of $2\ \text{V}/\Phi_0$ on the most sensitive scale [10.2,10.3,10.6,10.7].

The superconducting magnet utilized for the NAR experiments is designed for magnetic fields up to 1 T and to be stable to one part in 10^{10} per day. Single-strand TiNb superconducting wire with copper cladding is used to form the magnet winding. Each of the adjacent layers is separated by a 250μm thick sheet of oxygen-free copper. Copper cladding and the copper sheets keep the magnet in close contact with the liquid helium bath and thus help prevent local warming of the magnet. Although the magnet is designed to be homogeneous to one part in 10^4, in practice the presence of superconducting flux-transformer coils (described below) distort the magnetic field, making the actual homogeneity of the order of 1 part in 500.

The sample under investigation is located near the center of the superconducting magnet and inside one of the two astatically wound flux-transformer pick-up coils. The SQUID itself cannot be located next to the sample inside the magnet, since it will not function when exposed to large magnetic fields. The housing containing the SQUID is located outside the magnet, approximately 15 cm from the end of the magnet, where the magnetic field is small compared to that at the center. A change is magnetization in the sample is coupled to the SQUID via a magnetic flux-transformer circuit. One winding of the flux transformer consists of the two astatically wound pick-up coils and a third coil at the SQUID. All of the coils form a continuous superconducting circuit made of TiNb wire, as shown schematically in Fig. 10.3. Since the magnetic flux inside a closed loop of superconducting material remains constant as long as the critical current is not exceeded, a change in flux in one of the pick-up coils transfers a corresponding change in flux to the coil coupled to the SQUID. An expression for the amount of flux transferred to the SQUID, in terms of the inductances of the various coils, is given by the flux-transfer factor.

The magnet, pick-up coils and sample are all enclosed by a superconducting lead cylinder closed at one end. When superconducting, the lead shield provides magnetic shielding against extraneous magnetic fields. A lead shield and a niobium shield surround the SQUID itself to prevent extraneous magnetic fields from interacting directly with the SQUID.

The discussion in Section 10.2.2 of the effect of microphonic noise on the SQUID is to be recalled. Any movement of the pick-up coils with respect to an inhomogeneous magnetic field is seen as a flux change by the SQUID. It is therefore important to clamp tightly the pick-up coil form to the magnet form, thus minimizing the relative motion of the two. The sample probe is itself also tightly secured to the magnet form—again in an attempt to minimize the relative motion of the pick-up coils and sample. Not until these various pieces are anchored together, in the case of

10.3 THE SQUID ACOUSTOMAGNETIC SPECTROMETER

the Pickens spectrometer [10.3] or [10.6], is the microphonic noise reduced to a level which permits the observation of magneto–acoustic phenomena. The entire susceptometer is stationed on a stand weighing approximately 1000 kg. The stand, in turn, is isolated from the floor by springs, which serve to reduce the effects of building vibrations.

A typical data–acquisition system is also shown in Fig. 10.3. The frequency synthesizer, which provides the RF signal to drive the piezoelectric transducer bonded to the sample, is computer–controlled to sweep through a specified frequency range or to remain at a given frequency. Output of the frequency synthesizer, amplified by an RF amplifier, is sent through a section of adjustable transmission lines, enabling one to tune the spectrometer to the desired power level at a given frequency. A major problem is that one cannot keep the system tuned when frequency is being swept; this must be dealt with in the analysis of the data. A voltage proportional to the detected magnetization is put out by the SQUID system. This output voltage is fed to an analog–digital converter and stored digitally in computer memory. Signal averaging is possible by repeatedly sweeping through the region of interest and summing the data.

In order to study the SQUID–detected NAR signals one can vary frequency at fixed magnetic field; another way is to vary magnetic field for fixed frequency. A major disadvantage of the SQUID AME spectrometer described above is the inability to sweep magnetic field, since the changing flux due to the magnetic field change would couple into the SQUID, overwhelming any NAR signal or related acoustomagnetic effect. Further, the SQUID spectrometer baseline is not constant as the magnetic field changes value, since the flux–locked loop of the SQUID may lose its lock–point when the magnetic field is changed. Also, as mentioned above, in order to have spatially coherent acoustic radiation (so that acoustic waves have directional properties), it is necessary to be at the center frequency of a standing wave resonance. In order to distinguish SQUID–detected NAR signals, the following options are available:

(1) In SQUID–detected NMR experiments, the conventional high–Q RF coil is constructed such that $Q \simeq 1$; with frequency sweeping, an approximate SQUID–detected NMR line shape can be recorded. To obtain the proper H_1 for determining the SQUID–detected NMR signal, the amplitude of the RF voltage must be increased above that used for the high Q coil. Such an increase in RF power to an NMR resonance coil is not coupled to the NMR sample (I^2R heating) if the resonance coil is made superconducting. The equivalent in SQUID–detected NAR is the use of a non–mechanically resonant sample or one in which the mechanical Q has been intentionally re-

duced. Contrary to the NMR case, however, in a CW acoustic experiment reducing the Q while keeping the acoustic power fixed results in heating of the sample. The baseline between acoustic standing wave resonances is not flat, which results in a SQUID-detected NAR signal (as frequency is swept) with large distortion.

(2) One can set the frequency for the center of a standing wave resonance. For a long T_1, we can measure the SQUID-detected NAR signal at a fixed time after the acoustic pulse is turned off, and at the baseline of the recovery. The magnetic field is then changed and the SQUID system flux–locked loop returns to a lock point. At the new magnetic field, the SQUID-detected NAR signal is again measured for the same fixed time after the acoustic pulse turn–off, and again at the baseline. Doing this for a number of different fields will allow determination of the shape of the SQUID-detected NAR. For a short T_1, one measures the shape of a resonance by using a 100% amplitude modulation of the acoustic driving signal; after the change of H, the magnitude of the flux change associated with the amplitude modulation frequency can be determined as the output signal from a phase sensitive detector, which uses the modulation frequency as a reference frequency.

(3) To obtain a magnetic field change as in (2), an alternative is to use an additional superconducting coil to provide an incremental magnetic field on top of the magnetic field provided by the main superconducting coil. The current through this additional superconducting coil can be changed in small steps (not large enough to move the flux–locked loop away from its set–point). This system allows the NAR frequency to be set at the center of a standing wave resonance; and, under these conditions, the magnetic field changes should give accurate line shapes.

Considering option (3), since the changes in magnetic field needed to sweep over a typical NAR line may exceed the dynamic range of the SQUID system, the flux–locked loop will have lost its lock point when the field is changed. Modulating the acoustic drive, as in (b) above, is required to obtain a baseline. For T_1 long, a fixed point on the spin–lattice relaxation decay curve can be used to obtain the baseline. Some of the difficulties in implementing the stepped–field option are discussed by *Pickens* [10.3].

10.4 Detection of Acoustic Composite Resonator Responses

The SQUID acoustomagnetic technique can be used to display the acoustic responses of a composite resonator. A typical bulk sample has a finite concentration of paramagnetic impurities, which contribute a Curie–type paramagnetic susceptibility (χ_p) to the bulk susceptibility of the sample.

The paramagnetic susceptibility

$$\chi_p = Np^2 \left(\frac{\mu_B^2}{3k_B T}\right), \tag{10.8}$$

where

$$p = g[J(J+1)]^{\frac{1}{2}}$$

is the effective number of Bohr magnetons. N is the impurity concentration, μ_B is the Bohr magneton, g is the spectroscopic splitting factor, and J is the value of the azimuthal quantum number. When the sample temperature increases by ΔT, the corresponding magnetization change in the direction of the external field H is

$$\Delta M_z = -\frac{\chi_p H \Delta T}{T}. \tag{10.9}$$

An increase in temperature causes a negative M_z, consistent with the fact that SQUID-detected NMR or NAR corresponds to an increase of magnetization antiparallel to $\boldsymbol{H}$. An NMR magnetic field calibration sample in one of the two pick-up coils may be utilized to confirm that ΔM_z is negative upward.

A pattern of CW mechanical resonances is displayed by the SQUID AME spectrometer as the acoustic driving frequency is varied. Typical plots are shown in Fig. 10.4 for a composite resonator at 4 K consisting of a tantalum single crystal, a 7-MHz, X-cut quartz transducer and a bond of Nonaq stopcock grease. Whether ΔM_z corresponding to a mechanical resonance is negative- or positive-going depends on the time of sweep through the mechanical resonance, on the applied acoustic power and on the magnitude of the external magnetic field.

When the radial distribution of acoustic energy in the composite resonator is measured as a function of frequency, corresponding to a mechanical resonance, it is found to fall rapidly from a maximum value near the center to approximately 25% of the maximum value at R (πR^2 = the radiating area of the piezoelectric transducer) and approaches zero at $\simeq 1.3R$. The presence of this distribution of acoustic energy raises the temperature of the sample within the affected volume an amount ΔT compared to the sample boundary and the surrounding sample holder. The distinct behavior at frequencies corresponding to mechanical resonance is consistent with the spatial and temporal coherence that characterizes the acoustic radiation at a normal-mode frequency, in contrast to the spatial incoherence at frequencies removed from a normal-mode frequency.

When the acoustic frequency is shifted in the absence of NAR from the center of a mechanical resonance to an off-resonance position in μsec,

ΔM_z is measured to have a decay time of several seconds. This long decay suggests that the mechanism by which ΔM_z decays, as suggested above, is that of thermal conduction radially outward from the acoustic energy distribution to the sample boundary and sample holder. The mediated magnetization change is related to ΔT in the manner given in (10.9) above.

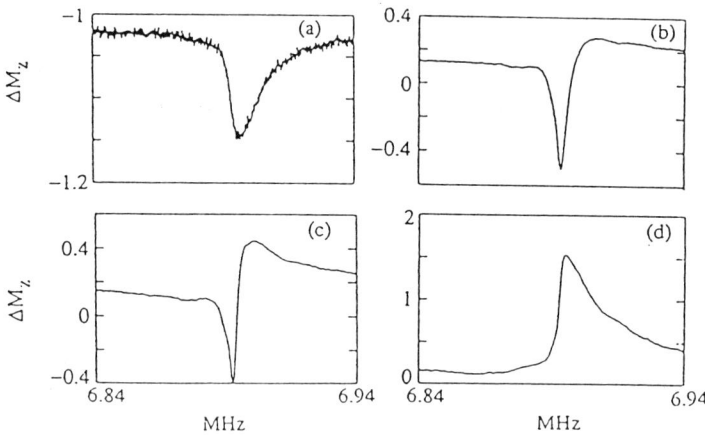

FIGURE 10.4. Plots of magnetization change versus acoustic frequency. RF power at the composite resonator is maintained constant, as is the frequency sweep range. (a) 195 G, 1 sec sweep, 2 sec delay, 300 sweep average; (b) 1046 G, 4 sec sweep, 5 sec delay, 100 sweep average; (c) 1046 G, 10 sec sweep, 5 sec delay, 30 sweep average; (d) 1046 G, 100 sec sweep, 10 sec delay, 3 sweep average. (From Ref. 10.8)

The signals observed in Figs. 10.4a–10.4d have been analyzed [10.8] by means of proposed competition among processes with the following characteristic times: (i) time of sweep through a fixed frequency range: 1 to 100 sec; (ii) thermal diffusion time in the sample: $\simeq 0.001$ sec; (iii) thermal conduction time between the sample thermal phonons and the liquid helium bath: $\simeq 24$ sec; (iv) time for 7 MHz acoustic phonons to decay into the thermal phonon spectra characteristic of the sample at a given temperature: $\simeq 0.15$ sec. Furthermore, as mentioned earlier, the observed signals depend upon the behavior of the (longitudinal) acoustic phonon distribution on and off the mechanical resonance. In the present instance, for example, the acoustic power in the sample is $\simeq 1000$ times greater at the center of the mechanical resonance than it is off the mechanical resonance.

In Figs. 10.4b–10.4d the low-frequency portion of the resonance response obeys a Lorentz line shape function with a half-width at half amplitude of

2.56 kHz; the high-frequency portion obeys an exponential decay function with time constant of 24 sec. In Figs. 10.4b and 10.4c, the center portion of the resonance response obeys an exponential decay with time constant of approximately 0.15 sec. In Fig. 10.4a, the low-frequency portion also obeys a Lorentzian line shape, with a half-width at half amplitude of 2.56 kHz; the high frequency exponential decay has a time constant of approximately 0.15 sec. As the sweep time is increases from Fig. 10.4a to 10.4c, the position of the minimum moves lower in frequency.

A plausible interpretation of these interesting acoustomagnetic effects observed with the SQUID follows. At the low-frequency end of the sweep in Fig. 10.4, the 7-MHz longitudinal phonons, which are spatially incoherent and fill the volume of the sample, decay into thermal phonons with a time constant of 0.15 sec. After many repetitive sweeps, with a delay between each sweep, the average sample temperature at the sweep beginning corresponds to a pseudo steady-state equilibrium between input acoustic power and heat energy per unit time transferred from the sample to its surroundings. As a particular sweep in frequency approaches a mechanical resonance center, the 7-MHz longitudinal acoustic phonon density becomes nonuniform, increasing density in the volume subsumed by the transducer, decreasing density elsewhere. For passage times through a mechanical resonance less than (or approximately the same as) the decay of 7-MHz longitudinal acoustic phonons to thermal phonons, the nonuniform 7-MHz acoustic phonon density results in an average temperature decrease of the sample, as in Fig. 10.4a. As the sweep frequency approaches the mechanical resonance center, a point is reached at which the spatially coherent 7-MHz acoustic phonons in the volume under the transducer achieve a high-enough density to decay into thermal phonons, which then diffuse rapidly through the sample volume, resulting in an increase of sample temperature. From this point in frequency, since the acoustic-phonon-to-thermal-phonon decay dominates, one observes an exponential 0.15 sec decay time. The qualitative model used in the analysis—0.15-sec acoustic-to-thermal-phonon decay time, 24-sec thermal conduction time, 1 msec thermal phonon diffusion time, and non-uniform 7-MHz acoustic phonon densities—do not, however, explain the observed line shape asymmetry in Figs. 10.4a–10.4c.

The energy in the 7-MHz acoustic-phonon standing-wave mode depends on the sweep time through the mechanical resonance. Thus, in Fig. 10.4b, the total sweep time of 4 sec permits the observation on the high frequency side of the resonance of the beginning of the sample's 24-sec thermal decay to the temperature of the liquid helium bath. The passage time through the mechanical resonance in Fig. 10.4d is 10 sec and the average energy at the mechanical resonance center is $\simeq 100$ times that in Fig. 10.4a. As a result of the slow sweep, the fast acoustic-phonon-to-thermal-phonon decay enables the average sample temperature to track the mechanical resonance as a positive temperature change up to the mechanical resonance center, at which point the 24-sec thermal decay become effective. The mechanical resonance center in Fig. 10.4d is 1 kHz higher in frequency than is the minimum in Fig. 10.4a.

The SQUID AME mechanical resonance responses shown in Fig. 10.4 for tantalum metal appear to characterize other solid samples. Similar line shapes are obtained, for example, for the insulating crystal MgO, for the single-crystal GaAs and for polycrystalline Cu. To describe acoustic responses, it is sufficient to specify the low-frequency half of the mechanical resonance structure, such as that shown in Fig. 10.4a. The passage time through a mechanical resonance then becomes much shorter than the time constant associated with the heat exchange with sample surroundings. In sum, measurements of CW acoustic absorption and dispersion can be carried out utilizing the very high sensitivity of the SQUID AME spectrometer.

Of equal interest, perhaps, the SQUID AME technique is applicable to the study of spatial coherence and incoherence in composite resonators. The impress of coherent and incoherent modes, indeed, is already evident in Fig. 10.4; in all four figures the line shape on the low-frequency side departs noticeably from a pure Lorentzian. The observed effect may be interpreted as due to a superposition of a broad mode (spatially incoherent) with a narrow standing-wave mode (spatially coherent).

10.5 Nuclear Spin–Phonon Interactions in Tantalum Metal

The SQUID AME technique has been used to study CW and transient nuclear spin–phonon interactions of ^{181}Ta nuclei in bulk tantalum metal at 4 K. Since NAR2 transitions are observed, the coupling is via the dynamic nuclear electric quadrupole interaction. In this case, the magnetization change can be written

$$\Delta M_z = \frac{\chi_0 H_0 \omega_1^2 T_1 T_2}{1 + \omega_1^2 T_1 T_1} , \qquad (10.10)$$

where ω_1^2 is the driving term for the dynamic quadrupole coupling. For the particular case of longitudinal acoustic wave propagation along a [110] direction in a cubic crystal, the magnetic field H_0 along the [1$\bar{1}$0] direction, and the frequency adjusted for $\Delta m = \pm 2$ and set at the center of a mechanical resonance, the driving term is

$$\omega_1^2 = \frac{1}{720} \frac{(I+1)^2(2I+3)}{(2I-1)\hbar^2} e^2 Q^2 S_{44}^2 \varepsilon^2 , \qquad (10.11)$$

where S_{44} is the appropriate S-tensor component and ε is the average acoustic strain amplitude in the sample. The strain amplitude can be approximated by

$$\varepsilon^2 = \frac{2P_0}{\alpha \rho v_a^3}, \qquad (10.12)$$

where P_0 is the incident acoustic power at the transducer per unit area and α is the sample background attenuation. For dynamic coupling with the frequency set at the center of a mechanical resonance at low temperatures, the appropriate driving term is

$$\omega_1^2 = \left(\frac{\gamma \varepsilon H_0}{2}\right)^2. \qquad (10.13)$$

In addition to the ^{181}Ta NAR2 in the bulk single crystal, when the frequency is adjusted between the mechanical resonances at constant magnetic field, ^{181}Ta $\Delta m = \pm 1$ resonance is observed at one-half the NAR resonant frequency. This $\Delta m = \pm 1$ resonance is expected to be that due to dynamic quadrupole coupling; our argument proceeds as follows:

(1) between mechanical resonances, the acoustic waves have spatial incoherence;
(2) the equations following Figs. 3.7 and 3.8 for longitudinal spatially coherent acoustic waves show that the transition probability for $\Delta m = \pm 1$ is zero for [110] propagation and $\boldsymbol{H}_0$ in the [1$\bar{1}$0], but a maximum when the angle between the [110] and $\boldsymbol{H}_0$ is 45° or when the angle between $\boldsymbol{H}_0$ and [001] is 45°; for spatially incoherent waves, on the other hand, there is a finite transition probability for those acoustic waves not traveling in either $\langle 001 \rangle$ or $\langle 110 \rangle$ directions;
(3) in tantalum single crystal, the magnetization change for dynamic quadrupole coupling is much larger than that for dynamic dipolar coupling [10.9];
(4) the $\Delta m = \pm 1$ magnetization change observed is too large to be explained by pick-up from the connecting wires to the piezoelectric transducer, which could result in NMR of ^{181}Ta in the skin depth of the tantalum sample.

10.5.1 CW SQUID–Detected NAR

The SQUID–detected CW NAR2 spectra is shown in Figs. 10.5 and 10.6 as a function of frequency at a constant magnetic field of 4.82 kG. The sweep period is 10 sec except in Fig. 10.5b, where it has been increased to 100 sec. To obtain the spectra shown, the sample, SQUID, pick–up coils and magnet are placed in a common liquid helium bath. The Ta crystal is annealed at ultra-high vacuum at temperatures approaching the melting temperature to remove interstitial impurities, including hydrogen. After the anneal, the end faces are reground flat and parallel to within 1.3μm, and hydrogen is again removed in a 1000 K vacuum annealing process. The end faces are within ±0.1° of the (110) planes, as measured by X-ray diffraction. Longitudinal acoustic waves are propagated along the [110] axis. The magnetic field is oriented along the [1$\bar{1}$0] direction.

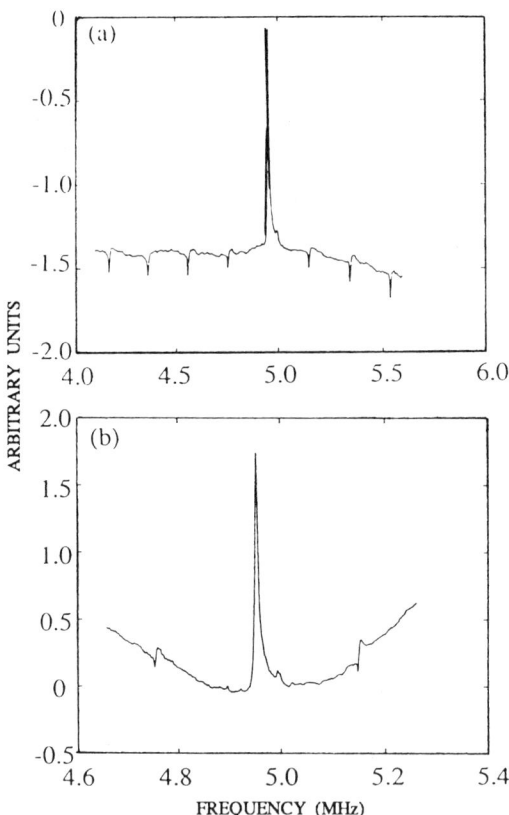

FIGURE 10.5. Negative-going comb of mechanical resonance responses, on which is superposed the positive-going enhanced ^{181}Ta NAR $\Delta m = \pm 2$ signal. (a) Sweep time 10 sec; (b) sweep time 100 sec. (From Ref. 10.10)

10.5 SPIN—PHONON INTERACTIONS IN TANTALUM

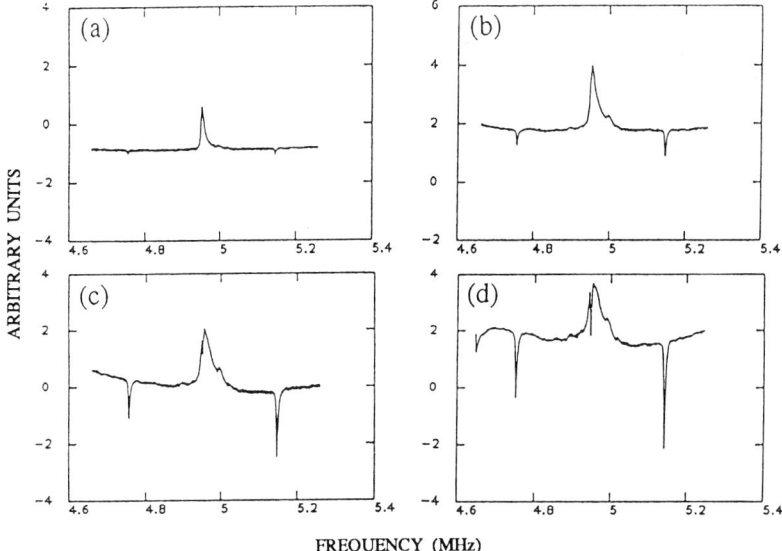

FIGURE 10.6. The enhanced ^{181}Ta NAR $\Delta m = \pm 2$ line, with accompanying mechanical resonance responses, as a function of acoustic power. Power level increases from (a) to (d): (a) −43 dB; (b) −36 dB; (c) −33 dB; (d) −31 dB. (Ref. 10.10)

The comb of narrow, negative–going peaks in Fig. 10.5a corresponds to the characteristic SQUID AME response of an acoustic composite resonator, described in Section 10.4 above. The negative–going (for tantalum at the sweep rates chosen) changes in the bulk susceptibility are due to temperature changes in the composite resonator induced by the CW standing acoustic waves. The positive–going peak at 5 MHz, however, is the SQUID–detected NAR2 resonance line of ^{181}Ta due to changes in the nuclear spin susceptibility caused by the acoustic waves. We observe that the positive peak is enhanced by the overlapping of a mechanical resonance peak with the CW NAR line. The overlapping is obtained by small adjustments in the magnetic field. The enhancement is achieved because of the increased acoustic energy available at a mechanical resonance frequency. Figure 10.5b, obtained using a slower sweep rate, shows the central structure in more detail: the positive–going enhanced ^{181}Ta NAR2 signal with the immediately adjacent, negative–going, SQUID–detected mechanical resonance responses.

Between Fig. 10.6a and Fig. 10.6d the acoustic power input is increased stepwise by a total of approximately 12 dB. The ^{181}Ta NAR signal (proportional to the change in nuclear spin susceptibility) increases in a positive sense, as power is increased until saturation occurs. The negative-going changes in the susceptibility—proportional to changes in temperature in the composite resonator as the acoustic energy density changes with frequency—corresponding to mechanical resonances, are measured to be more strongly dependent upon power; as a result, their magnitude increases more rapidly than the NAR signal. At some power level (*e.g.*, in Fig. 10.6c) the enhancement of the NAR signal that is induced by the high strains present at a mechanical resonance is overwhelmed by the competing, negative-going changes in the bulk susceptibility. Figures 10.6c and 10.6d are illustrative of this combination of bulk effect and the ^{181}Ta NAR signal. In Fig. 10.6, the NAR signal itself is not saturated.

The positive-going ^{181}Ta resonance line without enhancement is shown in Fig. 10.7. The magnetic field has been adjusted so that the NAR line falls between two mechanical resonances. The increase in RF power is 6 dB from (a) to (b) and 6 dB from (b) to (c). Figure 10.7d amplifies the central structure of Fig. 10.7c. The additional structure that appears on the NAR line is caused by slight variations in the coupling of the spin system that result from the frequency-dependent properties of the composite resonator.

The SQUID-detected NAR2 linewidths in Figs. 10.5–10.7 are consistent with the linewidths in the same Ta sample, reported in direct NAR experiments, described in Chapter 5.

Holding the magnetic field constant while varying the frequency, the SQUID technique—used to obtain the acoustomagnetic responses of Figs. 10.5, 10.6 and 10.7—results in a rather complex pattern of NAR resonances superposed on background acoustic resonant and nonresonant responses. In the event of the ability to rotate the sample, not available in the experiments of *Pickens et al.*, [10.2,10.7], one should be able to observe the additional NAR resonances—NAR(A-R) and NAR1—for the frequency position at the center of a mechanical resonances where the acoustic waves are spatially coherent. Indeed, there is no restriction in theory to the use of SQUID-detected NAR to observe nuclear spin–phonon effects via any one of the established coupling mechanisms (*e.g.*, in the case of Ta the dynamic dipole [Alpher-Rubin] and the dynamic electric quadrupole coupling). An alternative SQUID technique, as discussed above in Section 10.3, is to maintain the RF frequency fixed at a suitable mechanical resonance frequency, while the magnetic field is varied incrementally to display the spin–phonon responses. The use of this stepped field technique (i) would allow the operation of the system to be optimized for the highest required strain amplitude, (ii) would allow the use of only spatially coherent acoustic radiation and (iii) would minimize distortion of the CW NAR line shapes

10.5 SPIN—PHONON INTERACTIONS IN TANTALUM

by acoustic effects. When the sample has high acoustic absorption (*i.e.*, little or no measurable mechanical resonance response) as in Figs. 10.5–10.7, maintaining the magnetic field fixed and varying the signal frequency (the acoustic radiation is probably spatially incoherent) does not introduce the complications (or the enhancement) described above.

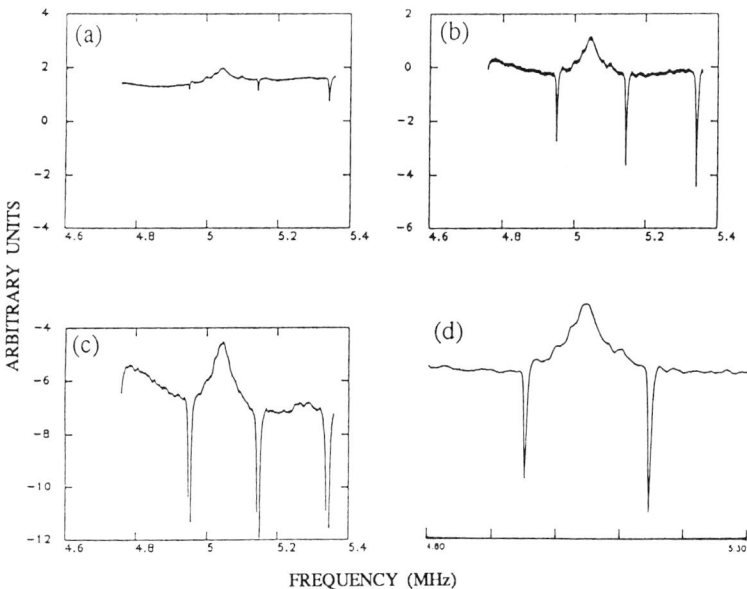

FIGURE 10.7. The enhanced ^{181}Ta NAR line without enhancement. RF power level increases 6 dB from (a) to (b), and 6 dB from (b) to (c). (d) amplifies the central structure of (c). (From Ref. 10.10)

10.5.2 Transient SQUID–Detected NAR

10.5.2.1 Spin–Lattice Relaxation Time

With the SQUID system, transient measurements of spin–lattice relaxation times T_1 are made by first applying acoustic power at a frequency corresponding to the center of the ^{181}Ta NAR2 line. On shifting frequency to a position off the NAR2 resonance, the resultant decay of the M_z component of the magnetization is monitored. (It is noted that the width in frequency of a mechanical resonance in the tantalum composite resonator is much smaller than the tantalum NAR line width.) Since the RF power level is approximately constant during this procedure, drift induced by thermal effects accompanying changes in power level is kept low. The acoustic ringdown after this frequency shift, both on and off a mechanical resonance, is less than 0.1 msec. (It is noted that this is approximately the estimated time of acoustic-to-thermal phonon decay—see Section 10.4.) The value of T_1 is the same whether the NAR $\Delta m = \pm 2$ line does or does not coincide with a mechanical resonance. A typical T_1 decay curve for ^{181}Ta at 1.6 K is shown in Fig. 10.8.

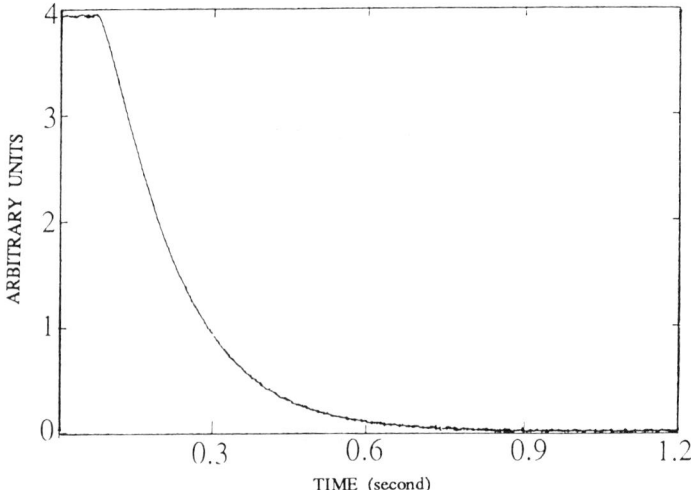

FIGURE 10.8. Typical T_1 decay curve for ^{181}Ta at 1.6 K. (From Ref. 10.2)

In Fig. 10.9, $1/T_1$ is plotted as a function of temperature in the range between 1.58 K and 4.2 K. In other metals with smaller nuclear electric quadrupole moments, $1/T_1$ approaches zero as $T \to 0$ K because of the domination of the conduction electron spin contribution to the nuclear relaxation rate. Figure 10.9 shows that the nuclear spin system of ^{181}Ta in tantalum metal has an experimental $1/T_1$ value that approaches a nonzero constant as $T \to 0$ K. An explanation of Fig. 10.9 is that for $T < 4.2$ K, a temperature-independent contribution, related to the extremely large ^{181}Ta nuclear electric quadrupole interaction, begins to dominate the conduction electron contribution to spin-lattice relaxation. The two contributions to this low temperature spin-lattice relaxation time can be written

$$\frac{1}{T_1} = \frac{T}{A} + B \ . \tag{10.14}$$

10.6 SPIN–PHONON INTERACTIONS IN ANTIMONY

A weighted least-squares fit to the data points enables the evaluation of the constants A and B in (10.14), where $A = 0.34 \pm 0.03$ sec K and $B = 2.7 \pm 0.5$ sec^{-1}. The temperature dependent relaxation rate due to dynamic coupling between ^{181}Ta nuclear spin and conduction electron spin gives $T_1 T = 0.34$ sec K, or $1/T_1 = 8.6 \pm 0.9$ sec^{-1} at 2 K. A temperature-independent relaxation rate of 2.7 sec^{-1} is also determined. By taking into account the large contribution of Raman–process relaxation at 78 K, the value of $1/T_1$ in Fig. 10.9 extrapolated to 78 K is consistent with the NMR measured value $1/T_1 = 670$ sec^{-1}.

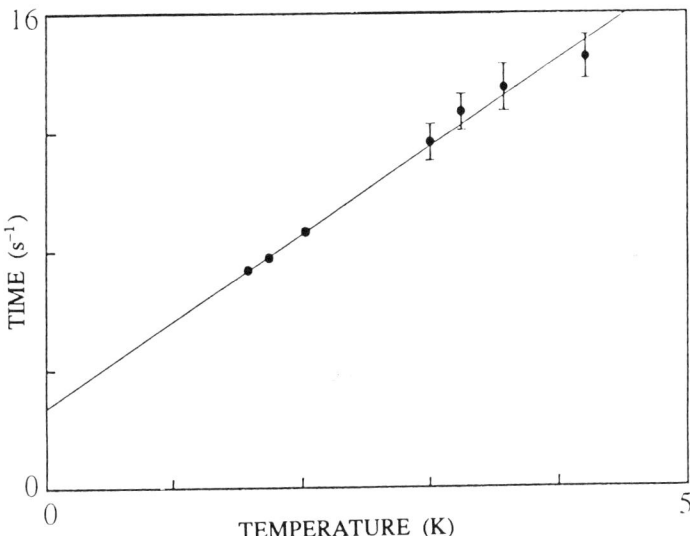

FIGURE 10.9. $1/T_1$ versus temperature for ^{181}Ta. Error bars on the three lower-temperature points are smaller than the dot size. $H_0 = 4.86$ kG. (From Ref. 10.2)

10.5.2.2 Spin–Spin Relaxation Time

The spin–spin relaxation time, T_2, is determined from the measurement of ΔM_z as a function of input acoustic power. Thus $1/T_2$ is determined if measurements are made of ΔM_z, $1/T_1$, the appropriate S–tensor components and the average acoustic strain in the sample.

10.6 Nuclear Spin–Phonon Interactions in Antimony Metal

A further application of the SQUID AME technique to NAR is the study of acoustic nuclear quadrupole resonance (ANQR) spectra and associated transient relaxation phenomena. An example is SQUID–detected ANQR and T_1 measurements of ^{121}Sb and ^{123}Sb in single–crystal antimony metal.

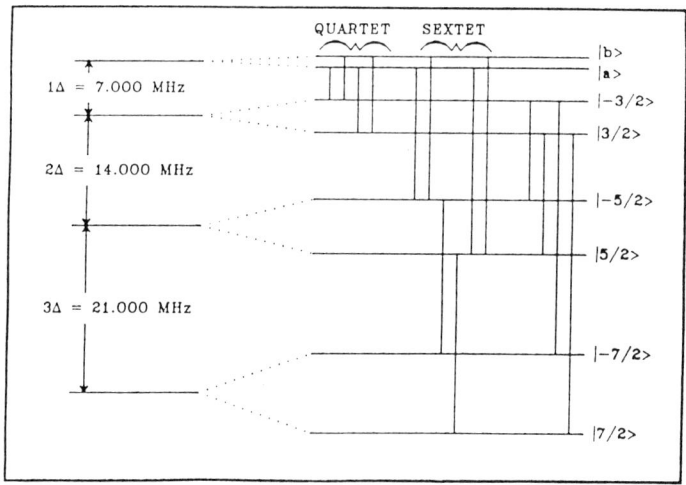

FIGURE 10.10. ^{123}Sb energy level configuration, with $H = 0$ (left) and $H = 500$ G (right). Spacings for $H = 500$ G are schematic. The transitions labelled *quartet* coincide with lines shown in Fig. 10.11. (From Ref. 10.7)

10.6.1 CW SQUID–Detected ANQR

Analogous to the quadrupole–shifted high–field energy level diagram of Fig. 3.2, transitions in the present instance occur between Zeeman–split pure quadrupole levels at low values of magnetic field. We choose to denote the transitions among the split levels, shown in Fig. 10.10, by the equivalent high–field energy changes, $\Delta m = \pm 1$ and $\Delta m = \pm 2$. Transitions between the energy levels may be observed and studied for either ^{121}Sb ($I = \frac{5}{2}$) or ^{123}Sb ($I = \frac{7}{2}$). Figure 10.10 represents the Zeeman–split pure quadrupole levels for ^{123}Sb for a magnetic field parallel to the trigonal axis; observed transitions are indicated by vertical lines. For an axially symmetric electric field gradient, as in antimony, and for the application of a weak ($\gamma \hbar H \ll e^2 qQ$) constant magnetic field $\boldsymbol{H}$ at an angle θ with respect to the EFG symmetry axis (z–axis), the energy levels for $m > \frac{1}{2}$ are given, analogously to (3.43b), by

$$E = \pm m\hbar\omega_0 \cos\theta = \frac{A}{4}[3m^2 - I(I+1)] , \qquad (10.15)$$

10.6 SPIN—PHONON INTERACTIONS IN ANTIMONY

where

$$A = \frac{e^2 qQ}{I(2I-1)} .$$

For weak magnetic fields, mixing between the originally unsplit energy levels is negligible except in the case of $m = \pm\frac{1}{2}$, in which zero–order mixing occurs. For magnetic fields ≤ 500 G, this approximation is valid; for higher fields, the approximation becomes poorer and must be replaced by an exact calculation of the intermediate field energy levels, as in the NAR study of Re (Section 3.6).

In Fig. 10.11 typical swept frequency spectra at 4 K are shown, corresponding to ANQR at 504 G for transverse waves between 6 MHz and 8 MHz that are propagated along the trigonal axis of the antimony sample. The observed ^{123}Sb resonance lines shown in Fig. 10.11 correspond to transitions labeled *quartet* in Fig. 10.10. Correspondence between the numbered spectra in Fig. 10.11 and the transitions of Fig. 10.10 is given in Table 10.2. The observed resonance lines have relatively high signal-to-noise ratios (for low magnetic fields). The sign of the observed changes in magnetization (ΔM_z) depends upon the sign change in m of the transition induced—an effect not observable by the non–SQUID-detected NAR or NMR techniques. The line widths at half amplitude are approximately 30 kHz for ^{121}Sb and 40 kHz for ^{123}Sb. The observed transitions between the levels $|a\rangle$ and $|b\rangle$, which are linear combinations of the Zeeman states $|+\frac{1}{2}\rangle$ and $|-\frac{1}{2}\rangle$, and the levels $|+\frac{3}{2}\rangle$ and $|-\frac{3}{2}\rangle$ are attributed to the dynamic quadrupole transitions $\Delta m = \pm 1$ and $\Delta m = \pm 2$. The magnetization changes for the observed transitions are proportional to RF driving power to the transducer, except at higher levels, where saturation occurs.

Table 10.2

Transitions for ^{123}Sb Quartet*

Figure 10.10	Figure 10.11
$\|-\frac{3}{2}\rangle - \|a\rangle$	1
$\|-\frac{3}{2}\rangle - \|b\rangle$	2
$\|+\frac{3}{2}\rangle - \|a\rangle$	3
$\|+\frac{3}{2}\rangle - \|b\rangle$	4

*(From Ref. 10.7)

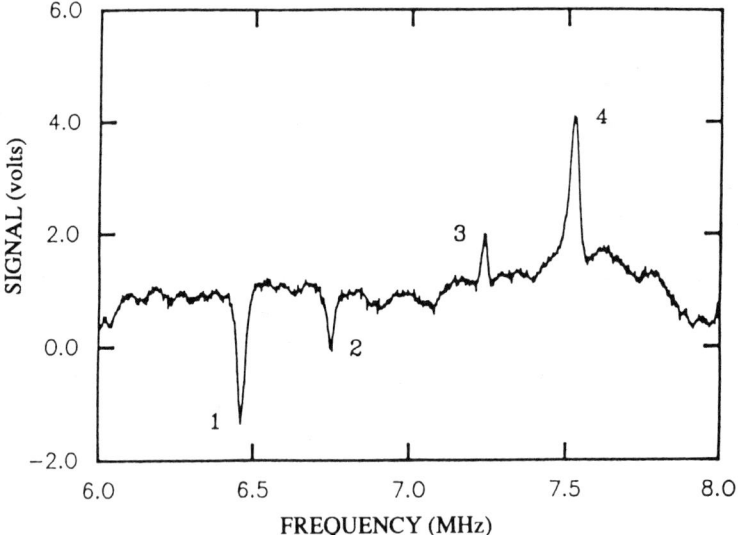

FIGURE 10.11. ^{123}Sb NAR quartet at $H = 504$ G for 7-MHz transverse waves with energy propagation at $28°40'$ (due to internal conical refraction) to the trigonal axis. Notation for transitions are as in Table 10.2. (From Ref. 10.7)

10.6.2 Spin–Lattice Relaxation Times

The transient SQUID technique has been used to measure T_1 in ^{121}Sb and ^{123}Sb for $\boldsymbol{H}$ parallel to the trigonal axis. Results for ^{123}Sb of T_1 at several values of H are given in Table 10.3. In Fig. 10.12 is plotted the nuclear spin relaxation rate $W_1(\equiv 1/T_1)$ as function of H^2. Following *Wolf* [10.11], the form of the field-dependent relaxation rate is

$$W_{1e} = W_0 \frac{H^2 + \delta H_L^2}{H^2 + H_L^2}, \qquad (10.16)$$

where W_0 is the relaxation rate at $H \geq H_L$. H_L is the effective local field. The factor δ is the ratio of the zero-field relaxation rate to the high-field relaxation rate. From an analysis of the fit of (10.16) to the data of Table 10.3, one obtains values of $\delta(^{123}Sb) = 1.49$ and of $H_L(^{123}Sb) = 410$ G. The solid curve in Fig. 10.12 represents the best fit of (10.16) to the data of Table 10.3.

10.6 SPIN—PHONON INTERACTIONS IN ANTIMONY

Table 10.3
Measured ^{123}Sb T_1 Values at 4.2 K for $\boldsymbol{H}$ Along the Trigonal Axis*

transition	frequency(MHz)	H(G)	T_1(sec)	W_1(sec^{-1})
$\|+\tfrac{3}{2}\rangle - \|b\rangle$	7.14	135	2.43	0.206
$\|+\tfrac{3}{2}\rangle - \|b\rangle$	7.16	147	2.48	0.202
$\|+\tfrac{3}{2}\rangle - \|b\rangle$	7.52	473	3.01	0.166
$\|+\tfrac{3}{2}\rangle - \|b\rangle$	7.56	504	2.93	0.171
$\|-\tfrac{3}{2}\rangle - \|b\rangle$	6.74	504	2.95	0.169
$\|+\tfrac{3}{2}\rangle - \|a\rangle$	7.26	504	2.86	0.175
$\|-\tfrac{3}{2}\rangle - \|a\rangle$	6.43	504	2.96	0.169
$\|-\tfrac{3}{2}\rangle - \|a\rangle$	5.36	1467	3.41	0.147

*The error is ±0.08 sec. (From Ref. 10.7)

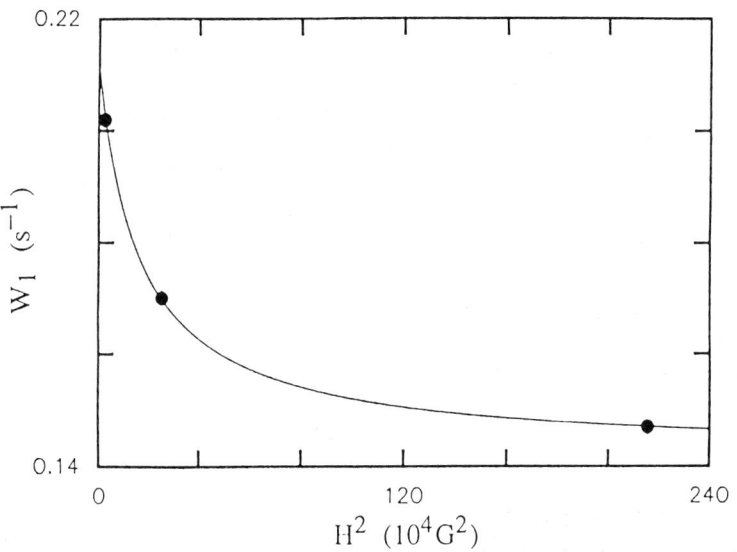

FIGURE 10.12. Magnetic field dependence of the nuclear spin relaxation rate W_1 of ^{123}Sb. The curve is the best fit of (10.16) to the data in Table 10.3. The plotted data points are averages of the W_1 data at each of three fields. (From Ref. 10.7)

10.6.3 T_1 Determination by Adiabatic Fast Passage

In Section 7.5.3, we give a brief discussion of acoustic adiabatic fast passage with SQUID–detection (SQUID AME AFP). We have seen in Section 10.5.2 how SQUID AME can be used to measure T_1's. For similar time ranges, SQUID AME AFP can also be used to measure T_1's, especially in the range of very long T_1's. Because of the distinctive adiabatic fast passage shape near the resonance center (see Fig. 7.20), SQUID AME AFP is also very useful in determining resonance center position for those spin systems in which T_1 is much larger than any frequency sweep time through the resonance.

10.7 SQUID–Detected Acoustic Magnetic Resonance: A Prognosis

SQUID–detected NAR is the most recent technique for coupling acoustic energy to nuclear spins in solids, particularly metals. It holds promise for being the most sensitive and flexible of the known nuclear spin–phonon measurement techniques. It bears reminding that SQUID–detected acoustic magnetic resonance (AMR), an inclusive term for SQUID–detected acoustic coupling to nuclear (NAR) and electron (Acoustic Paramagnetic Resonance = APR) spins, is not equivalent to conventional magnetic (NMR/EPR) resonance. The SQUID acoustic (NAR/APR) technique permits study of bulk metals, alloys and semiconductors; it also provides information complementary to NMR/EPR (whether conventional or SQUID–detected). For a bulk conductor, regardless of shape, one can utilize SQUID–detection at a particular temperature, magnetic field magnitude and angle to obtain all of the common magnetic resonance measurements: CW line shapes and line widths, relaxation times, quadrupole effects, Knight shifts, *etc.*

From the observed ΔM_z signal of ^{181}Ta in tantalum metal, one can estimate that the nuclear spin systems should be detectable by the SQUID technique for the case of dynamic nuclear electric quadrupole coupling when the product of

$$Q(1-\gamma_\infty)\frac{N}{N_{total}} > 0.1 \ ,$$

where Q is the quadrupole moment, γ_∞ is the Sternheimer antishielding factor appropriate for the nuclear spin system in the solid, and N/N_{total} is the fractional nuclear abundance of the nuclear spin system. Computation of the magnitude of ΔM_z for dynamic dipolar coupling yields a magnitude 10^{-3} of that found above for detectability of dynamic quadrupole coupling under similar conditions.

One may compute ΔM_z for both NAR(A-R) and NAR2 (with the assumption $\omega_1^2 T_1 T_2 \ll 1$) in the following way. The driving term for

10.7 SQUID—DETECTED AMR: A PROGNOSIS

NAR(A-R) is given by (10.13) as $\omega_1 = \gamma\varepsilon H_0/2$; the driving term for NAR2 for the case of acoustic shear waves propagated along a cube axis is

$$\omega_1^2 = \frac{1}{720}\frac{(I+1)^2(2I+3)}{(2I-1)}\frac{e^2 Q^2 S_{44}^2}{\hbar^2}\varepsilon^2 .$$

Substituting for ε from (10.12), one obtains for ΔM_z in the two cases

$$\Delta M_{z\,\text{A-R}} = \frac{\gamma^4 N T_1 T_2 I(I+1)}{\rho v^3}\left(\frac{H_0^3 \hbar^3}{6 k_B T}\right)\left(\frac{P_0}{\alpha}\right) \qquad (10.17)$$

and

$$\Delta M_{z\,2} = \frac{I(I+1)^3(2I+3)}{1080(2I-1)}\left(\frac{Q^2 S_{44}^2 T_1 T_2 N \gamma^2}{\rho v^3}\right)\frac{e^2 H_0}{k_B T}\left(\frac{P_0}{\alpha}\right). \qquad (10.18)$$

Substituting for known parameters in (10.17) and (10.18), one obtains the values given in Tables 10.4 and 10.5 for typical metal (and semiconductor) nuclei.

Table 10.4

Calculated Values of ΔM_z for NAR(A-R)*

Nucleus	$\Delta M_z [\times 10^{-12}(P_0/\alpha)]$
^{181}Ta	0.1
^{93}Nb	3.5
^{51}Va	2.0
^{63}Cu	3.2
^{27}Al	4.5

*(From Ref. 10.1)

Table 10.5

Calculated Values of ΔM_z for NAR 2*

Nucleus	$\Delta M_z [\times 10^{-9}(P_0/\alpha)]$
^{181}Ta	190
^{93}Nb	34
^{115}InAs	>10

*(From Ref. 10.1)

As mentioned in Chapter 5, the NAR line widths of some metals (*e.g.*, ^{183}W) have a large magnetic field–proportional contribution due to demagnetization effects. The true homogeneous line width is observable only in small magnetic fields; in these, however, the direct NAR signal-to-noise ratio is unacceptably small. For SQUID-detected NAR in such metals, on the other hand, the use of small magnetic fields is not necessarily restrictive. SQUID-detected NAR (as well as APR) at low magnetic fields lends itself to other applications:

(1) the methodical study of nuclear pure quadrupole resonance in metals and alloys;
(2) NAR in bulk conducting ferromagnets such as Fe, Ni and Co, in which temperature, for example, may be varied to display the resonance line;
(3) acoustic electron spin–phonon coupling, SQUID-detected APR, in conductors. In the majority of metals, electromagnetic conduction electron spin resonance is not observed, presumably because of extreme broadening by very fast spin–lattice relaxation. The implication of this rapid relaxation is that acoustic energy can be utilized efficiently to excite the spin transitions. A serious drawback, however, is that even in a sample with a narrow, electromagnetically excited line, one expects in an experiment to observe a line severely broadened by the motion of the conduction electrons ($v_{\text{Fermi}} \simeq 10^8$ cm/sec) relative to the acoustic wave ($v_{\text{AC}} \simeq 10^5$ cm/sec). In metals with finite conductivity, the electron motion is diffusive. The diffusive contribution to the line width Γ is given by $\Gamma = Dq^2$, where D is the electron diffusion coefficient, and q is the acoustic wave vector. For a 1-GHz experiment, a typical value of Γ is hundreds of MHz. Such a wide line would be difficult to attribute to a given mechanism. By using SQUID detection, however, a change in M_z will unambiguously identify the presence or absence of electron spin excitations.

A particularly promising possibility for observing APR in bulk metals is the *Fedders'* $\Delta m = 0$ resonance line [10.12], which is determined by the condition $\omega T_1 = 1$. This lends itself particularly well to investigation by the SQUID acoustomagnetic technique, since ω should be independent of magnetic field, and, for some values of T_1 in metals ($\simeq 10^{-7}$ sec at 4 K to $\simeq 10^{-10}$ sec at 300 K) ω falls in a particularly attractive frequency range ($\simeq 1$ to $\simeq 1000$ MHz). The SQUID acoustomagnetic spectrometer, indeed, has been used in $\Delta m = 0$ paramagnetic relaxation studies of Fe^{2+} ion in single crystal MgO [10.13].

CHAPTER 10 REFERENCES

10.1 G. Mozurkewich. "Acoustic Magnetic Resonance Investigations Utilizing Direct, Backward Wave, and SQUID Detection." Ph. D. Dissertation, Washington University, 1981 (unpublished). Also, written communications to D. I. Bolef on February 6, February 11 and March 2, 1980.

10.2 K. S. Pickens, G. Mozurkewich, D. I. Bolef, R. K. Sundfors. "SQUID Detection of Acoustomagnetic Effects: Nuclear Acoustic Resonance in Tantalum Metal." *Phys. Rev. Lett.* **52**, 156–159 (1984).

10.3 K. S. Pickens. "SQUID Detection of Acoustomagnetic Effects." Ph. D. Dissertation, Washington University, 1984 (unpublished).

10.4 C. Connor, J. Chang, A. Pines. "Magnetic Resonance Spectrometer with a DC SQUID Detector." *Rev. Sci. Instrum.* **61**, 1059–1063 (1990).

10.5 J. Clarke, R. H. Koch. "The Impact of High–Temperature Superconductivity on SQUID Magnetometers." *Science* **242**, 217–223 (1988).

10.6 M. R. Holland, R. K. Sundfors. "A SQUID–Based Acoustomagnetic Spectrometer." *J. de Phy. Colloque C10*, **46**, 779–782 (1985).

10.7 K. S. Pickens, D. I. Bolef, M. R. Holland, R. K. Sundfors. "Superconducting Quantum Interference Device Detection of Acoustic Nuclear Quadrupole Resonance of ^{121}Sb and ^{123}Sb in Antimony Metal." *Phys. Rev. B* **30**, 3644–3648 (1984).

10.8 D. I. Bolef, R. K. Sundfors. "Continuous–Wave Ultrasonics with Applications to Acoustic Magnetic Resonance: A Review." *Perspectives in Physical Acoustics*, 1–88, Ed. by Y. Fu, R. K. Sundfors, and P. Suntharothok (World Scientific, Singapore 1992).

10.9 P. A. Fedders, R. K. Sundfors. "Quadrupole Exchange Contributions to the ^{181}Ta Nuclear Acoustic Resonance Line Shapes in Pure Ta Single Crystal." *Phys. Rev. B* **19**, 1345–1350 (1979).

10.10 K. S. Pickens (unpublished).

10.11 D. Wolf. *Spin Temperature and Nuclear Spin–Relaxation in Matter* (Clarendon, London, 1979).

10.12 P .A. Fedders. "Resonant and Nonresonant Effects of Paramagnetic Spins on Acoustic Modes." *Phys. Rev.* **12**, 2045–2948 (1975).

10.13 R. K. Sundfors, M. R. Holland. "SQUID Acoustomagnetic Paramagnetic Relaxation Studies of MgO:Fe^{2+}." *J. de Phys. Colloque C10*, **46**, 783–785 (1985).

APPENDICES

A. Selected Physical Constants; Energy Conversion Factors

B. Properties of Selected Stable Nuclei; Pair Isotopes Suitable for Investigation of the Hexadecapole Interaction

C. S-Tensor Components for Various Crystal Structures

D. Dynamic Quadrupole Interaction: Transformation of the S-Tensor

E. T-Tensor Components for Cubic $\bar{4}$3m Symmetry

F. Dynamic Hexadecapole Interaction: Transformation of the T-Tensor

Table A–1
Selected Physical Constants

Quantity	Symbol and Definition	Value	Units
Planck's constant	h	6.626076 ×	$\begin{cases} 10^{-34} \text{ J-s} \\ 10^{-27} \text{ erg-s} \end{cases}$
$h/2\pi$	$\hbar$	1.054573 ×	$\begin{cases} 10^{-34} \text{ J-s} \\ 10^{-27} \text{ erg-s} \end{cases}$
Boltzmann constant	$k = R/N_A$	1.380658 ×	$\begin{cases} 10^{-23} \text{ J-K}^{-1} \\ 10^{-16} \text{ erg-K}^{-1} \end{cases}$
Speed of Light in Vacuum	c	2.997925 ×	$\begin{cases} 10^8 \text{ ms}^{-1} \\ 10^{10} \text{cm s}^{-1} \end{cases}$
Avogadro's Number	N_A	6.022137 ×	$\begin{cases} 10^{26} \text{ (kg-mol)}^{-1} \\ 10^{23} \text{ (g-mol)}^{-1} \end{cases}$
Elementary Charge	e		$\begin{cases} 1.602177 \times 10^{-19} \text{ C} \\ 4.803206 \times 10^{-10} \text{ esu} \end{cases}$
Electron Rest Mass	m_e	9.109390 ×	$\begin{cases} 10^{-31} \text{ kg} \\ 10^{-28} \text{ g} \end{cases}$
Bohr Magneton	$\mu_B = e\hbar/2m_e c$	9.274015 ×	$\begin{cases} 10^{-24} \text{ JT}^{-1} \\ 10^{-21} \text{ erg-G}^{-1} \end{cases}$
Nuclear Magneton	$\mu_N = e\hbar/2m_p c$	5.050787 ×	$\begin{cases} 10^{-27} \text{ JT}^{-1} \\ 10^{-24} \text{ erg-G}^{-1} \end{cases}$
Electron Magnetic Moment	μ_e	9.284770 ×	$\begin{cases} 10^{-24} \text{ JT}^{-1} \\ 10^{-21} \text{ erg-G}^{-1} \end{cases}$
Proton Magnetic Moment	μ_p	1.410608 ×	$\begin{cases} 10^{-26} \text{ JT}^{-1} \\ 10^{-23} \text{ erg-G}^{-1} \end{cases}$

Table A-2

Energy Conversion Factors

		J	erg	cm^{-1}	m^{-1}	K	Hz
1 J	=	1	10^7	5.0341×10^{22}	5.0341×10^{24}	7.2430×10^{22}	1.5092×10^{33}
1 erg	=	10^{-7}	1	5.0341×10^{15}	5.0341×10^{17}	7.2430×10^{15}	1.5092×10^{26}
1 cm^{-1}	=	1.9865×10^{-23}	1.9865×10^{-16}	1	10^2	1.4388	2.9979×10^{10}
1 m^{-1}	=	1.9865×10^{-25}	1.9864×10^{-18}	10^{-2}	1	1.4388×10^{-2}	2.9979×10^8
1 K	=	1.3807×10^{-23}	1.3807×10^{-16}	6.9501×10^{-1}	6.9501×10^1	1	2.0837×10^{10}
1 Hz	=	6.6261×10^{-34}	6.6261×10^{-27}	3.3356×10^{-11}	3.3356×10^{-9}	4.7993×10^{-11}	1

Table B-1

Properties of Selected Stable Nuclei

Isotope	Natural Abundance (atom %)	Spin Quantum (I)	Magnetic Moment (μ/μ_B)	Electric Q Moment ($e \times 10^{-24}$ cm^2)	NMR Frequency (MHz)
^{1}H	99.986	1/2	2.7928	-	42.5759
^{7}Li	92.70	3/2	3.2564	-0.0366	16.546
^{9}Be	100.	3/2	-1.1778	+0.053	5.9834
^{10}B	19.9	3	1.8007	+0.08472	4.5754
^{11}B	81.17	3/2	2.6886	+0.04065	13.660
^{14}N	99.635	1	0.4036	+0.0156	3.0756
^{19}F	100.	1/2	2.6287	-	40.0515
^{23}Na	100.	3/2	2.2175	+0.101	11.262
^{25}Mg	10.00	5/2	-0.85545	+0.22	2.6054
^{27}Al	100.	5/2	3.6414	+0.140	11.094
^{35}Cl	75.77	3/2	0.82187	-0.08249	4.1717
^{37}Cl	24.23	3/2	0.68412	-0.06493	3.472
^{39}K	93.258	3/2	0.39147	+0.049	1.9868
^{41}K	6.7302	3/2	0.21487	+0.060	1.0905
^{45}Sc	100.	7/2	4.75648	-0.22	10.343
^{47}Ti	7.3	5/2	-0.78848	+0.29	2.4000
^{49}Ti	5.5	7/2	-1.10417	+0.24	2.4005
^{51}V	99.750	7/2	5.1514	-0.052	11.19
^{53}Cr	9.501	3/2	-0.47454	0.022	2.4065
^{55}Mn	100.	5/2	3.4532	+0.40	10.501
^{57}Fe	2.2	1/2	0.0905	-	0.38
^{59}Co	100.	7/2	4.627	+0.404	10.054
^{61}Ni	1.13	3/2	0.75002	+0.162	3.8047
^{63}Cu	69.17	3/2	2.2233	-0.209	11.285
^{65}Cu	30.83	3/2	2.3817	-0.195	12.089
^{69}Ga	60.1	3/2	2.01659	+0.168	10.22
^{71}Ga	39.9	3/2	2.56227	+0.106	12.984
^{73}Ge	7.8	9/2	-0.87947	-0.173	1.4852
^{75}As	100.	3/2	1.43947	+0.29	7.2919

Table B-1 (continued)

Properties of Selected Stable Nuclei

Isotope	Natural Abundance (atom %)	Spin Quantum (I)	Magnetic Moment (μ/μ_B)	Electric Q Moment ($e \times 10^{-24}$ cm^2)	NMR Frequency (MHz)
^{79}Br	50.69	3/2	2.10639	+0.293	10.667
^{81}Br	49.31	3/2	2.27056	+0.27	11.498
^{83}Kr	11.55	9/2	-0.97066	+0.27	1.638
^{85}Rb	72.165	5/2	1.35303	+0.274	4.1108
^{87}Rb	27.835	3/2	2.75124	+0.132	13.931
^{87}Sr	7.00	9/2	-1.09282	0.15	1.8452
^{89}Y	100.	1/2	-0.13732	-	2.086
^{91}Zr	11.22	5/2	-1.30362	-0.21	3.9724
^{93}Nb	100.	9/2	6.1705	-0.36	10.407
^{95}Mo	15.92	5/2	-0.9142	-0.019	2.774
^{97}Mo	9.55	5/2	-0.9335	-0.102	2.832
^{99}Ru	12.7	5/2	-0.6413	+0.076	1.44
^{101}Ru	17.0	5/2	-0.7188	+0.44	2.1
^{103}Rh	100.	1/2	-0.0883	-	1.340
^{105}Pd	22.33	5/2	-0.642	+0.8	1.95
^{107}Ag	51.92	1/2	-0.11355	-	1.723
^{109}Ag	48.08	1/2	-0.13054	-	1.981
^{111}Cd	12.75	1/2	-0.59501	-	9.028
^{113}Cd	12.26	1/2	-0.62243	-	9.444
^{113}In	4.3	9/2	5.5289	+0.846	9.3099
^{115}In	95.7	9/2	5.5408	+0.861	9.301
^{117}Sn	7.51	1/2	-0.99983	-	15.17
^{119}Sn	8.45	1/2	-1.04621	-	15.87
^{121}Sb	57.3	5/2	2.5498	-0.20	10.189
^{123}Sb	42.7	7/2	2.81327	-0.26	5.5176
^{125}Te	6.98	1/2	-0.88715	-	13.45
^{127}I	100.	5/2	2.81327	-0.789	8.5183
^{129}Xe	26.44	1/2	-0.77686	-	11.78
^{131}Xe	21.18	3/2	0.69066	-0.120	3.4911

Table B-1 (continued)

Properties of Selected Stable Nuclei

Isotope	Natural Abundance (atom %)	Spin Quantum (I)	Magnetic Moment (μ/μ_B)	Electric Q Moment ($e \times 10^{-24}$ cm^2)	NMR Frequency (MHz)
^{133}Cs	100.	7/2	2.58202	-0.003	5.5847
^{135}Ba	6.592	3/2	0.83794	+0.18	4.2296
^{137}Ba	11.23	3/2	0.93736	+0.28	4.7315
^{139}La	99.91	7/2	2.7832	+0.22	6.0144
^{141}Pr	100.	5/2	4.136	-0.0589	12.5
^{143}Nd	12.18	7/2	-1.065	-0.484	2.315
^{145}Nd	8.30	7/2	-0.656	-0.253	1.42
^{147}Sm	15.0	7/2	-0.8148	-0.18	1.76
^{149}Sm	13.8	7/2	-0.6717	+0.052	1.40
^{151}Eu	47.8	5/2	3.4717	+1.15	10.559
^{153}Eu	52.2	5/2	1.5330	+2.94	4.6627
^{155}Gd	14.80	3/2	-0.2591	+1.59	1.6
^{157}Gd	15.65	3/2	-0.3398	+2.03	2.0
^{159}Tb	100.	3/2	2.014	1.18	9.66
^{161}Dy	18.9	5/2	-0.4805	+2.44	1.4
^{163}Dy	24.9	5/2	0.6726	+2.57	2.0
^{165}Ho	100.	7/2	4.173	+2.73	8.73
^{167}Er	22.95	7/2	-0.5665	+2.87	1.23
^{169}Tm	100.	1/2	-0.229	-	3.49
^{171}Yb	14.27	1/2	0.4930	-	7.51
^{173}Yb	16.12	5/2	-0.67989	+2.8	2.0659
^{175}Lu	97.41	7/2	2.2327	+5.68	4.86
^{176}Lu	2.59	6	3.19	+8.0	3.4
^{177}Hf	18.606	7/2	0.7935	+4.5	1.3
^{179}Hf	13.629	9/2	-0.6409	+5.1	0.80
^{181}Ta	99.988	7/2	2.370	+3.9	5.096
^{183}W	14.32	1/2	0.11722	-	1.75
^{185}Re	37.40	5/2	3.1871	2.36	9.5855

Table B–1 (continued)
Properties of Selected Stable Nuclei

Isotope	Natural Abundance (atom %)	Spin Quantum (I)	Magnetic Moment (μ/μ_B)	Electric Q Moment ($e \times 10^{-24}$ cm^2)	NMR Frequency (MHz)
^{187}Re	62.60	5/2	3.21970	2.24	9.6837
^{189}Os	16.1	3/2	0.65993	+0.91	3.3034
^{191}Ir	37.3	3/2	0.14610	0.78	0.7318
^{193}Ir	62.7	3/2	0.15910	0.70	0.7968
^{197}Au	100.	3/2	0.14816	0.594	0.7292
^{199}Mg	17.04	1/2	0.50270	-	7.612
^{201}Hg	13.22	3/2	-0.56023	+0.455	2.8099
^{203}Tl	29.46	1/2	1.61169	-	24.33
^{205}Tl	70.54	1/2	1.62754	-	24.57
^{207}Pb	22.62	1/2	0.58954	-	8.899
^{209}Bi	100.	9/2	4.1106	-0.46	6.8418

Table B-2

Paired Isotopes Suitable for Investigation of the Hexadecapole Interaction*

Nuclei	A	F	I	$\mu(\mu_N)$	Q (b)	$M^{(4)}(b^2)$	EC
^{47}Ti	7.75	2.400	5/2	-0.78848	l	*	$-3d^24s^2$
^{49}Ti	5.51	2.401	7/2	-1.10417	l	*	$-3d^24s^2$
^{95}Mo	15.78	2.774	5/2	-0.9142	l	*	$-4d^55s$
^{97}Mo	9.60	2.833	5/2	-0.9335	l	*	$-4d^55s$
^{99}Ru	12.81	1.9	5/2	-0.6413	ml	*	$-4d^75s$
^{101}Ru	16.98	2.1	5/2	-0.7188	ml	*	$-4d^75s$
^{113}In	4.16	9.310	9/2	5.5289	0.750	*	$-4d^{10}5s^25p$
^{115}In	95.84	9.329	9/2	5.5408	0.761	*	$-4d^{10}5s^25p$
^{121}Sb	57.25	10.19	5/2	2.5498	-0.53	*	$-4d^{10}5s^25p^3$
^{123}Sb	42.75	5.518	7/2	2.81327	-0.68	*	$-4d^{10}5s^25p^3$
^{143}Nd	12.20	2.72	7/2	-1.065	-0.57	~ 0.4	$-4f^46s^2$
^{145}Nd	8.30	1.7	7/2	-0.656	-0.30	~ 0.4	$-4f^46s^2$
^{147}Sm	15.07	1.5	7/2	-0.8148	0.72	~ 0.3	$-4f^66s^2$
^{149}Sm	13.84	1.2	7/2	-0.6717	0.72	~ 0.3	$-4f^66s^2$
^{151}Eu	47.77	10.49	5/2	3.4717	mh	~ 0.4	$-4f^76s^2$
^{153}Eu	52.23	4.638	5/2	1.5330	mh	~ 0.4	$-4f^76s^2$
^{161}Dy	18.73	1.2	5/2	-0.4805	h	~ 0.2	$-4f^96s^2$
^{163}Dy	24.97	1.6	5/2	0.6726	h	~ 0.2	$-4f^96s^2$
^{175}Lu	97.40	4.86	7/2	2.2327	5.7	~ 0.15	$-4f^{14}5d6s^2$
^{176}Lu	2.60	5.3	6	3.19	8.	~ 0.15	$-4f^{14}5d6s^2$
^{177}Hf	18.39	1.3	7/2	0.7935	3.	~ 0.05	$-4f^{14}5d^26s^2$
^{179}Hf	13.78	0.80	9/2	-0.6409	3.	~ 0.05	$-4f^{14}5d^26s^2$
^{185}Re	37.07	9.586	5/2	3.1871	2.8	~ 0.2	$-4f^{14}5d^56s^2$
^{187}Re	62.93	9.684	5/2	3.21970	2.6	~ 0.2	$-4f^{14}5d^56s^2$

*Nuclear abundance of the isotope in percent (A), resonance frequency in MHz in a magnetic field of 1 T (F), electronic configuration (EC), unavailability of the hexadecapole moment value (*), moderately low (ml), low (l) moderately high (mh), high (h).

APPENDIX C
S-Tensor Components for Various Crystal Structures

For the S-tensor relating the components of the electrostatic field gradient tensor to the elastic strain tensor components,

$$V_{ij} = S_{ijkl}\varepsilon_{kl} ,$$

where $ijkl$ take the values (1, 2 or 3), the following relationships are used in the computation of the S-tensor components given below (D. J. Barnes, Ph.D. dissertation, University of London, Imperial College, 1963 [unpublished]):

$$V_{ij} = V_{ji}$$
$$\varepsilon_{kl} = \varepsilon_{lk}$$
$$\sum_{i,j,k,l} V_{ii} = 0 .$$

The last equation above expresses the fact that the V_{ij} has a vanishing trace. This is the condition for purely ionic bonding, and the S-tensor components below are derived for this condition.

In addition, Voigt notation is employed so that the defining equation for the S-tensor becomes

$$V_a = S_{ab}\varepsilon_b ,$$

where

$$1 = xx \qquad 4 = yz(zy)$$
$$2 = yy \qquad 5 = zx(xz)$$
$$3 = zz \qquad 6 = xy(yx) .$$

For simplification in expressing the S-tensor results, $SP = S_{11} + S_{12}$ and $SM = S_{11} - S_{12}$.

Triclinic (class 1; 30 constants):

$$S_{\alpha\beta} = \begin{vmatrix} S_{11} & S_{12} & S_{13} & S_{14} & S_{15} & S_{16} \\ S_{21} & S_{22} & S_{23} & S_{24} & S_{25} & S_{26} \\ S_{31} & S_{32} & S_{33} & S_{34} & S_{35} & S_{36} \\ S_{41} & S_{42} & S_{43} & S_{44} & S_{45} & S_{46} \\ S_{51} & S_{52} & S_{53} & S_{54} & S_{55} & S_{56} \\ S_{61} & S_{62} & S_{63} & S_{64} & S_{65} & S_{66} \end{vmatrix} , \qquad (C-1)$$

with

$$S_{11} + S_{21} + S_{31} = 0$$
$$S_{12} + S_{22} + S_{32} = 0$$
$$S_{13} + S_{23} + S_{33} = 0$$
$$S_{14} + S_{24} + S_{34} = 0$$
$$S_{15} + S_{25} + S_{35} = 0$$
$$S_{16} + S_{26} + S_{36} = 0 \ . \tag{C-2}$$

Various substitutional patterns may be used for (C-2).

Monoclinic (classes 2, m and $2/m$; 16 constants):

$$S_{\alpha\beta} = \begin{vmatrix} S_{11} & S_{12} & S_{13} & 0 & S_{15} & 0 \\ S_{21} & S_{22} & S_{23} & 0 & S_{25} & 0 \\ S_{31} & S_{32} & S_{33} & 0 & S_{35} & 0 \\ 0 & 0 & 0 & S_{44} & 0 & S_{46} \\ S_{51} & S_{52} & S_{53} & 0 & S_{55} & 0 \\ 0 & 0 & 0 & S_{64} & 0 & S_{66} \end{vmatrix}, \tag{C-3}$$

with the relationships

$$S_{11} + S_{21} + S_{31} = 0$$
$$S_{12} + S_{22} + S_{32} = 0$$
$$S_{13} + S_{23} + S_{33} = 0$$
$$S_{15} + S_{25} + S_{35} = 0 \ . \tag{C-4}$$

Orthorhombic (classes 222, $mm2$ and mmm; 9 constants):

$$S_{\alpha\beta} = \begin{vmatrix} S_{11} & S_{12} & S_{13} & 0 & 0 & 0 \\ S_{21} & S_{22} & S_{23} & 0 & 0 & 0 \\ S_{31} & S_{32} & S_{33} & 0 & 0 & 0 \\ 0 & 0 & 0 & S_{44} & 0 & 0 \\ 0 & 0 & 0 & 0 & S_{55} & 0 \\ 0 & 0 & 0 & 0 & 0 & S_{66} \end{vmatrix}, \tag{C-5}$$

with

$$S_{11} + S_{21} + S_{31} = 0$$
$$S_{12} + S_{22} + S_{32} = 0$$
$$S_{13} + S_{23} + S_{33} = 0 \ . \tag{C-6}$$

Trigonal (classes 3 and $\bar{3}$; 10 constants):

$$S_{\alpha\beta} = \begin{vmatrix} S_{11} & S_{12} & -\tfrac{1}{2}S_{33} & S_{14} & -S_{25} & 2S_{62} \\ S_{12} & S_{11} & -\tfrac{1}{2}S_{33} & -S_{14} & S_{25} & -2S_{62} \\ -SP & -SP & S_{33} & 0 & 0 & 0 \\ S_{41} & -S_{41} & 0 & S_{44} & S_{45} & 2S_{52} \\ -S_{52} & S_{52} & 0 & -S_{45} & -S_{44} & 2S_{41} \\ -S_{62} & S_{62} & 0 & S_{25} & S_{14} & SM \end{vmatrix}$$

(C–7)

Trigonal (classes 32, 3m and $\bar{3}m$; 6 constants):

$$S_{\alpha\beta} = \begin{vmatrix} S_{11} & S_{12} & -\tfrac{1}{2}S_{33} & S_{14} & 0 & 0 \\ S_{12} & S_{11} & -\tfrac{1}{2}S_{33} & -S_{14} & 0 & 0 \\ -SP & -SP & S_{33} & 0 & 0 & 0 \\ S_{41} & -S_{41} & 0 & S_{44} & 0 & 0 \\ 0 & 0 & 0 & 0 & S_{44} & 2S_{41} \\ 0 & 0 & 0 & 0 & S_{14} & SM \end{vmatrix}$$

(C–8)

Tetragonal (classes 4, $\bar{4}$ and 4/m; 8 constants):

$$S_{\alpha\beta} = \begin{vmatrix} S_{11} & S_{12} & -\tfrac{1}{2}S_{33} & 0 & 0 & S_{16} \\ S_{12} & S_{11} & -\tfrac{1}{2}S_{33} & 0 & 0 & -S_{16} \\ -SP & -SP & S_{33} & 0 & 0 & 0 \\ 0 & 0 & 0 & S_{44} & 0 & 0 \\ 0 & 0 & 0 & 0 & S_{55} & 0 \\ S_{61} & -S_{61} & 0 & 0 & 0 & S_{66} \end{vmatrix}$$

(C–9)

Tetragonal (classes 4mm, $\bar{4}2m$, 422 and 4/mmm; 6 constants):

$$S_{\alpha\beta} = \begin{vmatrix} S_{11} & S_{12} & -\tfrac{1}{2}S_{33} & 0 & 0 & 0 \\ S_{12} & S_{11} & -\tfrac{1}{2}S_{33} & 0 & 0 & 0 \\ -SP & -SP & S_{33} & 0 & 0 & 0 \\ 0 & 0 & 0 & S_{44} & 0 & 0 \\ 0 & 0 & 0 & 0 & S_{55} & 0 \\ 0 & 0 & 0 & 0 & 0 & S_{66} \end{vmatrix}$$

(C–10)

APPENDIX C: S—TENSOR COMPONENTS

Hexagonal (classes 6, $\bar{6}$ and $6/m$; 6 constants):

$$S_{\alpha\beta} = \begin{vmatrix} S_{11} & S_{12} & -\frac{1}{2}S_{33} & 0 & 0 & 2S_{62} \\ S_{12} & S_{11} & -\frac{1}{2}S_{33} & 0 & 0 & -2S_{62} \\ -SP & -SP & S_{33} & 0 & 0 & 0 \\ 0 & 0 & 0 & S_{44} & S_{45} & 0 \\ 0 & 0 & 0 & S_{45} & S_{44} & 0 \\ S_{62} & S_{62} & 0 & 0 & 0 & SM \end{vmatrix}$$

(C-11)

Hexagonal (classes 622, $6mm$ and $\bar{6}m2$; 4 constants):

$$S_{\alpha\beta} = \begin{vmatrix} S_{11} & S_{12} & -\frac{1}{2}S_{33} & 0 & 0 & 0 \\ S_{12} & S_{11} & -\frac{1}{2}S_{33} & 0 & 0 & 0 \\ -SP & -SP & S_{33} & 0 & 0 & 0 \\ 0 & 0 & 0 & S_{44} & 0 & 0 \\ 0 & 0 & 0 & 0 & S_{44} & 0 \\ 0 & 0 & 0 & 0 & 0 & SM \end{vmatrix}$$

(C-12)

Isometric (classes 23 and m3; 3 constants):

$$S_{\alpha\beta} = \begin{vmatrix} S_{11} & S_{12} & -SP & 0 & 0 & 0 \\ -SP & S_{11} & S_{12} & 0 & 0 & 0 \\ S_{12} & -SP & S_{11} & 0 & 0 & 0 \\ 0 & 0 & 0 & S_{44} & 0 & 0 \\ 0 & 0 & 0 & 0 & S_{44} & 0 \\ 0 & 0 & 0 & 0 & 0 & S_{44} \end{vmatrix}$$

(C-13)

Cubic (classes $\bar{4}3m$ and $m3m$; 2 constants):

$$S_{\alpha\beta} = \begin{vmatrix} S_{11} & -\frac{1}{2}S_{11} & -\frac{1}{2}S_{11} & 0 & 0 & 0 \\ -\frac{1}{2}S_{11} & S_{11} & -\frac{1}{2}S_{11} & 0 & 0 & 0 \\ -\frac{1}{2}S_{11} & -\frac{1}{2}S_{11} & S_{11} & 0 & 0 & 0 \\ 0 & 0 & 0 & S_{44} & 0 & 0 \\ 0 & 0 & 0 & 0 & S_{44} & 0 \\ 0 & 0 & 0 & 0 & 0 & S_{44} \end{vmatrix}$$

(C-14)

Isotropic (1 constant):

$$S_{\alpha\beta} = \begin{vmatrix} S_{11} & -\frac{1}{2}S_{11} & -\frac{1}{2}S_{11} & 0 & 0 & 0 \\ -\frac{1}{2}S_{11} & S_{11} & -\frac{1}{2}S_{11} & 0 & 0 & 0 \\ -\frac{1}{2}S_{11} & -\frac{1}{2}S_{11} & S_{11} & 0 & 0 & 0 \\ 0 & 0 & 0 & \frac{3}{2}S_{11} & 0 & 0 \\ 0 & 0 & 0 & 0 & \frac{3}{2}S_{11} & 0 \\ 0 & 0 & 0 & 0 & 0 & \frac{3}{2}S_{11} \end{vmatrix}$$

(C–15)

APPENDIX D
Transformation of the S Tensor for Cubic $\bar{4}3m$ Symmetry

The S-tensor is defined in (3.47) through the relationship

$$V_{ij} = S_{ijkl}\,\varepsilon_{kl}\,, \qquad \text{(D-1)}$$

where $i,j,k,l = x,y,z$, and V_{ij} is the component of the electric field gradient tensor dependent on strain. S_{ijkl} is the defined fourth rank tensor, ε_{kl} is a component of the strain tensor, and x, y and z are mutually perpendicular directions, which we assume are along the axes of a cubic crystal. Summation over repeated indices is also assumed in (D-1). In terms of the Voigt notation, we show in Section 3.4 that (D-1) can be rewritten as

$$V_i = S_{ij}\,\varepsilon_j\,, \qquad \text{(D-2)}$$

where i and j now go from 1 to 6.

It is often necessary, however, to use coordinate systems that do not coincide with the cubic axes of the crystal. We write (D-2) in an orthogonal coordinate system, obtained by simple rotations from the coordinate system along the cubic axes, as

$$V'_l = S'_{lk}\,\varepsilon'_k \qquad l,k = 1,2,\cdots,6\,, \qquad \text{(D-3)}$$

where $x\prime$, y' and z' are the rotated axes. The relationship between the primed and unprimed strain tensor components can be written

$$\varepsilon_j = \alpha_{jk}\,\varepsilon'_k \qquad j,k = 1,2,\cdots,6\,; \qquad \text{(D-4)}$$

and between the primed and unprimed electric field gradient tensor components,

$$V'_l = \beta_{li}\,V_i \qquad l,i = 1,2,\cdots,6\,. \qquad \text{(D-5)}$$

By combining (D-2), (D-4) and (D-5), we find

$$V'_l = \beta_{li}\,V_i = \beta_{li}\,S_{ij}\,\varepsilon_j = \beta_{li}\,S_{ij}\,\alpha_{jk}\,\varepsilon'_k\,. \qquad \text{(D-6)}$$

If one compares (D-6) with (D-3), one finds

$$S'_{lk} = \beta_{li}\,S_{ij}\,\alpha_{jk}\,. \qquad \text{(D-7)}$$

In order to write (D-3) explicitly, one must first determine the transformation tensors whose components are α_{jk} and β_{li}. The primed and unprimed coordinate systems can be connected by the linear transformation

$$\begin{pmatrix} x' \\ y' \\ z' \end{pmatrix} = \begin{pmatrix} l_1 & m_1 & n_1 \\ l_2 & m_2 & n_2 \\ l_3 & m_3 & n_3 \end{pmatrix} \begin{pmatrix} x \\ y \\ z \end{pmatrix}, \quad \text{(D-8)}$$

where, for instance, $x' = l_1 x + m_1 y + n_1 z$. It is assumed that this transformation is unitary, and, therefore, the l, m and n are direction cosines.

To determine how the ε_j transform, it is convenient to define the *strain quadratic*. If $x + X$, $y + Y$ and $z + Z$ are the coordinates near point x, y, z in the crystal, then through this point there can pass one and only one quadratic surface of the family

$$\varepsilon_{xx} X^2 + \varepsilon_{yy} Y^2 + \varepsilon_{zz} Z^2 + \varepsilon_{yz} YZ + \varepsilon_{zx} ZX + \varepsilon_{xy} XY = C, \quad \text{(D-9)}$$

where C is a constant. Any one of these quadratics is called a strain quadratic; such a surface has the property that the reciprocal of the square of its central radius vector in any direction is proportional to the extension of a line in that direction due to the strain. Now, this property of the strain quadratic must be independent of the coordinate system in which it is described. Hence, in the primed system,

$$\varepsilon'_{xx} X'^2 + \varepsilon'_{yy} Y'^2 + \varepsilon'_{zz} Z'^2 + \varepsilon'_{yz} Y'Z' + \varepsilon'_{zx} Z'X' + \varepsilon_{xy} X'Y' = C. \quad \text{(D-10)}$$

The X's transform in the same way that the x's do. Therefore we can substitute in (D-8) to get

$$\begin{aligned} C = &\varepsilon'_{xx}(l_1 X + m_1 Y + n_1 Z)^2 + \varepsilon'_{yy}(l_2 X + m_2 Y + n_2 Z)^2 \\ &+ \varepsilon'_{zz}(l_3 X + m_3 Y + n_3 Z)^2 \\ &+ \varepsilon'_{yz}(l_2 X + m_2 Y + n_2 Z)(l_3 X + m_3 Y + n_3 Z) \\ &+ \varepsilon'_{zx}(l_3 X + m_3 Y + n_3 Z)(l_1 X + m_1 Y + n_1 Z) \\ &+ \varepsilon'_{xy}(l_1 X + m_1 Y + n_1 Z)(l_2 X + m_2 Y + n_2 Z), \end{aligned} \quad \text{(D-11)}$$

where the C's in (D-10) and (D-11) are the same. If we next set the left side of (D-9) to the right side of (D-11) and equate the coefficients of X^2, Y^2, XY and so forth, we obtain

$$\varepsilon_{xx} = l_1^2 \varepsilon'_{xx} + l_2^2 \varepsilon'_{yy} + l_3^2 \varepsilon'_{zz} + l_2 l_3 \varepsilon'_{yz} + l_3 l_1 \varepsilon'_{zx} + l_1 l_2 \varepsilon'_{xy}, \quad \text{(D-12)}$$

with similar expressions for ε_{yy} and ε_{zz}; and we obtain

$$\begin{aligned} \varepsilon_{yz} = &2 m_1 n_1 \varepsilon'_{xx} + 2 m_2 n_2 \varepsilon'_{yy} + 2 m_3 n_3 \varepsilon'_{zz} \\ &+ (m_3 n_2 + m_2 n_3) \varepsilon'_{yz} + (m_3 n_1 + m_1 n_3) \varepsilon'_{zx} \\ &+ (m_1 n_2 + n_1 m_2) \varepsilon'_{xy}, \end{aligned} \quad \text{(D-13)}$$

APPENDIX D: S—TENSOR TRANSFORMATION

with similar expressions for ε_{zx} and ε_{xy}. The tensor that transforms ε_k to ε_j, as per (D–4), is given in Table D–1.

Table D–1

A Representation of the Equation $\varepsilon_j = \alpha_{jk}\varepsilon'_k$*

	ε'_{xx}	ε'_{yy}	ε'_{zz}	ε'_{yz}	ε'_{zx}	ε'_{xy}
ε_{xx}	l_1^2	l_2^2	l_3^2	$l_2 l_3$	$l_3 l_1$	$l_1 l_2$
ε_{yy}	m_1^2	m_2^2	m_3^2	$m_2 m_3$	$m_3 m_1$	$m_1 m_2$
ε_{zz}	n_1^2	n_2^2	n_3^2	$n_2 n_3$	$n_3 n_1$	$n_1 n_2$
ε_{yz}	$2m_1 n_1$	$2m_2 n_2$	$2m_3 n_3$	$m_2 n_3 + m_3 n_2$	$m_1 n_3 + m_3 n_1$	$m_1 n_2 + m_2 n_1$
ε_{zx}	$2n_1 l_1$	$2n_2 l_2$	$2n_3 l_3$	$n_2 l_3 + n_3 l_2$	$n_1 l_3 + n_3 l_1$	$n_1 l_2 + n_2 l_1$
ε_{xy}	$2l_1 m_1$	$2l_2 m_2$	$2l_3 m_3$	$l_2 m_3 + l_3 m_2$	$l_1 m_3 + l_3 m_1$	$l_1 m_2 + l_2 m_1$

*In Voigt notation, where $j, k = 1, 2, \cdots, 6$.

In order to obtain the tensor components β_{li} given in (D–5), we make use of the fact that the inverse of a real orthogonal transformation is its transpose. Therefore the relationship that allows the unprimed coordinates to be expressed in terms of the primed coordinates is the following:

$$\begin{pmatrix} x \\ y \\ z \end{pmatrix} = \begin{pmatrix} l_1 & l_2 & l_3 \\ m_2 & m_2 & m_3 \\ n_1 & n_2 & n_3 \end{pmatrix} \begin{pmatrix} x' \\ y' \\ z' \end{pmatrix}. \quad \text{(D–14)}$$

By the laws of partial differentiation,

$$\frac{\partial V}{\partial x'} = \frac{\partial V}{\partial x}\frac{\partial x}{\partial x'} + \frac{\partial V}{\partial y}\frac{\partial y}{\partial x'} + \frac{\partial V}{\partial z}\frac{\partial z}{\partial x'}.$$

But from the relationship above between unprimed and primed coordinates,

$$\frac{\partial x}{\partial x'} = l_1 \qquad \frac{\partial y}{\partial x'} = m_1 \qquad \frac{\partial z}{\partial x'} = n_1.$$

Therefore

$$\frac{\partial V}{\partial x'} = l_1\frac{\partial V}{\partial x} + m_1\frac{\partial V}{\partial y} + n_1\frac{\partial V}{\partial z}.$$

In a similar way,

$$\frac{\partial V}{\partial y'} = l_2\frac{\partial V}{\partial x} + m_2\frac{\partial V}{\partial y} + n_2\frac{\partial V}{\partial z}$$

and

$$\frac{\partial V}{\partial z'} = l_3\frac{\partial V}{\partial x} + m_3\frac{\partial V}{\partial y} + n_3\frac{\partial V}{\partial z}.$$

We repeat this procedure for the second derivatives, obtaining

$$\frac{\partial^2 V}{\partial x'^2} = \left(l_1\frac{\partial}{\partial x} + m_1\frac{\partial}{\partial y} + n_1\frac{\partial}{\partial z}\right)\left(l_1\frac{\partial V}{\partial x} + m_1\frac{\partial V}{\partial y} + n_1\frac{\partial V}{\partial z}\right),$$

with similar expressions for $\partial^2 V/\partial y'^2$ and $\partial^2 V/\partial z'^2$; and

$$\frac{\partial^2 V}{\partial y' \partial z'} = \left(l_2\frac{\partial}{\partial x} + m_2\frac{\partial}{\partial y} + n_2\frac{\partial}{\partial z}\right)\left(l_3\frac{\partial V}{\partial x} + m_3\frac{\partial V}{\partial y} + n_3\frac{\partial V}{\partial z}\right),$$

with similar expressions for $\partial^2 V/\partial z'\partial x'$ and $\partial^2 V/\partial x'\partial y'$. The tensor that transforms V_i to V_l' is given in Table D-2.

Table D-2

A Representation of the Equation $V_l' = \beta_{li} V_i$ *

	V_{xx}	V_{yy}	V_{zz}	V_{yz}	V_{zx}	V_{xy}
V_{xx}'	l_1^2	m_1^2	n_1^2	$2m_1n_1$	$2n_1l_1$	$2l_1m_1$
V_{yy}'	l_2^2	m_2^2	n_2^2	$2m_2n_2$	$2n_2l_2$	$2l_2m_2$
V_{zz}'	l_3^2	m_3^2	n_3^2	$2m_3n_3$	$2n_3l_3$	$2l_3n_3$
V_{yz}'	l_2l_3	m_2m_3	n_2n_3	$m_2n_3 + m_3n_2$	$n_2l_3 + n_3l_2$	$l_2m_3 + l_3m_2$
V_{zx}'	l_3l_1	m_3m_1	n_3n_1	$m_1n_3 + m_3n_1$	$n_1l_3 + n_3l_1$	$l_1m_3 + l_3m_1$
V_{xy}'	l_1l_2	m_1m_2	n_1n_2	$m_1n_2 + m_2n_1$	$n_1l_2 + n_2l_1$	$l_1m_2 + l_2m_1$

*In Voigt notation, where $l, i = 1, 2, \cdots, 6$.

We note that the β_{li}-tensor is the transpose of the α_{jk}-tensor.

By using the tensor of (D-2) and of Tables D-1 and D-2, the matrix product

$$\beta_{li} S_{ij} \alpha_{jk} = S_{lk}' \qquad i,j,k,l = 1,2,\cdots,6 \qquad (D\text{-}15)$$

can be formed. From this, the relationship between ε_k' and V_l' in the new coordinate system can also be formed:

$$V_l' = S_{lk}' \varepsilon_k' \qquad l,k = 1,2,\cdots,6. \qquad (D\text{-}16)$$

As an application, we transform S_{ij} to the coordinate axes x', y' and z', in which x' lies along the [110] axis, and z and z' are common axes. From Fig. D-1, the transformation is given by

$$x' = x\cos\theta + y\sin\theta,$$
$$y' = -x\sin\theta + y\cos\theta,$$

and
$$z' = z . \tag{D-17}$$

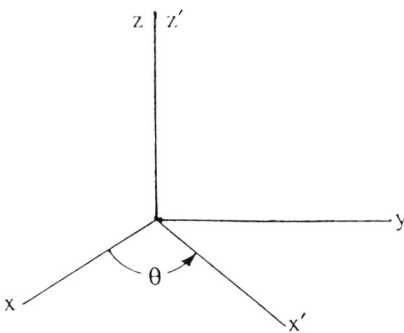

FIGURE D-1. Figure D-1. Cartesian coordinate axes for rotation from x, y, z to x', y', z' with a common zz' axis.

By using the relationships given in (D-8), one can identify
$$\begin{aligned} l_1 &= c & m_1 &= s & n_1 &= 0 \\ l_2 &= -s & m_2 &= c & n_2 &= 0 \\ l_3 &= 0 & m_3 &= 0 & n_3 &= 0 , \end{aligned} \tag{D-18}$$

where $c = \cos\theta$, and $s = \sin\theta$. From Table D-1, we then have for the α-tensor
$$\alpha_{jk} = \begin{pmatrix} c^2 & s^2 & 0 & 0 & 0 & -sc \\ s^2 & c^2 & 0 & 0 & 0 & sc \\ 0 & 0 & 1 & 0 & 0 & 0 \\ 0 & 0 & 0 & c & s & 0 \\ 0 & 0 & 0 & -s & c & 0 \\ 2sc & -2sc & 0 & 0 & 0 & c^2 - s^2 \end{pmatrix}, \tag{D-19}$$

and for the β-tensor
$$\beta_{li} = \begin{pmatrix} c^2 & s^2 & 0 & 0 & 0 & 2sc \\ s^2 & c^2 & 0 & 0 & 0 & -2sc \\ 0 & 0 & 1 & 0 & 0 & 0 \\ 0 & 0 & 0 & c & -s & 0 \\ 0 & 0 & 0 & s & c & 0 \\ -sc & sc & 0 & 0 & 0 & c^2 - s^2 \end{pmatrix} . \tag{D-20}$$

For x' along the [110], $\theta = 45°$,

$$\alpha_{jk} = \begin{pmatrix} \frac{1}{2} & \frac{1}{2} & 0 & 0 & 0 & -\frac{1}{2} \\ \frac{1}{2} & \frac{1}{2} & 0 & 0 & 0 & \frac{1}{2} \\ 0 & 0 & 1 & 0 & 0 & 0 \\ 0 & 0 & 0 & \sqrt{\frac{1}{2}} & \sqrt{\frac{1}{2}} & 0 \\ 0 & 0 & 0 & -\sqrt{\frac{1}{2}} & \sqrt{\frac{1}{2}} & 0 \\ 1 & -1 & 0 & 0 & 0 & 0 \end{pmatrix} \quad (D\text{-}21)$$

and

$$\beta_{li} = \begin{pmatrix} \frac{1}{2} & \frac{1}{2} & 0 & 0 & 0 & 1 \\ \frac{1}{2} & \frac{1}{2} & 0 & 0 & 0 & -1 \\ 0 & 0 & 10 & 0 & 0 & 0 \\ 0 & 0 & 0 & \sqrt{\frac{1}{2}} & -\sqrt{\frac{1}{2}} & 0 \\ 0 & 0 & 0 & \sqrt{\frac{1}{2}} & \sqrt{\frac{1}{2}} & 0 \\ -\frac{1}{2} & \frac{1}{2} & 0 & 0 & 0 & 0 \end{pmatrix}. \quad (D\text{-}22)$$

In the x, y, z coordinate system along the cubic axes,

$$S_{ij} = \begin{pmatrix} S_{11} & S_{12} & S_{12} & 0 & 0 & 0 \\ S_{12} & S_{11} & S_{12} & 0 & 0 & 0 \\ S_{12} & S_{12} & S_{11} & 0 & 0 & 0 \\ 0 & 0 & 0 & S_{44} & 0 & 0 \\ 0 & 0 & 0 & 0 & S_{44} & 0 \\ 0 & 0 & 0 & 0 & 0 & S_{44} \end{pmatrix}. \quad (D\text{-}23)$$

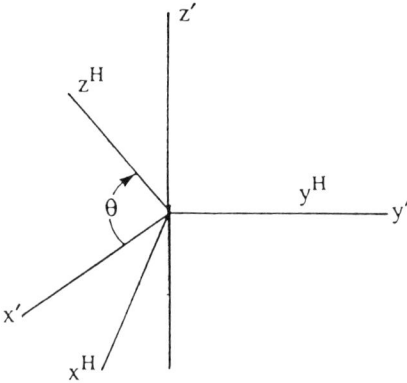

FIGURE D-2. Figure D-2. Cartesian coordinate axes for rotation from x', y', z' to x^H, y^H, z^H with a common $y'y^H$ axis.

APPENDIX D: S—TENSOR TRANSFORMATION

By using (D-20) through (D-22) in (D-7), we find the transformed S-tensor which is given in Table D-3.

Table D-3

A Representation of the Equation $V_l' = S_{lk}\varepsilon_k'$ *

	ε_{xx}'	ε_{yy}'	ε_{zz}'	ε_{yz}'	ε_{zx}'	ε_{xy}'
V_{xx}'	SP	SM	S_{12}	0	0	0
V_{yy}'	SM	SP	S_{12}	0	0	0
V_{zz}'	S_{12}	S_{12}	S_{11}	0	0	0
V_{yz}'	0	0	0	S_{44}	0	0
V_{zx}'	0	0	0	0	S_{44}	0
V_{xy}'	0	0	0	0	0	$\frac{1}{2}(S_{11} - S_{12})$

*In Voigt notation, where $l, k = 1, 2, \cdots, 6$. This equation refers to the S-tensor in a reference frame where x is along [110], y along [$\bar{1}$10], and z along [001]. To simplify the table, $SP = \frac{1}{2}(S_{11} + S_{12} + 2S_{44})$ and $SM = \frac{1}{2}(S_{11} + S_{12} - 2S_{44})$.

For the case of cubic symmetry, which we are considering, the magnetic field $\boldsymbol{H}_0$ could be rotated in a (110) plane or in a (1$\bar{1}$0) plane. For the latter case, shown in Fig. D-2, we take the angle θ to be measured from the propagation direction—the [110]—to the direction of $\boldsymbol{H}_0$, which defines the z^H direction of a rotated coordinate system:

$$x^H = x' \sin\theta - z' \cos\theta ,$$
$$y^H = y' ,$$

and

$$z^H = x' \cos\theta + z' \sin\theta . \tag{D-24}$$

With the use of (D-8), we identify

$$l_1 = s \quad m_1 = 0 \quad n_1 = -c$$
$$l_2 = 0 \quad m_2 = 1 \quad n_2 = 0$$
$$l_3 = c \quad m_3 = 0 \quad n_3 = s \quad .$$

We rewrite (D-5) as

$$V_l^H = \beta_{li} V_i' , \tag{D-25}$$

which allows representation of the field gradient components in the reference frame of the external magnetic field. The β-tensor of (D-25) is given in Table D-4.

Table D-4

A Representation of the Equation $V_l^H = \beta_{li} V_i'^*$

	V'_{xx}	V'_{yy}	V'_{zz}	V'_{yz}	V'_{zx}	V'_{xy}
V^H_{xx}	s^2	0	c^2	0	$-2sc$	0
V^H_{yy}	0	1	0	0	0	0
V^H_{zz}	c^2	0	s^2	0	$2sc$	0
V^H_{yz}	0	0	0	s	0	c
V^H_{zx}	sc	0	$-sc$	0	s^2-c^2	0
V^H_{xy}	0	0	0	$-c$	0	s

*In Voigt notation, where $i, l = 1, 2, \cdots, 6$. This equation gives the field gradient components V_l^H rotated in the $(1\bar{1}0)$ plane. The DC magnetic field $\boldsymbol{H}_0$ is in the z-direction of this rotated reference frame.

The field gradient operators for the dynamic quadrupole interaction can be written as

$$V_{\pm 1} = V^H_{zx} \pm i V^H_{yz}$$

and

$$V_{\pm 2} = \left(V^H_{xx} - V^H_{yy}\right) \pm 2i V^H_{xy} . \tag{D-26}$$

With the use of (D-26) together with Tables D-3 and D-4, we can now determine the dependence of these operators on particular dynamic quadrupole interaction transitions and on the chosen experimental strain.

For a volume conserving longitudinal strain in the direction of ε'_{xx}, we write that $\varepsilon'_{yy} = -(1/2)\varepsilon'_{xx}$ and $\varepsilon'_{zz} = -(1/2)\varepsilon'_{xx}$, giving

$$\begin{aligned}
V_{\pm 1, xx} &= cs V'_{xx} - cs V'_{zz} = c\, s(V'_{xx} - V'_{zz}) \\
&= c\, s \left[\frac{1}{2}(S_{11} + S_{12} - 2S_{44})\varepsilon'_{xx} \right. \\
&\quad + \frac{1}{2}(S_{11} + S_{12} - 2S_{44})\varepsilon'_{yy} \\
&\quad \left. + S_{12}\varepsilon'_{zz} - (S_{12}\varepsilon'_{xx} + S_{12}\varepsilon'_{yy} + S_{11}\varepsilon'_{zz}) \right] \\
&= \frac{3}{4}[(S_{11} - S_{12}) + 2S_{44}]\sin\theta\cos\theta\, \varepsilon'_{xx} \tag{D-27}
\end{aligned}$$

and

$$\begin{aligned}
V_{\pm 2,xx} &= (s^2 V'_{xx} + c^2 V'_{zz} - V'_{yy}) \\
&= s^2 \left[\frac{1}{2}(S_{11} + S_{12} + 2S_{44})\varepsilon'_{xx} \right. \\
&\quad \left. + \frac{1}{2}(S_{11} + S_{12} - 2S_{44})\varepsilon'_{yy} + S_{12}\varepsilon'_{zz} \right] \\
&\quad + c^2 [S_{12}\varepsilon'_{xx} + S_{12}\varepsilon'_{yy} + S_{11}\varepsilon'_{zz}] \\
&\quad - \left[\frac{1}{2}(S_{11} + S_{12} - 2S_{44})\varepsilon'_{xx} \right. \\
&\quad \left. + \frac{1}{2}(S_{11} + S_{12} + 2S_{44})\varepsilon'_{yy} + S_{12}\varepsilon'_{zz} \right] \\
&= \frac{3}{4}[2S_{44}(1 + \sin^2\theta) - (S_{11} - S_{12})\cos^2\theta]\varepsilon'_{xx} \; .
\end{aligned} \qquad \text{(D-28)}$$

For the case in which only $\varepsilon'_{xy} \neq 0$,

$$\begin{aligned}
V_{\pm 1,xy} &= ic V'_{xy} \\
&= \frac{1}{2}i(S_{11} - S_{12})\cos\theta \, \varepsilon'_{xy}
\end{aligned} \qquad \text{(D-29)}$$

and

$$\begin{aligned}
V_{\pm 2,xy} &= 2is V'_{xy} \\
&= \frac{1}{2}i(S_{11} - S_{12})2\sin\theta \, \varepsilon'_{xy} \\
&= i(S_{11} - S_{12})\sin\theta \, \varepsilon'_{xy} \; .
\end{aligned} \qquad \text{(D-30)}$$

For the case in which only $\varepsilon'_{xz} \neq 0$,

$$\begin{aligned}
V_{\pm 1,xz} &= (s^2 - c^2) V'_{xz} \\
&= S_{44}(\sin^2\theta - \cos^2\theta) \, \varepsilon'_{xz}
\end{aligned} \qquad \text{(D-31)}$$

and

$$\begin{aligned}
V_{\pm 2,xz} &= -2sc V'_{xz} \\
&= -2S_{44}\sin\theta\cos\theta \, \varepsilon'_{xz} \; .
\end{aligned} \qquad \text{(D-32)}$$

APPENDIX E
T-Tensor Components for Cubic $\bar{4}3m$ Symmetry

The fourth derivative of the electric potential at a nuclear position can be expanded as a Taylor series in strain. For the case of small strains, the relationship $V_{ijkl} = T_{ijklmn}\varepsilon_{mn}$ can be defined, where the sixth-rank T-tensor relates the elastic strain tensor to the fourth derivative of the electric potential tensor. From the definition of V_{ijkl} and ε_{mn}, the $ijkl$ indices commute, and the mn indices commute. For a crystal with cubic $\bar{4}3m$ symmetry, there are 6 independent components and a total of 183 nonzero components of the tensor T_{ijklmn}. These are given in Tables E-1 and E-2.

Table E-1

Nonzero Components of the Tensor T_{ijklmn}, $T_1 - T_4$

name (total no.)	components					
T_1 (3)	111111	222222	333333			
T_2 (18)	223311	113322	112233			
	332211	331122	221133			
	233211	133122	211233			
	322311	311322	122133			
	232311	313122	121233			
	323211	131322	212133			
T_3 (36)	221111	331111	223333	113333	332222	112222
	112211	113311	332233	331133	223322	221122
	121211	131311	323233	313133	232322	121222
	212111	313111	232333	131333	323222	212122
	122111	133111	133233	133133	322322	211222
	211211	311311	322333	311333	233222	122122
T_4 (6)	222211	333311	111122	333322	111133	222233

APPENDIX E: CUBIC T—TENSOR COMPONENTS

Table E-2
Nonzero Components of the Tensor T_{ijklmn}, T_5 and T_6

name (total no.)	components					
T_5	122221	211112	311113	133331	322223	233332
(48)	212221	121112	131113	313331	232223	323332
	221221	112112	113113	331331	223223	332332
	222121	111212	111313	333131	222323	333232
	211121	122212	133313	311131	233323	322232
	121121	212212	313313	131131	323323	232232
	112121	221212	331313	223131	332323	223232
	111221	222112	333113	111331	333223	222332
T_6	213321	123312	312231	132213	231123	321132
(72)	123321	213312	132231	312213	321123	231132
	331221	332112	221331	221313	112323	112332
	332121	331212	223131	223113	113223	113232
	321321	321312	213231	213213	123123	123132
	312321	312312	231231	231213	132123	132132
	313221	313212	212331	212313	121323	131232
	323121	323112	232131	232113	131223	121332
	132321	231312	123231	123213	213123	213132
	231321	132312	321231	321123	312123	312132
	133221	233112	122331	122313	311223	211332
	233121	133212	322131	322113	211323	311232

APPENDIX F
Dynamic Hexadecapole Interaction:
Transformation of the T-Tensor

As in Appendix E, we expand the fourth derivative of the electric potential, evaluated at a nuclear position, as a Taylor series in strain. The term linear in strain can be written

$$V_{ijkl} = T_{ijklmn}\varepsilon_{mn} , \qquad \text{(F-1)}$$

where the indices $ijklmn$ are equal to x, y or z. Since the $ijkl$ indices permute and the mn indices permute, it is convenient to rewrite (F-1) in terms of the Voigt notation

$$V_{ij} = T_{ijm}\varepsilon_m , \qquad \text{(F-2)}$$

where the indices i, j and $m = 1, \cdots 6$. In a coordinate system not coinciding with the cubic axes, we can write

$$V'_{rs} = T'_{rsv}\varepsilon'_v , \qquad \text{(F-3)}$$

where again the indices r, s and $m = 1, \cdots 6$.

As an example, we define the [100] reference frame with x along [100], y along [010], and z along [001]. It is in this reference frame that the specific nonzero tensor components in Table E-1 have been determined. We consider an elastic strain along [100]. Using volume conservation for this longitudinal strain, only four of the six independent components contribute to the terms of V_{ijkl}, and these occur in the combination of $T_1 - T_4$ and $T_3 - T_2$, as shown in Table F-1.

Table F-1

T-Tensor Components in the [100] Reference System for the ε_{xx} Volume–Conserving Longitudinal Strain.

Value of V/ε_{xx}					
$T_1 - T_4$	$(T_3 - T_2)/2$	$(T_3 - T_2)/2$	0	0	0
$(T_3 - T_2)/2$	$(T_4 - T_1)/2$	$T_2 - T_3$	0	0	0
$(T_3 - T_2)/2$	$T_2 - T_3$	$(T_4 - T_1)/2$	0	0	0
0	0	0	$T_2 - T_3$	0	0
0	0	0	0	$(T_3 - T_2)/2$	0
0	0	0	0	0	$(T_3 - T_2)/2$

We now determine the tensors that relate (F-2) to (F-3). The tensor V'_{rs} is found from V_{ij} by tensor transformations involving rotations of coordinate systems in which the rotation tensor components are direction cosines: l_{xy}.

$$V_{rs} = l_{ri}l_{sj}V_{ij} \tag{F-4}$$

and

$$\varepsilon_m = l_{vm}\varepsilon'_v . \tag{F-5}$$

Therefore

$$\begin{aligned} V'_{rs} &= l_{ri}l_{sj}V_{ij} \\ &= l_{ri}l_{sj}T_{ijm}\varepsilon_m \\ &= l_{ri}l_{sj}T_{ijm}l_{vm}\varepsilon'_v \\ &= T'_{rsv}\varepsilon'_v , \end{aligned} \tag{F-6}$$

where

$$T'_{rsv} = l_{ri}l_{sj}l_{vm}T_{ijm} , \tag{F-7}$$

and all indices range from $1, \cdots 6$.

We choose $V'_{rs} = T'_{rsv}\varepsilon'_v$ to represent the tensor relationship in the [110] reference system, where x is along [110], y along [$\bar{1}$10] and z along [001]. The actual components of V'_{rs} are determined by choosing a particular strain in the [110] reference frame that determines the indices v in (F-7). The various nonzero components of T'_{rsv} are determined from the known direction cosines and the values of the known components T_{ijm} in the [100] reference frame. These nonzero components of T'_{rsv} determine the nonzero components of V'_{rs} in the [110] reference frame.

To include the effect of a variable magnetic field direction, one can show that the tensor l_{vm} in (F-5) is identical with the tensor identified as α given in Table D-1. By extending the derivation following (D-14) in Appendix D, one can show that the transformation given by (F-4) is identical to the following matrix transformation:

$$V'' = \beta V' \alpha , \tag{F-8}$$

where V'' is the tensor in the reference system that contains the magnetic field, V' is the tensor in the [110] reference frame, β is the tensor given in Table D-2, and α is the tensor given in Table D-1.

To find the contributions to the fourth-order potential that occur in the transition probability, the components of V'' are used in the hexadecapole transitions described in (3.89) of Section 3.7, where the field gradient components are now written in Voigt notation:

$$V^{\pm 1} = \frac{1}{24}(V''_{35} \pm iV''_{34}) \tag{F-9}$$

$$V^{\pm 2} = \frac{1}{24}(V''_{31} - V''_{32} \pm 2iV''_{36}) \tag{F-10}$$

$$V^{\pm 3} = \frac{1}{12}[(V''_{51} - V''_{52}) \pm i(3V''_{56} - V''_{42})] \tag{F-11}$$

$$V^{\pm 4} = \frac{1}{48}[(V''_{11} + V''_{22} - 6V''_{12}) \pm 4i(V''_{16} - V''_{62})] \:. \tag{F-12}$$

INDEX

Absorption, 329,334,346, 350,352
Absorption acoustic, 46,364, 378,379,383
  coefficient, 40,330,340,350
Absorption dynamic Alpher–Rubin, 68
Absorption NAR, 32,33,43,49, 117,334–337,339,343
  line, 68,141,142,169,175,188, 337,360
  measurements, 208
  signal, 145,147,148
Absorption NAR(A-R)
  signal, 146,149
Absorption NAR1 first derivative, 190,192
Absorption NAR2 first derivative, 190,191
Absorption nuclear spin–phonon coefficient, 41
Acoustic absorption, 46,364, 378,379,383
  coefficient, 40,330,340,350
Acoustic adiabatic fast passage, 390
Acoustic dispersion, 46,364,378
Acoustic energy density, 35,41, 296,305,307,308,356,357,360, 379,382
Acoustic impedance, 33,46, 307,308
Acoustic nuclear adiabatic fast passage, 317
Acoustic nuclear quadrupole resonance (ANQR), 385
Acoustic nuclear spin echo, 311–313
Acoustic phonon to thermal phonon decay time, 377
Acoustic saturation experiment, 22

NMR (ASNMR), 37,232,263, 295,297,303,305,339
  of NMR in a rotating reference frame, 314
  technique, 24,28
Acoustic susceptibility, 33
Acoustically induced electric field, 150,158
Acoustically induced magnetic field, 137,150,157,158
Acoustically induced nuclear spin relaxation rate, 317
Acoustically induced RF magnetic field, 131,155
Acoustomagnetic effect (AME), 363
Adiabatic demagnetization in the rotating frame, 271
Adiabatic fast passage, 269,270, 288,289
Adiabatic limit, 87
Alpher–Rubin (A-R) attenuation, 117,127,138
Alpher–Rubin DC
  effect, 285
  interaction, 262
  mechanism, 366
Alpher–Rubin dynamic absorption, 68
  coupling, 39,42,49,92,115, 117,130,151,359,382
  interaction, 21,28,67,72,115, 149,231,264,366
Alpher–Rubin induced electric field, 118
Alpher–Rubin NAR susceptibility, 155
Alpher–Rubin nonresonant interaction, 126
Alpher–Rubin oscillating fields, 119

Alpher–Rubin pure dynamic
    susceptibility, 151
Alpher–Rubin velocity shift, 127
Amplitude attenuation
    coefficient, 41
Angular momentum
    operators, 63
Anisotropic exchange
    interaction, 55
Anisotropic spin–spin
    interaction, 165,195,196
Anomalous NAR line shapes, 164
ANQR, 387
ASNMR, 3,73,102,108,109,232,
    233,295,296,305
  continuous wave, 310
  spectrometer, 306
  transient, 311
Associated Legendre
    polynomial, 51
Asymmetry parameter, 9
Attenuation factor, 35
Axially symmetric EFG, 67
Axially symmetric electric field
    gradient, 65,87,386
Axially symmetric static
    quadrupole interaction, 88

Bloch–like equation, 178
Bond–orbital–model, 211

Characteristic impedance, 36
Chemical shift, 39
Clebsch-Gordon coefficients, 61
Coherent fractional phonon, 108
Composite resonator, 25–27
Computer–controlled reflection
    bridge spectrometer, 255,
    256,291
Conduction electron field
    gradient, 219
Coupling dynamic
    Alpher–Rubin, 39,42,49,92,
    115,117,130,151,382,359
Coupling dynamic
    dipolar, 115,129,364,379,390

Coupling dynamic
    dipole–dipole, 49
Coupling dynamic electric
    quadrupole, 39,48,382
Coupling dynamic multipole, 39
Coupling dynamic nuclear
    dipole, 42
  electric quadrupole, 42,72,221,
    224,390
Coupling dynamic
    quadrupole, 92,100,151,184,
    192,379
Coupling hexadecapole
    T-tensor, 287,288
Coupling
    magneto–elastic, 324,329
Coupling nuclear electric
    quadrupole, 171
Coupling quadrupole
    S-tensor, 287,288
Covalent contribution, 210
  to S_{11}, 211
  to S_{44}, 214
Crossed–coil NMR
    spectrometer, 303
Cubic $\bar{4}3m$ symmetry, 407,416
CW acoustic
    absorption, 379
    dispersion, 379
CW line shapes, 390
CW line widths, 390
CW NAR line shapes, 382

DC Alpher–Rubin
    effect, 285
    interaction, 262
    mechanism, 366
DC SQUID, 370
DC SQUID–based magnetic
    resonance spectrometer, 370
De Wette and Schachner
    method, 219,220
DEFG tensor, 221,222
Deformed spherical mode, 173
Demagnetization field
    effect, 353,359
Diagrammatic method by

Reiter, 52
Dipolar Hamiltonian, 14
Dipolar interaction, 55,64
Dipolar line width, 14
Dipole-dipole
  broadening, 170,171
Dipole-dipole dynamic
  coupling, 49
Dipole-dipole dynamic
  interaction, 107
Dipole-dipole dynamic magnetic
  interaction, 49,103
Dipole-dipole interaction, 16,
  18-20,164,165,168,170
Dipole-dipole nuclear magnetic
  exchange, 172
Dipole-dipole second
  moment, 168,170
Direct spin-lattice
  relaxation, 17,18
Direct spin-relaxation rate, 198
Dispersion, 329,346,348
  acoustic, 46,364,378,379
  NAR, 32,33,43,49,117,335-337
  NAR line, 142,337
  NAR signal, 145,148
  NAR(A-R) signal, 147
  resonant Alpher-Rubin, 127
Double acoustic magnetic
  resonance, 295
Double resonance techniques, 38
Double-tuned coupling
  network, 248
Double-tuned over-coupled RF
  transformer, 246,282
Dynamic Alpher-Rubin
  absorption, 68
Dynamic Alpher-Rubin
  coupling, 39,42,49,92,115,
  117,130,151,359,382
Dynamic Alpher-Rubin
  interaction, 21,28,67,72,115,
  149,231,264,366
Dynamic dipolar
  coupling, 115,129,364,379,
  390

Dynamic dipolar interaction, 231
Dynamic dipolar NAR, 116
Dynamic dipole-dipole
  coupling, 49
Dynamic dipole-dipole
  interaction, 107
Dynamic electric field gradient
  (DEFG), 163,188
Dynamic electric quadrupole
  coupling, 39,48,382
Dynamic electric quadrupole
  interaction, 149,267,269,270
Dynamic HDI, 287,290
Dynamic hexadecapole
  interaction (HDI), 98,99,418
Dynamic longitudinal
  magnetoresistivity, 159
Dynamic magnetic dipole-dipole
  interaction, 49,103
Dynamic multipole coupling, 39
Dynamic nuclear dipole
  coupling, 42
Dynamic nuclear electric
  hexadecapole interaction, 49,
  103
Dynamic nuclear electric
  quadrupole coupling, 73,221,
  224,390
Dynamic nuclear electric
  quadrupole interaction, 20,
  27,49,70,72,168,231,378
Dynamic nuclear hexadecapole
  interaction, 287,288
Dynamic nuclear quadrupole
  coupling, 42
Dynamic nuclear quadrupole
  interaction, 107,287
Dynamic quadrupolar
  interaction, 66,67,98
Dynamic quadrupolar
  transition, 66
Dynamic quadrupole
  coupling, 92,100,151,184,192,
  379
Dynamic quadrupole
  Hamiltonian, 49

Dynamic quadrupole
  interaction, 62,99,103,109,
  264,366,367,414
Dynamic quadrupole transition
  probabilities, 90
Dynamic transverse
  magnetoresistivity, 159
$d(i,j;m)$ transformation
  tensor, 56,57,185

Effective magnetization, 263,265
EFG
  fluctuation spectral
    function, 187
  operator, 62
  symmetry axis, 386
  tensor, 9,202
Electric field gradient (EFG)
  operator, 62
  in metals, 216
  tensor, 173,174,202
Electric hexadecapole
  moment, 54
Electric multipole, 60
Electric quadrupole tensor, 64
Electron-nuclear double
  resonance, 320,321
Electrostatic multipole, 59,60
Energy conversion factors, 396
Energy density of coherent
  phonons, 19
Enhanced magnetic field, 39
Enhanced NAR technique, 199
Enhanced nuclear acoustic
  resonance, 197,198,323,
  346–348, 351
Enhanced nuclear spin
  Hamiltonian, 348
Enhanced nuclear spin-strain
  Hamiltonian, 198
Enhancement factor, 347
Exchange interaction, 165,168
Exchange second moment, 168
Extra-ionic electric field
  gradients in metals, 218

Filling factor, 133,136,365

Fractional harmonic
  phonons, 102
Fractional phonons, 103,110,111

Gaussian first derivative, 169
Gaussian line shape, 141,142,168,
  170,172,192
Generalized Bloch
  equation, 164,172
Generalized Kubo
  susceptibility, 130
Generalized magnetic
  susceptibility, 179
Generalized NAR
  susceptibility, 150,154
Generalized susceptibility, 43,45,
  46,149
Gradient–elastic tensor
  components, 207

Harrison–Phillips model, 212,213
Heteropolar bonding, 213,215
Hexadecapole antishielding
  factor, 100
Hexadecapole coupling
  T-tensor, 287,288
Hexadecapole dynamic
  interaction, 98,99,418
Hexadecapole dynamic nuclear
  electric interaction, 49,103
Hexadecapole dynamic nuclear
  interaction, 287,288
Hexadecapole electric
  moment, 54
Hexadecapole energy shifts, 95,97
Hexadecapole Hamiltonian, 94
Hexadecapole interaction, 59,97
Hexadecapole moment, 50,60,
  61,110
Hexadecapole nuclear
  deformation parameter, 100
Hexadecapole nuclear electric
  moment, 49
Hexadecapole nuclear
  interaction, 101,103
Hexadecapole nuclear
  moment, 100,101

INDEX

Hexadecapole spin factors, 99
Hexadecapole static interaction, 95
Hexadecapole static nuclear electric interaction, 49
Hexagonal close-packed symmetry, 90
High-field magnetic energy levels, 11
High-field quadrupole Hamiltonian, 164
Homopolar bonding, 213,215
Hybrid junction, 253,255
Hybrid power splitter, 306

Impurity line broadening, 107
Incoherent fractional phonon, 108
Indirect exchange line broadening, 93
Indirect interaction, 38
Indirect nuclear spin interaction, 39
Indirect spin-lattice relaxation, 17
Inhomogeneous resonance line broadening, 354,359
Input impedance, 36
Interaction anisotropic
  exchange, 55
  spin-spin, 195,196
Interaction axially symmetric static quadrupole, 88
Interaction DC Alpher-Rubin, 262
Interaction dipolar, 55,64
Interaction dipole-dipole, 16, 18-20,165,168,170
Interaction dynamic Alpher-Rubin, 21,28,67, 72,115,149,264,366
Interaction dynamic dipolar, 231
Interaction dynamic dipole-dipole, 107
Interaction dynamic electric quadrupole, 149,267,269,270

Interaction dynamic hexadecapole, 98,99,418
Interaction dynamic magnetic dipole-dipole, 49,103
Interaction dynamic nuclear
  electric hexadecapole, 49,103
  electric quadrupole, 20,27,49, 72,168,231,378
  hexadecapole, 287,288
  quadrupole, 62,66,67,98,99, 103,109,264,287,366,367,414
Interaction exchange, 165
Interaction hexadecapole, 59,97
Interaction indirect, 38
  nuclear spin, 39
Interaction isotropic
  exchange, 55
  pseudo-exchange, 193
  spin-spin, 194,195
Interaction
  magneto-elastic, 324,325, 334,339,324
Interaction modulation
  of exchange, 325
  of hyperfine, 324
Interaction multipole, 59
Interaction nonresonant Alpher-Rubin, 126
Interaction nuclear electric quadrupole, 19,384
Interaction nuclear
  hexadecapole, 101,102
  quadrupole, 32
  spin-electron, 39
  spin-lattice, 39
  spin-phonon, 38
  spin-thermal phonon, 15,16
Interaction pseudo-dipolar, 170,171,193
Interaction pseudo-exchange, 168
Interaction quadrupole, 21,59,64,165,170
Interaction second-order quadrupole-quadrupole, 102
Interaction spin-phonon, 85,109
Interaction spin-spin, 177,183
Interaction static

hexadecapole, 95
nuclear electric
  hexadecapole, 49
  quadrupolar, 37,67,95
Interference susceptibility, 151
Intraspin cross relaxation, 163, 164,180–182,189
Intrinsic direct process, 198
Intrinsic direct spin–lattice relaxation process, 199
Intrinsic nuclear quadrupole moment, 63
Ionic contribution, 210,212,215
  to S_{11}, 212
  to S_{44}, 214
Ionic electric field gradients in metals, 218
Irreducible multipole operators, 83,84
Irreducible spherical tensor formalism, 104
Irreducible spherical tensor operator, 50,172
Irreducible tensor operator, 54, 183,184
Isothermal limit, 87
Isotropic distribution of electric field gradients, 196
Isotropic exchange interaction, 55
Isotropic pseudo–exchange, 192, 193
Isotropic quadrupolar exchange, 197
Isotropic spin–spin interaction, 194,195

Knight shift, 39,390
Kramers-Kronig relations, 138, 141,142
Kubo susceptibility, 43,44

Line broadening
  impurity, 107
  indirect exchange, 93
  spin–diffusion, 107
  static nuclear electric
    quadrupole, 93
Lorentz line shape, 86,236, 237,376,378
Lorentzian distribution of field gradients, 180
Lorentzian line shape, 123,128, 139,140,176,188,329,332,337
Lorentzian tail, 192

Magnetic dipole
  interaction Hamiltonian, 152
  moment, 116
Magnetic multipole, 60
Magnetic octupole moment, 54
Magnetic relaxation, 302
Magneto–elastic coupling, 324, 329
Magneto–elastic interaction, 324, 325,334,339,342
Magnetoconductivity tensor, 120
Magnetoresistivity tensor, 120, 122
Marginal oscillator, 243,245–248, 250,253
  ultrasonic spectrometer (MOUS), 232,238,243,245
Mechanical Q, 36
Mechanical resonance, 26,27,108
Mechanical resonator, 25
Method of Fumi and Ripamonti, 224
Method of moments, 163,164,165
Modified Bloch equation, 177
Modulation
  of anisotropy energy, 324
  of effective anisotropy field, 329
  of exchange interaction, 325
  of hyperfine interaction, 324
MOUS, 245,247,248,253,261, 282,289–291
  RF transformer, 261
Multiple quantum transitions, 49
Multipolar interaction, 59
Multipole moment, 61,215

NAR absorption, 32,33,43,49, 117,334–337,339,343

INDEX 427

line, 68,141,142,337
measurements, 208
signal, 145,147,148
NAR Alpher–Rubin
   susceptibility, 155
NAR dipole susceptibility, 134,
   136
NAR dispersion, 32,33,43,49,117,
   335–337
  line, 142,337
  signal, 145,148
NAR in antiferromagnetic
   insulators, 323
NAR in ferromagnets, 323,346
NAR interference
   susceptibility, 157
NAR Knight shift, 147
NAR line
  shapes in bulk metals, 323
  shapes, 140,164,165
  width, 147,164,165,392
NAR pure quadrupole
   resonance, 49,168
NAR quadrupole
   susceptibility, 155
NAR signal intensity, 159,160
NAR(A-R), 43,44,72,73,382,
   390,391
  absorption signal, 146,149
  dispersion signal, 147
  line shape, 163,193
  line width, 192,193
  second moment, 193
NAR1, 43,72,382
  absorption first derivative, 190,
   192
  first derivative line shape, 189
  line shape, 176,188,193–195
  line width, 188,192–195
  second moment, 166,193
NAR2, 43,72,378,379,390,391
  absorption first derivative, 190,
   191
  line shape, 176,188,194
  line width, 188,192–194
  second moment, 167,193
NMR, 73,101

line width, 193
second moment, 193
signal amplitude, 299
Nonaq grease bond, 274,283,
   284,290
Nonresonant Alpher–Rubin, 262
  interaction, 126
NQR, 101
Nuclear electric hexadecapole
   moment, 49
Nuclear electric quadrupole
  coupling, 171
  energy level, 364
  Hamiltonian, 73
  interaction, 19,384
  moment, 102
Nuclear electric
   quadrupole–quadrupole
   exchange, 172
Nuclear hexadecapole
  deformation parameter, 100
  interaction, 101,103
  moment, 100,101
Nuclear magnetic dipole
  energy level, 364
  moment, 102
Nuclear magnetic dipole–dipole
   exchange, 172
Nuclear magnetic octupole
   transition, 102
Nuclear multipole moments, 172
Nuclear quadrupole
  deformation parameter, 100
  interaction, 32
  moment operator, 62
  resonance (NQR), 101
Nuclear shell model, 71
Nuclear spin susceptibility, 136
Nuclear spin–electron
   interaction, 39
Nuclear spin–lattice
  interaction, 39
  relaxation, 3,7,13–17
Nuclear spin–phonon
  absorption coefficient, 41
  Hamiltonian, 40–42

428   INDEX

interaction, 38,109
relaxation time, 41
spectra, 109
Nuclear spin–thermal phonon
  interaction, 15,16
Nuclear structure model, 72
Nuclear–nuclear double
  resonance, 320,321

Orbach process, 198,199

Paired isotopes for HDI, 401
Parity, 60
Permeability tensor, 121
Plane–wise summation
  method, 215,219
Principal axes system, 66
Progressive saturation
  method, 267
Propagating wave model, 231,
  233,240
Properties of selected
  nuclei, 397–400
Pseduo-exchange, 164
  interaction, 164,168,170,171
Pseudo–dipolar broadening, 171
Pseudo–dipolar contribution, 192
Pseudo–dipolar interaction, 170,
  171,193
Pseudo–dipole, 164
Pseudo–exchange
  broadening, 170–172
Pseudo–quadrupolar
  exchange, 164
Pulse–echo
  response, 231,233
  spectrometer, 258,271
  technique, 258, 272
Pure dynamic Alpher–Rubin
  susceptibility, 151
Pure dynamic quadrupole
  susceptibility, 151
Pure nuclear electric quadrupole
  resonance, 87
Pure nuclear quadrupole energy
  levels, 13
Pure quadrupole energy
  levels, 12,13

Quadrupolar broadening, 171
Quadrupolar Hamiltonian, 62,66
Quadrupolar line width, 14
Quadrupole acoustic nuclear
  resonance, 385
Quadrupole antishielding
  factor, 100
Quadrupole axially symmetric
  static interaction, 88
Quadrupole coupling
  S-tensor, 287,288
  constant, 92
Quadrupole dynamic
  coupling, 39,48,92,100,151,
  184,192,379,382
Quadrupole dynamic electric
  interaction, 149,267,269,270
Quadrupole dynamic
  Hamiltonian, 49
Quadrupole dynamic
  interaction, 62,99,103,109,
  264,366,367,414
Quadrupole dynamic nuclear
  coupling, 42
Quadrupole dynamic nuclear
  electric coupling, 73,221,
  224,390
Quadrupole dynamic nuclear
  electric interaction, 20,27,49,
  70,72,168,231,378
Quadrupole dynamic nuclear
  interaction, 107,287
Quadrupole dynamic transition
  probabilities, 90
Quadrupole electric tensor, 64
Quadrupole Hamiltonian, 64
Quadrupole high–field
  Hamiltonian, 164
Quadrupole interaction, 21,59,
  64,165,170
Quadrupole intrinsic nuclear
  moment, 63
Quadrupole moment tensor, 8
Quadrupole NAR

INDEX 429

susceptibility, 155
Quadrupole nuclear deformation
  parameter, 100
Quadrupole nuclear electric
  coupling, 171
  energy level, 364
  Hamiltonian, 73
  interaction, 19,384
  moment, 102
Quadrupole nuclear
  interaction, 32
Quadrupole nuclear moment
  operator, 62
Quadrupole nuclear
  resonance, 101
Quadrupole pure dynamic
  susceptibility, 151
Quadrupole pure energy
  levels, 12,13
Quadrupole pure nuclear electric
  resonance, 87
Quadrupole pure nuclear energy
  levels, 13
Quadrupole pure
  SQUID-detected
  resonance, 392
Quadrupole relaxation, 301
Quadrupole second moment, 168
Quadrupole spectroscopic
  nuclear moment, 63
Quadrupole static axially
  symmetric interaction, 88
Quadrupole static interaction, 37
Quadrupole static nuclear
  electric broadening, 172
Quadrupole static nuclear
  electric line broadening, 93
Quadrupole S-tensor
  coupling, 287,288
Quadrupole–quadrupole
  nuclear electric exchange, 172
  second-order interaction, 102
  second-order transition, 108
Quadrupole-split energy
  levels, 67

Raman spin–lattice
  relaxation, 17
Raman–process relaxation, 385
Reflection bridge
  spectrometer, 253,261
Reiter's definitions, 53
Relaxation spin–lattice, 84
Resistivity tensor, 121
Resonant Alpher–Rubin
  dispersion, 127
RF acoustic field, 43
RF magnetic field, 43
RF phase–sensitive
  detection, 257,284
RF reflection spectrometer, 238
RF SQUID, 365,367,368,370
RF SQUID–based AME
  spectrometer, 370
RF transmission
  spectrometer, 238,251
RF–biased SQUID, 369,371
Rotated coordinate systems, 77,
  411,412

Salol bond, 274
Sampled–CW (SCW)
  response, 231,233
  spectrometer, 238,257,259
  technique, 259
Saturation and recovery, 268
SCW spectrometer, 259,260
Second moment
  determination, 170
Second-order quadrupolar, 110
Second-order
  quadrupole–quadrupole
  interaction, 102
  transition, 108
Selected physical constants, 395
Semi-phenomenological
  Bloch equation, 84
Sensitivity enhancement
  factor, 239–243
Shear distortions, 202,221
Shear wave propagation, 78–82
Signal-enhancement
  factor, 239–241,244

430 INDEX

Silicone grease bond, 274
Silverstein–Fedders theory, 323, 329
Sixth-rank tensor T, 111
Spatial coherence, 375
  in mechanical resonator, 378
Spatial incoherence, 375
  in mechanical resonator, 378
Spatially coherent radiation, 382
Spectral shape function, 164, 179,180,183
Spectrometer ASNMR, 306
Spectrometer
  computer–controlled reflection bridge, 255,256
Spectrometer DC SQUID–based magnetic resonance, 370
Spectrometer marginal oscillator, 238,243,245
Spectrometer NMR crossed–coil, 303
Spectrometer pulse–echo, 271
Spectrometer reflection bridge, 253,261
Spectrometer RF
  reflection, 238
  SQUID–based AME, 370
  transmission, 238,251
Spectrometer SCW, 238,257, 259,260
Spectrometer SQUID
  acoustomagnetic, 270,289, 370,371
  AME, 373,375
  NAR, 232
Spectrometer transmission, 238
  bridge, 254
  oscillator ultrasonic, 248
Spectroscopic nuclear quadrupole moment, 63
Spherical harmonic, 49,51,59,173
Spherical multipole tensor, 58
Spin–diffusion line broadening, 107
Spin–flopped antiferromagnet, 331

Spin–flopped case, 341
Spin–flopped configuration, 327
Spin–flopped regime, 332
Spin–lattice
  interaction, 55
  relaxation, 84,87
  relaxation time T_1, 364
Spin–phonon
  Hamiltonian, 85
  interaction, 85
Spin–spin relaxation time T_2, 364,385
SQUID acoustomagnetic spectrometer, 270,289, 370,371
SQUID AME, 363,366,378,390
  mechanical resonance response, 378
  spectrometer, 373,375
SQUID DC, 370
SQUID method of detecting NAR, 269
SQUID NAR spectrometer, 232
SQUID RF, 365,367,368,370
SQUID RF–biased, 369,371
SQUID–based
  DC magnetic resonance spectrometer, 370
  RF AME spectrometer, 370
SQUID–detected acoustic paramagnetic resonance, 392
SQUID–detected adiabatic fast passage, 288,289
SQUID–detected AME, 390
SQUID–detected ANQR, 386
SQUID–detected CW NAR2, 379
SQUID–detected magnetization, 270
SQUID–detected NAR, 240,364, 365,367,373,375,382,392
SQUID–detected NAR2 line widths, 382
SQUID–detected NMR, 364, 373,375
SQUID–detected pure quadrupole resonance, 392
SQUID–detected transient

NAR, 383
SQUID-detection of
  acoustic composite resonator
  responses, 374
Static electric field gradient, 55,
  163,164,188,224
Static hexadecapole
  interaction, 95
Static nuclear electric
  hexadecapole interaction, 49
Static nuclear electric quadrupole
  broadening, 93,172
Static quadrupole interaction, 37,
  67,95
Sternheimer antishielding factor
  $(1-\gamma_\infty)$, 204,206,210,
  215,217,390
Sternheimer antishielding
  term, 210
Sternheimer shielding factor
  $(1-R)$, 218
Strain tensor, 33
Strain-electric field gradient
  tensor S, 201
Strain-induced EFG, 202
Susceptibility acoustic, 33
Susceptibility Alpher-Rubin
  NAR, 155
susceptibility generalized, 43,45,
  46,149
  Kubo, 130
  magnetic, 179
  NAR, 150,154
Susceptibility interference, 151
Susceptibility Kubo, 43,44
Susceptibility NAR dipole, 134,
  136
Susceptibility NAR
  interference, 157
Susceptibility NAR
  quadrupole, 155
Susceptibility nuclear spin, 136
Susceptibility pure dynamic
  Alpher-Rubin, 151
Susceptibility pure dynamic
  quadrupole, 151

S-tensor, 49,73-76,90,202,364,
  379,385,407,413
S-tensor component, 202,
  203,206-208,210,219,221,
  224,402-406
S-tensor measurements, 219

Temporal coherence, 375
Tetragonal distortions, 224
Tetragonal shear, 221-223
Tetragonal strain, 223
Thermal phonon diffusion
  time, 377
Time-varying elastic strain, 70
Time-varying electric field
  gradients, 70
TOUS, 248-251
Transformation
  of S-tensor, 407
  of T-tensor, 418
Transient measurements of
  T_1, 383
Transient SQUID-detected
  NAR, 383
Transmission
  bridge spectrometer, 254
  oscillator ultrasonic
    spectrometer (TOUS), 248
  spectrometer, 238
Transmission-type acoustic
  resonator, 237
Transverse pure mode, 34
Trigonal distortions, 223,224
Trigonal shear, 221,222
Trigonal strain, 223
T-tensor, 100,103,416
  components, 416-418

Ultrasonic propagation
  constant, 36
Uniform background lattice
  model, 220
Universal correlation of
  Raghavan, 226

Van Vleck paramagnet, 39,197,
  323,347,348

Voigt notation, 34,73,76,402,407, 409,410,413,418,420
Volume–conserving longitudinal strain, 414,418
Volume–conserving linear dilatations, 202

Watson sphere model for ionic solids, 205
Weak field energy levels, 12

Wigner–Eckart theorem, 61, 62,178

Young's modulus, 36

Zeeman Hamiltonian, 66
Zero–field splitting, 87
Zinc–blende structure, 207, 212,213